UNDERSTANDING PHYSICS

PART 2

Karen Cummings
Rensselaer Polytechnic Institute
Southern Connecticut State University

Priscilla W. Laws
Dickinson College

Edward F. Redish
University of Maryland

Patrick J. Cooney
Millersville University

GUEST AUTHOR

Edwin F. Taylor
Massachusetts Institute of Technology

ADDITIONAL MEMBERS OF ACTIVITY BASED PHYSICS GROUP

David R. Sokoloff
University of Oregon

Ronald K. Thornton
Tufts University

Understanding Physics is based on *Fundamentals of Physics* by David Halliday, Robert Resnick, and Jearl Walker.

WILEY

John Wiley & Sons, Inc.

This book is dedicated to Arnold Arons, whose pioneering work in physics education and reviews of early chapters have had a profound influence on our work.

SENIOR ACQUISITIONS EDITOR	Stuart Johnson
SENIOR DEVELOPMENT EDITOR	Ellen Ford
MARKETING MANAGER	Bob Smith
SENIOR PRODUCTION EDITOR	Elizabeth Swain
SENIOR DESIGNER	Kevin Murphy
INTERIOR DESIGN	Circa 86, Inc.
COVER DESIGN	David Levy
COVER PHOTO	© Antonio M. Rosario/The Image Bank/Getty Images
ILLUSTRATION EDITOR	Anna Melhorn
PHOTO EDITOR	Hilary Newman

This book was set in 10/12 Times Ten Roman by Progressive and printed and bound by Von Hoffmann Press. The cover was printed by Von Hoffmann Press.

This book is printed on acid free paper. ∞

To order books or for customer service please, call 1(800)-CALL-WILEY (225-5945).

Library of Congress Cataloging in Publication Data:

Understanding physics / Karen Cummings . . . [et al.]; with additional members of the
    Activity Based Physics Group.
        p. cm.
    Includes index.
    ISBN 0-471-46436-8 (pt. 2 : pbk. : acid-free paper)
    1. Physics.   I. Cummings, Karen.   II. Activity Based Physics Group.

QC23.2.U54 2004		
530—dc21		2003053481
L.C. Call no.	Dewey Classification No.	L.C. Card No.
ISBN 0-471-46436-8		

Printed in the United States of America

10 9 8 7 6 5 4 3 2

Preface

Welcome to *Understanding Physics*. This book is built on the foundations of the 6th Edition of Halliday, Resnick, and Walker's *Fundamentals of Physics* which we often refer to as HRW 6th. The HRW 6th text and its ancestors, first written by David Halliday and Robert Resnick, have been best-selling introductory physics texts for the past 40 years. It sets the standard against which many other texts are judged. You are probably thinking, "Why mess with success?" Let us try to explain.

Why a Revised Text?

A physics major recently remarked that after struggling through the first half of his junior level mechanics course, he felt that the course was now going much better. What had changed? Did he have a better background in the material they were covering now? "No," he responded. "I started reading the book before every class. That helps me a lot. I wish I had done it in Physics One and Two." Clearly, this student learned something very important. It is something most physics instructors wish they could teach all of their students as soon as possible. Namely, no matter how smart your students are, no matter how well your introductory courses are designed and taught, your students will master more physics if they learn how to read an "understandable" textbook carefully.

We know from surveys that the vast majority of introductory physics students do not read their textbooks carefully. We think there are two major reasons why: (1) many students complain that physics textbooks are impossible to understand and too abstract, and (2) students are extremely busy juggling their academic work, jobs, personal obligations, social lives and interests. So they develop strategies for passing physics without spending time on careful reading. We address both of these reasons by making our revision to the sixth edition of *Fundamentals of Physics* easier for students to understand and by providing the instructor with more **Reading Exercises** (formerly known as Checkpoints) and additional strategies for encouraging students to read the text carefully. Fortunately, we are attempting to improve a fine textbook whose active author, Jearl Walker, has worked diligently to make each new edition more engaging and understandable.

In the next few sections we provide a summary of how we are building upon HRW 6th and shaping it into this new textbook.

A Narrative That Supports Student Learning

One of our primary goals is to help students make sense of the physics they are learning. We cannot achieve this goal if students see physics as a set of disconnected mathematical equations that each apply only to a small number of specific situations. We stress conceptual and qualitative understanding and continually make connections between mathematical equations and conceptual ideas. We also try to build on ideas that students can be expected to already understand, based on the resources they bring from everyday experiences.

In *Understanding Physics* we have tried to tell a story that flows from one chapter to the next. Each chapter begins with an introductory section that discusses why new topics introduced in the chapter are important, explains how the chapter builds on previous chapters, and prepares students for those that follow. We place explicit emphasis on basic concepts that recur throughout the book. We use extensive forward and backward referencing to reinforce connections between topics. For example, in the introduction of Chapter 16 on Oscillations we state: "Although your study of simple harmonic motion will enhance your understanding of mechanical systems it is also vital to understanding the topics in electricity and magnetism encountered in Chapters 30-37. Finally, a knowledge of SHM provides a basis for understanding the wave nature of light and how atoms and nuclei absorb and emit energy."

Emphasis on Observation and Experimentation

Observations and concrete everyday experiences are the starting points for development of mathematical expressions. Experiment-based theory building is a major feature of the book. We build ideas on experience that students either already have or can easily gain through careful observation.

Whenever possible, the physical concepts and theories developed in *Understanding Physics* grow out of simple observations or experimental data that can be obtained in typical introductory physics laboratories. We want our readers to develop the habit of asking themselves: What do our observations, experiences and data imply about the natural laws of physics? How do we know a given statement is true? Why do we believe we have developed correct models for the world?

Toward this end, the text often starts a chapter by describing everyday observations with which students are familiar. This makes *Understanding Physics* a text that is both relevant to students' everyday lives and draws on existing student knowledge. We try to follow Arnold Arons' principle "idea first, name after." That is, we make every attempt to begin a discussion by using everyday language to describe common experiences. Only then do we introduce formal physics terminology to represent the concepts being discussed. For example, everyday pushes, pulls, and their impact on the motion of an object are discussed before introducing the term "force" or Newton's Second Law. We discuss how a balloon shrivels when placed in a cold environment and how a pail of water cools to room temperature before introducing the ideal gas law or the concept of thermal energy transfer.

The "idea first, name after" philosophy helps build patterns of association between concepts students are trying to learn and knowledge they already have. It also helps students reinterpret their experiences in a way that is consistent with physical laws.

Examples and illustrations in *Understanding Physics* often present data from modern computer-based laboratory tools. These tools include computer-assisted data acquisition systems and digital video analysis software. We introduce students to these tools at the end of Chapter 1. Examples of these techniques are shown in Figs. P-1 and P-2 (on the left) and Fig. P-3 on the next page. Since many instructors use these computer tools in the laboratory or in lecture demonstrations, these tools are part of the introductory physics experience for more and more of our students. The use of real data has a number of advantages. It connects the text to the students' experience in other parts of the course and it connects the text directly to real world experience. Regardless of whether data acquisition and analysis tools are used in the student's own laboratory, our use of realistic rather that idealized data helps students develop an appreciation of the role that data evaluation and analysis plays in supporting theory.

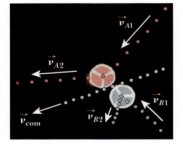

FIGURE P-1 ■ A video analysis shows that the center of mass of a two-puck system moves at a constant velocity.

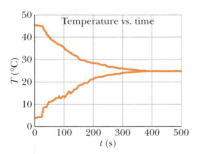

FIGURE P-2 ■ Electronic temperature sensors reveal that if equal amounts of hot and cold water mix the final temperature is the average of the initial temperatures.

FIGURE P-3 ■ A video analysis of human motion reveals that in free fall the center of mass of an extended body moves in a parabolic path under the influence of the Earth's gravitational force.

Using Physics Education Research

In re-writing the text we have taken advantage of two valuable findings of physics education research. One is the identification of concepts that are especially difficult for many students to learn. The other is the identification of active learning strategies to help students develop a more robust understanding of physics.

Addressing Learning Difficulties

Extensive scholarly research exists on the difficulties students have in learning physics.[1] We have made a concerted effort to address these difficulties. In *Understanding Physics,* issues that are known to confuse students are discussed with care. This is true even for topics like the nature of force and its effect on velocity and velocity changes that may seem trivial to professional physicists. We write about subtle, often counter-intuitive topics with carefully chosen language and examples designed to draw out and remediate common alternative student conceptions. For example, we know that students have trouble understanding passive forces such as normal and friction forces.[2] How can a rigid table exert a force on a book that rests on it? In Section 6-4 we present an idealized model of a solid that is analogous to an inner spring mattress with the repulsion forces between atoms acting as the springs. In addition, we invite our readers to push on a table with a finger and experience the fact that as they push harder on the table the table pushes harder on them in the opposite direction.

FIGURE P-4 ■ Compressing an innerspring mattress with a force. The mattress exerts an oppositely directed force, with the same magnitude, back on the finger.

Incorporating Active Learning Opportunities

We designed *Understanding Physics* to be more interactive and to foster thoughtful reading. We have retained a number of the excellent Checkpoint questions found at the end of HRW 6th chapter sections. We now call these questions **Reading Exercises.** We have created many new Reading Exercises that require students to reflect on the material in important chapter sections. For example, just after reading Section 6-2 that introduces the two-dimensional free-body diagram, students encounter Reading Exercise 6-1. This multiple-choice exercise requires students to identify the free-body diagram for a helicopter that experiences three non-collinear forces. The distractors were based on common problems students have with the construction of free-body diagrams. When used in "Just-In-Time Teaching" assignments or for in-class group discussion, this type of reading exercise can help students learn a vital problem solving skill as they read.

---

[1] L. C. McDermott and E. F. Redish, "Resource Letter PER-1: Physics Education Research," *Am. J. Phys.* **67**, 755-767 (1999)

[2] John J. Clement, "Expert novice similarities and instruction using analogies," *Int. J. Sci. Ed. 20*, 1271-1286 (1998)

We also created a set of **Touchstone Examples.** These are carefully chosen sample problems that illustrate key problem solving skills and help students learn how to use physical reasoning and concepts as an essential part of problem solving. We selected some of these touchstone examples from the outstanding collection of sample problems in HRW 6th and we created some new ones. In order to retain the flow of the narrative portions of each chapter, we have reduced the overall number of sample problems to those necessary to exemplify the application of fundamental principles. Also, we chose touchstone examples that require students to combine conceptual reasoning with mathematical problem-solving skills. Few, if any, of our touchstone examples are solvable using simple "plug-and-chug" or algorithmic pattern matching techniques.

Alternative problems have been added to the extensive, classroom tested end-of-chapter problem sets selected from HRW 6th. The design of these new problems are based on the authors' knowledge of research on student learning difficulties. Many of these new problems require careful qualitative reasoning. They explicitly connect conceptual understanding to quantitative problem solving. In addition, estimation problems, video analysis problems, and "real life" or "context rich" problems have been included.

The organization and style of *Understanding Physics* has been modified so that it can be easily used with other research-based curricular materials that make up what we call *The Physics Suite*. The *Suite* and its contents are explained at length at the end of this preface.

Reorganizing for Coherence and Clarity

For the most part we have retained the organization scheme inherited from HRW 6th. Instructors are familiar with the general organization of topics in a typical course sequence in calculus-based introductory physics texts. In fact, ordering of topics and their division into chapters is the same for 27 of the 38 chapters. The order of some topics has been modified to be more pedagogically coherent. Most of the reorganization was done in Chapters 3 through 10 where we adopted a sequence known as *New Mechanics*. In addition, we decided to move HRW 6th Chapter 25 on capacitors so it becomes the last chapter on DC circuits. Capacitors are now introduced in Chapter 28 in *Understanding Physics*.

The New Mechanics Sequence

HRW 6th and most other introductory textbooks use a familiar sequence in the treatment of classical mechanics. It starts with the development of the kinematic equations to describe constantly accelerated motion. Then two-dimensional vectors and the kinematics of projectile motion are treated. This is followed by the treatment of dynamics in which Newton's Laws are presented and used to help students understand both one- and two-dimensional motions. Finally energy, momentum conservation, and rotational motion are treated.

About 12 years ago when Priscilla Laws, Ron Thornton, and David Sokoloff were collaborating on the development of research-based curricular materials, they became concerned about the difficulties students had working with two-dimensional vectors and understanding projectile motion before studying dynamics.

At the same time Arnold Arons was advocating the introduction of the concept of momentum before energy.[3] Arons argued that (1) the momentum concept is simpler than the energy concept, in both historical and modern contexts and (2) the study

---

[3] Private Communication between Arnold Arons and Priscilla Laws by means of a document entitled "Preliminary Notes and Suggestions," August 19, 1990; and Arnold Arons, *Development of Concepts of Physics* (Addison-Wesley, Reading MA, 1965)

of momentum conservation entails development of the concept of center-of-mass which is needed for a proper development of energy concepts. Additionally, the impulse-momentum relationship is clearly an alternative statement of Newton's Second Law. Hence, its placement immediately after the coverage of Newton's laws is most natural.

In order to address these concerns about the traditional mechanics sequence, a small group of physics education researchers and curriculum developers convened in 1992 to discuss the introduction of a new order for mechanics.[4] One result of the conference was that Laws, Sokoloff, and Thornton have successfully incorporated a new sequence of topics in the mechanics portions of various curricular materials that are part of the Physics Suite discussed below.[5] These materials include *Workshop Physics*, the *RealTime Physics Laboratory Module in Mechanics*, and the *Interactive Lecture Demonstrations*. This sequence is incorporated in this book and has required a significant reorganization and revisions of HRW 6th Chapters 2 through 10.

The New Mechanics sequence incorporated into Chapters 2 through 10 of understanding physics includes:

- Chapter 2: One-dimensional kinematics using constant horizontal accelerations and vertical free fall as applications.

- Chapter 3: The study of one-dimensional dynamics begins with the application of Newton's laws of motion to systems with one or more forces acting along a single line. Readers consider observations that lead to the postulation of "gravity" as a constant invisible force acting vertically downward.

- Chapter 4: Two-dimensional vectors, vector displacements, unit vectors and the decomposition of vectors into components are treated.

- Chapter 5: The study of kinematics and dynamics is extended to two-dimensional motions with forces along only a single line. Examples include projectile motion and circular motion.

- Chapter 6: The study of kinematics and dynamics is extended to two-dimensional motions with two-dimensional forces.

- Chapters 7 & 8: Topics in these chapters deal with impulse and momentum change, momentum conservation, particle systems, center of mass, and the motion of the center-of-mass of an isolated system.

- Chapters 9 & 10: These chapters introduce kinetic energy, work, potential energy, and energy conservation.

Just-in-Time Mathematics

In general, we introduce mathematical topics in a "just-in-time" fashion. For example, we treat one-dimensional vector concepts in Chapter 2 along with the development of one-dimensional velocity and acceleration concepts. We hold the introduction of two- and three-dimensional vectors, vector addition and decomposition until Chapter 4, immediately before students are introduced to two-dimensional motion and forces in Chapters 5 and 6. We do not present vector products until they are needed. We wait to introduce the dot product until Chapter 9 when the concept of physical work is presented. Similarly, the cross product is first presented in Chapter 11 in association with the treatment of torque.

---

[4] The New Mechanics Conference was held August 6-7, 1992 at Tufts University. It was attended by Pat Cooney, Dewey Dykstra, David Hammer, David Hestenes, Priscilla Laws, Suzanne Lea, Lillian McDermott, Robert Morse, Hans Pfister, Edward F. Redish, David Sokoloff, and Ronald Thornton.

[5] Laws, P. W. "A New Order for Mechanics" pp. 125-136, *Proceedings of the Conference on the Introductory Physics Course*, Rensselaer Polytechnic Institute, Troy New York, May 20-23, Jack Wilson, Ed. 1993 (John Wiley & Sons, New York 1997)

Notation Changes

Mathematical notation is often confusing, and ambiguity in the meaning of a mathematical symbol can prevent a student from understanding an important relationship. It is also difficult to solve problems when the symbols used to represent different quantities are not distinctive. Some key features of the new notation include:

- We adhere to recent notation guidelines set by the U.S. National Institute of Standard and Technology Special Publication 811 (SP 811).

- We try to balance our desire to use familiar notation and our desire to avoid using the same symbol for different variables. For example, p is often used to denote momentum, pressure, and power. We have chosen to use lower case p for momentum and capital P for pressure since both variables appear in the kinetic theory derivation. But we stick with the convention of using capital P for power since it does not commonly appear side by side with pressure in equations.

- We denote vectors with an arrow instead of bolding so handwritten equations can be made to look like the printed equations.

- We label each vector component with a subscript that explicitly relates it to its coordinate axis. This eliminates the common ambiguity about whether a quantity represents a magnitude which is a scalar or a vector component which is not a scalar.

- We often use subscripts to spell out the names of objects that are associated with mathematical variables even though instructors and students will tend to use abbreviations. We also stress the fact that one object is exerting a force on another with an arrow in the subscript. For example, the force exerted by a rope on a block would be denoted as $\vec{F}_{\text{rope}\rightarrow\text{block}}$.

Our notation scheme is summarized in more detail in Appendix A4.

Encouraging Text Reading

We have described a number of changes that we feel will improve this textbook and its readability. But even the best textbook in the world is of no help to students who do not read it. So it is important that instructors make an effort to encourage busy students to develop effective reading habits. In our view the single most effective way to get students to read this textbook is to assign appropriate reading, reading exercises, and other reading questions after every class. Some effective ways to follow up on reading question assignments include:

1. Employ a method called "Just-In-Time-Teaching" (or JiTT) in which students submit their answers to questions about reading before class using just plain email or one of the many available computer based homework systems (Web Assign or E-Grade for example). You can often read enough answers before class to identify the difficult questions that need more discussion in class;

2. Ask students to bring the assigned questions to class and use the answers as a basis for small group discussions during the class period;

3. Assign multiple choice questions related to each section or chapter that can be graded automatically with a computer-based homework system; and

4. Require students to submit chapter summaries. Because this is a very effective assignment, we intentionally avoided doing chapter summaries for students.

Obviously, all of these approaches are more effective when students are given some credit for doing them. Thus you should arrange to grade all, or a random sample, of the submissions as incentives for students to read the text and think about the answers to Reading Exercises on a regular basis.

The Physics Suite

In 1997 and 1998, Wiley's physics editor, Stuart Johnson, and an informally constituted group of curriculum developers and educational reformers known as the *Activity Based Physics Group* began discussing the feasibility of integrating a broad array of curricular materials that are physics education research-based. This led to the assembly of an *Activity Based Physics Suite* that includes this textbook. The *Physics Suite* also includes materials that can be combined in different ways to meet the needs of instructors working in vastly different learning environments. The *Interactive Lecture Demonstration Series*[6] is designed primarily for use in lecture sessions. Other *Suite* materials can be used in laboratory settings including the *Workshop Physics Activity Guide*,[7] the *Real Time Physics Laboratory* modules,[8] and *Physics by Inquiry*.[9] Additional elements in the collection are suitable for use in recitation sessions such as the University of Washington *Tutorials in Introductory Physics* (available from Prentice Hall)[10] and a set of *Quantitative Tutorials*[11] developed at the University of Maryland. The *Activity Based Physics Suite* is rounded out with a collection of thinking problems developed at the University of Maryland. In addition to this **Understanding Physics** text, the Physics Suite elements include:

1. **Teaching Physics with the Physics Suite** by Edward F. Redish (University of Maryland). This book is not only the "Instructors Manual" for *Understanding Physics*, but it is also a book for anyone who is interested in learning about recent developments in physics education. It is a handbook with a variety of tools for improving both teaching and learning of physics—from new kinds of homework and exam problems, to surveys for figuring out what has happened in your class, to tools for taking and analyzing data using computers and video. The book comes with a Resource CD containing 14 conceptual and 3 attitude surveys, and more than 250 thinking problems covering all areas of introductory physics, resource materials from commercial vendors on the use of computerized data acquisition and video, and a variety of other useful reference materials. (Instructors can obtain a complimentary copy of the book and Resource CD, from John Wiley & Sons.)

2. **RealTime Physics** by David Sokoloff (University of Oregon), Priscilla Laws (Dickinson College), and Ronald Thornton (Tufts University). *RealTime Physics* is a set of laboratory materials that uses computer-assisted data acquisition to help students build concepts, learn representation translation, and develop an understanding of the empirical base of physics knowledge. There are three modules in the collection: Module 1: Mechanics (12 labs), Module 2: Heat and Thermodynamics (6 labs), and Module 3: Electric Circuits (8 labs). (Available both in print and in electronic form on *The Physics Suite CD*.)

---

[6]David R. Sokoloff and Ronald K. Thornton, "Using Interactive Lecture Demonstrations to Create an Active Learning Environment." *The Physics Teacher*, **35**, 340-347, September 1997.

[7]Priscilla W. Laws, *Workshop Physics Activity Guide*, Modules 1-4 w/ Appendices (John Wiley & Sons, New York, 1997).

[8]David R. Sokoloff, *RealTime Physics*, Modules 1-2, (John Wiley & Sons, New York, 1999).

[9]Lillian C. McDermott and the Physics Education Group at the University of Washington, *Physics by Inquiry* (John Wiley & Sons, New York, 1996).

[10]Lillian C. McDermott, Peter S. Shaffer, and the Physics Education Group at the University of Washington, *Tutorials in Introductory Physics*, First Edition (Prentice-Hall, Upper Saddle River, NJ, 2002).

[11]Richard N. Steinberg, Michael C. Wittmann, and Edward F. Redish, "Mathematical Tutorials in Introductory Physics," in, *The Changing Role Of Physics Departments In Modern Universities*, Edward F. Redish and John S. Rigden, editors, AIP Conference Proceedings **399**, (AIP, Woodbury NY, 1997), 1075-1092.

3. **Interactive Lecture Demonstrations** by David Sokoloff (University of Oregon) and Ronald Thornton (Tufts University). ILDs are worksheet-based guided demonstrations designed to focus on fundamental principles and address specific naïve conceptions. The demonstrations use computer-assisted data acquisition tools to collect and display high quality data in real time. Each ILD sequence is designed for delivery in a single lecture period. The demonstrations help students build concepts through a series of instructor led steps involving prediction, discussions with peers, viewing the demonstration and reflecting on its outcome. The ILD collection includes sequences in mechanics, thermodynamics, electricity, optics and more. (Available both in print and in electronic form on *The Physics Suite CD.*)

4. **Workshop Physics** by Priscilla Laws (Dickinson College). *Workshop Physics* consists of a four part activity guide designed for use in calculus-based introductory physics courses. Workshop Physics courses are designed to replace traditional lecture and laboratory sessions. Students use computer tools for data acquisition, visualization, analysis and modeling. The tools include computer-assisted data acquisition software and hardware, digital video capture and analysis software, and spreadsheet software for analytic mathematical modeling. Modules include classical mechanics (2 modules), thermodynamics & nuclear physics, and electricity & magnetism. (Available both in print and in electronic form on *The Physics Suite CD.*)

5. **Tutorials in Introductory Physics** by Lillian C. McDermott, Peter S. Shaffer and the Physics Education Group at the University of Washington. These tutorials consist of a set of worksheets designed to supplement instruction by lectures and textbook in standard introductory physics courses. Each tutorial is designed for use in a one-hour class session in a space where students can work in small groups using simple inexpensive apparatus. The emphasis in the tutorials is on helping students deepen their understanding of critical concepts and develop scientific reasoning skills. There are tutorials on mechanics, electricity and magnetism, waves, optics, and other selected topics. (Available in print from Prentice Hall, Upper Saddle River, New Jersey.)

6. **Physics by Inquiry** by Lillian C. McDermott and the Physics Education Group at the University of Washington. This self-contained curriculum consists of a set of laboratory-based modules that emphasize the development of fundamental concepts and scientific reasoning skills. Beginning with their observations, students construct a coherent conceptual framework through guided inquiry. Only simple inexpensive apparatus and supplies are required. Developed primarily for the preparation of precollege teachers, the modules have also proven effective in courses for liberal arts students and for underprepared students. The amount of material is sufficient for two years of academic study. (Available in print.)

7. **The Activity Based Physics Tutorials** by Edward F. Redish and the University of Maryland Physics Education Research Group. These tutorials, like those developed at the University of Washington, consist of a set of worksheets developed to supplement lectures and textbook work in standard introductory physics courses. But these tutorials integrate the computer software and hardware tools used in other Suite elements including computer data acquisition, digital video analysis, simulations, and spreadsheet analysis. Although these tutorials include a range of classical physics topics, they also include additional topics in modern physics. (Available only in electronic form on *The Physics Suite CD.*)

8. **The Understanding Physics Video CD for Students** by Priscilla Laws, et. al.: This CD contains a collection of the video clips that are introduced in *Understanding Physics* narrative and alternative problems. The CD includes a number of Quick-Time movie segments of physical phenomena along with the QuickTime player

software. Students can view video clips as they read the text. If they have video analysis software available, they can reproduce data presented in text graphs or complete video analyses based on assignments designed by instructors.

9. **WPTools** by Priscilla Laws and Patrick Cooney: These tools consist of a set of macros that can be loaded with Microsoft Excel software that allow students to graph data transferred from computer data acquisition software and video analysis software more easily. Students can also use the *WPTools* to analyze numerical data and develop analytic mathematical models.

10. **The Physics Suite CD.** This CD contains a variety of the Suite Elements in electronic format (Microsoft Word files). The electronic format allows instructors to modify and reprint materials to better fit into their individual course syllabi. The CD contains much useful material including complete electronic versions of the following: *RealTime Physics, Interactive Lecture Demonstrations, Workshop Physics, Activity Based Physics Tutorials.*

A Final Word to the Instructor

Over the past decade we have learned how valuable it is for us as teachers to focus on what most students actually need to do to learn physics, and how valuable it can be for students to work with research-based materials that promote active learning. We hope you and your students find this book and some of the other *Physics Suite* materials helpful in your quest to make physics both more exciting and understandable to your students.

Supplements for Use with Understanding Physics

Instructor Supplements

1. **Instructor's Solution Manual** prepared by Anand Batra (Howard University). This manual provides worked-out solutions for most of the end-of-chapter problems.

2. **Test Bank** by J. Richard Christman (U. S. Coast Guard Academy). This manual includes more than 2500 multiple-choice questions adapted from HRW 6th. These items are also available in the *Computerized Test Bank* (see below).

3. **Instructor's Resource CD.** This CD contains:

   - The entire *Instructor's Solutions Manual* in both Microsoft Word© (IBM and Macintosh) and PDF files.
   - A *Computerized Test Bank,* for use with both PCs and Macintosh computers with full editing features to help you customize tests.
   - All text illustrations, suitable for classroom projection, printing, and web posting.

4. **Online Homework and Quizzing:** *Understanding Physics* supports WebAssign and eGrade, two programs that give instructors the ability to deliver and grade homework and quizzes over the Internet.

Student Supplements

1. **Student Study Guide** by J. Richard Christman (U. S. Coast Guard Academy). This student study guide provides chapter overviews, hints for solving selected end-of-chapter problems, and self-quizzes.

2.   **Student Solutions Manual** by J. Richard Christman (U. S. Coast Guard Academy). This manual provides students with complete worked-out solutions for approximately 450 of the odd-numbered end-of-chapter problems.

Acknowledgements

Many individuals helped us create this book. The authors are grateful to the individuals who attended the weekend retreats at Airlie Center in 1997 and 1998 and to our editor, Stuart Johnson and to John Wiley & Sons for sponsoring the sessions. It was in these retreats that the ideas for *Understanding Physics* crystallized. We are grateful to Jearl Walker, David Halliday and Bob Resnick for graciously allowing us to attempt to make their already fine textbook better.

The authors owe special thanks to Sara Settlemyer who served as an informal project manager for the past few years. Her contributions included physics advice (based on her having completed Workshop Physics courses at Dickinson College), her use of Microsoft Word, Adobe Illustrator, Adobe Photoshop and Quark XPress to create the manuscript and visuals for this edition, and skillful attempts to keep our team on task—a job that has been rather like herding cats.

Karen Cummings: I would like to say "Thanks!" to: Bill Lanford (for endless advice, use of the kitchen table and convincing me that I really could keep the same address for more than a few years in a row), Ralph Kartel Jr. and Avery Murphy (for giving me an answer when people asked why I was working on a textbook), Susan and Lynda Cummings (for the comfort, love and support that only sisters can provide), Jeff Marx, Tim French and the poker crew (for their friendship and laughter), my colleagues at Southern Connecticut and Rensselaer, especially Leo Schowalter, Jim Napolitano and Jack Wilson (for the positive influence you have had on my professional life) and my students at Southern Connecticut and Rensselaer, Ron Thornton, Priscilla Laws, David Sokoloff, Pat Cooney, Joe Redish, Ken and Pat Heller and Lillian C. McDermott (for helping me learn how to teach).

Priscilla Laws: First of all I would like thank my husband and colleague Ken Laws for his quirky physical insights, for the Chapter 11 Kneecap puzzler, for the influence of his physics of dance work on this book, and for waiting for me countless times while I tried to finish "just one more thing" on this book. Thanks to my daughter Virginia Jackson and grandson Adam for all the fun times that keep me sane. My son Kevin Laws deserves special mention for sharing his creativity with us—best exemplified by his murder mystery problem, *A(dam)nable Man,* reprinted here as problem 5-68. I would like to thank Juliet Brosing of Pacific University who adapted many of the Workshop Physics problems developed at Dickinson for incorporation into the alternative problem collection in this book. Finally, I am grateful to my Dickinson College colleagues Robert Boyle, Kerry Browne, David Jackson, and Hans Pfister for advice they have given me on a number of topics.

Joe Redish: I would like to thank Ted Jacobsen for discussions of our chapter on relativity and Dan Lathrop for advice on the sources of the Earth's magnetic field, as well as many other of my colleagues at the University of Maryland for discussions on the teaching of introductory physics over many years.

Pat Cooney: I especially thank my wife Margaret for her patient support and constant encouragement and I am grateful to my colleagues at Millersville University: John Dooley, Bill Price, Mike Nolan, Joe Grosh, Tariq Gilani, Conrad Miziumski, Zenaida Uy, Ned Dixon, and Shawn Reinfried for many illuminating conversations.

We also appreciate the absolutely essential role many reviewers and classroom testers played. We took our reviewers very seriously. Several reviewers and testers deserve special mention. First and foremost is Arnold Arons who managed to review 29 of the 38 chapters either from the original HRW 6th material or from our early drafts before he passed away in February 2001. Vern Lindberg from the Rochester

Institute of Technology deserves special mention for his extensive and very insightful reviews of most of our first 18 chapters. Ed Adelson from Ohio State did a particularly good job reviewing most of our electricity chapters. Classroom tester Maxine Willis from Gettysburg Area High School deserves special recognition for compiling valuable comments that her advanced placement physics students made while class testing Chapters 1-12 of the preliminary version. Many other reviewers and class testers gave us useful comments in selected chapters.

Class Testers

Gary Adams
Rensselaer Polytechnic Institute

Marty Baumberger
Chestnut Hill Academy

Gary Bedrosian
Rensselaer Polytechnic Institute

Joseph Bellina,
Saint Mary's College

Juliet W. Brosing
Pacific University

Shao-Hsuan Chiu
Frostburg State

Chad Davies
Gordon College

Hang Deng-Luzader
Frostburg State

John Dooley
Millersville University

Diane Dutkevitch
Yavapai College

Timothy Hayes
Rensselaer Polytechnic Institute

Brant Hinrichs
Drury College

Kurt Hoffman
Whitman College

James Holliday
John Brown University

Michael Huster
Simpson College

Dennis Kuhl
Marietta College

John Lindberg
Seattle Pacific University

Vern Lindberg
Rochester Institute of Technology

Stephen Luzader
Frostburg State

Dawn Meredith
University of New Hampshire

Larry Robinson
Austin College

Michael Roth
University of Northern Iowa

John Schroeder
Rensselaer Polytechnic Institute

Cindy Schwarz
Vassar College

William Smith
Boise State University

Dan Sperber
Rensselaer Polytechnic Institute

Roger Stockbauer
Louisiana State University

Paul Stoler
Rensselaer Polytechnic Institute

Daniel F. Styer
Oberlin College

Rebecca Surman
Union College

Robert Teese
Muskingum College

Maxine Willis
Gettysburg Area High School

Gail Wyant
Cecil Community College

Anne Young
Rochester Institute of Technology

David Ziegler
Sedro-Woolley High School

Reviewers

Edward Adelson
Ohio State University

Arnold Arons
University of Washington

Arun Bansil
Northeastern University

Chadan Djalali
University of South Carolina

William Dawicke
Milwaukee School of Engineering

Robert Good
California State University-Hayware

Harold Hart
Western Illinois University

Harold Hastings
Hofstra University

Laurent Hodges
Iowa State University

Robert Hilborn
Amherst College

Theodore Jacobson
University of Maryland

Leonard Kahn
University of Rhode Island

Stephen Kanim
New Mexico State University

Hamed Kastro
Georgetown University

Debora Katz
U. S. Naval Academy

Todd Lief
Cloud Community College

Vern Lindberg
Rochester Institute of Technology

Mike Loverude
California State University-Fullerton

Robert Luke
Boise State University

Robert Marchini
Memphis State University

Tamar More
Portland State University

Gregor Novak
U. S. Air Force Academy

Jacques Richard
Chicago State University

Cindy Schwarz
Vassar College

Roger Sipson
Moorhead State University

George Spagna
Randolf-Macon College

Gay Stewart
University of Arkansas-Fayetteville

Sudha Swaminathan
Boise State University

We would like to thank our proof readers Georgia Mederer and Ernestine Franco, our copyeditor Helen Walden, and our illustrator Julie Horan.

Last but not least we would like to acknowledge the efforts of the Wiley staff; Senior Acquisitions Editor, Stuart Johnson, Ellen Ford (Senior Development Editor), Justin Bow (Program Assistant), Geraldine Osnato (Project Editor), Elizabeth Swain (Senior Production Editor), Hilary Newman (Senior Photo Editor), Anna Melhorn (Illustration Editor), Kevin Murphy (Senior Designer), and Bob Smith (Marketing Manager). Their dedication and attention to endless details was essential to the production of this book.

Brief Contents

Contents

Appendices

13 | Equilibrium and Elasticity

Rock climbing may be the ultimate physics exam. Failure can mean death, and even "partial credit" can mean severe injury. For example, in a long chimney climb, in which your torso is pressed against one wall of a wide vertical fissure and your feet are pressed against the opposite wall, you need to rest occasionally or you will fall due to exhaustion. Here the exam consists of a single question: What can you do to relax your push on the walls in order to rest? If you relax without considering the physics, the walls will not hold you up.

What is the answer to this life-and-death, one-question exam?

The answer is in this chapter.

13-1 Introduction

In this chapter we consider objects that remain motionless in the presence of external forces and torques. In particular we address the questions: (1) Under what conditions can objects that experience external forces and torques remain motionless (a state we will come to call static equilibrium)? (2) Under what conditions can objects that experience external forces and torques deform in order to remain in static equilibrium? The answers to these questions are of vital importance to engineers in the design of structures such as buildings, dams, roads, and bridges. The design engineer must identify all the external forces and torques that may act on a structure and, by good design and wise choice of materials, ensure that the structure will remain stable under these loads.

We begin this chapter by defining static equilibrium. We then use Newton's laws to summarize the conditions needed to keep structures in static equilibrium. For simplicity, the conditions for static equilibrium are developed assuming that objects do not change shape in the presence of external forces.

In Section 6-4 we discussed how the contact forces that an object can exert, such as normal and tension forces, result from the deformation of tiny spring-like bonds that separate atoms. For this reason no material is perfectly rigid. "Perfect rigidity" is an idealization, just like the assumptions that air resistance is negligible or that a surface is frictionless. Even a table sags under the load of a sheet of paper. Often, the change in shape is so small that we cannot observe it directly. However, sometimes the change in shape is easily noticeable. Engineers need to predict how an object of a given composition, shape, and size will deform as a function of the external forces on it for two reasons. First, its change of shape may, in turn, change the nature of the external forces it experiences and thereby force it out of static equilibrium. Second, the object could be deformed to the breaking point. For these reasons, the remainder of the chapter deals with an introduction to how structures deform in the presence of external forces.

13-2 Equilibrium

Consider these objects: (1) a book resting on a table, (2) a hockey puck sliding across a frictionless surface with constant velocity, (3) the rotating blades of a ceiling fan, and (4) the wheel of a bicycle that is traveling along a straight path at constant speed. For each of these four objects:

1. The translational momentum $\vec{p}_{com}$ of its center of mass is constant.

2. The rotational momentum $\vec{L}_{com}$ about its center of mass, or about any other point, is also constant.

Even though all four of these objects are moving, we say that they are in **equilibrium** because in each case both the translational momentum and rotational momentum of the object's center of mass are constant. Thus, the two requirements for equilibrium are

$$\vec{p}_{com} = \text{a constant} \quad \text{and} \quad \vec{L}_{com} = \text{a constant.} \tag{13-1}$$

Static Equilibrium and Stability

Our primary concern in this chapter is with objects that are not moving in any way—either in translation or in rotation—in the reference frame from which we observe them. Such objects are defined as being in **static equilibrium** whenever both the translational momentum and rotational momentum of the center of mass of the system is zero. In other words, if an object is in static equilibrium the constants in Eq. 13-1 must be zero. Of the four objects mentioned at the beginning of this section, only one — the book resting on the table — is in static equilibrium.

The balancing rock of Fig. 13-1 is another example of an object that, for the present at least, is in static equilibrium. It shares this property with countless other structures, such as cathedrals, houses, filing cabinets, and taco stands, that remain stationary over time.

As we discussed in Chapter 10, if a body returns to a state of static equilibrium after having been displaced from it by a force, the body is said to be in *stable* static equilibrium. A marble placed at the bottom of a hemispherical bowl is an example. However, if a small force can displace the body and end the equilibrium, the body is in *unstable* static equilibrium.

As an example of unstable static equilibrium, suppose we balance a domino with the domino's center of mass vertically above the supporting edge as in Fig. 13-2a. The torque about the supporting edge due to the gravitational force $\vec{F}^{\text{grav}}$ on the domino is zero, because the line of action of $\vec{F}^{\text{grav}}$ is through that edge. Thus, the domino is in equilibrium. Of course, even a slight force on it due to some chance disturbance ends the equilibrium. As the line of action of $\vec{F}^{\text{grav}}$ moves to one side of the supporting edge (as in Fig. 13-2b), the torque due to $\vec{F}^{\text{grav}}$ will cause the domino's rotation. Thus, the domino in Fig. 13-2a is in unstable static equilibrium.

FIGURE 13-1 ■ A balanced rock in the Arches National Park, Utah. Although its perch seems precarious, the rock is in static equilibrium.

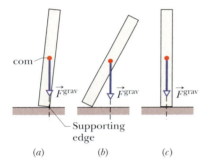 

FIGURE 13-2 ■ (a) A domino balanced on one edge, with its center of mass vertically above that edge. The gravitational force $\vec{F}^{\text{grav}}$ on the domino is directed through the supporting edge. (b) If the domino is rotated even slightly clockwise from the balanced orientation, then $\vec{F}^{\text{grav}}$ causes a torque that increases the rotation. (c) A domino upright on a narrow side is somewhat more stable than the domino in (a). (d) A cubical block is even more stable.

The domino in Fig. 13-2c is slightly more stable. To topple this domino, a force would have to rotate it through and then beyond the balance position of Fig. 13-2a, in which the center of mass is above a supporting edge. A slight force will not topple this domino, but a vigorous flick of the finger against the domino certainly will. (If we arrange a chain of such upright dominos, a finger flick against the first can cause the whole chain to fall.)

The child's cubical block in Fig. 13-2d is even more stable because its center of mass would have to be moved even farther to get it to pass above a supporting edge. A flick of the finger may not topple the block. (This is why you never see a chain of toppling blocks.) The worker in Fig. 13-3 is like both the domino and the square block. Parallel to the beam, his stance is wide and he is stable. Perpendicular to the beam, his stance is narrow and he is unstable (and at the mercy of a chance gust of wind).

FIGURE 13-3 ■ A construction worker balanced above New York City is in static equilibrium but is more stable parallel to the beam than perpendicular to it.

The Conditions for Static Equilibrium

The translational motion of a body is governed by Newton's Second Law. In its translational momentum form, this relation is given as

$$\vec{F}^{\text{net}} = \frac{d\vec{p}}{dt}.$$

(13-2)

If the body is in translational equilibrium—that is, if $\vec{p}$ is a constant—then $d\vec{p}/dt = 0$ and we must have

$$\vec{F}^{\text{net}} = 0 \qquad \text{(balance of forces)}.$$

(13-3)

The rotational motion of a body is governed by Newton's Second Law in its rotational momentum form, given by Eq. 12-32 as

$$\vec{\tau}^{\text{net}} = \frac{d\vec{L}}{dt}.$$

(13-4)

If the body is in rotational equilibrium—that is, if $\vec{L}$ is a constant—then $d\vec{L}/dt = 0$ and we must have

$$\vec{\tau}^{\text{net}} = 0 \qquad \text{(balance of torques)}.$$

(13-5)

Thus, two requirements for a body to be in equilibrium are as follows:

> **If a body is in equilibrium:** (1) The vector sum of all the external forces that act on it must be zero; and (2) the vector sum of all the external torques that act on it, measured about *any* possible point, must also be zero.

Although these requirements obviously hold for *static* equilibrium, they also hold for the more general equilibrium in which $\vec{p}$ and $\vec{L}$ are constant but not zero.

We can express the vector form of equilibrium represented by Eqs. 13-3 and 13-5 in terms of three independent component equations, one for each axis in the chosen coordinate system:

Balance of Force Components	Balance of Torque Components	
$F_x^{\text{net}} = \Sigma F_x = 0$	$\tau_x^{\text{net}} = \Sigma \tau_x = 0$	
$F_y^{\text{net}} = \Sigma F_y = 0$	$\tau_y^{\text{net}} = \Sigma \tau_y = 0$	(13-6)
$F_z^{\text{net}} = \Sigma F_z = 0$	$\tau_z^{\text{net}} = \Sigma \tau_z = 0$	

We shall simplify matters by considering only situations in which the forces that act on the body lie in the *x-y* plane. This means that the only torques that can act on the body must tend to cause rotation around an axis parallel to the *z* axis. With this assumption, we eliminate one force equation and two torque equations leaving

$$F_x^{\text{net}} = 0 \qquad \text{(balance of forces)},$$

(13-7)

$$F_y^{\text{net}} = 0 \qquad \text{(balance of forces)},$$

(13-8)

$$\tau_z^{\text{net}} = 0 \qquad \text{(balance of torques)}.$$

(13-9)

Here, τ_z^{net} is the net torque that the external forces produce either about the z axis or about *any* axis parallel to the z axis.

The conditions for static equilibrium are more stringent than those for general equilibrium. For example, a hockey puck that is sliding at constant velocity over ice while spinning about its center of mass with a constant rotational velocity satisfies the conditions for general equilibrium. But the puck is *not in static* equilibrium. The requirements for static equilibrium are that:

> **In static equilibrium** all parts of a body must be at rest in an inertial (nonaccelerating) frame of reference with no net force and no net torque acting on it.

READING EXERCISE 13-1: The figure gives six overhead views of a uniform rod on which two or more forces act perpendicular to the rod. If the magnitudes of the forces are adjusted properly (but kept nonzero), in which situations can the rod be in static equilibrium?

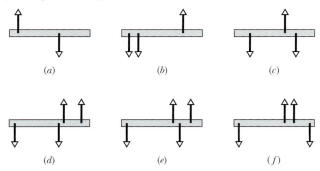

13-3 The Center of Gravity

Consider an extended body that is close to the surface of the Earth. The gravitational force on this body is the vector sum of the gravitational forces acting on the individual elements (the atoms) of the body. Instead of considering all those individual elements, we can say:

> The gravitational force $\vec{F}^{\text{grav}}$ on a body effectively acts at a single point, called the **center of gravity** (cog) of the body.

Here the word "effectively" means that if the gravitational forces on the individual elements were somehow turned off and force $\vec{F}^{\text{grav}}$ at the center of gravity were turned on, the net force and the net torque (about any point) acting on the body would not change.

Until now, we have assumed that the gravitational force $\vec{F}^{\text{grav}}$ acts at the center of mass (com) of the body. This is equivalent to assuming that the center of gravity is at the center of mass. Considering Fig. 13-4, it can be shown mathematically that

> If the local gravitational strength, g, is the same for all elements of a body, then the body's center of gravity (cog) is coincident with the body's center of mass (com).

This constancy is a very good approximation for everyday objects because, as we explained in Section 3-9, g varies only slightly along Earth's surface and with altitude. Thus, for objects like a mouse or a skyscraper, we can assume that the gravitational force acts at the center of mass.

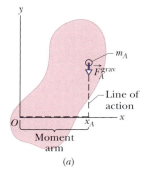

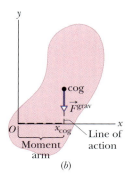

FIGURE 13-4 ■ (a) An element of mass m_A in an extended body. The y-component of the gravitational force F_{Ay}^{grav} on it has moment arm x_A about the origin O of the coordinate system. (b) The gravitational force $\vec{F}^{\text{grav}}$ on a body is said to act at the center of gravity (cog) of the body. Here it has moment arm x_{cog} about origin O.

READING EXERCISE 13-2: Suppose that you skewer an apple with a thin rod, missing the apple's center of gravity. When you hold the rod horizontally and allow the apple to rotate freely, where does the center of gravity end up and why? ■

READING EXERCISE 13-3: A ladder is leaning against a wall as shown in the figure. If you were going to climb the ladder, are you better off if there is no friction between the ladder and the wall or no friction between the ladder and the floor? Explain your answer in terms of the physics you have learned. *Hint:* It will be helpful to draw an extended free-body diagram like that shown in Touchstone Example 13-2 and then consider the conditions that must be met to keep the ladder from moving.

TOUCHSTONE EXAMPLE 13-1: Cat on a Plank

Two workmen are carrying a 6.0-m-long plank as shown in Fig. 13-5. The plank has a mass of 15 kg. A cat, with a mass of 5.0 kg, jumps on the plank and hangs on, 1.0 m from the end of the plank. Assuming that the workers are walking at a constant velocity, how much force does each workman have to exert to hold the plank up?

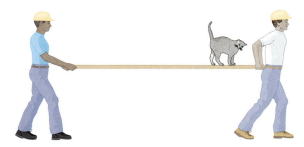

FIGURE 13-5 ■ Workmen holding up a cat and a plank.

SOLUTION ■ The first **Key Idea** here is that for an extended object such as the plank and cat to move at a constant velocity without rotating, it must be in equilibrium. Two conditions must be satisfied for equilibrium: (1) The net force on the plank must be zero to ensure that its center of mass is not accelerating; and (2) the net torque on it must be zero to ensure that it is not rotating. The second **Key Idea** here is that we can treat the downward gravitational forces on each of the mass elements that make up the plank as a single force acting at the plank's center of gravity.

The first step of our solution is to identify and locate the forces acting on the plank and display them as an extended free-body diagram as shown in Fig. 13-6. There are two downward forces on the plank: the gravitational force assumed to be acting as its center of gravity and the force exerted by the cat due to its weight. These must be counterbalanced by the upward forces exerted by worker A on the left and worker B on the right.

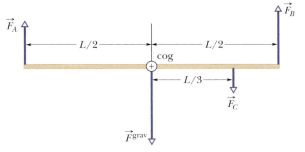

FIGURE 13-6 ■ An extended free-body diagram.

Next we can set up equations to balance the force and torque components shown in Eq. 13-6. Let us choose a conventional rectangular coordinate system with the origin at the center of gravity of the plank. The y axis is up, the x axis is in the plane of the page, and the z axis points out of the page. In this case we find that all the forces are parallel to the y axis, so there is only one force component balance equation,

$$F_y^{\text{net}} = F_{Ay} + F_y^{\text{grav}} + F_{By} + F_{Cy} = 0 \quad \text{(force component balance).}$$
(13-10)

In this case, the equation for balancing torque components must be taken about the z axis (or any axis parallel to it). Let's stick with the z axis since it rather conveniently passes through the center of gravity of the plank. The torque balance equation that follows has a positive (counterclockwise) z-component of torque due to worker B's force and two negative (clockwise) z-components of torque due to worker A's force and the force exerted by the cat. We can express this as

$$\tau_z^{\text{net}} = +\tfrac{L}{2}|F_{By}| - \tfrac{L}{2}|F_{Ay}| - \tfrac{L}{3}|F_{Cy}| = 0.$$

Since we know that the y-components of the workers' forces are both positive and the cat's force component is negative, we can rewrite the torque balance equation as

$$\tau_z^{\text{net}} = +\tfrac{L}{2}F_{By} - \tfrac{L}{2}F_{Ay} + \tfrac{L}{3}F_{Cy} = 0 \qquad \text{(torque component balance)}.$$

If we then eliminate the length of the plank by dividing each term by $L/2$, we get

$$F_{By} - F_{Ay} + \tfrac{2}{3}F_{Cy} = 0. \qquad (13\text{-}11)$$

Recall that we are trying to use our balance equations to determine $\vec{F}_A = F_{Ay}\,\hat{j}$ and $\vec{F}_B = F_{By}\,\hat{j}$. We have been given the information needed to find the gravitational force of the center of gravity of the plank and the force the cat exerts on the plank. In particular,

$$F_y^{\text{grav}} = -Mg = -(15\,\text{kg})(9.8\,\text{N/kg}) = -147\,\text{N}, \quad (13\text{-}12)$$

and

$$F_{Cy} = -m_{\text{cat}}g = -(5.0\,\text{kg})(9.8\,\text{N/kg}) = -49.0\,\text{N}. \quad (13\text{-}13)$$

Thus we have two equations (Eqs. 13-10 and 13-11) with two unknowns, so we should be able to find our unknowns F_{Ay} and F_{By}. If we add Eq. 13-10 to Eq. 13-11 and solve for F_{By}, we get

$$F_{By} = \tfrac{1}{2}(-F_y^{\text{grav}} - \tfrac{5}{3}F_{Cy}) = \tfrac{1}{2}(147\,\text{N} + (\tfrac{5}{3})49\,\text{N})$$
$$= 114.333\,\text{N} \cong 114\,\text{N}. \qquad \text{(Answer)}$$

Finally, we can solve Eq. 13-11 for F_{Ay} to get

$$F_{Ay} = F_{By} + \tfrac{2}{3}F_{Cy} = 114\,\text{N} + \tfrac{2}{3}(-49\,\text{N}) = 81.666\,\text{N} \cong 82\,\text{N}.$$
$$\text{(Answer)}$$

As you might have predicted, worker B, who is closer to the cat, has to exert a larger force than worker A does. However, note that we were able to factor the length L out of the torque balance equation, so the forces exerted by the workers do not depend on the length of the plank but only on the fraction of the distance that the cat is from the ends of the plank.

TOUCHSTONE EXAMPLE 13-2: Fireman on a Ladder

In Fig. 13-7a, a ladder of length $L = 12$ m and mass $m = 45$ kg leans against a slick (frictionless) wall. Its upper end is at height $h = 9.3$ m above the pavement on which the lower end rests (the pavement is not frictionless). The ladder's center of mass is $L/3$ from the lower end. A firefighter of mass $M = 72$ kg climbs the ladder until her center of mass is $L/2$ from the lower end. What then are the magnitudes of the forces on the ladder from the wall and the pavement?

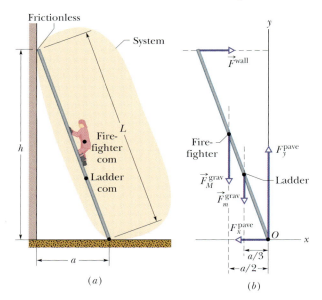

FIGURE 13-7 ■ (a) A firefighter climbs halfway up a ladder that is leaning against a frictionless wall. The pavement beneath the ladder is not frictionless. (b) A free-body diagram, showing the forces that act on the firefighter–ladder system. The origin O of a coordinate system is placed at the point of application of the unknown force $\vec{F}^{\text{pave}}$ (whose components F_x^{pave} and F_y^{pave} are shown).

SOLUTION ■ First, we choose our system as being the firefighter and ladder together, and then we draw the free-body diagram shown in Fig. 13-7b. The firefighter is represented with a dot within the boundary of the ladder. The gravitational force on her, $\vec{F}_M^{\text{grav}}$, has been shifted along its line of action, so that its tail is on the dot. (The shift does not alter a torque due to $\vec{F}_M^{\text{grav}}$ about any axis perpendicular to the figure.)

The only force on the ladder from the wall is the horizontal force $\vec{F}^{\text{wall}}$ (there cannot be frictional force along a frictionless wall). The force $\vec{F}^{\text{pave}}$ on the ladder from the pavement has a horizontal component F_x^{pave} that is a static frictional force and a vertical component F_y^{pave} that is a normal force.

A **Key Idea** here is that the system is in static equilibrium, so the balancing equations (Eqs. 13-7 through 13-9) apply to it. Let us start with Eq. 13-9 ($\tau_z^{\text{net}} = 0$). To choose an axis about which to calculate the torques, note that we have unknown forces ($\vec{F}^{\text{wall}}$ and $\vec{F}^{\text{pave}}$) at the two ends of the ladder. To eliminate, say, $\vec{F}^{\text{pave}}$ from the calculation, we place the axis at point O, perpendicular to the figure. We also place the origin of an xy coordinate system at O. We can find torques about O with any of Eqs. 11-29 through 11-31, but Eq. 11-31 ($|\vec{\tau}| = r_\perp |\vec{F}|$) is easiest to use here.

To find the moment arm $r_\perp$ of $\vec{F}^{\text{wall}}$, we draw a line of action through that vector (Fig. 13-7b). Then $r_\perp$ is the perpendicular distance between O and the line of action. In Fig. 13-7b, it extends along the y axis and is equal to the height h. Similarly, we draw lines of action for $\vec{F}_M^{\text{grav}}$ and $\vec{F}_m^{\text{grav}}$ and see that their moment arms extend along the x axis. For the distance a shown in Fig. 13-7a, the moment arms are $a/2$ (the firefighter is halfway up the ladder) and $a/3$ (the ladder's center of mass is one-third of the way up the ladder), respectively. The moment arms for F_x^{pave} and F_y^{pave} are zero.

Now, the torques can be written in the form $r_\perp F$. The balancing equation $\tau_z^{\text{net}} = 0$ becomes

$$-(h)(|\vec{F}^{\,\text{wall}}|) + (a/2)(Mg) + (a/3)(mg) + (0)(|F_x^{\text{pave}}|)$$
$$+ (0)(|F_y^{\text{pave}}|) = 0. \qquad (13\text{-}14)$$

(Recall our rule: A positive torque corresponds to counterclockwise rotation and a negative torque corresponds to clockwise rotation.)
Using the Pythagorean theorem, we find that

$$a = \sqrt{L^2 - h^2} = 7.58 \text{ m}.$$

Then Eq. 13-14 gives us

$$|\vec{F}^{\,\text{wall}}| = \frac{ga(M/2 + m/3)}{h}$$
$$= \frac{(9.8 \text{ m/s}^2)(7.58 \text{ m})(72/2 \text{ kg} + 45/3 \text{ kg})}{9.3 \text{ m}} \quad \text{(Answer)}$$
$$= 4.1 \times 10^2 \text{ N}.$$

But since $\vec{F}^{\,\text{wall}}$ points to the right, its component F_x^{wall} is also $+410$ N.

Now we need to use the force-balancing equations. The equation $F_x^{\text{net}} = 0$ gives us

$$F_x^{\text{wall}} + F_x^{\text{pave}} = 0,$$

so

$$F_x^{\text{pave}} = -F_x^{\text{wall}} = -4.1 \times 10^2 \text{ N}, \qquad \text{(Answer)}$$

where the minus sign tells us F_x^{pave} points to the left.

Since gravitational forces are negative whereas the local gravitational strength g and the force component F_y^{pave} are positive, the equation $F_y^{\text{net}} = 0$ gives us

$$F_y^{\text{pave}} - Mg - mg = 0.0 \text{ N} \quad \text{(force component balance)},$$

so

$$F_y^{\text{pave}} = (M + m)g = (72 \text{ kg} + 45 \text{ kg})(9.8 \text{ m/s}^2)$$
$$= 1146.6 \text{ N} \approx 1.15 \times 10^3 \text{ N}. \qquad \text{(Answer)}$$

TOUCHSTONE EXAMPLE 13-3: Safe on a Boom

Figure 13-8a shows a safe, of mass $M = 430$ kg, hanging by a rope from a boom with dimensions $a = 1.9$ m and $b = 2.5$ m. The boom consists of a hinged beam and a horizontal cable that connects the beam to a wall. The uniform beam has a mass m of 85 kg; the mass of the cable and rope are negligible.

(a) What is the tension T^{cable} in the cable? In other words, what is the magnitude of the force $\vec{T}^{\text{cable}}$ on the beam from the cable?

SOLUTION ■ The system here is the beam alone, and the forces on it are shown in the free-body diagram of Fig. 13-8b. The force from the cable is $\vec{T}^{\text{cable}}$. The gravitational force on the beam acts at the beam's center of mass (at the beam's center) and is represented by its equivalent $\vec{F}_m^{\text{grav}}$. The vertical component of the force on the beam from the hinge is F_y^{hinge} and the horizontal component of the force from the hinge is F_x^{hinge}. The force from the rope supporting the safe is $\vec{T}^{\text{rope}}$. Because beam, rope, and safe are stationary, the magnitude of $\vec{T}^{\text{rope}}$ is equal to the weight of the safe: $|\vec{T}^{\text{rope}}| = Mg$. We place the origin O of an xy coordinate system at the hinge.

One **Key Idea** here is that our system is in static equilibrium, so the balancing equations apply to it. Let us start with Eq. 13-9 $\tau_z^{\text{net}} = 0$. Note that we are asked for the magnitude of force $\vec{T}^{\text{cable}}$ and not of force components F_x^{hinge} and F_y^{hinge} acting at the hinge, at point O. Thus, a second **Key Idea** is that, to eliminate F_x^{hinge} and F_y^{hinge} from the torque calculation, we should calculate torques about an axis that is perpendicular to the figure at point O. Then F_x^{hinge} and F_y^{hinge} will have moment arms of zero. The lines of action for $\vec{T}^{\text{cable}}$, $\vec{T}^{\text{rope}}$, and $\vec{F}_m^{\text{grav}}$ are dashed in Fig. 13-8b. The corresponding moment arms are $a, b,$ and $b/2$.

Writing torques in the form of $r_\perp |\vec{F}^{\text{hinge}}|$ and using our rule about signs for torques, the balancing equation $\tau_z^{\text{net}} = 0$ becomes

$$(a)(|\vec{T}^{\text{cable}}|) - (b)(|\vec{T}^{\text{rope}}|) - (\tfrac{1}{2}b)(mg) = 0.$$

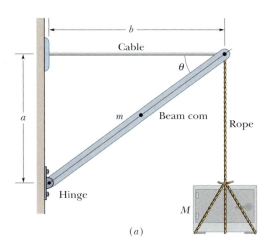

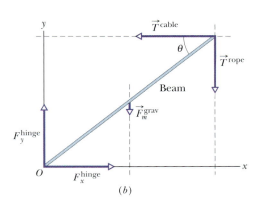

(a) (b)

FIGURE 13-8 ■ (a) A heavy safe is hung from a boom consisting of a horizontal steel cable and a uniform beam. (b) A free-body diagram for the beam.

Substituting $|\vec{F}_M^{\,grav}|$ for $|\vec{T}^{\,rope}|$ and solving for $|\vec{T}^{\,cable}|$, we find that

$$|\vec{T}^{\,cable}| = \frac{gb(M + \frac{1}{2}m)}{a}$$

$$= \frac{(9.8 \text{ m/s}^2)(2.5 \text{ m})(430 \text{ kg} + 85/2 \text{ kg})}{1.9 \text{ m}}$$

$$= 6093 \text{ N} \approx 6100 \text{ N}. \qquad \text{(Answer)}$$

Since $\vec{T}^{\,cable}$ points along the negative x axis, its component T_x^{cable} is negative, so

$$T_x^{cable} = -|\vec{T}^{\,cable}| = -6.1 \times 10^3 \text{ N}.$$

(b) Find the magnitude $|\vec{F}^{\,hinge}|$ of the net force on the beam from the hinge.

SOLUTION ■ Now we want to know the values of the force components F_x^{hinge} and F_y^{hinge} so we can combine them to get $|\vec{F}^{\,hinge}|$. Because we know T_x^{cable}, our **Key Idea** here is to apply the force-balancing equations to the beam. For the horizontal balance, we write $F_x^{\,net} = 0$ as

$$F_x^{hinge} + T_x^{cable} = 0,$$

and so

$$F_x^{hinge} = -T_x^{cable} = +6093 \text{ N}.$$

For the vertical balance, we write $F_y^{\,net} = 0$ as

$$F_y^{hinge} - mg + T_y^{rope} = 0.$$

Substituting $-Mg$ for T_y^{rope} and solving for F_y^{hinge}, we find that

$$F_y^{hinge} = (m + M)g = (85 \text{ kg} + 430 \text{ kg})(9.8 \text{ m/s}^2)$$

$$= +5047 \text{ N} \approx 5.0 \times 10^3 \text{ N}.$$

The net force vector is then

$$\vec{F}^{\,hinge} = F_x^{hinge}\,\hat{i} + F_y^{hinge}\,\hat{j}$$

$$= (6093 \text{ N})\hat{i} + (5047 \text{ N})\hat{j}.$$

From the Pythagorean theorem, we now have

$$|\vec{F}^{\,hinge}| = \sqrt{(F_x^{hinge})^2 + (F_y^{hinge})^2}$$

$$= \sqrt{(6093 \text{ N})^2 + (5047 \text{ N})^2} \approx 7.9 \times 10^3 \text{ N}.$$

$$\text{(Answer)}$$

Note that $|\vec{F}^{\,hinge}|$ is substantially greater than either the combined weights of the safe and the beam, which is 5.0×10^3 N, or the tension in the horizontal cable, which is 6.1×10^3 N.

TOUCHSTONE EXAMPLE 13-4: Chimney Climbing

In Fig. 13-9, a rock climber with mass $m = 55$ kg rests during a "chimney climb," pressing only with her shoulders and feet against the walls of a fissure of width $w = 1.0$ m. Her center of mass is a horizontal distance $d = 0.20$ m from the wall against which her shoulders are pressed. The coefficient of static friction between her shoes and the wall is $\mu_{shoes}^{stat} = 1.1$, and between her shoulders and the wall it is $\mu_{shoulders}^{stat} = 0.70$. To rest, the climber wants to minimize her horizontal push on the walls. The minimum occurs when her feet and her shoulders are both on the verge of sliding.
(a) What is that minimum horizontal push on the walls?

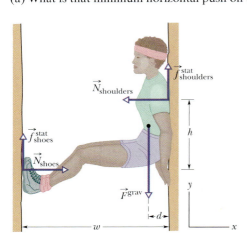

FIGURE 13-9 ■ The forces on a climber resting in a rock chimney. The push of the climber on the chimney walls results in a rise of the normal forces $\vec{N}_{shoes}$ and $\vec{N}_{shoulders}$ and the static frictional forces $\vec{f}_{shoes}^{\,stat}$ and $\vec{f}_{shoulders}^{\,stat}$.

SOLUTION ■ Our system is the climber, and Fig. 13-9 shows the forces that act on her. The only horizontal forces are the normal forces $\vec{N}_{shoes}$ and $\vec{N}_{shoulders}$ on her from the walls, at her feet and shoulders. The static frictional forces on her are $\vec{f}_{shoes}^{\,stat}$ and $\vec{f}_{shoulders}^{\,stat}$, directed upward. The gravitational force $\vec{F}^{\,grav}$ acts downward at her center of gravity.

A **Key Idea** is that, because the system is in static equilibrium, we can apply the force-balancing equations (Eqs. 13-7 and 13-8) to it. The equation $F_x^{\,net} = 0$ tells us that the two normal forces on her must be equal in magnitude and opposite in direction. We seek the magnitude $|\vec{N}|$ of these two forces, which is also the magnitude of her push against either wall.

The balancing equation $F_y^{\,net} = 0$ gives us

$$F_y^{\,net}\hat{j} = \vec{f}_{shoes}^{\,stat} + \vec{f}_{shoulders}^{\,stat} + \vec{F}^{\,grav} = 0, \quad \text{where } \vec{F}^{\,grav} = -mg\hat{j}.$$
$$(13\text{-}15)$$

We want the climber to be on the verge of sliding at both her feet and her shoulders. That means we want the static frictional forces there to be at their maximum values. Those maximum magnitudes are, from Eq. 6-11, $(|\vec{f}_{max}^{\,stat}| = \mu^{stat}|\vec{N}|)$,

$$|\vec{f}_{shoes}^{\,stat}| = \mu_{shoes}^{stat}|\vec{N}_{shoes}| \quad \text{and} \quad |\vec{f}_{shoulders}^{\,stat}| = \mu_{shoulders}^{stat}|\vec{N}_{shoulders}|,$$
$$(13\text{-}16)$$

where $|\vec{N}| = |\vec{N}_{shoes}| = |\vec{N}_{shoulders}|$. Substituting these expressions into Eq. 13-15 and solving for the magnitude of $|\vec{N}|$ gives us

$$|\vec{N}| = \frac{mg}{\mu_{\text{shoes}}^{\text{stat}} + \mu_{\text{shoulders}}^{\text{stat}}} = \frac{(55\text{ kg})(9.8\text{ m/s}^2)}{1.1 + 0.70}$$

$$= 299\text{ N} \approx 3.0 \times 10^2\text{ N}.$$

Thus, her minimum horizontal push must be about 300 N.
(b) For that push, what must be the vertical distance h between her feet and her shoulder if she is to be stable?

SOLUTION ■ A **Key Idea** here is that the climber will be stable if the torque-balancing equation giving the z-component of torque ($\tau_z^{\text{net}} = 0$) applies to her. This means that the forces on her must not produce a net torque about any rotation axis. Another **Key Idea** is that we are free to choose a rotation axis that helps simplify the calculation. We shall write the torques in the form $r_\perp |\vec{F}|$, where $r_\perp$ is the moment arm of force $\vec{F}$. In Fig. 13-9, we choose a rotation axis at her shoulders, perpendicular to the figure's plane. Then the moment arms of the forces acting there ($\vec{N}_{\text{shoulders}}$ and $\vec{f}_{\text{shoulders}}^{\text{stat}}$) are zero. Frictional force $\vec{f}_{\text{shoes}}^{\text{stat}}$, the normal force $\vec{N}_{\text{shoes}}$ at her feet, and the gravitational force $\vec{F}^{\text{grav}} = -mg\hat{j}$ have the corresponding moment arms w, h, and d.

Recalling our rule about the signs of torques and the corresponding directions, we can now write the torque component $\tau_z^{\text{net}} = 0$ as

$$-(w)(|\vec{f}_{\text{shoes}}^{\text{stat}}|) + (h)(\vec{N}_{\text{shoes}}) + (d)(mg)$$
$$+ (0)(|\vec{f}_{\text{shoulders}}^{\text{stat}}|) + (0)(|\vec{N}_{\text{shoulders}}|) = 0. \qquad (13\text{-}17)$$

(Note how the choice of rotation axis neatly eliminates $|\vec{f}_{\text{shoulders}}^{\text{stat}}|$ from the calculation.) Next, solving Eq. 13-17 for h, setting $|\vec{f}_{\text{shoes}}^{\text{stat}}| = \mu_{\text{shoes}}^{\text{stat}} |\vec{N}_{\text{shoes}}|$, and substituting $|\vec{N}| = |\vec{N}_{\text{shoes}}| = |\vec{N}_{\text{shoulders}}| = 299$ N and other known values, we find that

$$h = \frac{|\vec{f}_{\text{shoes}}^{\text{stat}}|w - mgd}{|\vec{N}|} = \frac{\mu_{\text{shoes}}^{\text{stat}}|\vec{N}|w - mgd}{|\vec{N}|} = \mu_{\text{shoes}}^{\text{stat}}w - \frac{mgd}{|\vec{N}|}$$

$$= (1.1)(1.0\text{ m}) - \frac{(55\text{ kg})(9.8\text{ m/s}^2)(0.20\text{ m})}{299\text{ N}} \qquad \text{(Answer)}$$

$$= 0.739\text{ m} \approx 0.74\text{ m}.$$

We would find the same required value of h if we wrote the torques about any other rotation axis perpendicular to the page, such as one at her feet.

If h is more than *or* less than 0.74 m, she must exert a force greater than 299 N on the walls to be stable. Here, then, is the advantage of knowing the physics before you climb a chimney. When you need to rest, you will avoid the (dire) error of novice climbers who place their feet too high or too low. Instead, you will know that there is a "best" distance between shoulders and feet, requiring the least push, and giving you a good chance to rest.

13-4 Indeterminate Equilibrium Problems

We decided earlier in the chapter to reduce the complication of equilibrium calculations by only working with situations in which forces that act on a body all lie in the x-y plane. In these cases we have only three independent equations at our disposal. These are the two balance of force components equations (typically for the x and y axis force components) and the one balance of torque components equation about a rotation axis (typically the z axis). If a problem has more than three unknowns, we cannot solve it.

It is easy to find such problems. For example, the seesaw shown in Fig. 13-10 with its board weighing 100 N will remain in static equilibrium if the magnitudes of the three forces $\vec{F}_B$, $\vec{F}_C$, and $\vec{F}_D$ are 150 N, 100 N, and 50 N, respectively. But it will also remain in equilibrium for a force combination of 160 N, 80 N, 60 N, and so on. A variant on Touchstone Example 13-2 provides another example. We could have assumed that there is friction between the wall and the top of the ladder. Then there would have been a vertical frictional force acting where the ladder touches the wall, making a total of four unknown forces. With only three equations, we could not have solved this problem. Problems like these, in which there are more unknowns than equations, are called **indeterminate.**

Yet solutions to indeterminate problems exist in the real world even in cases where the forces on an object do not necessarily lie in one plane. If you rest the tires of the car on four platform scales, each scale will register a definite reading, the sum of the readings being the weight of the car. What is eluding us in our efforts to find the individual forces by solving equations?

The problem is that we have assumed—without making a great point of it—that the bodies to which we apply the equations of static equilibrium are perfectly rigid. By this we mean that they do not deform when forces are applied to them. Strictly,

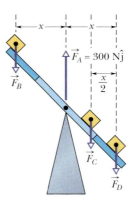

FIGURE 13-10 ■ The seesaw with three small boxes of unknown weight placed on its rough board is in static equilibrium. The board weighing 100 N is pivoted at its center of mass. However the system forms an indeterminate structure even when the total downward force on the seesaw pivot is known to be $(-400\text{ N})\hat{j}$. This is because the weights of the three boxes cannot be found from the conditions for static equilibrium alone.

there are no such bodies. The tires of the car, for example, deform easily under this load until the car settles into a position of static equilibrium.

We have all had experience with a wobbly restaurant table, which we usually level by putting folded paper under one of the legs. If a big enough elephant sat on such a table, however, you may be sure that if the table did not collapse, it would deform just like the tires of a car. Its legs would all touch the floor, the forces acting upward on the table legs would all assume definite (and different) values, and the table would no longer wobble. How do we find the values of those forces acting on the legs?

To solve such indeterminate equilibrium problems, we must supplement equilibrium equations with some knowledge of deformation or *elasticity,* the branch of physics and engineering that describes how real bodies deform when forces are applied to them. The next section provides an introduction to this subject.

READING EXERCISE 13-4: A horizontal uniform bar of weight 10 N is to hang from a ceiling by three wires that exert upward forces $\vec{F}_A$, $\vec{F}_B$, and $\vec{F}_C$ on the bar. The figure shows three arrangements for the wires. Which arrangements, if any, are indeterminate (so that we cannot solve for numerical values of $\vec{F}_A$, $\vec{F}_B$, and $\vec{F}_C$)?

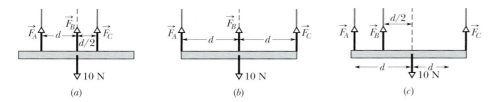

(a)   (b)   (c)

13-5 Elasticity

In Section 6-4 we introduced an idealized model based on atomic physics to help us explain the behavior of contact forces. This model can also be used to help us understand the elastic properties of solids. When a large number of atoms come together to form a metallic solid, such as an iron nail, they settle into equilibrium positions in a three-dimensional *lattice,* a repetitive arrangement in which each atom has a well-defined equilibrium distance from its nearest neighbors. The atoms are held together by interatomic forces that act like tiny springs. A two-dimensional drawing of this model is shown in Fig. 6-7 and a three-dimensional picture of this model is shown in Figs. 6-5 and Fig. 13-11. As shown in Fig. 6-19c, when an object that is in equilibrium is stretched or compressed by forces acting at opposite ends, each atom or molecule within the material feels oppositely directed forces on it that balance.

The lattice of a metallic solid is remarkably rigid. This is another way of saying that the "interatomic springs" are extremely stiff. It is for this reason that we perceive many ordinary objects such as metal ladders, tables, and spoons as perfectly rigid. Of course, some ordinary objects, such as garden hoses or rubber gloves, do not strike us as rigid at all. The atoms that make up these objects *do not* form a rigid lattice like that of Fig. 13-11 but are aligned in long, flexible molecular chains, each chain being only loosely bound to its neighbors.

Given the atomic-molecular picture presented above, we can visualize all objects as being made up of discrete particles. Can we feel or observe the spring-like interactions between particles that make up an object? Consider what happens when we pull on an object. For example, suppose that you pull on your finger. Your finger does not immediately come apart, but you feel stretching forces along the entire length of your finger. These forces along the finger feel greater as you increase the magnitude of your pull. What you are feeling is an opposing, balancing force that arises within your

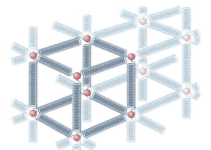

FIGURE 13-11 ■ The atoms of a metallic solid are distributed on a repetitive three-dimensional lattice. The interatomic forces behave like tiny springs.

finger to oppose your pull. This opposing force works to keep the particles that make up your finger from moving away from one another (which ultimately would result in your finger coming apart). This experience can be generalized by saying that the particles that make up your finger (or other object) are held together by forces that are *attractive* forces when a stretching force is applied. This attractive force increases as stretching occurs and the particles that make up the object move apart. Stretching forces like your pull are called **tensile forces.**

If instead, you push on your finger (in trying to shorten it), you will again sense balancing, opposing forces along the length of your finger. Push harder on your finger. The feeling will change and likely convince you that these opposing forces also increase with increasing applied force. From this experience, we gather that the particles that make up our finger (or other objects) will encounter *repulsive* forces if we try to push them closer together. Furthermore, the repulsive forces increase with decreasing separation. Squeezing forces like the one you applied to your finger (in an attempt to shorten it) are called **compressive forces.** As it turns out, the attractive and repulsive forces between the particles that make up an object are associated with the electrical nature of the particles that comprise atoms as well as their interactions on a microscopic scale. However, the nature of the forces between the particles that make up an object is very much like the nature of a spring force. Hence, we can model these forces quite well as spring forces. This is why Fig. 13-11 shows the particles connected with springs.

As you surely have experienced, neither tensile nor compressive forces can be increased indefinitely. Eventually the object simply breaks. In general, solid materials tend to be strongest under compression forces and weakest under stretching (tensile) forces. There is another way to break a long thin object—by *bending* it. If you have ever tried to break a dry stick, you probably observed that it breaks more easily when it is bent than when it is stretched or compressed. This makes sense when you visualize what is happening to the atoms or molecules that makeup an object. When bending occurs, these tiny particles are compressed on one side while they are simultaneously stretched on the opposite side. Figure 13-12 shows combinations of forces that can lead to compression, stretching, and bending.

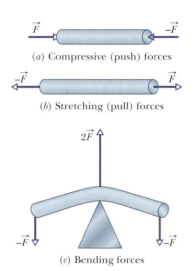

FIGURE 13-12 ■ Forces leading to (*a*) compression, (*b*) stretching, and (*c*) bending.

Tensile and Compressive Forces

Let's consider what can happen when we pull on each end of a long, thin, apparently rigid object with forces of greater and greater magnitude. If you fix one end of a steel rod that is 1 m long and 1 cm in diameter and hang a subcompact car from the end, the rod will stretch. However, it stretches only about 0.5 mm, or 0.05% as shown in Fig. 13-13. In this case the rod acts like an *elastic* spring that obeys Hooke's law, $\vec{F}_x^{\text{spring}}(x) = -k\Delta\vec{x}$(Eq. 9-16), so that the magnitude of the applied force on it is proportional to the displacement of the spring.

> If the deformation of an object is proportional to the magnitude of the applied forces on its ends and if the object returns to its original length when the forces are removed, then the deformation is **elastic.**

If you hang two cars from the rod, the rod will be permanently stretched and will not recover its original length when you remove the load. The stretching is now **inelastic.** If you hang three cars from the rod, the rod will break or **rupture.** Just before breaking, the elongation of the rod will be less than 0.2%. Although deformations of this size seem small, they are important in engineering practice. (Whether a wing under load will stay on an airplane is obviously important.)

We want to quantify the relationship between applied tensile or compressive forces and deformations. Let's think about elastic deformations (those that are pro-

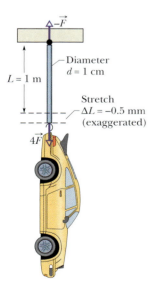

FIGURE 13-13 ■ A subcompact car is hanging from a 1-m-long steel rod that is 1 cm in diameter. Even though the thin rod supports the entire weight of the car, the rod will only stretch about 0.5 mm.

portional to the applied forces that cause them). How do these deformations depend on the particular geometry of the object and nature of the material being placed under tension or compression forces? Let's see.

If we pull on the ends of a rod of length L parallel to it and with equal but opposite forces of magnitude F, we can measure the change in the rod's length, ΔL. In practice we can do this by pulling on one end while the other end is attached to an immovable object. What happens if we now apply the same force to the end of a rod of the same material and diameter that is twice as long? By how much will it stretch? We can figure this out by imagining that the double-length rod is just two rods connected in the middle as shown in Figure 13-14.

Let's assume that the magnitude of pulling force on the ends of our double-length rod is much greater than the rod's weight. In static equilibrium, the support point at the other end of the rod exerts a force of the same magnitude in the opposite direction. What happens to a small cross section of the rod at its midpoint? As we saw in our discussion of tension in Chapter 6, every cross-sectional element along the rod will experience the same pair of opposite pull forces. So the pull force from the lower half of the rod must be equal in magnitude and opposite in direction to the pull force from the upper half of the rod. Therefore, each half of the rod will stretch by an amount ΔL as shown in Fig. 13-14. As a result, the total stretch for the double length rod will be $2\,\Delta L$. We can easily generalize this argument to show that for a rod of arbitrary length L, the amount of stretch ΔL produced by a given force will be proportional to the length of the rod—assuming that we keep everything else about the rod and situation the same.

Similarly, we can consider a rod of the same length as our original rod but with double the cross-sectional area. We can picture this as two rods with our original diameter placed right next to each other but not touching as shown in Fig. 13-15. Now we see that our two rods would have to *share* the force $\vec{F}$ between them. This means that each effectively feels only half of the total force. As a result, the double diameter rods will only stretch half the amount of the single rod. This implies that the change in length should be *inversely* proportional to the cross-sectional area of the rod. That is, the deformation ΔL is proportional to $1/A$ where A is the cross-sectional area of the rod.

Recapping what we discussed in the paragraphs above: The deformation of the rod ΔL is proportional to the magnitude of the applied force F, proportional to the length of the rod L, and inversely proportional to the cross-sectional area of the rod A. If we combine these results, we get

$$\Delta L \propto \frac{LF}{A} \quad \text{or} \quad \Delta L = c\frac{LF}{A},$$

where c is a constant of proportionality that may well be present. It is customary to express this proportionality so that we can relate the force magnitude to the length change for a given rod. Thus, we can write

$$F = \left(\frac{1}{c}\right)\left(\frac{A}{L}\right)\Delta L. \tag{13-18}$$

Let's examine the factors in parentheses in the equation shown above. First, the $(1/c)$ factor is a constant, which turns out to depend on the material an object is made of but not on its shape. This factor is a kind of stiffness constant that characterizes the tensile and compressive strength of a material. This factor $(1/c)$ is represented in engineering practice by the symbol E and is called the **Young's modulus.** So, we can write

$$F = E\left(\frac{A}{L}\right)\Delta L. \tag{13-19}$$

The second factor given by (A/L) depends only on the geometry of the rod.

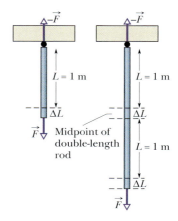

FIGURE 13-14 ■ If the forces applied to the ends of a double-length rod are the same as those applied to a single-length rod the double-length rod will experience twice the deformation. The extent of the deformation is exaggerated.

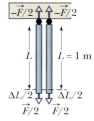

FIGURE 13-15 ■ A rod that has twice the cross-sectional area but has the same forces applied to its ends will experience half the deformation. The deformation is exaggerated.

As we mentioned in the beginning of this section, whenever the magnitude of the force applied to an object is proportional to its deformation in the direction of the force, the object is behaving like a Hooke's law spring. Note that if we represent $E(A/L)$ for a given rod by k, then we can write Eq. 13-19 in the form $F = k\Delta L$. In this form, Eq. 13-19 and Eq. 9-16 describing Hooke's law for a spring in one dimension are essentially the same. However, a length of rod or wire in general is much stiffer (that is, more resistant to deformation) than the same rod or wire would be if it were coiled into a spring.

It is customary to write Eq. 13-19 in a form known as the stress–strain equation:

$$\frac{F}{A} = E\left(\frac{\Delta L}{L}\right) \qquad \text{(stress–strain equation).} \qquad (13\text{-}20)$$

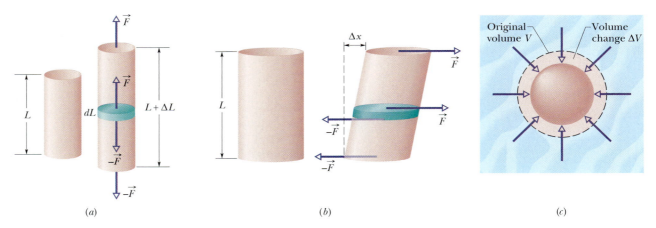

(a)   *(b)*   *(c)*

FIGURE 13-16 ▪ When solid matter is in static equilibrium, it can experience oppositely directed tensile, compressive, or shearing forces that deform it. These deformation forces are experienced by every tiny element that the matter is composed of, such as the cross-sectional disks of infinitesimal thickness *dl* depicted here. (*a*) A cylinder subject to *tensile stress* stretches by an amount ΔL. (*b*) A cylinder subject to *shearing stress* deforms by an amount Δx, somewhat like a pack of playing cards would. (*c*) A solid sphere that is immersed in a fluid is subject to compressive, uniform *hydraulic stress*. It shrinks in volume by an amount ΔV. All the deformations shown are greatly exaggerated.

Figure 13-16 shows three ways in which a solid might change its dimensions when forces act on it. In Fig. 13-16*a*, a cylinder is stretched. In Fig. 13-16*b*, a cylinder is deformed by a force perpendicular to its axis, much as we might deform a pack of cards or a book. In Fig. 13-16*c*, a solid object, placed in a fluid under high pressure, is compressed uniformly on all sides. What the three deformation types have in common is that a deforming force per unit area F/A produces a unit deformation $\Delta L/L$. The deforming force per unit area F/A is called a **stress** and the unit deformation $\Delta L/L$ is called a **strain.** So, the stresses and strains take different forms in the three situations of Fig. 13-16. When engineers design structures, they usually work with strains that are small enough that the materials are elastic (so that stress and strain are proportional to each other). That is,

$$\text{stress} = \text{modulus} \times \text{strain.} \qquad (13\text{-}21)$$

As we already mentioned, the modulus is a stiffness factor. Obviously Eq. 13-21 tells us that when the modulus is very large, it takes a lot of force per unit of cross-sectional area (stress) to produce a small deformation (strain). In Fig. 13-16, *tensile stress* (associated with stretching) is illustrated in (*a*), *shearing stress* in (*b*), and *hydraulic stress* in (*c*).

To get a better feel for the relationship between the applied force and the force per unit area (or stress), consider the normal force exerted by the table on a chunk of clay. Suppose that the clay starts as a tall cylinder. It is then squashed down into a flat pancake of clay, without any change in mass. Comparing these two geometries for the clay, we note that the total normal force does not change (the weight has not changed). However, the area of the clay does change. Therefore, the force per unit area or stress must also change. As Eq. 13-21 shows, thinking about deformations is more direct when we consider stress rather than force.

In a typical test of tensile (stretching) properties, the stress on a test specimen (like that in Figs. 13-17b and 13-18) is slowly increased until the cylinder fractures, and the stress vs. strain are carefully measured and plotted. For metals, the result is a graph like those shown in Fig. 13-17 or Fig. 13-19. For a substantial range of applied stresses, the stress–strain relation is linear, and the specimen recovers its original dimensions when the stress is removed. Here stress = modulus × strain (Eq. 13-21) applies. If the stress is increased beyond the **yield strength** S^{yield} of the specimen, the specimen becomes permanently deformed. If the stress continues to increase, the specimen eventually ruptures, at a stress called the **ultimate strength** S^{ultimate}.

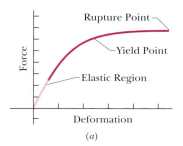

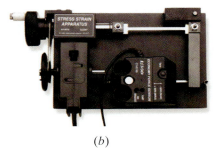

(b)

FIGURE 13-17 ■ (a) A graph showing the deformation of a hard steel bar as a function of the magnitude of the forces applied to its ends. The applied forces are increased until the bar ruptures. (b) The measurements are made using an electronic force sensor and a rotary motion sensor to gauge the deformation of the bar. (Courtesy of PASCO scientific.)

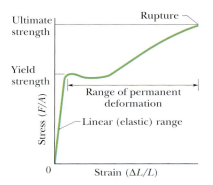

FIGURE 13-19 ■ A stress–strain curve for a soft steel test specimen such as that of Fig. 13-18. Young's modulus is the slope of the first linear portion of the graph. The specimen deforms permanently when the stress is equal to the *yield strength* of the material. It ruptures when the stress is equal to the *ultimate strength* of the material.

For simple tension or compression, the stress on an object is defined as F/A, where F represents the magnitude of the force applied perpendicular to the cross-sectional area A of the object. The strain, or unit deformation, is then the dimensionless quantity $\Delta L/L$. This is the fractional (or sometimes percentage) change in the length of the specimen. Because the strain is dimensionless, and stress = modulus × strain (Eq. 13-21), the modulus has the same dimensions as the stress—namely, force per unit area.

The strain $\Delta L/L$ in a specimen can often be measured conveniently with a *strain gage* (Fig. 13-20). This simple and useful device, which can be attached directly to operating machinery with an adhesive, is based on the principle that its electrical properties are dependent on the strain it undergoes.

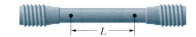

FIGURE 13-18 ■ A test specimen, used to determine a stress–strain curve such as that of Fig. 13-19. The change ΔL that occurs in a certain length L is measured in a tensile stress–strain test.

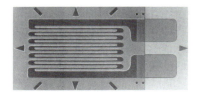

FIGURE 13-20 ■ A strain gage of overall dimensions 9.8 mm by 4.6 mm. The gage is fastened with adhesive to the object whose strain is to be measured; it experiences the same strain as the object. The electrical resistance of the gage varies with the strain, permitting deformations up to about 3 mm to be measured.

Although the Young's modulus for an object is typically almost the same for tension and compression, the object's ultimate strength or yield strength may well be different for the two types of stress. Concrete, for example, is very strong in compression but is so weak in tension that it is almost never used in that manner. In fact, steel bars are often embedded in concrete to enhance the tensile strength of a concrete structure. Table 13-1 shows the Young's modulus and other elastic properties for some materials of engineering interest.

TABLE 13-1
Some Elastic Properties of Selected Materials of Engineering Interest

Material	Density ρ (kg/m³)	Young's Modulus E (10⁹ N/m²)	Ultimate Strength S^{ultimate} (10⁶ N/m²)	Yield Strength S^{yield} (10⁶ N/m²)
Steel[a]	7860	200	400	250
Aluminum	2710	70	110	95
Glass	2190	65	50[b]	—
Concrete[c]	2320	30	40[b]	—
Wood[d]	525	13	50[b]	—
Bone	1900	9[b]	170[b]	—
Polystyrene	1050	3	48	—

[a]Structural steel (ASTM-A36).　　　[b]In compression.　　　[c]High strength.　　　[d]Douglas fir.

Shearing

Consider Fig. 13-16b. Here the force is applied parallel to the cross-sectional area. We call this force orientation *shearing stress*. In the case of shearing, the stress is still a force per unit area, but the force vector lies in the plane of the area rather than perpendicular to it. The strain is effectively the angle of the shear in radians as shown in Fig. 13-16b. It is given by the dimensionless ratio $\Delta x/L$, with the quantities defined as shown in Fig. 13-16b. The corresponding modulus, which is given the symbol G in engineering practice, is called the **shear modulus.** For shearing, the stress = modulus × strain equation (Eq. 13-21) is written as

$$\frac{F^{\text{perp}}}{A} = G\frac{\Delta x}{L}, \tag{13-22}$$

where F^{perp} is the magnitude of the component of an applied force that is applied perpendicular to the length of the rod. Shearing stresses play a critical role in the buckling of shafts that rotate under load and in bone fractures caused by bending.

Hydraulic Stress

In Fig. 13-16c, the stress is the fluid pressure P on the object. We will discuss fluid pressure in more detail in Chapter 15. For example, as you will see in Eq. 15-1, if the forces that act on an area A are uniform, then pressure is the ratio of the force component perpendicular to the area and to the area itself. The strain is then $|\Delta V|/V$, where V is the original volume of the specimen and ΔV is the absolute value of the change in volume. The corresponding modulus, with symbol B, is called the **bulk modulus** of the material. The object is said to be under *hydraulic compression,* and the pressure can be called the *hydraulic stress.* For this situation, we write stress = modulus × strain (Eq. 13-21) as

$$P = B\frac{|\Delta V|}{V}. \tag{13-23}$$

The bulk modulus is 2.2×10^9 N/m² for water and 16×10^{10} N/m² for steel. The pressure at the bottom of the Pacific Ocean, at its average depth of about 4000 m, is 4.0×10^7 N/m². The fractional compression $\Delta V/V$ of a volume of water due to this pressure is 1.8%; that for a steel object is only about 0.025%. In general, solids—with their rigid atomic lattices—are less compressible than liquids, in which the atoms or molecules are less tightly coupled to their neighbors.

READING EXERCISE 13-5: Four cylindrical rods are stretched as in Fig. 13-16a. The force magnitudes, the cross-sectional areas, the initial lengths, and the changes in length are shown in the table below. Rank the rods from largest to smallest Young's modulus.

Rod	Force Magnitude	Area	Length Change	Initial Length
1	F	A	ΔL	L
2	$2F$	$2A$	$2\Delta L$	L
3	F	$2A$	$2\Delta L$	$2L$
4	$2F$	A	ΔL	$2L$

■

READING EXERCISE 13-6: Visualizing the microscopic particles that make up objects as shown in Fig. 13-11, discuss the difference between the bending shown in Fig. 13-12c and the shearing shown in Fig. 13-16b.

■

TOUCHSTONE EXAMPLE 13-5: Stretching a Rod

A structural steel rod has a radius R of 9.5 mm and a length L of 81 cm. A 6.2×10^4 N force of magnitude F stretches it along its length. What are the stress on the rod and the elongation and strain of the rod?

SOLUTION ■ The first **Key Idea** here has to do with what is meant by the second sentence in the problem statement. We assume the rod is held stationary by, say, a clamp or vise at one end. Then force $\vec{F}$ is applied at the other end, parallel to the length of the rod and thus perpendicular to the end face there. Therefore, the situation is like that in Fig. 13-16a.

The next **Key Idea** is that we assume the force is applied uniformly across the end face and thus over an area $A = \pi R^2$. Then the stress on the rod is given by the left side of Eq. 13-20,

$$\text{stress} = \frac{F}{A} = \frac{F}{\pi R^2} = \frac{6.2 \times 10^4 \,\text{N}}{(\pi)(9.5 \times 10^{-3} \,\text{m})^2} \quad \text{(Answer)}$$

$$= 2.2 \times 10^8 \,\text{N/m}^2.$$

The yield strength for structural steel is 2.5×10^8 N/m², so this rod is dangerously close to its yield strength.

Another **Key Idea** is that the elongation of the rod depends on the stress, the original length L, and the type of material in the rod. The last determines which value we use for Young's modulus E (from Table 13-1). Using the value for steel, Eq. 13-20 gives us

$$\Delta L = \frac{(F/A)L}{E} = \frac{(2.2 \times 10^8 \,\text{N/m}^2)(0.81 \,\text{m})}{2.0 \times 10^{11} \,\text{N/m}^2} \quad \text{(Answer)}$$

$$= 8.9 \times 10^{-4} \,\text{m} = 0.89 \,\text{mm}.$$

The last **Key Idea** we need here is that strain, which is the dimensionless ratio of the change in length to the original length, is

$$\frac{\Delta L}{L} = \frac{8.9 \times 10^{-4} \,\text{m}}{0.81 \,\text{m}} \quad \text{(Answer)}$$

$$= 1.1 \times 10^{-3} = 0.11\%.$$

Problems

SEC. 13-3 ■ THE CENTER OF GRAVITY

1. Brady Bunch A physics Brady Bunch, whose weights in newtons are indicated in Fig. 13-21, is balanced on a seesaw. What is the number of the person who causes the largest torque, about the rotation axis at *fulcrum f*, directed (a) out of the page and (b) into the page?

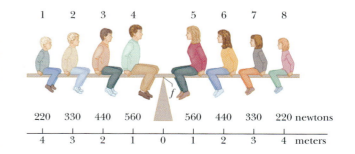

FIGURE 13-21 ■ Problem 1.

2. Tower of Pisa The leaning Tower of Pisa (Fig. 13-22) is 55 m high and 7.0 m in diameter. The top of the tower is displaced 4.5 m from the vertical. Treat the tower as a uniform, circular cylinder. (a) What additional displacement, measured at the top, would bring the tower to the verge of toppling? (b) What angle would the tower then make with the vertical?

FIGURE 13-22 ■
Problem 2.

3. Particle Acted on A particle is acted on by forces given by $\vec{F}_A = (10 \text{ N})\hat{i} + (-4 \text{ N})\hat{j}$ and $\vec{F}_B = (17 \text{ N})\hat{i} + (2 \text{ N})\hat{j}$. (a) What force $\vec{F}_C$ balances these forces? (b) What direction does $\vec{F}_C$ have relative to the x axis?

4. A Bow is Drawn A bow is drawn at its midpoint until the tension in the string is equal to the force exerted by the archer. What is the angle between the two halves of the string?

5. Rope of Negligible Mass A rope of negligible mass is stretched horizontally between two supports that are 3.44 m apart. When an object of weight 3160 N is hung at the center of the rope, the rope is observed to sag by 35.0 cm. What is the tension in the rope?

6. Scaffold A scaffold of mass 60 kg and length 5.0 m is supported in a horizontal position by a vertical cable at each end. A window washer of mass 80 kg stands at a point 1.5 m from one end. What is the tension in (a) the nearer cable and (b) the farther cable?

7. Uniform Sphere In Fig. 13-23 a uniform sphere of mass m and radius r is held in place by a massless rope attached to a frictionless wall a distance L above the center of the sphere. Find (a) the tension in the rope and (b) the force on the sphere from the wall.

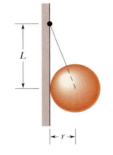

FIGURE 13-23 ■
Problem 7.

8. Automobile An automobile with a mass of 1360 kg has 3.05 m between the front and rear axles. Its center of gravity is located 1.78 m behind the front axle. With the automobile on level ground, determine the magnitude of the force from the ground on (a) each front wheel (assuming equal forces on the front wheels) and (b) each rear wheel (assuming equal forces on the rear wheels).

9. Diver A diver of weight 580 N stands at the end of a 4.5 m diving board of negligible mass (Fig. 13-24). The board is attached to two pedestals 1.5 m apart. What are the magnitude and direction of the force on the board from (a) the left pedestal and (b) the right pedestal? (c) Which pedestal is being stretched, and (d) which compressed?

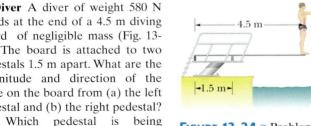

FIGURE 13-24 ■ Problem 9.

10. Car in the Mud In Fig. 13-25, a man is trying to get his car out of mud on the shoulder of a road. He ties one end of a rope tightly around the front bumper and the other end tightly around a utility pole 18 m away. He then pushes sideways on the rope at its midpoint with a force of 550 N, displacing the center of the rope 0.30 m

from its previous position, and the car barely moves. What is the magnitude of the force on the car from the rope? (The rope stretches somewhat.)

FIGURE 13-25 ■ Problem 10.

11. Meter Stick A meter stick balances horizontally on a knife-edge at the 50.0 cm mark. With two 5.0 g coins stacked over the 12.0 cm mark, the stick is found to balance at the 45.5 cm mark. What is the mass of the meter stick?

12. Uniform Cubical A uniform cubical crate is 0.750 m on each side and weighs 500 N. It rests on a floor with one edge against a very small, fixed obstruction. At what least height above the floor must a horizontal force of magnitude 350 N be applied to the crate to tip it?

13. Window Cleaner A 75 kg window cleaner uses a 10 kg ladder that is 5.0 m long. He places one end on the ground 2.5 m from a wall, rests the upper end against a cracked window, and climbs the ladder. He is 3.0 m up along the ladder when the window breaks. Neglecting friction between the ladder and window and assuming that the base of the ladder does not slip, find (a) the magnitude of the force on the window from the ladder just before the window breaks and (b) the magnitude and direction of the force on the ladder from the ground just before the window breaks.

14. Lower Leg Figure 13-26 shows the anatomical structures in the lower leg and foot that are involved in standing tiptoe with the heel raised off the floor so the foot effectively contacts the floor at only one point, shown as P in the figure. Calculate, in terms of a person's weight W, the forces on the foot from (a) the calf muscle (at A) and (b) the lower-leg bones (at B) when the person stands tiptoe on one foot. Assume that $a = 5.0$ cm and $b = 15$ cm.

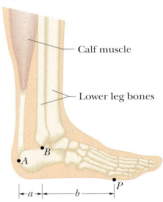

FIGURE 13-26 ■ Problem 14.

15. Construction In Fig. 13-27, an 817 kg construction bucket is suspended by a cable A that is attached at O to two other cables B and C, making angles of 51.0° and 66.0° with the horizontal. Find the tensions in (a) cable A, (b) cable B, and (c) cable C. (*Hint:* To avoid solving two equations in two unknowns, position the axes as shown in the figure.)

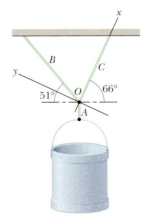

FIGURE 13-27 ■ Problem 15.

16. System in Equilibrium The system in Fig. 13-28 is in equilibrium, with the string in the center exactly horizontal. Find (a) tension T_A, (b) tension T_B, (c) tension T_C, and (d) angle θ.

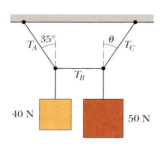

FIGURE 13-28 ▪ Problem 16.

17. Three Pulleys The force $\vec{F}$ in Fig. 13-29 keeps the 6.40 kg block and the pulleys in equilibrium. The pulleys have negligible mass and friction. Calculate the tension T in the upper cable. (*Hint:* When a cable wraps halfway around a pulley as here, the magnitude of its net force on the pulley is twice the tension in the cable.)

18. Triceps A 15 kg block is being lifted by the pulley system shown in Fig. 13-30. The upper arm is vertical, whereas the forearm makes an angle of 30° with the horizontal. What are the forces on the forearm from (a) the triceps muscle and (b) the upper-arm bone (the humerus)? The forearm and hand together have a mass of 2.0 kg with a center of mass 15 cm (measured along the arm) from the point where the forearm and upper-arm bones are in contact. The triceps muscle pulls vertically upward at a point 2.5 cm behind that contact point.

FIGURE 13-29 ▪ Problem 17.

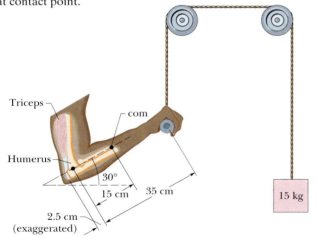

FIGURE 13-30 ▪ Problem 18.

19. Forces on Structure Forces $\vec{F}_A$, $\vec{F}_B$, and $\vec{F}_C$ act on the structure of Fig. 13-31 shown in an overhead view. We wish to put the structure in equilibrium by applying a fourth force, at a point such as P. The fourth force has vector

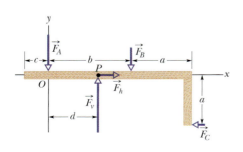

FIGURE 13-31 ▪ Problem 19.

components $\vec{F}_h$ and $\vec{F}_v$. We are given that $a = 2.0$ m, $b = 3.0$ m, $c = 1.0$ m, $|\vec{F}_A| = 20$ N, $|\vec{F}_B| = 10$ N, and $|\vec{F}_C| = 5.0$ N. Find (a) $|\vec{F}_h|$, (b) $|\vec{F}_v|$, and (c) d.

20. Square Sign In Fig. 13-32, a 50.0 kg uniform square sign, 2.00 m on a side, is hung from a 3.00 m horizontal rod of negligible mass. A cable is attached to the end of the rod and to a point on the wall 4.00 m above the point where the rod is hinged to the wall. (a) What is the tension in the cable? What are the magnitudes and directions of the (b) horizontal and (c) vertical components of the force on the rod from the wall?

21. Wheel and Obstacle In Fig. 13-33, what magnitude of force $\vec{F}$ applied horizontally at the axle of the wheel is necessary to raise the wheel over an obstacle of height h? The wheel's radius is r and its mass is m.

22. Rock Climber In Fig. 13-34, a 55 kg rock climber is in a lie-back climb along a fissure, with hands pulling on one side of the fissure and feet pressed against the opposite side. The fissure has width $w = 0.20$ m, and the center of mass of the climber is a horizontal distance $d = 0.40$ m from the fissure. The coefficient of static friction between hands and rock is $\mu_{\text{hands}}^{\text{stat}} = 0.40$, and between boots and rock it is $\mu_{\text{boots}}^{\text{stat}} = 1.2$. (a) What is the least horizontal pull by the hands and push by the feet that will keep the climber stable? (b) For the horizontal pull of (a), what must be the vertical distance h between hands and feet? (c) If the climber encounters wet rock, so that $\mu_{\text{hands}}^{\text{stat}}$ and $\mu_{\text{boots}}^{\text{stat}}$ are reduced, what happens to the answers to (a) and (b), respectively?

23. Beam and Hinge In Fig. 13-35, one end of a uniform beam that weighs 222 N is attached to a wall with a hinge. The other end is supported by a wire. (a) Find the tension in the wire. What are the (b) horizontal and (c) vertical components of the force of the hinge on the beam?

24. Four Bricks Four bricks of length L, identical and uniform, are stacked on top of one another (Fig. 13-36) in such a way that part of each extends beyond the one beneath.

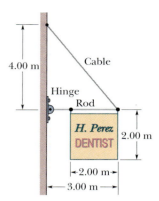

FIGURE 13-32 ▪ Problem 20.

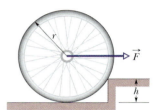

FIGURE 13-33 ▪ Problem 21.

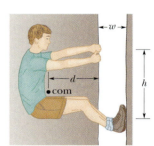

FIGURE 13-34 ▪ Problem 22.

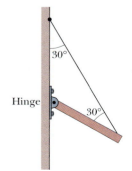

FIGURE 13-35 ▪ Problem 23.

Find, in terms of L, the maximum values of (a) a_A, (b) a_B, (c) a_C, (d) a_D, and (e) h, such that the stack is in equilibrium.

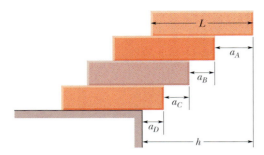

FIGURE 13-36 ■ Problem 24.

25. Concrete Block The system in Fig. 13-37 is in equilibrium. A concrete block of mass 225 kg hangs from the end of the uniform strut whose mass is 45.0 kg. Find (a) the tension T in the cable and the (b) horizontal and (c) vertical force components on the strut from the hinge.

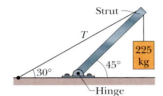

FIGURE 13-37 ■ Problem 25.

26. A Door A door 2.1 m high and 0.91 m wide has a mass of 27 kg. A hinge 0.30 m from the top and another 0.30 m from the bottom each support half the door's mass. Assume that the center of gravity is at the geometrical center of the door, and determine the (a) vertical and (b) horizontal components of the force from each hinge on the door.

27. Nonuniform Bar A nonuniform bar is suspended at rest in a horizontal position by two massless cords as shown in Fig. 13-38. One cord makes the angle $\theta = 36.9°$ with the vertical; the other makes the angle $\phi = 53.1°$ with the vertical. If the length L of the bar is 6.10 m, compute the distance x from the left-hand end of the bar to its center of mass.

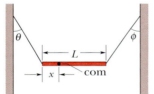

FIGURE 13-38 ■ Problem 27.

28. Thin Horizontal Bar In Fig. 13-39 a thin horizontal bar AB of negligible weight and length L is hinged to a vertical wall at A and supported at B by a thin wire BC that makes an angle θ with the horizontal. A load of weight W can be moved anywhere along the bar; its position is defined by the distance x from the wall to its center of mass. As a function of x, find (a) the tension in the wire, and the (b) horizontal and (c) vertical components of the force on the bar from the hinge at A.

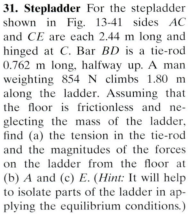

FIGURE 13-39 ■
Problems 28 and 30.

29. Uniform Plank In Fig. 13-40, a uniform plank, with a length L of 6.10 m and a weight of 445 N, rests on the ground and against a frictionless roller at the top of a wall of height $h = 3.05$ m. The plank remains in equilibrium for any value of $\theta \geq 70°$ but slips if $\theta < 70°$. Find the coefficient of static friction between the plank and the ground.

30. Max Tension In. Fig. 13-39, suppose the length L of the uniform bar is 3.0 m and its weight is 200 N. Also, let the load's weight $W = 300$ N and the angle $\theta = 30°$. The wire can withstand a maximum tension of 500 N. (a) What is the maximum possible distance x before the wire breaks? With the load placed at this maximum x, what are the (b) horizontal and (c) vertical components of the force on the bar from the hinge at A?

31. Stepladder For the stepladder shown in Fig. 13-41 sides AC and CE are each 2.44 m long and hinged at C. Bar BD is a tie-rod 0.762 m long, halfway up. A man weighting 854 N climbs 1.80 m along the ladder. Assuming that the floor is frictionless and neglecting the mass of the ladder, find (a) the tension in the tie-rod and the magnitudes of the forces on the ladder from the floor at (b) A and (c) E. (*Hint:* It will help to isolate parts of the ladder in applying the equilibrium conditions.)

32. Two Beams Two uniform beams, A and B, are attached to a wall with hinges and then loosely bolted together as in Fig. 13-42. Find the x- and y-components of the force on (a) beam A due to its hinge, (b) beam A due to the bolt, (c) beam B due to its hinge, and (d) beam B due to the bolt.

33. Box of Sand A cubical box is filled with sand and weighs 890 N. We wish to tip the box by pushing horizontally on one of the upper edges. (a) What minimum force is required? (b) What minimum coefficient of static friction between box and floor is required? (c) Is there a more efficient way to tip the box? If so, find the smallest possible force that would have to be applied directly to the box to tip it. (*Hint:* At the onset of tipping, where is the normal force located?)

34. Two Arrangements Four bricks of length L, identical and uniform, are stacked on a table in two ways, as shown in Fig. 13-43 (compare

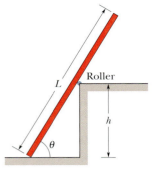

FIGURE 13-40 ■ Problem 29.

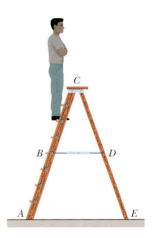

FIGURE 13-41 ■ Problem 31.

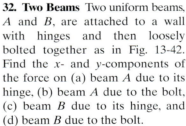

FIGURE 13-42 ■ Problem 32.

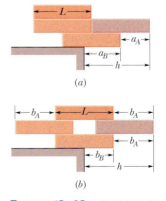

FIGURE 13-43 ■ Problem 34.

with Problem 24). We seek to maximize the overhang distance h in both arrangements. Find the optimum distances a_A, a_B, b_A, b_B, and calculate for the two arrangements. [See "The Amateur Scientist," *Scientific American*, June 1985, pp. 133–134, for a discussion and an even better version of arrangement (*b*).]

35. Cubical Crate A crate, in the form of a cube with edge lengths of 1.2 m, contains a piece of machinery; the center of mass of the crate and its contents is located 0.30 m above the crate's geometrical center. The crate rests on a ramp that makes an angle θ with the horizontal. As θ is increased from zero, an angle will be reached at which the crate will either start to slide down the ramp or tip over. Which event will occur (a) when the coefficient of static friction between ramp and crate is 0.60 and (b) when it is 0.70? In each case, give the angle at which the event occurs. (*Hint:* At the onset of tipping, where is the normal force located?)

SEC. 13-5 ■ ELASTICITY

36. Young's Modulus Figure 13-44 shows the stress–strain curve for quartzite. What are (a) the Young's modulus and (b) the approximate yield strength for this material?

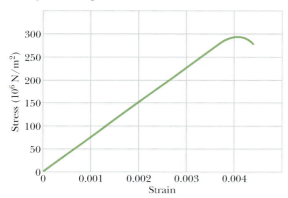

FIGURE 13-44 ■ Problem 36.

37. Aluminum Rod A horizontal aluminum rod 4.8 cm in diameter projects 5.3 cm from a wall. A 1200 kg object is suspended from the end of the rod. The shear modulus of aluminum is 3.0×10^{10} N/m². Neglecting the rod's mass, find (a) the shear stress on the rod and (b) the vertical deflection of the end of the rod.

38. Lead Brick In Fig. 13-45, a lead brick rests horizontally on cylinders A and B. The areas of the top faces of the cylinders are related by $A_A = 2A_B$; the Young's moduli of the cylinders are related by $E_A = 2E_B$. The cylinders had identical lengths before the brick was placed on them. What fraction of the brick's mass is supported (a) by cylinder A and (b) by cylinder B? The horizontal distances between the center of mass of the brick and the centerlines of the cylinders are d_A for cylinder A and d_B for cylinder B. (c) What is the ratio d_A/d_B?

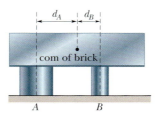

FIGURE 13-45 ■ Problem 38.

39. Uniform Log In Fig. 13-46, 103 kg uniform log hangs by two steel wires, A and B, both of radius 1.20 mm. Initially, wire A was 2.50 m long and 2.00 mm shorter than wire B. The log is now horizontal. What are the magnitudes of the forces on it from (a) wire A and (b) wire B? (c) What is the ratio d_A/d_B?

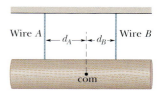

FIGURE 13-46 ■ Problem 39.

40. Tunnel A tunnel 150 m long, 7.2 m high, and 5.8 m wide (with a flat roof) is to be constructed 60 m beneath the ground. (See Fig. 13-47.) The tunnel roof is to be supported entirely by square steel columns, each with a cross-sectional area of 960 cm². The density of the ground material is 2.8 g/cm³. (a) What is the total mass of the material that the columns must support? (b) How many columns are needed to keep the compressive stress on each column at one-half its ultimate strength?

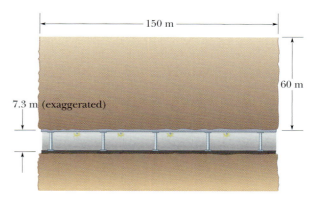

FIGURE 13-47 ■ Problem 40.

41. Cylindrical Aluminum Rod A cylindrical aluminum rod, with an initial length of 0.8000 m and radius 1000.0 μm, is clamped in place at one end and then stretched by a machine pulling parallel to its length at its other end. Assuming that the rod's density (mass per unit volume) does not change, find the force magnitude that is required of the machine to decrease the radius to 999.9 μm. (The yield strength is not exceeded.)

42. Stress Versus Strain Figure 13-48 shows the stress versus strain plot for an aluminum wire that is stretched by a machine pulling in opposite directions at the two ends of the wire. The wire has an initial length of 0.800 m and an initial cross-sectional area of 2.00×10^{-6} m². How much work does the force from the machine do on the wire to produce a strain of 1.00×10^{-3}?

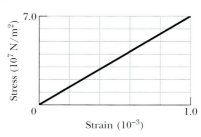

FIGURE 13-48 ■ Problem 42.

Additional Problems

43. Preventing Velociraptors In the movie Jurassic Park, there is a scene in which some members of the visiting group are trapped in the kitchen with dinosaurs outside the door. The paleontologist is pressing his shoulder near the center of the door, trying to keep out the dinosaurs who are on the other side. The botanist throws herself against the door at the edge right next to the hinge. A pivotal point in the film is that she cannot reach a gun on the floor because she is trying to help hold the door closed. Would they improve or worsen their situation if the paleontologist moved to the outer edge of the door and the botanist went for the gun? Estimate the change in the torque they are exerting on the door due to the change in their positions.

44. Rollerboards In the past few years, luggage carts that are rolling suitcases with handles, called "rollerboards," have become commonplace in airports around the country. Often, you will see people with a briefcase or additional small bag hung on the cart in one of the two ways shown, either hanging over the front of the cart (Fig. 13-49*a*) or resting on the handle (Fig. 13-49*b*). In this problem, we will figure out which way is easier for the traveler.

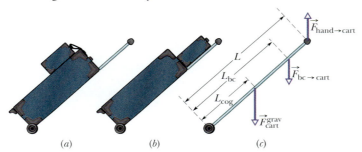

(a)   (b)   (c)

FIGURE 13-49 ■ Problem 44.

In Fig. 13-49*c* we have sketched a simplified idealization of the cart as a thin rod with forces acting on it. Three of the forces are shown, the gravitational attraction of the earth on the cart ($\vec{F}_{cart}^{grav}$), the force of the briefcase on the cart ($\vec{F}_{bc}$), and the force of the traveler's hand holding the cart up ($\vec{F}_{hand \rightarrow cart}$). Take the angle the cart makes with the ground to be θ, the mass of the cart to be M, and the mass of the briefcase to be m. Assume that the total length of the cart, handle and all, is L, the center of gravity of the cart is a distance L_{cog} from the wheel, and the center of gravity of the briefcase is a distance L_{bc} from the wheel.

(a) Find an expression for the force the hand has to exert in order to hold up the luggage cart. Express your answer in terms of the symbols given above.
(b) Does the force the cart exerts on the floor depend on the positioning of the briefcase? Explain.
(c) Estimate how different the force the hand has to exert would be in the two cases shown in Figs. 13-49*a* and 13-49*b*.

45. TV on a Handtruck You are working as a staff person on the Internet chat program "Ask Dr. Science." The following e-mail message comes in and needs a quick answer.

My wife just called me at the office and asked the following question. We had a large computer monitor delivered to her home office this morning. The delivery person was kind enough to put the box on our hand truck (see attached picture) but he put it on while the truck was lying flat. My wife has some back problems and doesn't want to have to exert more than 50 pounds of force. The monitor in its box weighs about 85 pounds. She's having a business meeting at the house later and would like to get the box out of the front room. What I want to know is, can she stand the truck upright safely without hurting her back?

A schematic diagram of the hand truck with the box on it is shown in Fig. 13-50. Is it safe for the man's wife to pull the hand truck upright so she can roll the box into the back room? Be sure to explain why you think so.

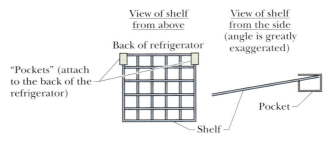

FIGURE 13-50 ■ Problem 45.

46. Refrigerator Shelf The shelves in your refrigerator are metal lattices that are held up by being slipped into two small (about 1 inch long) hollow boxes or "pockets" attached to the interior back wall of the refrigerator. See Fig. 13-51. If you put a full gallon of milk on the shelf, is it more likely to break the pocket if you place it near the back of the refrigerator or near the front? Explain your answer in terms of the physics you have learned. If the milk is the only thing on the shelf, estimate the downward force that the shelf exerts on the front of the pocket when the milk is placed at the front of the shelf.

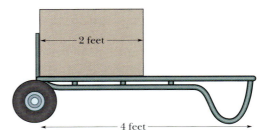

FIGURE 13-51 ■ Problem 46.

47. Weighing a Big Suitcase When preparing to travel to Australia last summer, a friend was concerned that her suitcase was too heavy. (There is a 20 kg limit on suitcases for international travel). Unfortunately, she only had a small bathroom scale. When she placed the suitcase directly on the scale, it covered the dial. She tried standing on the scale, measuring her weight, and then standing on the scale holding the suitcase. Unfortunately, when she was holding the suitcase, she couldn't see the markings on the scale, and it was too heavy to hold behind her. Design a way for her to measure the weight of the suitcase.

48. Indoor Playground You have been hired to build a large sand mound in an indoor playground and must be careful about the stress that the sand will put on the floor. Consulting research literature, you are surprised to find that the greatest stress occurs, not directly beneath the apex (top) of the mound, but at points that are a distance r^{max} from that central point (Fig. 13-52a). This outward displacement of the maximum stress is presumably due to the sand grains forming arches within the mound. For a mound of height $H = 3.00$ m and angle $\theta = 33°$, and with sand of density $\rho = 1800$ kg/m³, Fig. 13-52b gives the stress σ as a function of radius r from the central point of the mound's base. In that figure, $\sigma^{center} = 40\ 000$ N/m², $\sigma^{max} = 40\ 024$ N/m², and $r^{max} = 1.82$ m.

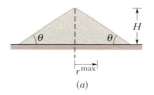

(a)

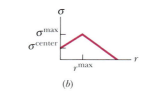

(b)

FIGURE 13-52 ■
Problem 48.

(a) What is the volume of sand contained in the mound for $r \leq r^{max}/2$? (*Hint:* The volume is that of a vertical cylinder plus a cone on top of the cylinder. The volume of the cone is $\pi R^2 h/3$, where R is the cone's radius and h is the cone's height.) (b) What is the weight W of that volume of sand? (c) Use Fig. 13-52b to write an expression for the stress σ on the floor as a function of radius r, for $r \leq r^{max}$. (d) On the floor, what is the area dA of a thin ring of radius r centered on the mound's central axis and with radial width dr? (e) What then is the magnitude dF of the downward force on the ring due to the sand? (f) What is the magnitude F of the net downward force on the floor due to all the sand contained in the mound for $r \leq r^{max}/2$? [*Hint:* Integrate the expression of (e) from $r = 0$ to $r = r^{max}/2$.] Now note the surprise: This force magnitude F on the floor is less than the weight W of the sand above the floor, as found in (b). (g) By what fraction is F reduced from W; that is, what is $(F - W)/W$?

49. Moving a Heavy Log Here is a way to move a heavy log through a tropical forest. Find a young tree in the general direction of travel; find a vine that hangs from the top of the tree down to ground level; pull the vine over to the log; wrap the vine around a limb on the log: pull

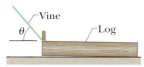

FIGURE 13-53 ■
Problem 49.

hard enough on the vine to bend the tree over; and then tie off the vine on the limb. Repeat this procedure with several trees; eventually the net force of the vines on the log moves the log forward. Although tedious, this technique allowed workers to move heavy logs long before modern machinery was available. Figure 13-53 shows the essentials of the technique. There, a single vine is shown attached to a branch at one end of a uniform log of mass M. The coefficient of static friction between the log and the ground is 0.80. If the log is on the verge of sliding, with the left end raised slightly by the vine, what are (a) the angle θ and (b) the magnitude T of the force on the log from the vine?

50. Uniform Ramp Figure 13-54a shows a uniform ramp between two buildings that allows for motion between the buildings due to

strong winds. At its left end, it is hinged to the building wall; at its right end, it has a roller that can roll along the building wall. There is no vertical force on the roller from the building, only a horizontal force with magnitude F^{horiz}. The horizontal distance between the buildings is $D = 4.00$ m. The rise of the ramp is $h = 0.490$ m. A man walks across the ramp from the left. Figure 13-54b gives F^{horiz} as a function of the horizontal distance x of the man from the building at the left. What are the masses of (a) the ramp and (b) the man?

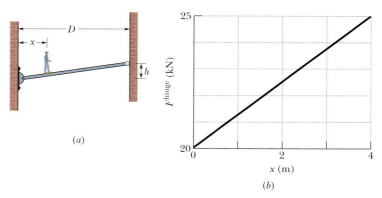

(a)

(b)

FIGURE 13-54 ■ Problem 50.

51. Diving Board In Fig. 13-55, a uniform diving board (mass = 40 kg) is 3.5 m long and is attached to two supports. When a diver stands on the end of the board, the support on the other end exerts a downward force of 1200 N on the board. Where on the board should the diver stand in order to reduce that force to zero?

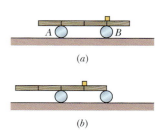

FIGURE 13-55 ■ Problem 51.

52. Rollers In Fig. 13-56a, a uniform 40 kg beam is centered over two rollers. Vertical lines across the beam mark off equal lengths. Two of the lines are centered over the rollers; a 10 kg package of tamale is centered over roller B. What are the magnitudes of the forces on the beam from (a) roller A and (b) roller B? The beam is then rolled to the left until the right-hand end is centered over roller B (Fig. 13-56b). What now are the magnitudes of the forces on the beam from (c) roller A and (d) roller B? Next, the beam is rolled to the right. Assume that it has a length of 0.800 m. (e) What horizontal distance between the package and roller B puts the beam on the verge of losing contact with roller A?

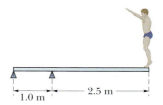

(a)

(b)

FIGURE 13-56 ■ Problem 52.

53. Horizontal Uniform Beam Figure 13-57a shows a horizontal uniform beam of mass m_{beam} and length L that is supported on the left by a hinge with a wall and on the right by a cable at angle θ with the horizontal. A package of mass m_{pack} is positioned on the beam at a distance x from the left end. The total mass is $m_{beam} +$

$m_{\text{pack}} = 61.22$ kg. Figure 13-57b gives the tension T in the cable as a function of the package's position given as a fraction x/L of the beam length. Evaluate (a) angle θ, (b) mass m_{beam}, and (c) mass in m_{pack}.

54. Vertical Uniform Beam Figure 13-58a shows a vertical uniform beam of length L that is hinged at its lower end. A horizontal force $\vec{F}^{\text{app}}$ is applied to the beam at a distance y from the lower end. The beam remains vertical because of a cable attached at the upper end, at angle θ with the horizontal. Figure 13-58b gives the tension T in the cable as a function of the position of the applied force given as a fraction y/L of the beam length. Figure 13-58c gives the magnitude F^{hinge} of the horizontal force on the beam from the hinge, also as a function of y/L. Evaluate (a) angle θ and (b) the magnitude of $\vec{F}^{\text{app}}$.

55. Makeshift Swing A makeshift swing is constructed by making a loop in one end of a rope and tying the other end to a tree limb. A child is sitting in the loop with the rope hanging vertically when an adult pulls on the child with a horizontal force and displaces the child to one side. Just before the child is released from rest, the rope makes an angle of 15° with the vertical and the tension in the rope is 280 N. (a) How much does the child weigh? (b) What is the magnitude of the (horizontal) force of the adult on the child just before the child is released? (c) If the maximum horizontal force that the adult can exert on the child is 93 N, what is the maximum angle with the vertical that the rope can make while the adult is pulling horizontally?

56. Emergency Stop A car on a horizontal road makes an emergency stop by applying the brakes so that all four wheels lock and skid along the road. The coefficient of kinetic friction between tires and road is 0.40. The separation between the front and rear axles is 4.2 m, and the center of mass of the car is located 1.8 m behind the front axle and 0.75 m above the road; see Fig. 13-59. The car weighs 11 kN. Calculate (a) the braking acceleration of the car, (b) the normal force on each wheel, and (c) the braking force on each wheel. (*Hint:* Although the car is not in translational equilibrium, it *is* in rotational equilibrium.)

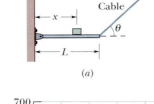
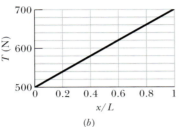

(a)

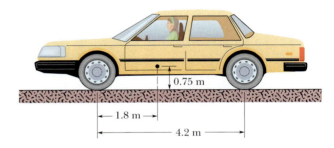

(b)

FIGURE 13-57 ■ Problem 53.

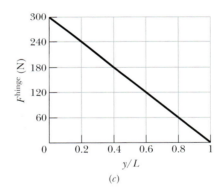

FIGURE 13-59 ■ Problem 56.

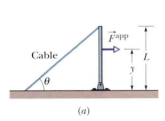

(a)

FIGURE 13-58 ■ Problem 54.

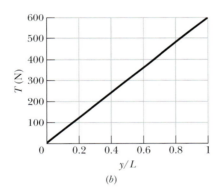

(b)

(c)

14 | Gravitation

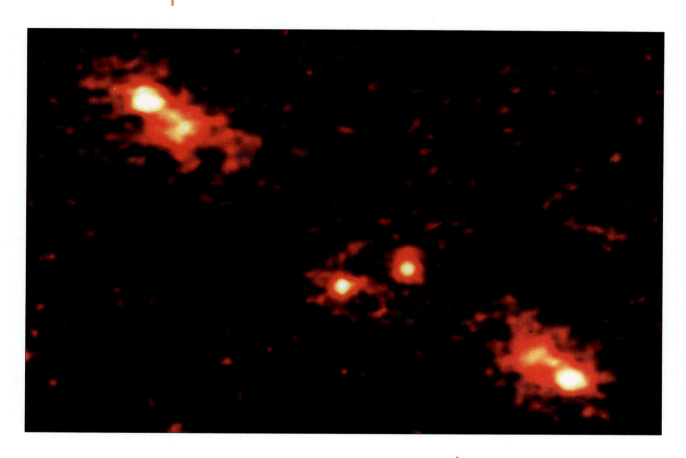

Quasars (or quasi-stellar objects) are highly luminous compact objects that are larger than stars and smaller than whole galaxies. Although some quasars emit energy at a trillion times the rate of our Sun and are bigger than our entire solar system, they cannot be detected with the naked eye. This is because quasars are the most distant objects yet detected in our universe. In 1979, astronomers using a powerful telescope were astonished to discover two similar-looking quasars that have the same spectrum of light coming from them. These are shown in the upper left and lower right corners of the photograph. Is this a rare coincidence or is there another explanation for the similarity of the two images?

How can Einstein's theory of gravitation be used to explain this coincidence?

The answer is in this chapter.

14-1 Our Galaxy and the Gravitational Force

Our Milky Way galaxy is a disk-shaped collection of gas, dust, and billions of stars, including our Sun and solar system. Figure 14-1 shows how our galaxy would look if we could view it from outside. Earth is near the edge of the disk of the galaxy, about 26 000 light-years (2.5×10^{20} m) from its central bulge. Our galaxy is a member of the Local Group of galaxies, which includes the Andromeda galaxy at a distance of 2.5×10^6 light-years, and several closer dwarf galaxies, such as the Large Magellanic Cloud.

FIGURE 14-1 ■ A scientifically constructed image of our Milky Way galaxy from the perspective of an observer outside the galaxy. Painting by Jon Lomberg, taken from a mural in the *Where Next, Columbus?* exhibit at the National Air and Space Museum.

The Local Group is part of the Local Supercluster of galaxies. Measurements taken during and since the 1980s suggest that the Local Supercluster and the supercluster consisting of the clusters Hydra and Centaurus are all moving toward an exceptionally massive region called the Great Attractor. This region appears to be about 300 million light-years away, on the opposite side of the Milky Way from us, past the clusters Hydra and Centaurus.

The force that binds together these progressively larger structures, from star to galaxy to supercluster, and may be drawing them all toward the Great Attractor, is known as the **gravitational force.** This force not only holds you on Earth but also reaches out across intergalactic space and acts between galaxies. It is our focus in this chapter.

14-2 Newton's Law of Gravitation

One of the strengths of physics as a scientific discipline is that physicists can often find connections between seemingly unrelated phenomena. The physicist's search for unification has been going on for centuries, and we continue the tradition in this chapter. We will search for connections between what we have already learned about gravitation close to the Earth's surface and a more general theory of gravitation.

In order to get started on this, we now turn our attention to the universe beyond Earth and see what Newton's laws reveal about the motion of a heavenly body like the Moon. Based on careful astronomical observation, we know the Moon orbits the Earth in an approximately circular path. Newton's laws of motion say that since the Moon is not going in a straight line there must be a force acting on it. This is because the direction of the speed (and so the velocity) is changing.

What force keeps the Moon in its circular orbit? Well, we know everything is pulled to Earth by a gravitational force, so this is a logical place to start. Suppose we assume that the magnitude of the force exerted on the moon by the Earth is given by the mass of the Moon, m, times the local gravitational strength at the Earth's surface given by $g = 9.8$ N/kg. If we also assume that this magnitude of the gravitational force on the Moon results in a centripetal acceleration, then the following equation should hold:

$$mg = \frac{mv^2}{r} \qquad \text{(proposed relationship)},$$

where v is the Moon's orbital speed, and r is its distance from the Earth. However, when we calculate the Moon's speed based on its orbit around the Earth, we discover that the actual magnitude of the Moon's centripetal acceleration, v^2/r, is thousands of times smaller than this equation would suggest. The gravitational force that we know and love (after all, it keeps our feet on the ground and our air from leaking away) is much too big to keep the Moon in its orbit. In fact, the gravitational force is about 3600 times too big. Such a gravitational force would cause the Moon to spiral in toward the Earth very rapidly.

Rather than abandon the idea that the gravitational force holds objects (including the Moon) near the Earth's surface, Newton suggested that perhaps gravity got weaker as you got farther from the center of the Earth. Since the Moon is about 60 Earth radii away, a gravitational force that decreased as the inverse square of the distance r between the Earth and the Moon would be just strong enough to be the centripetal force holding the Moon in its orbit. Furthermore, in our everyday life we only observe objects falling extremely short distances compared with the radius of the Earth. For this reason, a gravitational force that decreases as $1/r^2$ would appear constant to us. So, such a form for the gravitational force would be consistent with both astronomical observations and those made of objects close to the Earth's surface.

Suppose we pull together what we can observe and infer in an effort to construct a general statement regarding the gravitational force. We know from $|\vec{F}^{\text{grav}}| = mg$ that the gravitational force from the Earth on a mass m is proportional to the mass. Therefore we expect for two interacting masses m_A and m_B that the gravitational force on m_A is proportional to m_A and the force on m_B is proportional to m_B. However, Newton's Third Law tells us that these two forces must have equal magnitudes. So, the force magnitudes must be proportional to both masses. If the masses are separated by a distance r, our discussion of a $1/r^2$ dependence above implies that the force magnitudes should be given by

$$F_{B \to A}^{\text{grav}} = F_{A \to B}^{\text{grav}} \propto \frac{m_A m_B}{r^2}. \qquad (14\text{-}1)$$

In 1665, the 23-year-old Isaac Newton figured this out and made a historic contribution to physics. He showed that the force that holds the Moon in its orbit is the same force that makes an apple fall. We take this so much for granted now that it is not easy for us to comprehend the ancient belief that the motions of Earth-bound and heavenly bodies were governed by different laws. Furthermore, Newton determined that not only does Earth attract an apple and the Moon, but every body in the universe attracts every other body; this generalized tendency of bodies to move toward each other is called **gravitation.**

Newton's conclusion takes a little getting used to, because the familiar attraction of Earth for Earth-bound bodies is so great that it overwhelms the attraction that

Earth-bound bodies have for each other. For example, Earth attracts an apple with a force magnitude of about 0.8 N. You also attract a nearby apple (and it attracts you), but this force of attraction has less magnitude than the weight of a speck of dust, so we don't notice it.

However, if you consider the expression above (Eq. 14-1), you may notice that the units don't match. We have a unit of Newtons on the left, but not on the right. Newton suggested using a constant of proportionality, G, in the expression to create the equation describing the magnitude of the gravitational force between two particles:

$$F_{B \to A}^{\text{grav}} = F_{A \to B}^{\text{grav}} = G\frac{m_A m_B}{r^2} \qquad \text{(Newton's law of gravitation for particles).} \qquad (14\text{-}2)$$

This equation is the symbolic form of Newton's law of gravitation, which is expressed in words as follows:

Newton's law of gravitation: Every particle attracts any other particle with a *gravitational force*. This force has (1) a magnitude that is directly proportional to the product of the masses of the two particles and inversely proportional to the square of the distance between them; and (2) a direction that points along a line connecting the centers of the interacting particles.

The constant of proportionality G is known as the **gravitational constant.** Careful measurements show that in SI units G has a value of

$$G = 6.67 \times 10^{-11} \text{ N} \cdot \text{m}^2/\text{kg}^2$$
$$= 6.67 \times 10^{-11} \text{ m}^3/\text{kg} \cdot \text{s}^2. \qquad (14\text{-}3)$$

As Fig. 14-2 shows, a particle m_B attracts a particle m_A with a gravitational force $\vec{F}_{B \to A}^{\text{grav}}$ that is directed toward particle m_B. Particle m_A attracts particle m_B with a gravitational force $\vec{F}_{A \to B}^{\text{grav}}$ that is directed toward m_A. The forces $\vec{F}_{B \to A}^{\text{grav}}$ and $\vec{F}_{A \to B}^{\text{grav}}$ form a third-law force pair. So, we know that they must be opposite in direction but equal in magnitude. Thus,

$$\vec{F}_{B \to A}^{\text{grav}} = -\vec{F}_{A \to B}^{\text{grav}}.$$

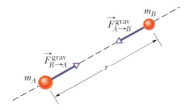

FIGURE 14-2 ■ Two particles, of masses m_A and m_B and with separation r, attract each other according to Newton's law of gravitation described in Eq. 14-2. The mutual forces of attraction are equal in magnitude and opposite in direction so that $\vec{F}_{B \to A}^{\text{grav}} = -\vec{F}_{A \to B}^{\text{grav}}.$

These interaction forces depend on the separation of the two particles, but not on their location: the particles could be in a deep cave or in deep space. Also forces $\vec{F}_{B \to A}^{\text{grav}}$ and $\vec{F}_{A \to B}^{\text{grav}}$ are not altered by the presence of other bodies, even if those bodies lie between the two particles we are considering. We know this because we observe that interposing an object (like a table) between the Earth and a book does not affect the gravitational force that the Earth exerts on the book.

Applying the Law of Gravitation to Spherical Objects

Although Newton's law of gravitation applies strictly to particles, we can also apply it to real objects as long as the sizes of the objects are small compared to the distance between them. The Moon and Earth are far enough apart so that, to a good approximation, we can treat them both as particles. But, what about an apple and Earth? From the point of view of the apple, the broad and level Earth, stretching out to the horizon beneath the apple, certainly does not look like a particle.

Newton solved the apple–Earth problem by proving an important theorem called the *shell theorem:*

> A uniform spherical shell of matter attracts a particle that is outside the shell as if all the shell's mass were concentrated at its center.*

Earth can be thought of as a nest of such shells, one within another, with the outer shells having less density. The shell theorem tells us that each shell will attract a particle outside Earth's surface as if the mass of that shell were at the center of the shell. If we also invoke the principle of superposition, then from the apple's point of view, the Earth *does* behave like a particle, one that is located at the center of Earth and has a total mass equal to that of Earth. So for spherical objects that don't overlap each other,

> The value of r in the expression $Gm_A m_B/r^2$ is always the center-to-center separation of two objects provided they do not overlap.

Suppose, as shown in Fig. 14-3, the Earth pulls down on an apple with a force of magnitude 0.80 N. The apple must then pull up on Earth with a force of magnitude 0.80 N, which we take to act at the center of Earth. Although the forces are matched in magnitude, they produce different accelerations due to the difference in the masses of the two objects. For the apple, the magnitude of the acceleration is about 9.8 m/s², the familiar acceleration of a falling body near Earth's surface. For Earth, the acceleration magnitude measured in a reference frame attached to the center of mass of the apple–Earth system is only about 1×10^{-25} m/s².

It is important to remember that Newton's theory of gravitation is an "action-at-a-distance" theory. That is, two objects exert gravitational forces on one another even if they do not touch one another. Newton's Third Law holds, so the forces between the two interacting bodies are equal and opposite *instant by instant*. This is interesting in that changes in the forces occur instantaneously at two different locations. So, variations cannot be propagated through intervening space at finite velocity or there would be an elapsed time interval. Massive objects do manage to interact instantaneously over huge intervening spaces. In the "General Scholium" section at the end of Book III of his *Principia*, Newton acknowledged this difficulty:

> *"But hitherto I have not been able to discover the cause of those properties of gravity from phenomena, and I frame no hypotheses . . . . to us it is enough that gravity does really exist, and acts according to the laws which we have explained, and abundantly serves to account for all the motions of the celestial bodies and of our sea."†*

Newton was clearly pleased that his theory could unify known astronomical observations. Ultimately, it was Einstein who successfully undertook a deeper inquiry into the very troubling and fundamental problem of how action-at-a-distance forces "travel" across space instantaneously. The answer he found lies in the fact that forces are not instantaneously transmitted. Rather, they are just transmitted very fast—at the speed of light. Although we introduce Einstein's theory of gravitation in Section 14-7, his explanation of action-at-a-distance phenomena is beyond the scope of this text.

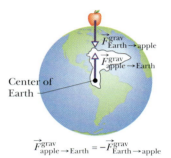

FIGURE 14-3 ■ The apple pulls up on Earth just as hard as Earth pulls down on the apple.

READING EXERCISE 14-1: In order to calculate the relative amount of force the Earth exerts on the Moon as compared to a mass close to the Earth's surface having the same mass as the Moon, you do not need to know the mass of the Moon. Why? ■

---

* This is an example of Gauss' law applied to gravitational forces. We introduce Gauss' law in more detail in Chapter 24.

† Isaac Newton. *Principia*, Vol 2: The System of the World. Translated by Andrew Motte and revised by Florian Cajori. (University of California Press Berkeley: 1962), p. 547.

READING EXERCISE 14-2: In 1666, the following facts stood by themselves without additional ramifications or supporting evidence: (1) The centripetal acceleration of the Moon is 3600 times smaller than the gravitational acceleration near the Earth's surface and (2) the square of the ratio of the Earth's radius to the mean radius of the Moon's orbit is 1/3600. How would you interpret the meaning of these facts? Do they "prove" that the Moon is held in its orbit by gravity? ■

14-3 Gravitation and Superposition

Suppose that we are given a group of particles and we want to find the net (or resultant) gravitational force on any one of them due to the others. How would we go about doing this? Previously, we found by observation that we could get the net force on an object by finding the vector sum of all of the forces acting on the object. This straightforward vector addition procedure is called "superposition." However, we should keep in mind that there are instances in which a simple linear superposition does *not* work. For example, superposition doesn't work on the atomic level. If we bring a neutron and proton together to form a heavy hydrogen nucleus, the mass of our nucleus is less than the sum of the neutron and proton masses.

When we test this idea of using the **principle of superposition** for gravitational forces, we find that it does work. Thus, we can compute the gravitational force that acts on our selected particle due to each of the other particles, in turn, by adding these forces vectorially. For n interacting particles, the force on the first particle is given by the vector addition

$$\vec{F}_A^{\text{net}} = \vec{F}_{B \to A}^{\text{grav}} + \vec{F}_{C \to A}^{\text{grav}} + \vec{F}_{D \to A}^{\text{grav}} + \vec{F}_{E \to A}^{\text{grav}} + \cdots + \vec{F}_{n \to A}^{\text{grav}}. \tag{14-4}$$

Here $\vec{F}_A^{\text{net}}$ is the net gravitational force on particle A, $\vec{F}_{B \to A}^{\text{grav}}$ is the gravitational force on particle A from particle B, and so on. We can express this equation more compactly as a vector sum:

$$\vec{F}_A^{\text{net}} = \sum_{i=B}^{n} \vec{F}_{i \to A}^{\text{grav}}. \tag{14-5}$$

What about the gravitational force on a particle from a real extended object? The force can be found by dividing the object into small particle-like parts, and then calculating the vector sum of the forces on the particle from all the parts. In the limiting case, we can divide the extended object into differential parts of mass dm, each of which produces a differential gravitational force $d\vec{F}^{\text{grav}}$ on particle A. In this limit, the net gravitational force on the particle is given by an integral

$$\vec{F}_A^{\text{net}} = \int d\vec{F}^{\text{grav}}, \tag{14-6}$$

taken over the entire extended object. If the extended object has spherical symmetry and if particle A lies outside of the sphere, we can avoid the integration by assuming that the extended object's mass is concentrated at its center. In this case we can use Newton's law of gravitation for particles described in Eq. 14-2,

$$F_{B \to A}^{\text{grav}} = F_{A \to B}^{\text{grav}} = G\,\frac{m_A m_B}{r^2}.$$

READING EXERCISE 14-3: A particle is to be placed, in turn, at the same distance, r, from the center of the four objects, shown in the figure, each of mass m: (1) a small uniform solid sphere, (2) a small uniform spherical shell (3) a large uniform spherical shell and (4) a large uniform solid sphere. In each situation, the distance between the particle and the center of the object is r. Rank the objects according to the magnitude of the gravitational force they exert on the particle, greatest first.

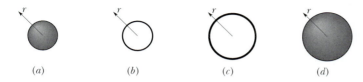

(a)   (b)   (c)   (d)

■

TOUCHSTONE EXAMPLE 14-1: Three Particles

Figure 14-4 shows an arrangement of three particles, particle A having mass $m_A = 6.0$ kg and particles B and C having mass $m_B = m_C = 4.0$ kg, and with distance $a = 2.0$ cm. What is the net gravitational force $\vec{F}_A^{\text{net}}$ that acts on particle A due to the other particles?

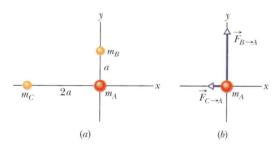

(a)   (b)

FIGURE 14-4 ■ (a) An arrangement of three particles. (b) The forces acting on the particle of mass m_A due to the other particles.

SOLUTION ■ One **Key Idea** here is that, because we have particles, the magnitude of the gravitational force on particle A due to particle B is given by Eq. 14-2 ($F_{B \to A} = Gm_Am_B/r^2$). Thus, the magnitude of the force $\vec{F}_{B \to A}$ on particle A from particle B is

$$F_{B \to A} = \frac{Gm_Am_B}{a^2}$$

$$= \frac{(6.67 \times 10^{-11}\ \text{m}^3/\text{kg} \cdot \text{s}^2)(6.0\ \text{kg})(4.0\ \text{kg})}{(0.020\ \text{m})^2}$$

$$= 4.00 \times 10^{-6}\ \text{N}.$$

Similarly, the magnitude of force $\vec{F}_{C \to A}$ on particle A from particle C is

$$F_{C \to A} = \frac{Gm_Am_C}{(2a)^2}$$

$$= \frac{(6.67 \times 10^{-11}\ \text{m}^3/\text{kg} \cdot \text{s}^2)(6.0\ \text{kg})(4.0\ \text{kg})}{(0.040\ \text{m})^2}$$

$$= 1.00 \times 10^{-6}\ \text{N}.$$

To determine the directions of $\vec{F}_{B \to A}$ and $\vec{F}_{C \to A}$ we use this **Key Idea**: Each force on particle A is directed toward the particle responsible for that force. Thus, $\vec{F}_{B \to A}$ is directed in the positive direction of y (Fig. 14-4b) and has only the y-component $F_{Ay} = F_{B \to A}$. Similarly, $\vec{F}_{C \to A}$ is directed in the negative direction of x and has only the x-component $F_{Ax} = -F_{C \to A}$.

To find the net force $\vec{F}_A^{\text{net}}$ on particle A, we first use this very important **Key Idea**: Because the forces are not directed along the same line, we *cannot* simply add or subtract their magnitudes or their components to get their net force. Instead, we must add them as vectors.

We can do so on a vector-capable calculator. However, here we note that $-F_{C \to A}$ and $F_{B \to A}$ are actually the x- and y-components of $\vec{F}_A^{\text{net}}$. Therefore, we shall follow the guide of Eq. 4-6 to find first the magnitude and then the direction of $\vec{F}_A^{\text{net}}$. The magnitude is

$$F_A^{\text{net}} = \sqrt{(F_{B \to A})^2 + (-F_{C \to A})^2}$$

$$= \sqrt{(4.00 \times 10^{-6}\ \text{N})^2 + (-1.00 \times 10^{-6}\ \text{N})^2} \quad \text{(Answer)}$$

$$= 4.1 \times 10^{-6}\ \text{N}.$$

Relative to the positive direction of the x axis, Eq. 4-6 gives the direction of $\vec{F}_A^{\text{net}}$ as

$$\theta = \tan^{-1}\frac{F_{B \to A}}{-F_{C \to A}} = \tan^{-1}\frac{4.00 \times 10^{-6}\ \text{N}}{-1.00 \times 10^{-6}\ \text{N}} = -76°.$$

Is this a reasonable direction? No, the direction of $\vec{F}_A^{\text{net}}$ must be between the directions of $\vec{F}_{B \to A}$ and $\vec{F}_{C \to A}$. A calculator displays only one of the two possible answers to a $\tan^{-1}$ function. We find the other answer by adding 180°. That gives us

$$-76° + 180° = 104°, \quad \text{(Answer)}$$

which is a reasonable direction for $\vec{F}_A^{\text{net}}$.

14-4 Gravitation in the Earth's Vicinity

The Earth has a mean radius of just over 6000 km. In this section we will consider the gravitational acceleration constants of objects located at various altitudes between 0 km (at the Earth's surface) and about 36 000 km (at the greatest altitude communications satellites achieve). We will also consider how the Earth's rotation and nonuniformity can cause relatively small changes in the measured weight for objects at the Earth's surface.

Gravitational Forces in the Vicinity of a Spherical Earth

Let's assume that Earth is spherically symmetric and has a total mass M and radius R. The magnitude of the gravitational force from the Earth on a particle of mass m, located outside Earth a distance $r > R$ from Earth's center can be expressed by modifying Eq. 14-2 as

$$F^{\text{grav}} = G\frac{Mm}{r^2}. \tag{14-7}$$

Let's focus on the particle of mass m. In Chapter 3 we introduced the local gravitational strength, g, as the ratio of the particle's gravitational force magnitude and its mass. In symbols this ratio is expressed as $g = F^{\text{grav}}/m$ (Eq. 3-7). If we combine Eq. 3-7 with Eq. 14-7 we see that the local gravitational strength can be expressed as

$$g = G\frac{M}{r^2}, \tag{14-8}$$

where the units for g are N/kg.

If our mass m experiences no other forces when it is released, it will fall toward the center of Earth under the influence of the only gravitational force $\vec{F}^{\text{grav}}$. As we saw in Chapter 3, according to Newton's Second Law the particle's **gravitational acceleration constant, $\vec{a}$,** is given by

$$\frac{\vec{F}^{\text{grav}}}{m} = \vec{a}.$$

By combining $F^{\text{grav}} = G\,Mm/r^2$ (Eq. 14-7) and the equation immediately above, the magnitude of the gravitational acceleration constant can be expressed in terms of the gravitational constant, G, the mass of the Earth, M, and the distance of the particle from the Earth's center, r, as

$$a = G\frac{M}{r^2} \quad \text{(gravitational acceleration constant)}, \tag{14-9}$$

where the units for a are m/s². The similarity of Eqs. 14-8 and 14-9 remind us that the *gravitational acceleration constant* and the Earth's *local gravitational strength* have the same value. However, as we observed in Section 3-9, we use different but dimensionally equivalent units to describe these two quantities.

Table 14-1 shows the calculated values of the magnitude of the gravitational acceleration constant of an object as a function of altitude. The calculations are made using $a = G\,(M/r^2)$ (Eq. 14-9) along with the known value for the mass of the Earth of $M = 5.98 \times 10^{24}$ kg. We note from the calculations in the table that anywhere on the Earth's land surface, from the bottom of the Dead Sea to the top of Mt. Everest, the gravitational acceleration constant calculated from the expression GM/r^2 is the same to two significant figures. In other words, the principles developed in this chapter

TABLE 14-1

Calculated Values of Gravitational Acceleration Constant with Altitude

Altitude* (km)	a (m/s^2)	Altitude Example
0.0	9.8	Mean Earth radius
8.8	9.8	Mt. Everest
36.6	9.7	Highest manned balloon
400	8.7	Space shuttle orbit
35 700	0.2	Communications satellite

*Altitude $= r - R$, where the radius of the Earth $R = 6370$ km.

reduce to the same familiar gravitational acceleration constant 9.8 m/s^2 that has been measured countless times in physics laboratories throughout the world.

The *reduction* of our "new" theory of gravitation discussed above to what we already had found to be true for the specific case of gravitation close to the surface of the Earth is an example of a general requirement for any "new" scientific model. When a model is developed to explain new, more general or more complicated phenomena, the "new" model must provide correct predictive information for the set of phenomena it describes and it must also be consistent with any simpler, more specific, or previously investigated phenomena. For example, in future chapters we will see that relativistic physics (for very high velocities) reduces to the nonrelativistic physics we have been studying so far when velocities are much less than the speed of light. Quantum physics reduces to classical physics for large, low-energy objects.

It is interesting to note that we can use our measurement of the gravitational acceleration constant at the Earth's surface (Eq. 14-9) along with our knowledge of G and r to calculate the mass of Earth!

Variations of Gravitational Forces over the Earth's Surface

In Section 6-3 we made two assumptions. First, we ignored the variations of the gravitational force and gravitational acceleration constant for an object at different locations on the Earth's surface. Second, we assumed that weight of an object as measured on a scale and the gravitational force on it, given by Eq. 14-7, are the same. Although these two assumptions are approximately true, geophysicists have measured slight variations or anomalies in the Earth's local gravitational field strength at different locations. There are many reasons for these anomalies. Some of the most significant are:

1. ***The Earth has an uneven surface.*** The Earth is covered with hills and mountains that rise above sea level and some valleys that are below sea level. The gravitational acceleration constant depends on altitude. When an object is closer to the dense core at a low altitude, it experiences a greater gravitational acceleration than it would at a high altitude. The distance between an object on dry land at the Earth's surface and the Earth's mean sea level varies from -0.414 km at the Dead Sea to $+8.85$ km at the summit of Mt. Everest. There is a difference of 0.03 m/s^2 in gravitational acceleration constant between sea level and the top of Everest.

2. ***The Earth bulges at its equator.*** Even if we were to "sand down" the bumps that represent hills and mountains rising above sea level and fill in the valleys below sea level, the Earth has the shape of an oblate spheroid. In other words, its shape is that of a sphere flattened at the poles and bulging at the equator. The equatorial radius is greater than its polar radius by 21 km. Thus, an object at the poles is

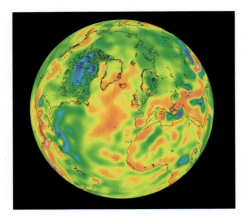

FIGURE 14-5 ■ Colored free-air gravity anomaly map of the Earth, centered on the Atlantic Ocean. These anomalies are the differences between the theoretical value for the gravity at the surface and the measured value. The colors range from purple (low gravity), through blue, green (normal), yellow, red, to white (high). The anomalies are only a tiny fraction of the gravitational field strength, but they can provide information on the Earth's internal structure. The gravity low in Hudson Bay, Canada (upper left), occurs partly because the area's rocks are recovering from being compressed during the ice age.

closer to the dense core of Earth than an object at the equator is. This is one of the reasons why the gravitational acceleration constant increases when we move it from the equator (where the latitude is 0°) toward either pole (where the latitude is 90°). This difference is about one-half of one percent or about 0.05 m/s^2.

3. **The Earth is rotating.** The rotation axis runs through the north and south poles of Earth. An object located on Earth's surface anywhere except at those poles must rotate in a circle about the rotation axis and thus must have a centripetal acceleration directed toward the center of the circle. This centripetal acceleration is caused by a centripetal force that is zero at the poles and most pronounced at the equator. This centripetal acceleration causes the apparent weight and measured gravitational acceleration of an object at the equator to be about 0.35% smaller than it would be at a pole. This latitude-dependent reduction in gravitational acceleration constant is about 0.03 m/s^2. (See below for more details.)

4. **The Earth has a crust of uneven thickness and density.** Even after geophysicists make corrections for the effects of altitude and latitude, measurements show that the gravitational force the Earth exerts on an object varies from location to location for other reasons. This is attributed to variations in: (1) the thickness of the Earth's crust (or outer section) and (2) the density of the rocks at the surface. Measurements in gravitational variations are useful in locating oil and mineral deposits. An example of regional variations is shown in Fig. 14-5.

Calculating the Effects of the Earth's Rotation

Recall from Chapter 6 that the weight we perceive for a mass, the object's apparent weight, is associated with the normal force on the object and can vary from the value mg if other forces act on the object. To see how Earth's rotation causes the apparent weight of an object at the Earth's surface to differ from the magnitude of the gravitational force on it, let us analyze a simple situation in which a crate of mass m is on a scale at the equator. Figure 14-6a shows this situation as viewed from a point in space above the north pole.

FIGURE 14-6 ■ (a) A crate lies on a scale at Earth's equator, as seen along Earth's rotation axis from above the north pole. (b) A free-body diagram for the crate, with a radially outward r axis. The gravitational force on the crate is represented by $\vec{F}^{\text{grav}}$. The normal force (or apparent weight) on the crate as read on a scale is represented by $\vec{N}$. Because of Earth's rotation, the crate also has a centripetal acceleration and hence a net centripetal force directed toward Earth's center.

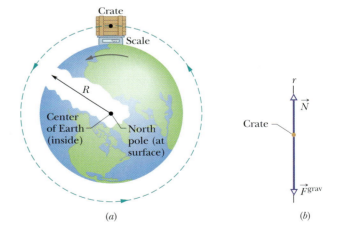

Figure 14-6b, a free-body diagram for the crate, shows the two forces on the crate, both acting along a radial axis r that extends from Earth's center. The normal force $\vec{N}$ on the crate exerted on it by the scale is directed outward, in the positive direction of axis r. The gravitational force, $\vec{F}^{\text{grav}}$, acts inward on the object of mass m. Because the crate travels in a circle about the center of Earth as the Earth turns, the crate also experiences a centripetal acceleration directed inward with its radial component given by a_r. From Eq. 11-20, we know the magnitude of this acceleration is equal to $\omega^2 R$, where ω is Earth's angular speed and R is the circle's radius (approximately Earth's radius). Thus, we can write Newton's Second Law in component form along the r axis ($F_r^{\text{net}} = ma_r$) as

$$N_r + F_r^{\text{grav}} = ma_r \quad \text{or} \quad |\vec{N}| - |\vec{F}^{\text{grav}}| = -m(\omega^2 R). \tag{14-10}$$

As we determined in Chapter 6, the magnitude N of the normal force is equal to the weight read on the scale. Solving Eq. 14-10 for the apparent (or measured) weight as the magnitude of the normal force gives us

$$|\vec{N}| = |\vec{F}^{\text{grav}}| - m(\omega^2 R) \quad \text{(apparent weight at the equator)}. \tag{14-11}$$

We see that the measured weight of the crate is actually slightly less than the magnitude of the gravitational force on the crate, because of Earth's rotation,

$$(\text{the measured weight}) = \begin{pmatrix} \text{the magnitude of} \\ \text{the gravitational force} \end{pmatrix} - \begin{pmatrix} \text{the object's mass times} \\ \text{its centripetal acceleration} \end{pmatrix}.$$

To find the difference at the equator, we can use Eq. 11-5 ($\langle\omega\rangle = \Delta\theta/\Delta t$) and Earth's radius $R = 6.37 \times 10^6$ m. For one Earth rotation ($\theta = 2\pi$ rad) the period is $\Delta t = 24$ h. Using these values (and converting hours to seconds), we find that $|\vec{N}|$ differs from $|\vec{F}^{\text{grav}}|$ by less than four-tenths of a percent (0.35%). Therefore, neglecting the difference between the apparent or measured weight and gravitational force magnitude is usually justified.

As we already mentioned, the difference between measured weight and the gravitational force magnitude is greatest on the equator (for one reason, the radius of the circle traveled by the crate is greatest there). At latitudes other than 0° it can be shown that Eq. 14-11 can be modified, to a very good approximation, to take the more general form,

$$|\vec{N}| = |\vec{F}^{\text{grav}}| - m(\omega^2 r) \quad \text{(apparent weight at any latitude)}, \tag{14-12}$$

where r is the perpendicular distance from a location on the Earth's surface to the axis of rotation and varies from R at the equator to zero at the poles.

READING EXERCISE 14-4: In the discussion above, we talked about the impact of the Earth's rotation on an object's measured weight as compared to the magnitude of the gravitational force on it. Do the factors that we discussed affect the direction of $\vec{F}^{\text{grav}}$ due to the Earth? Is $\vec{F}^{\text{grav}}$ directed toward the center of the Earth at all points on Earth? To answer this question, consider the arguments made above in regard to the effect of the Earth's rotation on the object's apparent weight. Then, without full algebraic analysis, do some visualization based on a relevant force diagram. ■

TOUCHSTONE EXAMPLE 14-2: Floor to Ceiling

How much less will a mass of 1.000 kg weigh at a ceiling of height $h = 3.00$ m compared to its weight on the floor? Assume that the local gravitational strength at floor level has the internationally adopted standard value of 9.80665 N/kg.

SOLUTION ■ The **Key Idea** here is that the mass is slightly further from the center of the Earth on the ceiling than on the floor so its local gravitational strength is slightly smaller.

The first step in our solution is to find the ratio of the local gravitational strength on the ceiling to that on the floor. We can start by using the equation $g = GM/R^2$ (Eq. 14-8) where R is the distance from the center of the Earth to the location of an object near its surface. If we represent the distance from the Earth's center to the ceiling as R_{ceiling} then we can represent the distance to the floor as

$$R_{\text{floor}} = R_{\text{ceiling}} - h.$$

Next we substitute the symbols representing distances to the floor and to ceiling from the center of the Earth into Eq. 14-8 and take a ratio. This gives

$$\frac{g_{\text{ceiling}}}{g_{\text{floor}}} = \frac{GM/R_{\text{ceiling}}^2}{GM/(R_{\text{ceiling}} - h)^2}$$

$$= \frac{(R_{\text{ceiling}} - h)^2}{R_{\text{ceiling}}^2} = \frac{R_{\text{ceiling}}^2 - 2hR_{\text{ceiling}} + h^2}{R_{\text{ceiling}}^2}.$$

We can then note in Appendix B that the mean radius of the Earth, which we can take as the Earth center to floor distance, is millions of meters. More precisely, it is 6.4×10^6 m. The ceiling height is so much less than R_{ceiling} that we can ignore h^2. So we can express the ratio to very good approximation as

$$\frac{g_{\text{ceiling}}}{g_{\text{floor}}} = \frac{R_{\text{ceiling}}^2 - 2hR_{\text{ceiling}} + h^2}{R_{\text{ceiling}}^2} \approx 1 - \frac{2h}{R_{\text{ceiling}}}.$$

Since the weight of our mass m is just $F^{\text{grav}} = mg$ in either location, the ratio of its weight at the ceiling to its weight at the floor is just

$$\frac{F_{\text{ceiling}}^{\text{grav}}}{F_{\text{floor}}^{\text{grav}}} = \frac{mg_{\text{ceiling}}}{mg_{\text{floor}}} \approx 1 - \frac{2h}{R_{\text{ceiling}}}.$$

So the difference in the weight of the mass between the ceiling and the floor is

$$F_{\text{ceiling}}^{\text{grav}} - F_{\text{floor}}^{\text{grav}} = -\frac{2hF_{\text{floor}}^{\text{grav}}}{R_{\text{ceiling}}}$$

$$= \frac{-(2)(3.00 \text{ m})(1.000 \text{ kg})(9.80665 \text{ m/s}^2)}{6.37 \times 10^6 \text{ m}}$$

$$= -9.24 \times 10^{-6} \text{ N}.$$

Hanging out on the ceiling is certainly not a viable way to lose a measurable amount of weight!

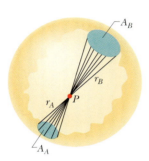

FIGURE 14-7 ■ The gravitational force at a point P due to an element of mass of area A is directly proportional to the area and inversely proportional to the square of the distance between the area and point P. The area of intersection increases with the square of the distance from it to point P. The gravitational forces of areas A_A and A_B on either side of any cone pair cancel each other out. This proves that there is no gravitational force acting anywhere inside a sphere of uniform mass.

14-5 Gravitation Inside Earth

The *shell theorem*, discussed in Section 14-2, states that "a uniform spherical shell of matter attracts a particle that is *outside* the shell as if all the shell's mass were concentrated at its center." However, we are still left with the question of what the force is on a particle that is *inside* the spherical shell of mass. In order to answer this question, let us refer to Fig. 14-7 and develop a geometric argument to show that a spherical shell with uniformly distributed mass exerts no net gravitational force on a particle inside itself.

Consider a mass m placed at a point P somewhere in the interior of this shell. Let's construct two cones that come together at P and have equal vertex angles. The cones will intercept patches of mass with areas A_A and A_B on opposite sides of the shell. The area of patch A_A and hence its mass m_A will be proportional to r_A^2, so that $m_A = Cr_A^2$ where C is a constant. Similarly the area of patch A_B and its mass m_B will be proportional to r_B^2 and $m_B = Cr_B^2$. Let's consider the magnitude of the gravitational force on an object of mass m at point P due to a patch having an area A_A and a mass m_A. This force magnitude is given by

$$F_{m_A \to m}^{\text{grav}} = G\frac{mm_A}{r_A^2} = G\frac{mCr_A^2}{r_A^2} = GmC.$$

We end up with a force that is independent of *both* the distance to the point (r) and the area of the patch (A). The same can obviously be said for the force on m from the mass m_B in the patch of area A_B. So, the forces exerted at P by patches A_A and A_B will be equal in magnitude. As we can see from Fig. 14-7, these forces are also opposite in direction. So, the forces from the two patches sum to zero and cancel out. We can cover the entire shell with such opposing patches, so overall the net force at P must be zero. Using this geometric argument, we can state a shell theorem as follows:

> A uniform shell of matter exerts no *net* gravitational force on a particle located anywhere inside it.

Be careful though: This statement does *not* mean that the gravitational forces on the particle from the various elements of the shell magically disappear. Rather, it means that the *sum* of the force vectors on the particle from all the elements is zero. In addition, if the gravitational force did not obey the inverse square law, the areas would not be canceled patch by patch by the radial distances, and the net force at P would not be zero. Therefore, this theorem is valid *only* for a force that obeys an inverse square law.

If we think of Earth as a series of concentric shells of uniform density, this shell theorem implies that the gravitational force acting on a particle would be a maximum at Earth's surface. If the particle were to move inward, perhaps down a deep mine shaft, the gravitational force would change for two reasons. (1) It would tend to increase because the particle would be moving closer to the center of Earth. (2) It would tend to decrease because the thickening shell of material lying outside the particle's radial position would not exert any net force on the particle. For a uniform Earth, the second influence would prevail and the force on the particle would steadily decrease to zero as the particle approached the center of Earth. However, for the real (nonuniform) Earth, the force on the particle actually increases as the particle begins to descend. The force reaches a maximum at a certain depth: Only then does it begin to decrease as the particle descends farther. This is because the Earth's crust is low density as compared to the average density of the Earth.

READING EXERCISE 14-5: How would the force of gravity from the Earth on a particle change in the following three cases? Case A: The particle starts at the surface of the Earth and moves outward from its center. Case B: The particle starts at the surface of the Earth and moves in toward the center of the Earth (assumed to have a uniform density). Case C: The particle starts at the surface of the Earth and moves in toward the center of the Earth (with the Earth's real density distribution). ■

TOUCHSTONE EXAMPLE 14-3: Pole to Pole

In *Pole to Pole*, an early science fiction story by George Griffith, three explorers attempt to travel by capsule through a naturally formed tunnel between the south pole and the north pole (Fig. 14-8). According to the story, as the capsule approaches Earth's center, the gravitational force on the explorers becomes alarmingly large and then, exactly at the center, it suddenly but only momentarily disappears. Then the capsule travels through the second half of the tunnel, to the north pole.

Check Griffith's description by finding the gravitational force on the capsule of mass m when it reaches a distance r from Earth's center. Assume that Earth is a sphere of uniform density ρ (mass per unit volume).

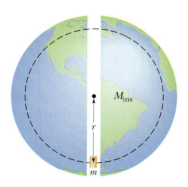

FIGURE 14-8 ■ A capsule of mass m falls from rest through a tunnel that connects Earth's south and north poles. When the capsule is at distance r from Earth's center, the portion of Earth's mass that is contained in a sphere of that radius is M_{ins}.

SOLUTION ■ Newton's shell theorem gives us three **Key Ideas** here:

1. When the capsule is at a radius r from Earth's center, the portion of Earth that lies outside a sphere of radius r does *not* produce a net gravitational force on the capsule.

2. The portion that lies inside that sphere *does* produce a net gravitational force on the capsule.

3. At a given location inside the Earth, we can treat the mass M_{ins} of the inside portion of Earth at that location as being the mass of a particle located at Earth's center.

All three ideas tell us that we can write Eq. 14-2, for the magnitude of the gravitational force on the capsule, as

$$F^{\text{grav}} = \frac{GmM_{\text{ins}}}{r^2}. \quad (14\text{-}13)$$

To write the mass M_{ins} in terms of the radius r, we note that the volume V_{ins} containing this mass is $\frac{4}{3}\pi r^3$. Also, its density is Earth's density ρ. Thus, we have

$$M_{\text{ins}} = \rho V_{\text{ins}} = \rho \frac{4\pi r^3}{3}. \quad (14\text{-}14)$$

Then, after substituting this expression into Eq. 14-13 and canceling, we have

$$F^{\text{grav}} = \frac{4\pi Gm\rho}{3}\,r. \qquad \text{(Answer)} \quad (14\text{-}15)$$

This equation tells us that the gravitational force magnitude F^{grav} depends linearly on the capsule's distance r from Earth's center. Thus, as r decreases, F^{grav} also decreases (the opposite of Griffith's description), until it is zero at Earth's center. At least Griffith got zero-at-the-center correct. However, forces near the Earth's center are not large, but "alarmingly" small instead.

Equation 14-15 can also be written in terms of the force vector $\vec{F}^{\text{grav}}$ and the capsule's position vector $\vec{r}$ along a radial axis extending from Earth's center. Let K represent the collection of constants $4\pi Gm\rho/3$. Then Eq. 14-15 becomes

$$\vec{F}^{\text{grav}} = -K\vec{r}, \qquad (14\text{-}16)$$

in which we have inserted a minus sign to indicate that $\vec{F}^{\text{grav}}$ and $\vec{r}$ have opposite directions, since $\vec{r}$ represents the displacement of an object from the Earth's center. Equation 14-15 has the form of Hooke's law (Eq. 9-16). Thus, under the idealized conditions of the story, the capsule would oscillate like a block on a spring, with the center of the oscillation at Earth's center. After the capsule had fallen from the south pole to Earth's center, it would travel from the center to the north pole (as Griffith said) and then back again.

14-6 Gravitational Potential Energy

In Chapter 10, we defined potential energy and derived an expression for the change in potential energy, ΔU, associated with any conservative force when a system of two objects is reconfigured as shown in Fig. 10-11. We did this by finding the internal work done when one of the objects exerts a force on the other during the reconfiguration. Our expression for the gravitational force was

$$\Delta U = -W^{\text{int}} = -\int_{r_1}^{r_2} \vec{F}^{\text{grav}}_{A\to B}(r)\cdot d\vec{r}, \qquad \text{(Eq. 10-13)}$$

where r_1 is the original separation between objects in the system and r_2 is the separation of the objects in the system at some later time. We then used this equation to find changes in gravitational potential energy (GPE) for an Earth–object system. Because we did not yet have a general expression for the gravitational forces between two objects, we only considered the special case in which the object is close to the Earth's surface and found $\Delta U = mg\Delta y$ (Eq. 10-6). However, we cannot use this expression to determine how much energy it would take to launch a rocket that escapes the gravitational pull of the Earth.

Gravitational Potential Energy Changes for Any Two-Particle System

A key objective of this section is to find an equation for the gravitational potential energy for a two-particle system and use it to describe an Earth–object system for objects at any distance from the Earth's surface. In Section 10-3 we determined the potential energy change for a two-particle system that interacts by means of a conservative force. This is where we will start (with Eq. 10-13) and use the gravitational force as our conservative force. So,

$$\Delta U = -\int_{r_1}^{r_2} \vec{F}^{\text{grav}}_{A\to B}(r)\cdot d\vec{r},$$

for a general equation of the magnitude of the gravitational force between any two particles of masses m_A and m_B (represented in both Fig. 14-2 and Eq. 14-2). This equation, which is simply Newton's law of gravitation, is given by

$$F^{\text{grav}}_{B\to A} = F^{\text{grav}}_{A\to B} = G\,\frac{m_A m_B}{r^2}. \qquad \text{(Eq. 14-2)}$$

If we increase the separation of the two masses from an initial separation r_1 to a final separation r_2, the direction of $\vec{F}_{A \to B}^{\,\text{grav}}$ is opposite to the direction of dr. If we take the direction of dr to be positive, then the r-component of the gravitational force exerted on particle B by particle A, $(F_{A \to B}^{\text{grav}})_r$, must be negative. Thus,

$$(F_{A \to B}^{\text{grav}})_r = -G\,\frac{m_A m_B}{r^2}.$$

We can substitute this expression for the gravitational force component along r into the integral and evaluate as we separate the particles from r_1 to r_2. This gives

$$\int_{r_1}^{r_2} (F_{A \to B}^{\text{grav}})_r\, dr = \int_{r_1}^{r_2} \left(\frac{-Gm_A m_B}{r^2} \right) dr = \left[\frac{+Gm_A m_B}{r} \right]_{r_1}^{r_2} = \frac{Gm_A m_B}{r_2} - \frac{Gm_A m_B}{r_1}.$$

We now substitute the value of this integral into Eq. 10-13 to find the gravitational potential energy change,

$$\Delta U = U(r_2) - U(r_1) = -\int_{r_1}^{r_2} (F_{A \to B}^{\text{grav}})_r\, dr = -\frac{Gm_A m_B}{r_2} + \frac{Gm_A m_B}{r_1}. \quad (14\text{-}17)$$

Defining an Absolute Gravitational Potential Energy for a Two-Particle System

In Chapter 10, we discussed the gravitational potential energy of a particle–Earth system where the particle is close to the Earth's surface. For that special case we found it useful to choose a "zero potential energy" configuration in which the particle was located at the surface of the Earth (or some other convenient height near the Earth's surface). In this more general situation in which the particles can be very far apart, we find it more useful to define a different reference configuration for which the potential energy is equal to zero. Since gravitational forces decrease rapidly to zero with separation, (in fact as $1/r^2$), it is very convenient to define our potential energy to be zero when separation distance r between particles is *infinite*. We can then define the gravitational potential energy as minus the internal work done on particle B by particle A as the separation of the two particle changes from an initial separation of infinity (denoted by ∞) to a final separation of r.

For this situation, Eq. 14-17 tell us

$$\Delta U = U(r) - U(\infty) = -\frac{Gm_A m_B}{r} + \frac{Gm_A m_B}{\infty} = -\frac{Gm_A m_B}{r} + 0.$$

Since we have defined our reference potential to be zero at infinity so $U(\infty) = 0$, the equation above reduces further to

$$U(r) = -\frac{Gm_A m_B}{r} \qquad \text{(gravitational PE relative to infinite separation).} \qquad (14\text{-}18)$$

Here $G = 6.67 \times 10^{-11}\ \text{N} \cdot \text{m}^2/\text{kg}^2$ is the gravitational constant, m_A is the mass of one object, m_B is the mass of the other object, and r is the center-to-center separation of the two particle-like masses. Note that $U(r)$ approaches zero as r approaches infinity and that for any finite value of r, the value of $U(r)$ is negative (Fig. 14-9).

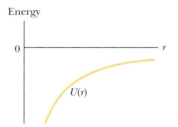

FIGURE 14-9 ■ The gravitational potential energy of a two-mass system. Note that the PE is negative everywhere. It has a very large magnitude as the distance r between the masses approaches 0, but it approaches 0 as r approaches infinity.

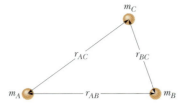

FIGURE 14-10 ■ Three particles form a system. (The separation for each pair of particles is labeled with a double subscript to indicate the particles.) The gravitational potential energy *of the system* is the sum of the gravitational potential energies of all three pairs of particles.

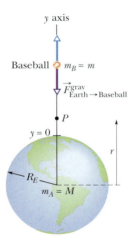

FIGURE 14-11 ■ A baseball is moved along a *y* axis to a point *P* which is a distance $r > R_E$ from the center of the Earth. Since the gravitational force is attractive, the Earth–baseball system loses gravitational PE as the particles get closer together.

Gravitational Potential Energy for a Many-Particle System

The potential energy given by this expression is a property of the system of two particles rather than of either particle alone. If our system contains more than two particles, we consider each pair of particles separately. We calculate the gravitational potential energy of that pair with this equation as if the other particles were not there. We then algebraically sum the results (energy is a scalar). Applying Eq. 14-18 to each of the three pairs of charges in Fig. 14-10, for example, gives the potential energy of this system as

$$U = -\left(\frac{Gm_A m_B}{r_{AB}} + \frac{Gm_A m_C}{r_{AC}} + \frac{Gm_B m_C}{r_{BC}} \right) \quad \text{(3-particle system).} \quad (14\text{-}19)$$

Gravitational Potential Energy for an Earth–Object System

Suppose we bring a baseball of mass *m* from infinity along the *y* axis to a point *P*, a distance *r* from the center of the Earth, which has mass *M* as shown in Fig. 14-11. If $r \geq R_E$, where R_E is the Earth's radius, what is the *general* expression for the gravitational potential energy of the Earth–baseball system? We simply substitute our new symbols into Eq. 14-18 shown above. This gives us

$$U(r) = -\frac{GMm}{r} \quad \text{(Earth–object system gravitational PE relative to } r = \infty \text{).} \quad (14\text{-}20)$$

However, this general expression must be consistent with what we derived in Chapter 10 for the special case of an object *close* to the Earth's surface. When we chose to define $y = 0$ at some convenient height near the Earth's surface, we found that

$$U(y) = mgy \quad \text{(near Earth gravitational PE relative to } y = 0 \text{).} \quad \text{(Eq. 10-8)}$$

Although the two expressions look quite different at first glance, we see that in both cases the potential energy decreases (becomes progressively more negative) as the Earth and the baseball move closer together. Our two expressions for gravitational potential energy are consistent in this regard.

However, if our general expression for gravitational potential energy in Eq. 14-20 is valid for all separations, it must be consistent with the more specific expression $U(y) = mgy$. To see that this is the case, suppose the object starts at the surface of the Earth. Its potential energy at this location is

$$U(R_E) = -\frac{GMm}{R_E}.$$

The object then moves upward to a height Δy above the Earth's surface. The potential energy of the object at this location is

$$U(R_E + \Delta y) = -\frac{GMm}{R_E + \Delta y}.$$

So, the change in potential energy between these two configurations is

$$\Delta U = U_2 - U_1 = U(R_E + \Delta y) - U(R_E)$$

$$= -\frac{GMm}{R_E + \Delta y} - \left(-\frac{GMm}{R_E} \right)$$

$$= -\frac{GMm}{R_E + \Delta y} + \frac{GMm}{R_E}.$$

Simplifying this expression by finding a common denominator and factoring gives us

$$\Delta U = \frac{GMm\Delta y}{R_E(R_E + \Delta y)}.$$

However, the radius of the Earth, R_E, is *orders of magnitude* (much, much) larger than the additional height Δy for situations in which the object is close to the surface of the Earth. So,

$$R_E + \Delta y \approx R_E,$$

and

$$\Delta U = \frac{GMm\Delta y}{R_E(R_E + \Delta y)} \approx \frac{GMm\Delta y}{R_E^2}.$$

Equation 14-9 tells us that the magnitude of the gravitational acceleration constant at the surface of the Earth (with radius R_E and mass M) is

$$a = \frac{GM}{R_E^2} \qquad \text{(gravitational acceleration constant).}$$

If we ignore small variations in the gravitational acceleration constant due to the Earth's rotation and its small deviations from a spherical shape, then the gravitational acceleration constant and the local gravitational strength have the same magnitude so that $a = g$.

Substituting these last two equations into the equation for the change in the gravitational potential energy expression gives

$$\Delta U \approx \frac{GMm\,\Delta y}{R_E^2} = mg\,\Delta y.$$

In other words, if the height Δy above the surface of the Earth is small compared to the radius of the Earth, the gravitational acceleration constant and the local gravitational strength g are essentially constant, and our general expression for gravitational potential energy allows us to predict the same changes in gravitational potential energy as the more specific one we used in Chapter 10. This is true even though we have chosen very different zero points for our general and near-Earth potential energies. The two expressions are consistent because they allow us to calculate the same changes in gravitational potential energy as long as we are near the Earth's surface.

Path Independence

In the equations derived in this section, we have made the simplifying assumption that our particles move apart or come together along a line connecting their centers. But, because the gravitational force is conservative, potential energy changes of the system are path independent as discussed in Section 10-2. Thus, our equations hold even when we allow the interacting particles to separate along any crazy path.

An example of this path independence is shown in Fig. 14-12. We imagine moving a baseball from point A to point G along a path consisting of three radial lengths and three circular arcs (centered on Earth). We are interested in the total work W done by Earth's gravitational force on the ball as it moves from A to G. The work done along each circular arc is zero, because the direction of the force is perpendicular to the arc at every point. Thus, the only work done by the force is along the three radial lengths, and the total work W is the sum of the work done along the radial lengths.

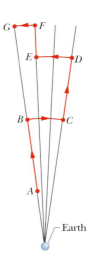

FIGURE 14-12 ■ Near Earth, a baseball is moved from point A to point G along a path consisting of radial lengths and circular arcs.

Now suppose we mentally shrink the arcs to zero. We would then be moving the ball directly from A to G along a single radial length. Does that change the total work done? No. Because no work was done along the arcs, eliminating them does not change the work. The path taken from A to G now is clearly different, but the work done by the force on the baseball is the same. This example reminds us that the internal work done by a system can be independent of the actual path taken. In that case, the change ΔU in the system's gravitational potential energy is path independent as well. This example of path independence for the gravitational force is consistent with what we have shown in Section 10-2.

Escape Speed

When launching a rocket, how fast does it need to be moving to escape the gravitational pull of the Earth? If you throw an object upward, it will usually slow, turn around, and then speed up as it travels back toward the Earth. There is, however, a minimum initial speed that will allow an object to move upward forever. In this case, the rocket or any other object launched upward slows and approaches zero speed as the object gets farther and farther away from the Earth, but never reverses direction and returns. This special initial speed is called the (Earth) **escape speed.**

In order to calculate the escape speed from Earth (or some other spherical astronomical body), consider a projectile of mass m, leaving the surface of a planet of radius R with escape speed v. Assume the rocket has an initial kinetic energy K given by $\frac{1}{2}mv^2$ and a potential energy U given by Eq. 14-20,

$$U = -\frac{GMm}{R},$$

where M is the mass of the planet, and R is its radius.

As the rocket gets far away, its speed approaches zero and the kinetic energy of the Earth–rocket system approaches zero. In addition, the system potential energy approaches zero because the planet–rocket separation is very large. So, total energy of the system at an "infinite" separation is zero. If we ignore the relatively small amount of energy lost to air drag, then mechanical energy is approximately conserved. In this case the rocket's total energy at the planet's surface must also have been zero, so we can use Eq. 10-19 to get

$$E^{\text{mec}} = K + U = \tfrac{1}{2}mv^2 + \left(-\frac{GMm}{R}\right) = 0.$$

This yields

$$v = \sqrt{\frac{2GM}{R}}. \tag{14-21}$$

The escape speed v does not depend on the direction in which a projectile is fired from a planet. However, attaining that speed is easier if the projectile is fired in the direction the launch site is moving as the planet rotates about its axis. For example, rockets are launched eastward at Cape Canaveral to take advantage of the Cape's eastward speed of 0.5 km/s due to Earth's rotation.

Equation 14-21 can be applied to find the escape speed of a projectile from any astronomical body, provided we substitute the mass of the body for M and the radius of the body for R. Table 14-2 shows escape speeds from some astronomical bodies. It is interesting to note that objects of any size can escape from an astronomical body. For example, gas molecules in planetary atmospheres sometimes reach escape speeds.

TABLE 14-2
Some Escape Speeds

Body	Mass (kg)	Radius (m)	Escape Speed (km/s)
Ceres[a]	1.17×10^{21}	3.8×10^5	0.64
Earth's Moon	7.36×10^{22}	1.74×10^6	2.38
Earth	5.98×10^{24}	6.37×10^6	11.2
Jupiter	1.90×10^{27}	7.15×10^7	59.5
Sun	1.99×10^{30}	6.96×10^8	618
Sirius B[b]	2×10^{30}	1×10^7	5200
Neutron star[c]	6×10^{30}	1×10^4	3.4×10^5

[a]The most massive of the asteroids.
[b]A *white dwarf* (a very compact old star that has burned its nuclear fuel) that is in orbit around the bright star Sirius. Sirius B has roughly the mass of our Sun and a radius close to that of the Earth.
[c]The collapsed core of a massive star that remains after that star has exploded in a *supernova* event and is more compact than a white dwarf (with 3 times the mass of our Sun and a diameter of only a few kilometers).

READING EXERCISE 14-6: You move a ball of mass m away from a sphere of mass M. (a) Does the gravitational potential energy of the ball-sphere system increase or decrease? (b) Is positive or negative work done by the gravitational force between the ball and the sphere?   ∎

TOUCHSTONE EXAMPLE 14-4: Asteroid

An asteroid, headed directly toward Earth, has a speed of 12 km/s relative to the planet when it is at a distance of 10 Earth radii from Earth's center. Neglecting the effects of Earth's atmosphere on the asteroid, find the asteroid's speed v_2 when it reaches Earth's surface.

SOLUTION ∎ One **Key Idea** is that, because we are to neglect the effects of the atmosphere on the asteroid, the mechanical energy of the asteroid-Earth system is conserved during the fall. Thus, the final mechanical energy (when the asteroid reaches Earth's surface) is equal to the initial mechanical energy. We can write this as

$$E^{mec} = K_1 + U_1 = K_2 + U_2, \qquad \text{(Eq. 10-19)}$$

where K is kinetic energy and U is gravitational potential energy.

A second **Key Idea** is that, if we assume the system is isolated, the system's linear momentum must be conserved during the fall. Therefore, the momentum change of the asteroid and that of Earth must be equal in magnitude and opposite in sign. However, because Earth's mass is so great relative to the asteroid's mass, the change in Earth's speed is negligible relative to the change in the asteroid's speed. So, the change in Earth's kinetic energy is also

negligible. Thus, we can assume that the kinetic energies in Eq. 10-19 are those of the asteroid alone.

Let m represent the asteroid's mass and M represent Earth's mass (5.98×10^{24} kg). The asteroid is initially at the distance $10R_E$ and finally at the distance R_E, where R_E is Earth's radius (6.37×10^6 m). Substituting Eq. 14-20 for U and $\frac{1}{2}mv^2$ for K, we rewrite Eq. 10-19 as

$$\tfrac{1}{2}mv_2^2 - \frac{GMm}{R_E} = \tfrac{1}{2}mv_1^2 - \frac{GMm}{10R_E}.$$

Rearranging and substituting known values, we find

$$v_2^2 = v_1^2 + \frac{2GM}{R_E}\left(1 - \frac{1}{10}\right)$$

$$= (12 \times 10^3 \text{ m/s})^2$$

$$+ \frac{2(6.67 \times 10^{-11} \text{ m}^3/\text{kg}\cdot\text{s}^2)(5.98 \times 10^{24} \text{ kg})}{6.37 \times 10^6 \text{ m}} 0.9$$

$$= 2.567 \times 10^8 \text{ m}^2/\text{s}^2,$$

and thus the magnitude of the impact velocity is

$$v_2 = 1.60 \times 10^4 \text{ m/s} = 16 \text{ km/s}. \qquad \text{(Answer)}$$

At this speed, the asteroid would not have to be particularly large to do considerable damage at impact. As an example, if it were only 5 m across, the impact could release about as much energy as the nuclear explosion at Hiroshima. Alarmingly, about 500 million asteroids of this size are near Earth's orbit, and in 1994 one of them apparently penetrated Earth's atmosphere and exploded at an altitude of 20 km near a remote South Pacific island (setting off nuclear-explosion warnings on six military satellites). The impact of an asteroid 500 m across (there may be a million of them near Earth's orbit) could end modern civilization and almost eliminate humans worldwide.

TOUCHSTONE EXAMPLE 14-5: Escape Speeds

(a) Suppose a particle in space is the same distance from the Sun as the Earth is. At this distance, what is the particle's escape speed from the Sun?

SOLUTION ■ We can use Eq. 14-21 to find the escape speed relative to the Sun. The **Key Idea** here is that the particle is not on the surface of the Sun but at a distance equivalent to the mean distance between the Earth and the Sun given by $R = 1.5 \times 10^{11}$ m. Since the mass of the Sun is $M = 1.99 \times 10^{30}$ kg, we get

$$v = \sqrt{\frac{2GM}{R}} = \sqrt{\frac{2(6.67 \times 10^{-11}\,\text{N}\cdot\text{m}^2/\text{kg}^2)(1.99 \times 10^{30}\,\text{kg})}{1.5 \times 10^{11}\,\text{m}}}$$

$$= v = 4.2 \times 10^4\,\text{m/s}. \qquad \text{(Answer)}$$

(b) How does the escape speed you just calculated compare to the particle's escape speed from the surface of the Earth?

SOLUTION ■ We can look up the escape speed from the Earth's surface in Table 14-2. The value is given by

$$v = 11.2\,\text{km/s} = 1.12 \times 10^4\,\text{m/s}.$$

The escape speed from the Earth's surface is about one-fourth (=1.12/4.2) of that needed to escape the Sun at an "Earth orbit distance" from it. The **Key Idea** here is that though the particle at the surface of the Earth is much closer to the center of the Earth than it is to the center of the Sun in situation a, the Sun is much more massive than the Earth is. (Answer)

(c) How does the escape speed calculated in part (a) compare to the particle's escape speed from the surface of the Sun?

SOLUTION ■ Once again we can look up the escape speed in Table 14-2. This time we need to list the speed needed to escape from the surface of the Sun,

$$v = 618\,\text{km/s} = 61.8 \times 10^4\,\text{km/s}.$$

The escape speed from the Sun's surface is about 15 times larger (= 61.8/4.2) that needed to escape the Sun at an "Earth orbit distance" from it. The **Key Idea** here is that the particle in this case is much closer to the Sun. (Answer)

14-7 Einstein and Gravitation

Principle of Equivalence

When we casually discuss gravitational forces, we often say things like "we can feel the pull of gravity" or "we can feel the pull of the Earth." However, careful observation will convince you that what we actually "feel" is the upward push of the floor or a chair. If we hang from a rope, we feel the upward pull of the rope. We do not feel any push or pull if the floor, chair, or rope is taken away.

In contrast, if we jump off a ladder or cliff, we are in free fall, and we feel no forces at all even though we are subject to the uncomfortable sensation that is (unfortunately) called "weightlessness." This is what Albert Einstein was referring to when he said: "I was ... in the patent office at Bern when all of a sudden a thought occurred to me: 'If a person falls freely, he will not feel his own weight.' I was startled. This simple thought made a deep impression on me. It impelled me toward a theory of gravitation."

Thus Einstein tells us how he began to form his **general theory of relativity.** The fundamental postulate of this theory about gravitation (the gravitating of objects toward each other) is called the **principle of equivalence,** which says that gravitation and acceleration are equivalent. If a physicist were locked up in a small box as in Fig. 14-13, he would not be able to tell whether the box was at rest on Earth (and subject only to Earth's gravitational force), as in Fig. 14-13a, or accelerating through interstellar space at 9.8 m/s^2 (and subject only to the force producing that acceleration), as in Fig. 14-13b. In both situations he would feel the same and would read the same value for his weight on a scale. Moreover, if he watched an object fall past him, the object would have the same acceleration relative to him in both situations.

Curvature of Space

We have introduced the concept of gravitation to explain the interaction forces between masses. Einstein introduced an alternative explanation of gravitation as a curvature (or shape) of space that is caused by masses. (As we will discuss in Chapter 38, space and time are entangled so the curvature of which Einstein spoke is really a curvature of *spacetime,* the combined four dimensions of our universe.)

Picturing how space (such as vacuum) can have curvature is difficult, but an analogy might help. Suppose that from orbit we watch a race in which two boats begin on the equator with a separation of 20 km and head due south (Fig. 14-14a). To the sailors, the boats travel along flat, parallel paths. However, with time the boats draw together until, nearer the south pole, they touch. The sailors in the boats can interpret this drawing together in terms of a force acting on the boats. However, we can see that the boats draw together simply because of the curvature of Earth's surface. We can see this because we are viewing the race from "outside" that surface.

(a)

(b)

FIGURE 14-13 ■ (a) A physicist in a box resting on Earth sees a cantaloupe falling with acceleration $a = 9.8$ m/s^2. (b) If he and the box accelerate in deep space at 9.8 m/s^2, the cantaloupe has the same acceleration relative to him. It is not possible, by doing experiments within the box, for the physicist to tell which situation he is in. For example, the platform scale on which he stands reads the same weight in both situations.

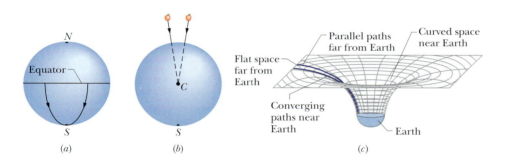

(a)   (b)   (c)

FIGURE 14-14 ■ (a) Two objects moving along lines of longitude toward the south pole converge because of the curvature of Earth's surface. (b) Two objects falling freely near Earth move along lines that converge toward the center of Earth because of the curvature of space near Earth. (c) Far from Earth (and other masses), space is flat and parallel paths remain parallel. Close to Earth, the parallel paths begin to converge because space is curved by Earth's mass.

Figure 14-14b shows a similar race: Two horizontally separated apples are dropped from the same height above Earth. Although the apples may appear to travel along parallel paths, they actually move toward each other because they both fall toward Earth's center. We can interpret the motion of the apples in terms of the gravitational force on the apples from Earth. We can also interpret the motion in terms of a curvature of the space near Earth, due to the presence of Earth's mass. This time we cannot see the curvature because we cannot get "outside" the curved space, as we got "outside" the curved Earth in the boat example. However, we can depict the curvature with a drawing like Fig. 14-14c. There the apples would move along a surface that curves toward Earth because of Earth's mass.

When light passes near Earth, its path bends slightly because of the curvature of space there, an effect called *gravitational lensing.* When it passes a more massive

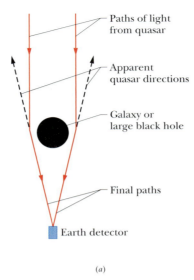

FIGURE 14-15 ■ (a) Light from a distant quasar named AC 114 follows curved paths around a galaxy because the mass of the galaxy has curved the adjacent space. If the light is detected, it appears to have originated along the backward extensions of the final paths (dashed lines).
(b) An image showing identical quasars. The source of the light is far behind a large, unseen "lensing" galaxy that has just the right shape and orientation to produce two images of the quasar. The two objects near the center of the image are believed to be unrelated galaxies in front of the lensing galaxy.

Paths of light from quasar

Apparent quasar directions

Galaxy or large black hole

Final paths

Earth detector

(a)

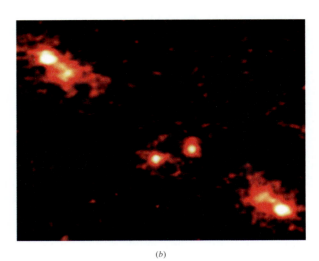

(b)

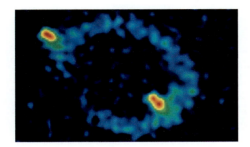

FIGURE 14-16 ■ The Einstein ring known as MG1131+0456 as it appeared on the computer screen of a radio telescope.

structure like a galaxy, its path can be bent more. If such a massive structure is between us and a quasar (an extremely bright, distant source of light), the light from the quasar can bend around the massive structure and toward us (Fig. 14-15a). Then, because the light seems to be coming to us from a number of slightly different directions in the sky, *we see the same quasar appearing to be located in two different directions* (Fig. 14-15b). In situations where the light from a distant quasar lies precisely behind the center of the lensing galaxy, the images of a single quasar can blend together to form a full ring of light known as an *Einstein ring* (Fig. 14-16).

Should we attribute gravitation to the curvature of spacetime due to the presence of masses or to a force between masses? Or should we attribute it to the actions of a type of fundamental particle called a *graviton,* as conjectured in some modern physics theories? We do not know.

Problems

SEC. 14-2 ■ NEWTON'S LAW OF GRAVITATION

1. What Separation? What must the separation be between a 5.2 kg particle and a 2.4 kg particle for their gravitational attraction to have a magnitude of 2.3×10^{-12} N?

2. Horoscopes Some believe that the positions of the planets at the time of birth influence the newborn. Others deride this belief and claim that the gravitational force exerted on a baby by the obstetrician is greater than that exerted by the planets. To check this claim, calculate and compare the magnitude of the gravitational force exerted on a 3 kg baby (a) by a 70 kg obstetrician who is 1 m away and roughly approximated as a point mass, (b) by the massive planet Jupiter ($m = 2 \times 10^{27}$ kg) at its closest approach to Earth ($= 6 \times 10^{11}$ m), and (c) by Jupiter at its greatest distance from Earth ($= 9 \times 10^{11}$ m). (d) Is the claim correct?

3. Echo Satellites One of the *Echo* satellites consisted of an inflated spherical aluminum balloon 30 m in diameter and of mass 20 kg. Suppose a meteor having a mass of 7.0 kg passes within 3.0 m of the surface of the satellite. What is the magnitude of the gravita-

tional force on the meteor from the satellite at the closest approach?

4. Sun and Earth The Sun and Earth each exert a gravitational force on the Moon. What is the ratio $F_{\text{Sun}\to\text{Moon}}/F_{\text{Earth}\to\text{Moon}}$ of the magnitudes of these two forces? (The average Sun–Moon distance is equal to the Sun–Earth distance.)

5. Split into Two A mass M is split into two parts, m and $M - m$, which are then separated by a certain distance. What ratio m/M maximizes the magnitude of the gravitational force between the parts?

SEC 14-3 ■ GRAVITATION AND SUPERPOSITION

6. Zero Net Force A spaceship is on a straight-line path between Earth and its moon. At what distance from Earth is the net gravitational force (due to the Earth and the Moon only) on the spaceship zero?

7. Space Probe How far from Earth must a space probe be along a line toward the Sun so that the Sun's gravitational pull on the probe balances Earth's pull?

8. Three Spheres Three 5.0 kg spheres are located in the xy plane as shown in Fig. 14-17. What is the magnitude of the net gravitational force on the sphere at the origin due to the other two spheres?

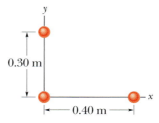

0.30 m

0.40 m

FIGURE 14-17 ■ Problem 8.

9. Four Spheres In Fig. 14-18a, four spheres form the corners of a square whose side is 2.0 cm long. What are the magnitude and direction of the net gravitational force from them on a central sphere with mass $m_A = 250$ kg?

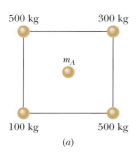

500 kg   300 kg

m_A

100 kg   500 kg

(a)

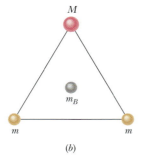

M

m_B

m m

(b)

FIGURE 14-18 ■ Problems 9 and 10.

10. Two Spheres In Fig. l4-18b, two spheres of mass m and a third sphere of mass M form an equilateral triangle, and a fourth sphere of mass m_B is at the center of the triangle. The net gravitational force on that central sphere from the three other spheres is zero. (a) What is M in terms of m? (b) If we double the value of m_B, what then is the magnitude of the net gravitational force on the central sphere?

11. Masses and Coordinates Given The masses and coordinates of three spheres are as follows: 20 kg, $x = 0.50$ m, $y = 1.0$ m; 40 kg, $x = -1.0$ m, $y = -1.0$ m; 60 kg, $x = 0$ m, $y = -0.50$ m. What is the magnitude of the gravitational force on a 20 kg sphere located at the origin due to the other spheres?

12. Four Uniform Spheres Four uniform spheres, with masses $m_A = 400$ kg, $m_B = 350$ kg, $m_C = 2000$ kg, and $m_D = 500$ kg, have (x, y) coordinates of (0, 50) cm, (0, 0) cm, (−80, 0) cm, and (40, 0) cm, respectively. What is the net gravitational force on sphere B due to the other spheres?

13. Spherical Hollow Figure l4-19 shows a spherical hollow inside a lead sphere of radius R; the surface of the hollow passes through the center of the sphere and "touches" the right side of the sphere. The mass of the sphere before hollowing was M. With what gravitational force does the hollowed-out lead sphere attract a small sphere of mass m that lies at a distance d from the center of the lead sphere, on the straight line connecting the centers of the spheres and of the hollow?

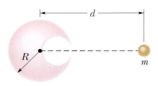

d

R m

FIGURE 14-19 ■ Problem 13.

SEC. 14-4 ■ GRAVITATION IN THE EARTH'S VICINITY.

14. Empire State Building You weigh 530 N at sidewalk level outside the Empire State Building in New York City. Suppose that you ride from this level to the 102nd floor tower, a height of 373 m. Ig-

noring Earth's rotation, how much less would you weigh there (because you are slightly farther from the center of Earth)?

15. $g = 4.9$ m/s² At which altitude above Earth's surface would the gravitational acceleration be 4.9 m/s²?

16. Moon's Surface (a) What will an object weigh on the Moon's surface if it weighs 100 N on Earth's surface? (b) How many Earth radii must this same object be from the center of Earth if it is to weigh the same as it does on the Moon?

17. Rate of Rotation The fastest possible rate of rotation of a planet is that for which the gravitational force on material at the equator just barely provides the centripetal force needed for the rotation. (Why?) (a) Show that the corresponding shortest period of rotation is

$$T = \sqrt{\frac{3\pi}{G\rho}},$$

where ρ is the uniform density of the spherical planet. (b) Calculate the rotation period assuming a density of 3.0 g/cm³, typical of many planets, satellites, and asteroids. No astronomical object has ever been found to be spinning with a period shorter than that determined by this analysis.

18. Model of a Planet One model for a certain planet has a core of radius R and mass M surrounded by an outer shell of inner radius R, outer radius $2R$, and mass $4M$. If $M = 4.1 \times 10^{24}$ kg and $R = 6.0 \times 10^6$ m, what is the gravitational acceleration of a particle at points (a) R and (b)$3R$ from the center of the planet?

19. Spring Scale A body is suspended from a spring scale in a ship sailing along the equator with speed v. (a) Show that the scale reading will be very close to W_0 $(1 \pm 2\ \omega v/g)$, where ω is the rotational speed of Earth and W_0 is the scale reading when the ship is at rest. (b) Explain the $\pm$ sign.

20. Neutron Stars Certain neutron stars (extremely dense stars) are believed to be rotating at about 1 rev/s. If such a star has a radius of 20 km, what must be its minimum mass so that material on its surface remains in place during the rapid rotation?

SEC. 14-5 ■ GRAVITATION INSIDE EARTH

21. Apple and Tunnel Assume that a planet is a sphere of radius R with a uniform density and (somehow) has a narrow radial tunnel through its center. Also assume that we can position an apple anywhere along the tunnel or outside the sphere. Let F_R be the magnitude of the gravitational force on the apple when it is located at the planet's surface. How far from the surface is a point where the magnitude of the gravitational force on the apple is $\frac{1}{2}F_R$ if we move the apple (a) away from the planet and (b) into the tunnel?

22. Two Concentric Shells Two concentric shells of uniform density having masses M_1 and M_2 are situated as shown in Fig. 14-20. Find the magnitude of the net gravitational force on a particle of mass m, due to the shells, when the particle is located at (a) point A, at distance $r = a$ from the center, (b) point B at $r = b$, and (c) point C at $r = c$. The distance r is measured from the center of the shells.

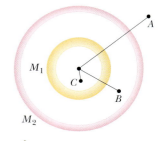

A

M_1

C

B

M_2

FIGURE 14-20 ■ Problem 22.

23. Solid Sphere A solid sphere of uniform density has a mass of 1.0×10^4 kg and a radius of 1.0 m. What is the magnitude of the gravitational force due to the sphere on a particle of mass m located at a distance of (a) 1.5 m and (b) 0.50 m from the center of the sphere? (c) Write a general expression for the magnitude of the gravitational force on the particle at a distance $r \leq 1.0$ m from the center of the sphere.

24. Uniform Solid Sphere A uniform solid sphere of radius R has a gravitational strength g_{local} at its surface. At what two distances from the center of the sphere is the gravitational strength $g_{local}/3$? (*Hint:* Consider distances both inside and outside the sphere.)

25. Crust, Mantle, Core Figure 14-21 shows, not to scale, a cross section through the interior of Earth. Rather than being uniform throughout, Earth is divided into three zones: an outer *crust*, a *mantle*, and an inner *core*. The dimensions of these zones and the masses contained within them are shown on the figure. Earth has a total mass of 5.98×10^{24} kg and a radius of 6370 km. Ignore rotation and assume that Earth is spherical. (a) Calculate the local gravitational strength g at the surface. (b) Suppose that a bore hole (the *Mohole*) is driven to the crust–mantle interface at a depth of 25 km. What would be the value of g at the bottom of the hole? (c) Suppose that Earth were a uniform sphere with the same total mass and size. What would be the value of g at a depth of 25 km? (Precise measurements of g are sensitive probes of the interior structure of Earth, although results can be clouded by local density variations.)

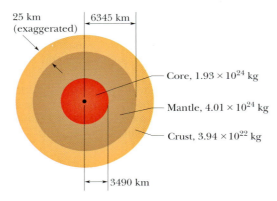

25 km (exaggerated)   6345 km

Core, 1.93×10^{24} kg

Mantle, 4.01×10^{24} kg

Crust, 3.94×10^{22} kg

3490 km

FIGURE 14-21 ■ Problem 25.

SEC. 14-6 ■ GRAVITATIONAL POTENTIAL ENERGY

26. Potential Energy (a) What is the gravitational potential energy of the two-particle system in Problem 1? If you triple the separation between the particles, how much work is done (b) by the gravitational force between the particles and (c) by you?

27. Remove Sphere A (a) In Problem 12, remove sphere A and calculate the gravitational potential energy of the remaining three-particle system. (b) If A is then put back in place, is the potential energy of the four-particle system more or less than that of the system in (a)? (c) In (a), is the work done by you to remove A positive or negative? (d) In (b), is the work done by you to replace A positive or negative?

28. Ratio *m/M* In Problem 5, what ratio m/M gives the least gravitational potential energy for the system?

29. Mars and Earth The mean diameters of Mars and Earth are 6.9×10^3 km and 1.3×10^4 km, respectively. The mass of Mars is 0.11 times Earth's mass. (a) What is the ratio of the mean density of Mars to that of Earth? (b) What is the value of the gravitational acceleration on Mars? (c) What is the escape speed on Mars?

30. Escape Calculate the amount of energy required to escape from (a) Earth's moon and (b) Jupiter relative to that required to escape from Earth.

31. Three Other Spheres The three spheres in Fig. 14-22, with masses $m_A = 800$ g, $m_B = 100$ g, and $m_C = 200$ g, have their centers on a common line, with $L = 12$ cm and $d = 4.0$ cm. You move sphere B along the line until its center-to-center separation from C is $d = 4.0$ cm. How much work is done on sphere B (a) by you and (b) by the net gravitational force on B due to spheres A and C?

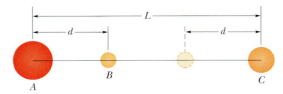

FIGURE 14-22 ■ Problem 31.

32. Zero Zero, a hypothetical planet, has a mass of 5.0×10^{23} kg, a radius of 3.0×10^6 m, and no atmosphere. A 10 kg space probe is to be launched vertically from its surface. (a) If the probe is launched with an initial energy of 5.0×10^7 J, what will be its kinetic energy when it is 4.0×10^6 m from the center of Zero? (b) If the probe is to achieve a maximum distance of 8.0×10^6 m from the center of Zero, with what initial kinetic energy must it be launched from the surface of Zero?

33. Rocket Accelerated A rocket is accelerated to speed $v = 2\sqrt{gR_E}$ near Earth's surface (where Earth's radius is R_E), and it then coasts upward. (a) Show that it will escape from Earth. (b) Show that very far from Earth its speed will be $v = \sqrt{2gR_E}$.

34. Roton Planet Roton, with a mass of 7.0×10^{24} kg and a radius of 1600 km, gravitationally attracts a meteorite that is initially at rest relative to the planet, at a great enough distance to take as infinite. The meteorite falls toward the planet. Assuming the planet is airless, find the speed of the meteorite when it reaches the planet's surface.

35. Escape Speed (a) What is the escape speed on a spherical asteroid whose radius is 500 km and whose gravitational acceleration at the surface is 3.0 m/s²? (b) How far from the surface will a particle go if it leaves the asteroid's surface with a radial speed of 1000 m/s? (c) With what speed will an object hit the asteroid if it is dropped from 1000 km above the surface?

36. Rocket Moving Radially A 150.0 kg rocket moving radially outward from Earth has a speed of 3.70 km/s when its engine shuts off 200 km above Earth's surface. (a) Assuming negligible air drag, find the rocket's kinetic energy when the rocket is 1000 km above Earth's surface. (b) What maximum height above the surface is reached by the rocket?

37. Two Neutron Stars Two neutron stars are separated by a distance of 10^{10} m. They each have a mass of 10^{30} kg and a radius of 10^5 m. They are initially at rest with respect to each other. As measured from that rest frame, how fast are they moving when (a) their separation has decreased to one-half its initial value and (b) they are about to collide?

38. Deep Space In deep space, sphere A of mass 20 kg is located at the origin of an x axis and sphere B of mass 10 kg is located on the axis at $x = 0.80$ m. Sphere B is released from rest while sphere A is held at the origin. (a) What is the gravitational potential energy of the two-sphere system as B is released? (b) What is the kinetic energy of B when it has moved 0.20 m toward A?

39. Projectile A projectile is fired vertically from Earth's surface with an initial speed of 10 km/s. Neglecting air drag, how far above the surface of Earth will it go?

SEC. 14-7 ■ EINSTEIN AND GRAVITATION

40. Cantaloupe In Fig. 14-13b, the scale on which the 60 kg physicist stands reads 220 N. How long will the cantaloupe take to reach the floor if the physicist drops it from rest (relative to himself), 2.1 m from the floor?

Additional Problems

41. Frames of Reference Figure 14-23 shows two identical spheres, each with mass 2.00 kg and radius $R = 0.0200$ m, that initially touch, somewhere in deep space. Suppose the spheres are blown apart such that they initially separate at the relative speed 1.05×10^{-4} m/s. They then slow due to the gravitational force between them.

FIGURE 14-23 ■
Problem 41.

  Center-of-mass frame: Assume that we are in an inertial reference frame that is stationary with respect to the center of mass of the two-sphere system. Use the principle of conservation of mechanical energy $(K_2 + U_2 = K_1 + U_1)$ to find the following when the center-to-center separation is $10R$: (a) the kinetic energy of each sphere and (b) the speed of sphere B relative to sphere A.

  Sphere frame: Next assume that we are in a reference frame attached to sphere A (we ride on the body). Now we see sphere B move away from us. From this reference frame, again use $K_2 + U_2 = K_1 + U_1$ to find the following when the center-to-center separation is $10R$: (c) the kinetic energy of sphere B and (d) the speed of sphere B relative to sphere A. (e) Why are the answers to (b) and (d) different? Which answer is correct?

42. Black Hole The radius R_h of a black hole is the radius of a mathematical sphere, called the event horizon, that is centered on the black hole. Information from events inside the event horizon cannot reach the outside world. According to Einstein's general theory of relativity, $R_h = 2GM/c^2$, where M is the mass of the black hole and c is the speed of light.

  Suppose that you wish to study black holes near them, at a radial distance of $50R_h$. However, you do not want the difference in gravitational acceleration between your feet and your head to exceed 10 m/s^2 when you are feet down (or head down) toward the black hole. (a) As a multiple of our sun's mass, what is the limit to the mass of the black hole you can tolerate at the given radial distance? (You need to estimate your height.) (b) Is the limit an upper limit (you can tolerate smaller masses) or a lower limit (you can tolerate larger masses)?

43. Romeo and Juliet Two schoolmates, Romeo and Juliet, catch each other's eye across a crowded dance floor at a school dance. Estimate the gravitational attraction they exert on each other.

44. The Alignment of the Planets Some authors seeking public attention have suggested that when many planets are "aligned" (i.e., are close together in the sky) their gravitational pull on the Earth all acting together might produce earthquakes and other disasters. To get an idea of whether this is plausible, set up the following calculation: (a) Draw a sketch of the solar system and arrange the planets so that Mars, Jupiter, and Saturn are on the same side of the Sun as the Earth. Look up (there is a table in the back of *Understanding Physics*) the radii of the planetary orbits and their masses. (b) Infer the distances these planets would be from Earth in this arrangement. (c) Without doing all the calculations, decide which of the three planets would exert the strongest gravitational force on the Earth. (*Hint:* Use the dependence of Newton's universal gravitation law on mass and distance.) (d) Calculate the gravitational force of the most important planet on the Earth. (e) Calculate how this compares to the gravitational force the Moon exerts on the Earth.

Note: In fact, it is not the gravitational force itself that produces the possibly dangerous effects, but the tidal forces—the derivative of the gravitational force. This reduces the effect by another factor of the distance. That is, the tidal force goes like $1/r^3$ instead of like $1/r^2$. This weakens the planet's gravitational effect compared to the Moon's by an additional factor of $r_{\text{Earth–moon}}/r_{\text{Earth–planet}}$, a number much less than 1.

45. Is Newton's Law of Gravity Wrong? A professional scientist (not a physicist) stops you in the hall and says: "I can prove Newton's theory of gravity is wrong. The Sun is 320,000 times as massive as the Earth, but only 400 times as far from the Moon as is the Earth. Therefore, the force of the Sun's gravity on the Moon should be twice as big as the Earth's and the Moon should go around the Sun instead of around the Earth. Since it doesn't, Newton's theory of gravity must be wrong!" What's the matter with this reasoning?

46. In the Shuttle When we see the astronauts in orbit in the space shuttle on TV, they seem to float. If they let go of something, it just stays where they put it. It doesn't fall. What happens to gravity for objects in orbit? Does gravity stop at the Earth's atmosphere? Explain what's happening in terms of the physics you have learned.

15 | Fluids

The force exerted by water on the body of a descending diver increases noticeably, even for a relatively shallow descent to the bottom of a swimming pool. However, in 1975, using scuba gear with a special gas mixture for breathing, William Rhodes emerged from a chamber that had been lowered 300 m into the Gulf of Mexico, and he then swam to a record depth of 350 m. Strangely, a novice scuba diver practicing in a swimming pool might be in more danger from the force exerted by the water than was Rhodes. Occasionally, novice scuba divers die because they have neglected that danger.

What is this potentially lethal risk?

The answer is in this chapter.

15-1 Fluids and the World Around Us

Fluids—which include both liquids and gases—play a central role in our daily lives. We breathe and drink them, and a rather vital fluid circulates in the human cardiovascular system. The Earth's oceans and atmosphere consist of fluids.

Cars and jet planes need many different fluids to operate, including fluids in their tires, fuel tanks, and engine combustion chambers. They also need fluids for their air conditioning, lubrication, and hydraulic systems. Windmills transform the kinetic energy in air to electrical energy, and hydroelectric plants convert the gravitational potential energy in water to electrical energy. Over long time periods, air and water carve out and reshape the Earth's landscape.

In our study of fluids, we will start by examining simple physical situations that we encounter every day. First, we will study the forces acting on fluids that are in static equilibrium and consider the forces on objects in fluids. Then we will examine how a hydraulic system can be used as a lever. Later in the chapter, we will study the motions of fluids as they flow through pipes and around objects.

15-2 What Is a Fluid?

To understand what we mean by the term "fluid," let us compare solids, liquids, and gases. A solid vertical column that rests on a table can retain its shape without external support. Since a gravitational force is acting on each of the columns shown in Fig. 15-1, each exerts a downward normal force on the table that is equal in magnitude to its weight. What happens if we try to make a column out of a liquid? Without external support, the gravitational forces on the liquid will cause it to collapse and flow into a puddle. (In the more formal terms introduced in Section 13-5, liquids cannot withstand shear stresses.)

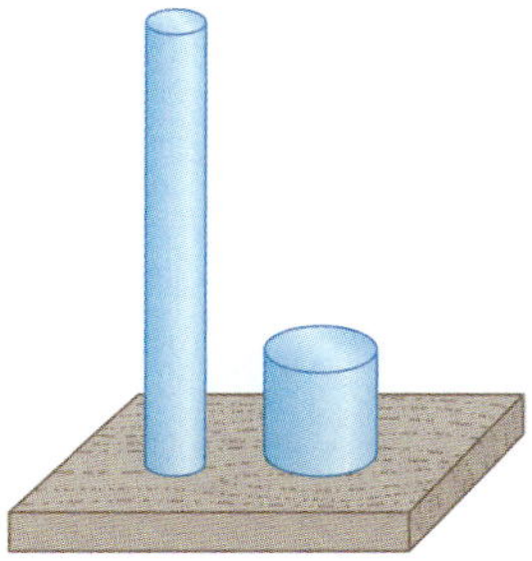

However, we can maintain a vertical column of liquid if we provide it with solid walls. In this case, the liquid presses sideways against the walls and the walls press back against the liquid. Thus, the vertical columns of the liquid and solid differ in that the column of liquid needs external forces acting on it to maintain its shape whereas the solid does not. However, both a solid column and a container full of liquid will exert normal forces on a table.

When external forces are present, a **fluid,** unlike a solid, can flow until it conforms to the boundaries of its container. Obviously gases, such as the air that surrounds us, are also fluids, because they can conform to the shape of a container quite rapidly. Some gooey materials, such as heavy syrup and silly putty, take a longer time to conform to the boundaries of a container. But since they can do so eventually, we also classify them as fluids.

FIGURE 15-1 ■ Two columns resting on a table have the same mass, so they each exert the same downward normal force on the table.

15-3 Pressure and Density

Defining Pressure for Uniform Forces

Let us consider the properties of the two solid columns shown in Fig. 15-1. Since they both have the same weight, they exert the same downward forces on the table. However, if you placed your hand under each of the columns, you would *feel* a difference. Why? Because the forces are spread out over different areas. It is this difference in *pressure* you feel when placing each column on your hand. If a force is evenly distributed over every point of an area (as is the case for the normal forces exerted by the

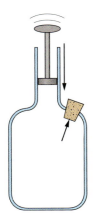

FIGURE 15-2 ■ Suppose a hole is drilled at some random place on a bottle and plugged with a cork. If an airtight plunger is thrust down the bottle's neck, the increased pressure in the bottle can cause the cork to pop out no matter what direction it faces. This can happen whenever the bottle contains either a gas (such as air) or a liquid (such as water).

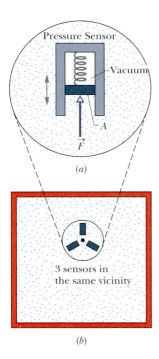

(a)

(b)

FIGURE 15-3 ■ (a) A tiny pressure sensor that uses spring compression to measure the net force normal to the area, A, of a piston. (b) When an array of pressure sensors pointing in different directions are placed in the vicinity of a single point in a fluid, their pressure measurements are identical.

cylinders), we say that it is *uniform* over the area. For a force that is both uniformly distributed over an area and perpendicular or normal to it, the **pressure** P on a surface is defined as the magnitude of the net force acting on the surface divided by its area. Thus, it can be expressed by the equation

$$P \equiv \frac{|\vec{F}_\perp|}{A} \quad \text{(uniform forces normal to area } A\text{),} \qquad (15\text{-}1)$$

where, as usual, the symbol $\equiv$ is used to signify that the equation holds "by definition." The column on the left in Fig. 15-1 has one-fourth the area that the right-hand column does. Even though the left and right columns exert the same net force on the table, the left column exerts four times more pressure on the table. Later on in this section we will refine our discussion of pressure to handle situations in which the normal forces acting on a surface are not uniform.

Is Pressure a Vector or Scalar?

In order to think about the idea of pressure exerted by a fluid, consider a bottle full of a fluid that has a piston on top. Although there is a hole in the bottle, it is plugged with a cork as shown in Fig. 15-2. If we press on the piston, the cork will pop out. This indicates that the fluid exerts a perpendicular force on the face of the cork that is sticking into the bottle. What's remarkable is that no matter where the hole and cork are located on the bottle, the cork would still pop out! Somehow, the downward force we apply with the piston to one part of the fluid is translated into "internal forces" that act in all directions. Thus, *the fluid pressure acting at the surface of a container appears to have no preferred direction.*

Let us consider a fluid that is not moving so that we can define it as being in a state of static equilibrium. What is the pressure like inside the fluid? We can consider this question both experimentally and theoretically.

Experimental Results: If we want to measure the pressure exerted by a fluid at a point inside a container of fluid, we can design a small pressure sensor like that shown in Fig. 15-3a. The sensor consists of a piston with a small cross-sectional area A. The piston fits snugly in an evacuated cylinder, so the cylinder contains no matter other than a coiled spring that is lodged behind the piston. By measuring the spring compression we can determine the normal force the fluid exerts on the piston.

Suppose we place an array of three tiny pistons at the point of interest inside the container as shown in Fig. 15-3. We find that the magnitude of force on each of the pistons is the same independent of the directions the pistons are facing. Thus, we only need to place a single pressure sensor at a point of interest and measure the force on its piston to calculate the pressure using Eq. 15-1.

> Experiments reveal that at a given point in a fluid that is in static equilibrium, the pressure P has the same value in all directions. In other words, pressure is a scalar, having no directional properties.

Agreement between Experiment and Theory: We can use the fact that we have chosen to examine a fluid that is in static equilibrium to see why we should indeed expect the pressure near a point in the fluid to be nondirectional. Let us simplify the situation by assuming that a container of fluid is located where there are no gravitational forces on it. In Section 15-4, we will revisit this idea for the more common case of nonzero gravitational forces. Next we can draw an imaginary cubical boundary around a tiny parcel of fluid (Fig. 15-4) centered on some point in the container. Since the parcel of fluid is in equilibrium it cannot be accelerating, and we must conclude that the net force on its boundaries is zero as shown in Fig. 15-4. Since the force vectors on opposite faces of the cubical parcel must be equal in magnitude and opposite in direction, the pressure on opposite faces must be the same. Furthermore, there are no gravitational forces acting in our special case and thus there is no preferred direction. These facts allow us to conclude that if no part of the fluid is accelerating, the pressure must be the same in all directions throughout the entire container. So far we have considered a very special shape for our parcel in the absence of gravitational forces. In Section 15-4 we consider what happens to the pressure in fluids in static equilibrium close to the Earth's surface or at other locations where gravitational forces must be taken into account.

Defining Pressure for Nonuniform Forces and Surfaces

If the forces on an area are not uniform or if the area is curved, we can still use our basic definition of pressure by breaking area A into segments. The area segments must be small enough so that the normal forces acting on each segment are uniform and each area segment is essentially flat (Fig. 15-5). If we do this, the pressure at the location of the ith segment of the area can be defined as

$$P_i \equiv \lim_{A_i \to 0} \frac{|\vec{F}_i|}{A_i} \qquad \text{(pressure at a point, nonuniform forces),} \qquad (15\text{-}2)$$

where $|\vec{F}_{i\perp}|$ is the magnitude of the net force normal to the ith area. That is, the pressure at any point is the limit of this ratio as the area A_i centered on that point is made smaller and smaller. Obviously, the net force acting on smaller areas will be smaller so the ratio is still physically meaningful.

The SI unit of pressure is the newton per square meter, which is given a special name, the **pascal** (Pa). In countries using the metric system, tire pressure gauges are calibrated in kilopascals (kPa). The pascal is related to some other common (non-SI) pressure units as follows:

$$1.00 \text{ atm} \equiv 101\,325 \text{ Pa} \equiv 760 \text{ Torr} \approx 14.7 \text{ lb/in}^2.$$

The *atmosphere* (atm) is, as the name suggests, the approximate average pressure of the atmosphere at sea level. The *torr* (named for Evangelista Torricelli, who invented the mercury barometer in 1674) was formerly called the *millimeter of mercury* (mm Hg). The pound per square inch is often abbreviated psi. Table 15-1 shows some pressures.

Density

Let us return one more time to the column of solid we discussed above. Clearly, the weight of a column of a given size and height depends on what the column is made of. If a certain column was constructed of a material like Styrofoam® it would be much lighter than if it was made of lead. Hence, it would be convenient to have a way to predict the weight of an object that has a certain size and shape.

For this purpose, we invent a new quantity that is a measure of the mass of one cubic meter of a material. To determine this value for a given substance, we measure the total mass M in a measured volume V of the material and calculate M/V. This quantity is called *density*. In general, density is a measure of mass per unit of volume. The standard symbol for density is ρ.

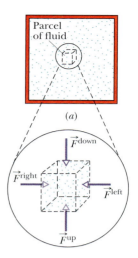

FIGURE 15-4 ■ (*a*) A tiny parcel of fluid with no gravitational forces acting on it in static equilibrium. (*b*) Since the parcel does not accelerate, the net force on it due to the surrounding fluid must be zero, so the pressure on the parcel is the same in all directions. For clarity, force vectors on the front and back parcel faces are not shown.

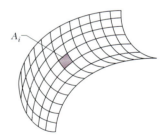

FIGURE 15-5 ■ To calculate the pressure exerted by fluids on curved surfaces or surfaces that have nonuniform forces acting on them, a surface area must be divided into a large number of small area elements, A_1, A_2, and so on. The ith area is shown in the diagram.

TABLE 15-1
Some Pressures

	Pressures (Pa)
Center of the Sun	2×10^{16}
Center of Earth	4×10^{11}
Highest sustained laboratory pressure	1.5×10^{10}
Deepest ocean trench (bottom)	1.1×10^{8}
Spike heels on a dance floor	1×10^{6}
Automobile tire[a]	2×10^{5}
Atmosphere at sea level	1.0×10^{5}
Normal blood pressure[a,b]	1.6×10^{4}
Best laboratory vacuum	10^{-12}

[a] Pressure in excess of atmospheric.

[b] The systolic pressure (120 torr on a physician's pressure gauge).

TABLE 15-2
Some Densities

Material or Object	Density (kg/m³)
Stray atoms in interstellar space	10^{-20}
Air remaining in the best laboratory vacuum	10^{-17}
Air: 20°C and 1 atm pressure	1.21
20°C and 50 atm pressure	60.5
Styrofoam	1×10^{2}
Ice	0.917×10^{3}
Water: 20°C and 1 atm	0.998×10^{3}
20°C and 50 atm	1.000×10^{3}
Seawater: 20°C and 1 atm	1.024×10^{3}
Whole blood	1.060×10^{3}
Iron	7.9×10^{3}
Mercury (the metal)	13.6×10^{3}
Earth: average	5.5×10^{3}
core	9.5×10^{3}
crust	2.8×10^{3}
Sun: average	1.4×10^{3}
core	1.6×10^{5}
White dwarf star (core)	10^{10}
Uranium nucleus	3×10^{17}
Neutron star (core)	10^{18}
Black hole (1 solar mass)	10^{19}

Table 15-2 shows the densities of several substances and the average densities of some objects. Notice that the density of a gas (see Air in the table) varies considerably with pressure, but the density of a liquid (see Water) does not. That is, gases are readily *compressible* but liquids are not.

The density of a fluid is not always uniform. For example, the density of the gas molecules and other particles that make up the Earth's atmosphere is much greater close to the surface of the Earth than in the stratosphere. As is the case for pressure, we can find the density ρ of any fluid at point i if we isolate a small volume element V_i around that point and measure the mass m_i of the fluid contained within that element. The **density** is then

$$\rho_i = \frac{m_i}{V_i} \qquad \text{(density in the vicinity of a point } i\text{). (15-3)}$$

In theory, the density at any point in a fluid is the limit of this ratio as the volume element V at that point is made smaller and smaller. In practice, many fluid samples are large compared to atomic dimensions and are thus "smooth" rather than "lumpy" with atoms. If it is reasonable to assume further that the sample has a uniform density, we can simplify Eq. 15-3 to

$$\rho = \frac{m}{V} \qquad \text{(uniform density), (15-4)}$$

where m and V are the total mass and volume of the sample. Density is a scalar property; its SI unit is the kilogram per cubic meter.

Density and pressure are fundamental concepts in regard to fluids. When we discuss solids, we are concerned with particular lumps of matter, such as wooden blocks, baseballs, or metal rods. Physical quantities that we find useful, and in whose terms we express Newton's laws, are *mass* and *force*. We might speak, for example, of a 3.6 kg block acted on by a 25 N force. With fluids, we are more interested in the extended substance, and in properties that can vary from point to point in that substance. In these cases, it is more useful to speak of *density* and *pressure* than of mass and force.

READING EXERCISE 15-1: Estimate the pressure in pascals exerted on a dance floor by just the spike heels worn by a 125 lb woman who is standing on both feet. Assume that half of her weight is on the spike heels and half is on the front soles of her shoes. How does your estimate compare with the number given in Table 15-1? Discuss why it is possible for the spike heels to exert more pressure on the floor than the pressure exerted on the road by a single tire holding up its share of a 2500 lb automobile. ■

READING EXERCISE 15-2: Examine the densities listed in Table 15-2. Use the fact that air and water have significantly different densities to develop a plausible explanation for the fact that air is a much more compressible fluid than is water. ■

READING EXERCISE 15-3: Consider a book of dimensions 8 in. by 10 in. Show that the downward force on the book by the atmosphere is about 1200 lb. The downward force on a smooth thick rubber mat of the same dimensions is also 1200 lb. You find that you cannot lift the rubber mat when it is placed on a smooth Formica table so no air can get under it. However, you can easily lift the book. Can you explain this phenomenon? ■

TOUCHSTONE EXAMPLE 15-1: Force Due to Air Pressure

A living room has floor dimensions of 3.5 m and 4.2 m and a height of 2.4 m.

(a) What does the air in the room weigh when the air pressure is 1.0 atm?

SOLUTION ■ The **Key Ideas** here are these: (1) The air's weight is equal to mg, where m is its mass. (2) Mass m is related to the air density ρ and the air's volume V by Eq. 15-4 ($\rho = m/V$). Putting these two ideas together and taking the density of air at 1.0 atm from Table 15-1, we find

$$mg = (\rho V)g$$

$$= (1.21 \text{ kg/m}^3)(3.5 \text{ m} \times 4.2 \text{ m} \times 2.4 \text{ m})(9.8 \text{ m/s}^2)$$

$$= 418 \text{ N} \approx 420 \text{ N}.$$

This is the weight of about 126 cans of soda.

(b) What is the magnitude of the atmosphere's force on the ceiling of the room?

SOLUTION ■ The **Key Idea** here is that the atmosphere pushes up on the ceiling with a force of magnitude $F_\perp = |\vec{F}_\perp|$ that is uniform over the ceiling. Thus, it produces a pressure that is related to F and the flat area A of the ceiling by Eq. 15-1 ($P = F_\perp/A$), which gives us

$$F_\perp = PA = (1.0 \text{ atm})\left(\frac{1.01 \times 10^5 \text{ N/m}^2}{1.0 \text{ atm}}\right)(3.5 \text{ m})(4.2 \text{ m})$$

$$= 1.5 \times 10^6 \text{ N}. \qquad \text{(Answer)}$$

This enormous force is equal to the weight of the column of air that has as its base the horizontal area of the room and extends all the way to the top of the atmosphere.

15-4 Gravitational Forces and Fluids at Rest

In the last section we considered the forces on a parcel of fluid that was "at rest" and does not experience gravitational forces. We found that the pressure in any fluid in a container was the same in every direction at every location in the container. This is not so in fluids that experience gravitational forces. In this section we will consider how the presence of gravitational forces leads to pressure differences at different levels in a container of fluid. As every diver knows, the pressure *increases* with depth below the air–water interface. The diver's depth gauge, in fact, is a pressure sensor much like that of Fig. 15-3a. As every mountaineer knows, the pressure *decreases* with altitude as one ascends into the atmosphere. The pressures encountered by the diver and the mountaineer are usually called **hydrostatic pressures** because they are pressures due to fluids that are static (at rest). Here we want to find an expression for hydrostatic pressure as a function of depth or altitude.

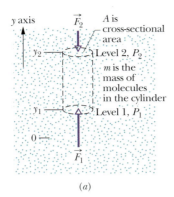

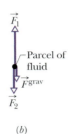

(b)

FIGURE 15-6 ■ (a) A parcel of fluid such as air or water is contained in an imaginary cylinder of cross-sectional area A. Forces $\vec{F}_1$ and $\vec{F}_2$ act, respectively, on the bottom and top of the cylinder. The gravitational force of the parcel of fluid is $\vec{F}^{\text{grav}} = -mg\,\hat{\jmath}$. (b) A free-body diagram of the forces that act on the parcel of fluid in the cylinder.

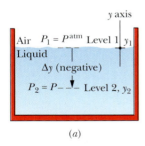

(a)

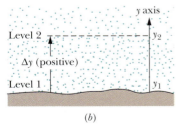

(b)

FIGURE 15-7 ■ Equation 15-7 can be used to determine (a) either the pressure underwater or (b) the atmospheric pressure above the surface of the Earth.

Let us consider a fluid such as air or water near the surface of the Earth. What happens to its pressure when one changes from an initial level y_1 to a final level y_2? Since a fluid is made up of lots of molecules, we can pick any subset of them as our "object" and apply Newton's laws to that parcel of fluid. So let us imagine a parcel of the fluid consisting of all the molecules contained in a cylindrical column of cross-sectional area A that extends between the two levels y_1 and y_2, as shown in Fig. 15-6a. The total mass of the molecules in the cylinder is m.

If the parcel of fluid is at rest, in *static equilibrium*, the horizontal forces on it from the sides must add up to zero. Similarly, static equilibrium requires that the vector sum of the vertical forces on the parcel of fluid must be zero too. There are three vertical forces that act on the parcel of fluid in the cylinder. Figure 15-6b shows a free-body diagram of these forces. Force $\vec{F}_1$ acts at the bottom surface of the cylinder and is due to the water below the cylinder. Similarly, force $\vec{F}_2$ acts at the top surface of the cylinder and is due to the fluid above the cylinder. The gravitational force on the water in the cylinder can be represented by $\vec{F}^{\text{grav}} = -mg\,\hat{\jmath}$. The net force comprised of these forces must be zero, so that

$$\vec{F}^{\text{net}} = \vec{F}_1 + \vec{F}_2 + \vec{F}^{\text{grav}} = 0.$$

Since we know that all the forces act in the vertical direction, we can rewrite this equation in terms of the y-components of the vectors as

$$F_y^{\text{net}} = F_{1y} + F_{2y} + F_y^{\text{grav}} = F_{1y} + F_{2y} + (-mg) = 0. \tag{15-5}$$

We know that force $\vec{F}_1$ acts in an upward direction and is inherently positive while force $\vec{F}_2$ acts in a downward direction and is inherently negative. So we can replace the force *components*, with the corresponding pressures and areas, with

$$F_{1y} = +P_1 A \quad \text{and} \quad F_{2y} = -P_2 A. \tag{15-6}$$

We use the explicit minus sign in front of the inherently positive P_2A term to signify the fact that the force component F_{2y} must be negative. The mass m of the fluid in the cylinder is $m = \rho V$, where ρ represents the density of the fluid and where V represents the volume of the cylinder. Since the volume is the product of its face area A and its height $\Delta y = y_2 - y_1$, the mass m is equal to $\rho A(y_2 - y_1)$. Using these facts and substituting Eq. 15-6 into Eq. 15-5, we get

$$P_2 = P_1 - \rho g(y_2 - y_1) = P_1 - \rho g\,\Delta y \quad \text{(only if } \rho \text{ is uniform),} \tag{15-7}$$

or since the pressure difference $\Delta P = P_2 - P_1$, we can also write

$$\Delta P = -\rho g\,\Delta y. \tag{15-8}$$

If SI units are used for ρ, g, and Δy in calculations, then the pressure will be in pascals.

Equation 15-7 is a general expression that can be used to find pressure changes in either a liquid (as a function of depth) or in the atmosphere (as a function of altitude or height) (Fig. 15-7). However, it is only valid when the density of the fluid and the local gravitational strength factors are essentially constant between the levels under consideration.

Special Case 1: Pressure in the Earth's Atmosphere

Equation 15-7 can be used to determine the pressure of the atmosphere at a given distance above a reference point in terms of the atmospheric pressure P_1 at that level. If we denote the density as $\rho = \rho_{air}$, we can write Eq. 15-7 as

$$P_2 = P_1 - \rho_{air} g \, \Delta y \qquad \text{(only if } \rho_{air} \text{ and } g \approx \text{const).} \tag{15-9}$$

Since we are *above* the reference level, we know that $y_2 > y_1$ so that Δy is *positive*. Thus, we obtain an expression for pressure that predicts that pressure decreases in a linear manner with altitude.

Observations like those shown in Fig. 15-8 verifies that pressure does indeed decrease with altitude. Data in Table 14-1 indicate that the local gravitational strength is essentially constant at any altitude found on Earth, including Mt. Everest at 8.8 km. However, the density of air decreases with altitude. Fortunately the density of air ρ_{air} is reasonably constant up to about 5000 ft (or 1500 m). This means that Eq. 15-9 can be used to calculate pressures for the range of altitudes encountered in the trip across the cascades described in Fig. 15-8. However, Eq. 15-9 would not be accurate when considering higher elevations.

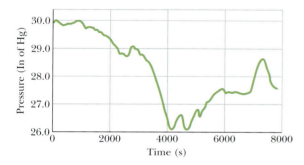

FIGURE 15-8 ■ Here is a graph of pressure data collected while driving a car from Portland, Oregon (altitude of about 3 m above sea level) over the Cascade Mountain Range to Madras, Oregon. The first pressure minimum came at about 4200 s when going over the pass near Government Camp. The second minimum came at about 4700 s while going over Blue Box pass, which is almost as high. Both passes are just over 1200 m above sea level. (Data courtesy of David Vernier.)

Special Case 2: Underwater Pressure

Equation 15-7 can also be used to determine the pressure underwater at a given distance below a reference point in terms of the pressure P_1 at the reference level. If we denote the density as $\rho = \rho_{water}$, we can write Eq. 15-7 as

$$P_2 = P_1 - \rho_{water} g \, \Delta y \qquad \text{(only if } \rho_{water} \text{ and } g \approx \text{const).} \tag{15-10}$$

Since we are *below* the reference level, we know that $y_2 < y_1$ so that Δy is *negative*. Thus, we obtain an expression for pressure that predicts that pressure increases in a linear manner with depth.

As you can see in Fig. 15-9, our theory, which predicts a linear increase in pressure as a function of depth (with a slope given by the factor $\rho_{water} g$), compares nicely with experiment. Although we observe that the pressure in a liquid increases with depth, it does not depend on the horizontal location of the parcel of liquid. If we have a liquid other than water, obviously we need to use the density of that liquid in place of the density of water.

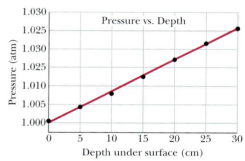

FIGURE 15-9 ■ Pressure as a function of depth for distances between 0 cm and 30 cm below the surface of a container of water. Measurements were made using a computer data acquisition system outfitted with a gas pressure sensor like that shown in the next section. The slope of the graph when expressed in SI units should be equal to the factor $\rho_{water} g$ shown in Eq. 15-10. Within the limits of experimental uncertainty, the data shown are compatible with theory.

> When gravitational forces are present, the pressure at a point in a fluid in static equilibrium depends on the depth of that point but not on any horizontal dimension of the fluid or the shape of its container.

READING EXERCISE 15-4: Scuba divers know that if they descend to a depth of 10 m the pressure they experience doubles. However, alpine mountain climbers must ascend to about 5.5 km to cut the atmospheric pressure in half. What factor in Eq. 15-7 accounts for the fact that the pressure change with distance is so much smaller in air than in water? ■

READING EXERCISE 15-5: The figure shows four containers of olive oil. Rank them according to the pressure at y_2, greatest first.

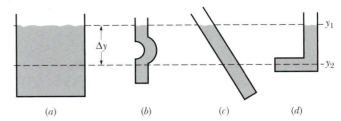

(a)   (b)   (c)   (d)

TOUCHSTONE EXAMPLE 15-2: Exhale!

A novice scuba diver practicing in a swimming pool takes enough air from his tank to fully expand his lungs before abandoning the tank at depth L and swimming to the surface. He ignores instructions and fails to exhale during his ascent. When he reaches the surface, the difference between the external pressure on him and the air pressure in his lungs is 9.3 kPa. From what depth does he start? What potentially lethal danger does he face?

SOLUTION ■ The **Key Idea** here is that when he fills his lungs at depth Δy, the external pressure on him (and thus the air pressure within his lungs) is greater than normal.

Assuming that P_1 equals the atmospheric pressure at the water's surface, we can rewrite Eq. 15-10 as

$$P_2 = P_1 - \rho_{water}\, g\, \Delta y = P^{atm} - \rho_{water}\, g\, \Delta y,$$

ρ_{water} is the water's density (998 kg/m³), given in Table 15-2. As the diver ascends, the external pressure on him decreases, until it is at mospheric pressure P^{atm} at the surface. His blood pressure also decreases, until it is normal. However, because he does not exhale, the air pressure in his lungs remains at the value it had at depth Δy. At the surface, the pressure difference between the higher pressure in his lungs and the lower pressure on his chest is

$$\Delta P = P_2 - P^{atm} = -\rho_{water}\, g\, \Delta y,$$

from which we find

$$\Delta y = \frac{\Delta P}{-\rho g} = -\frac{9300\ \text{Pa}}{(998\ \text{kg/m}^3)(9.8\ \text{m/s}^2)} \quad \text{(Answer)}$$
$$= -0.95\ \text{m}.$$

This is not deep! Yet, the pressure difference of 9.3 kPa (about 9% of atmospheric pressure) is sufficient to rupture the diver's lungs and force air from them into the depressurized blood, which then carries the air to the heart, killing the diver. If the diver follows instructions and gradually exhales as he ascends, he allows the pressure in his lungs to equalize with the external pressure, and then there is no danger.

TOUCHSTONE EXAMPLE 15-3: U-Tube

The U-tube in Fig. 15-10 contains two liquids in static equilibrium: Water of density ρ_{water} (= 998 kg/m³) is in the right arm, and oil of unknown density ρ_x is in the left. Measurement gives $l = 135$ mm and $d = 12.3$ mm. What is the density of the oil?

SOLUTION ■ One **Key Idea** here is that the pressure P^{int} at the oil–water interface in the left arm depends on the density ρ_x and height of the oil above the interface. A second **Key Idea** is

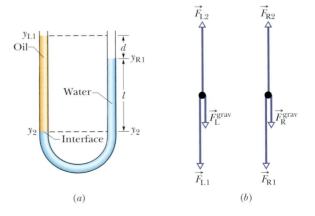

FIGURE 15-10 ■ (a) The oil in the left arm stands higher than the water in the right arm because the oil is less dense than the water. Both fluid columns produce the same pressure P^{int} at the level of the interface. (b) Free-body diagrams showing the forces on the parcel of oil on the left and the forces on the parcel of water above the interface on the right.

that the water in the right arm *at the same level* must be at the same pressure P^{int}. The reason is that, because the water is in static equilibrium, pressures at points in the water at the same level must be the same even if the points are separated horizontally.

In the right arm, the interface is a distance l below the free surface of the *water* and we have, from Eq. 15-10, $P_2 = P_1 - \rho_{water}\, g\, \Delta y$ where $P_2 = P^{int}$ and $\Delta y = y_2 - y_1 = -l$, so

$$P^{int} = P^{atm} - \rho_{water}\, g(-l) \qquad \text{(right arm).}$$

In the left arm, the interface is a distance $l + d$ below the free surface of the *oil* and we have, again from Eq. 15-10,

$$P^{int} = P^{atm} - \rho_x g[-(l + d)] \qquad \text{(left arm).}$$

Equating these two expressions and solving for the unknown density yield

$$\rho_x = \rho_{water}\left(\frac{l}{l + d}\right) = (998 \text{ kg/m}^3)\,\frac{135 \text{ mm}}{135 \text{ mm} + 12.3 \text{ mm}}$$

$$= 915 \text{ kg/m}^3 \qquad \text{(Answer)}$$

Note that the answer does not depend on the atmospheric pressure P^{atm} or the local gravitational strength g.

15-5 Measuring Pressure

Electronic Pressure Sensors

One of the most popular methods for measuring absolute pressure in a gas is to use an electronic sensor. Many electronic sensors work in much the same way as the test sensor shown in Fig. 15-3. A common electronic pressure sensor has a flexible membrane with a vacuum chamber on one side of it, examples of which are shown in Fig. 15-11. The flexing of the membrane under pressure is sensed electronically. These devices are used in recording barometers found in weather stations and in physics laboratories.

FIGURE 15-11 ■ Two popular gas pressure sensors used in contemporary physics laboratories can record between 0 atm and about 7 atm of pressure. (*a*) Vernier pressure sensor. (*b*) PASCO pressure sensor. (Photos used with permission of Vernier Software and Technology and PASCO scientific.)

Typically an electronic gas sensor will be damaged when immersed in a liquid. However, when air-filled tubing connected to a gas sensor is immersed in a liquid, the pressure at various depths in the liquid can also be measured.

The Mercury Barometer

For historical and practical reasons, the *mercury barometer* and the *open tube manometer* are still popular methods for measuring atmospheric pressure and pressures near atmospheric pressure.

Figure 15-12a shows a very basic *mercury barometer,* a device used to measure the pressure of the atmosphere. The long glass tube is filled with mercury and inverted with its open end in a dish of mercury, as the figure shows. The space above the mercury column contains only mercury vapor, whose pressure is so small at ordinary temperatures that it can be neglected.

We can use Eq. 15-10 to find the local atmospheric pressure P^{atm} in terms of the height $\Delta y = y_2 - y_1$ of the mercury column. Since the chemical symbol for mercury is

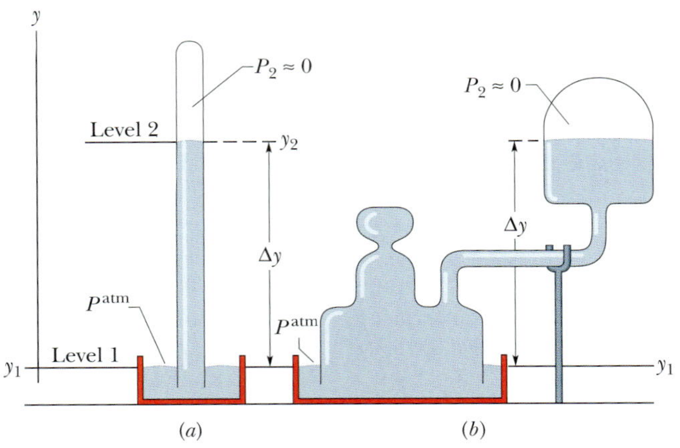

FIGURE 15-12 ■ (a) A mercury barometer. (b) Another mercury barometer. The difference Δy between liquid levels is the same in both cases.

Hg, we denote the density of the mercury by ρ_{Hg}. If we choose level 1 of Fig. 15-12a to be that of the air–mercury interface and level 2 to be that of the top of the mercury column, as labeled in Fig. 15-12a, we can substitute

$$P_1 = P^{atm}, \quad P_2 = 0, \quad \text{and} \quad \rho = \rho_{Hg},$$

into Eq. 15-10 to get

$$P^{atm} = \rho_{Hg} g \, \Delta y \tag{15-11}$$

where $\Delta y = y_2 - y_1$ is positive.

For a given pressure, the height Δy of the mercury column does not depend on the cross-sectional area of the vertical tube. The fanciful mercury barometer of Fig. 15-12b gives the same reading as that of Fig. 15-12a; all that counts is the vertical distance Δy between the mercury levels.

Equation 15-11 shows that, for a given pressure, the height of the column of mercury depends on the value of the local gravitational constant g at the location of the barometer and on the density of mercury, which varies only slightly with temperature. The column height (in millimeters) is numerically equal to the pressure (in torr) *only* if the barometer is at a place where g has its accepted average value of $+9.80665$ N/kg *and* the temperature of the mercury is 0°C. If these conditions do not prevail (and they rarely do), small corrections must be made before the height of the mercury column can be transformed into a pressure.

The Open-Tube Manometer

An *open-tube manometer* (Fig. 15-13) measures the pressure P_{gas} of a gas. It consists of a U-tube containing a liquid, with one end of the tube connected to the vessel whose pressure we wish to measure and the other end open to the atmosphere. Looking at the figure, we see that the "U" of fluid below the line marked "Level 2" is being pushed on the left by the force from the pressure of the gas in the tank and is being pushed on the right by the force arising from the pressure built up by everything above it, including the weight of the column between levels 1 and 2 and the pressure of the atmosphere. When the column is in equilibrium (no rising or falling), these forces must balance. Keeping this in mind, we can use Eq. 15-10 to find the pressure in terms of the distance from level 1 to level 2, Δy, shown in Fig. 15-13. Let us choose levels 1 and 2 as shown in Fig. 15-13. We then substitute

$$P_1 = P^{atm} \quad \text{and} \quad P_2 = P_{gas}$$

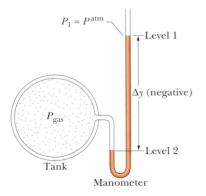

FIGURE 15-13 ■ An open-tube manometer, connected to measure the pressure of the gas in the tank on the left. The right arm of the U-tube is open to the atmosphere.

into Eq. 15-9, finding that

$$P_{gas} = P^{atm} - \rho g \, \Delta y, \qquad (15\text{-}12)$$

where ρ is the density of the liquid in the tube.

In Fig. 15-13, the Δy factor is negative since level 2 is lower than level 1. So this figure depicts a gas pressure that is greater than the atmospheric pressure. Inflated tires and the human circulatory network are examples of systems with pressures that are greater than atmospheric. As you can see from Fig. 15-13 if level 2 were above level 1, Δy would be positive. In this case the gas pressure calculated using Eq. 15-12 would be less than the atmospheric pressure. For example, when you suck on a straw to pull fluid up the straw, the (absolute) pressure in your lungs is actually less than atmospheric pressure.

Gauge Pressure

Often when measuring pressure we are interested in knowing the difference between the pressure at some point such as level 2 in Fig. 15-13, which we call the **absolute pressure,** and a reference pressure P_1, which is taken to be the atmospheric pressure. In general, the difference between an absolute pressure and an atmospheric pressure is called the **gauge pressure.** (The name comes from the use of a gauge to measure this difference in pressures.) For example, most tire pressure devices measure gauge pressure.

The gauge pressure for water or another liquid used in a barometer or open-tube manometer is given by Eq. 15-8, which is $\Delta P = P_2 - P_1 = -\rho g \, \Delta y$. Since the gauge pressure is simply ΔP in the case where $P_1 = P^{atm}$, we can write

$$P^{gauge} = P_2 - P_1 = P_2 - P^{atm} = -\rho g \, \Delta y. \qquad (15\text{-}13)$$

15-6 Pascal's Principle

When you push down the plunger in the bottle shown in Fig. 15-2 and the cork pops out, you are watching **Pascal's principle** in action. This principle is also the reason why toothpaste comes out of the open end of a tube when you squeeze on the other end, and it is the basis of the Heimlich maneuver. In that maneuver a sharp pressure increase properly applied to the abdomen is transmitted to the throat, forcefully ejecting food lodged there. Pascal's principle was first stated clearly in 1652 by Blaise Pascal (for whom the unit of pressure is named):

> A change in the pressure applied to an enclosed fluid is transmitted undiminished to every portion of the fluid and to the walls of its container.

This is just what we explained at the start of our discussion of pressure when we applied a force to the piston in Fig. 15-2. A cork inserted in the side of a bottle pops out no matter where it is located.

A Mathematical "Proof" of Pascal's Principle

Consider the case in which the incompressible fluid is a liquid contained in a tall cylinder, as in Fig. 15-14. The cylinder is fitted with a piston on which a container of lead shot rests. The atmosphere, container, and shot put pressure P^{ext} on the piston and thus on the liquid. The pressure P at any point in the liquid a distance Δy below the piston is then

$$P_2 = P_1 - \rho g \, \Delta y = P^{ext} - \rho g \, \Delta y, \qquad (15\text{-}14)$$

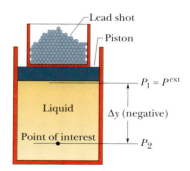

FIGURE 15-14 ■ The atmosphere and container with lead shot (small balls of lead) combine to create a pressure P^{ext} at the top of the enclosed (incompressible) liquid. If P^{ext} is increased, by adding more lead shot, the pressure increases by the same amount at all points within the liquid.

where, as usual, when level 2 is below level 1, Δy is negative. Let us add a little more lead shot to the container to increase P^{ext} by an amount ΔP^{ext}. Since the fluid is assumed to be incompressible, the quantities ρ, g, and Δy in Eq. 15-14 are unchanged. Thus the pressure change at any point is

$$\Delta P = \Delta P^{\text{ext}}. \tag{15-15}$$

This pressure *change* is independent of Δy, so it must hold for all points within the liquid, as Pascal's principle states.

Pascal's Principle and the Hydraulic Lift

Figure 15-15 shows how Pascal's principle can be made the basis of a hydraulic lift. In operation, let an external force of magnitude F^{in} be directed downward on the left-hand (or input) piston, whose area is A^{in}. An incompressible liquid in the device then produces an upward force of magnitude F^{out} on the right-hand (or output) piston, whose area is A^{out}. There will also be a downward force on the output piston with a magnitude equal to the weight of the external load (not shown). To keep the system in equilibrium, the weight of the external load must have the same magnitude as the upward output force so that $F^{\text{grav}}_{\text{L-load}} = F^{\text{out}}$.

The magnitude of the input force F^{in} applied on the left and the magnitude of the downward force from the load on the right, $F^{\text{grav}}_{\text{R-load}}$, both serve to produce a change ΔP in the pressure of the liquid. Since $F^{\text{grav}}_{\text{R-load}} = F^{\text{out}}$, this pressure change is given by

$$\Delta P = \frac{F^{\text{in}}}{A^{\text{in}}} = \frac{F^{\text{out}}}{A^{\text{out}}},$$

so

$$F^{\text{out}} = F^{\text{in}} \frac{A^{\text{out}}}{A^{\text{in}}}. \tag{15-16}$$

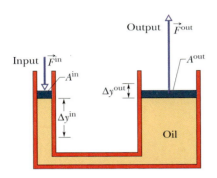

FIGURE 15-15 ■ A hydraulic arrangement that can be used to magnify a force $\vec{F}^{\text{in}}$. The work done is, however, not magnified and is the same for both the input and output forces.

Equation 15-16 shows that the magnitude of the output force F^{out} on the load must be greater than the magnitude of the input force F^{in} if $A^{\text{out}} > A^{\text{in}}$, as is the case in Fig. 15-15.

If we move the input piston downward a distance equal to the magnitude of Δy^{in}, the output piston moves upward a distance equal to the magnitude of Δy^{out}, such that the same volume V of the incompressible liquid is displaced at both pistons. Then

$$V = A^{\text{in}} |\Delta y^{\text{in}}| = A^{\text{out}} |\Delta y^{\text{out}}|,$$

which we can write as

$$A^{\text{out}} = A^{\text{out}} \frac{|\Delta y^{\text{in}}|}{|\Delta y^{\text{out}}|}. \tag{15-17}$$

This shows that, if $A^{\text{out}} > A^{\text{in}}$ (as in Fig. 15-15), the output piston moves a smaller distance than the input piston moves.

By combining Eqs. 15-16 and 15-17 and noting that the displacements on both the input and output sides of the lift are in the *same* direction as the forces that cause them, we get the following expression relating work out to work in,

$$W^{\text{out}} = F^{\text{out}} |\Delta y^{\text{out}}| = \left(F^{\text{out}} \frac{A^{\text{out}}}{A^{\text{in}}} \right) \left(|\Delta y^{\text{in}}| \frac{A^{\text{in}}}{A^{\text{out}}} \right) = F^{\text{in}} |\Delta y^{\text{in}}| = W^{\text{in}}. \tag{15-18}$$

This equation shows that in a hydraulic lift, the work W^{in} done *on* the input piston by the applied force should be equal to the work W^{out} done on the load by the output piston in lifting it.

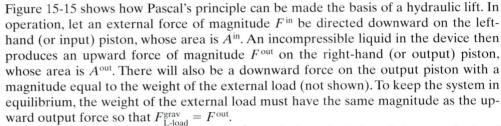

The advantage of a hydraulic lever is this:

> With a hydraulic lift, a given force applied over a given distance can be transformed to a greater force applied over a smaller distance.

The product of force and distance remains unchanged so that the same work is done. A small version of the hydraulic lift we have described is the jack used to change automobile tires. Most of us, for example, cannot lift an automobile directly but can with a hydraulic jack, even though we have to pump the handle farther than the automobile rises. In this device, the displacement is accomplished not in a single stroke but over a series of small strokes.

READING EXERCISE 15-6: Consider the cylinder of a real hydraulic jack filled with oil. The oil is slightly compressible. Discuss how the relations presented in this section are affected by the compressibility of the oil. For example, is the work put in still equal to the work out? If not, which is larger? What (if any) are the energy transformations that take place? ■

READING EXERCISE 15-7: The pressure on the bottom of a container with sloping walls is determined by the height Δy of the central column. Relate this observation to the concepts addressed in this section.

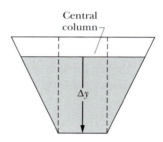

■

TOUCHSTONE EXAMPLE 15-4: Car Lift

You are a new employee at an auto repair shop. The exhaust system on a car you are repairing is leaking. Since you don't want to crawl under the car to work on the problem, you tell your boss that you'll use Pascal's principle to design a hydraulic lift that anyone can use to lift cars by hand that weigh up to 2000 kg using a lifting post with a 50 cm diameter. Describe your design.

SOLUTION ■ The **Key Ideas** here are these: (1) From Pascal's principle you know that the pressure is constant at the same level everywhere in a container of fluid, and (2) since $F = PA$ if two ends of a container have different cross-sectional areas, then the force exerted at the end with the small area can be much less than the force exerted at the end with the much larger area. Basically a hydraulic lift like that shown in Fig. 15-16 is a device that enables you to exert a small force over a large distance on one side of a container of fluid and to transmit a large force acting over a small distance on the other side.

Assume the lift is in equilibrium and the level of the liquid on the input and output of the hydraulic lift shown in Fig. 15-16 is the same. Then $P^{\text{in}} = P^{\text{out}}$ where $P^{\text{in}} = F^{\text{in}}/A^{\text{in}}$ and $P^{\text{out}} = F^{\text{out}}/A^{\text{out}}$, so

$$F^{\text{out}} = F^{\text{in}}\frac{A^{\text{out}}}{A^{\text{in}}}, \qquad (15\text{-}19)$$

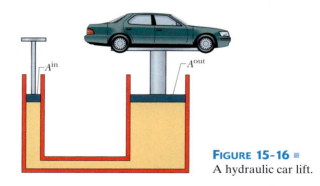

FIGURE 15-16 ■
A hydraulic car lift.

where F^{in} and F^{out} are the magnitudes of the forces pressing down on the pistons at the two ends of the hydraulic lift. You know that the magnitude of the force F^{out} needed to hold a car up is given by

$$F^{\text{out}} = F^{\text{grav}}_{\text{car}} = mg = 2000 \text{ kg} \times 9.8 \text{ m/s} = 2.0 \times 10^4 \text{ N}.$$

Most people can manage to lift a 20 kg mass without too much difficulty, so you assume that the input force you exert is given by

$$F^{\text{in}} = 20 \text{ kg} \times 9.8 \text{ m/s} = 2.0 \times 10^2 \text{ N},$$

which is 100 times less than the force you need to exert on a car to hold it up. So we can rewrite Eq. 15-19 to get

$$\frac{A^{out}}{A^{in}} = \frac{F^{out}}{F^{in}} = \frac{2.0 \times 10^4 \text{ N}}{2.0 \times 10^2 \text{ N}} = 100.$$

Using the fact that $A = \pi r^2$ and that $A^{out} = \pi r_{out}^2 = \pi(50 \text{ cm}/2)^2$, we can rearrange Eq. 15-19 to calculate the size of the plunger you can use in the input side to exert the force needed to hold the car up,

$$\frac{A^{out}}{A^{in}} = \frac{\pi r_{out}^2}{\pi r_{in}^2} = 100$$

so that

$$r_{in} = \sqrt{\frac{r_{out}^2}{100}} = \frac{r_{out}}{10} = \frac{50 \text{ cm}/2}{10} = 2.5 \text{ cm}.$$

Thus, in your design, you construct a column of diameter $2 \times 2.5 \text{ cm} = 5.0 \text{ cm}$ at the input end and push a piston into it with a force of just over

$$F^{in} = 2.0 \times 10^2 \text{ N}. \qquad \text{(Answer)}$$

FIGURE 15-17 ■ A thin-walled plastic sack of water is in static equilibrium in the pool. Its weight must be balanced by a net upward force on the water in the sack from the surrounding water.

15-7 Archimedes' Principle

Let us think about what happens when we immerse an object in a fluid. Your first guess might be that the weight of the water above the object would push it to the bottom. But think about what happens when we try to push a beach ball down under the surface in a swimming pool. It will be hard to push it down. If we do get it under the surface and release it, it will go shooting up into the air. What's going on here? It turns out that the critical issue is that pressure is a scalar. You can get forces in all directions—up as well as down. Let us see how this works.

Figure 15-17 shows a student in a swimming pool, manipulating a very thin plastic sack (of negligible mass) that is filled with water. She finds that the sack and its contained water are in static equilibrium, tending neither to rise nor to sink. The downward gravitational force $\vec{F}^{grav}$ on the contained water must be balanced by a net upward force from the water surrounding the sack.

This net upward force on an object in a fluid is called a **buoyant force** $\vec{F}^{buoy}$. It exists because the pressure in the surrounding water increases with depth below the surface. Thus, the pressure near the bottom of the sack is greater than the pressure near the top. Then the forces on the sack due to this pressure are greater in magnitude near the bottom of the sack than near the top. Some of the forces are represented in Fig. 15-18a, where the space occupied by the sack has been left empty. As you can see, the force vectors drawn near the bottom of that space (with upward components) have longer lengths than those drawn near the top of the sack (with downward components). If we vectorially add all the forces on the sack from the water, the horizontal components cancel and the vertical components add to yield the upward buoyant force $\vec{F}^{buoy}$ on the sack. (Force $\vec{F}^{buoy}$ is shown to the right of the pool in Fig. 15-18a.)

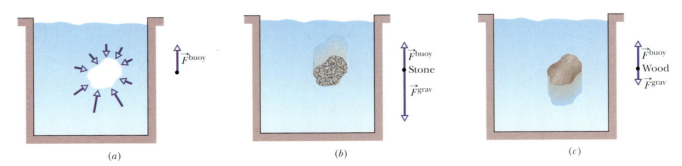

(a)   (b)   (c)

FIGURE 15-18 ■ (a) The water surrounding the hole in the water produces a net upward buoyant force on whatever fills the hole. (b) For a stone of the same volume as the hole, the gravitational force exceeds the buoyant force in magnitude. (c) For a lump of wood of the same volume, the gravitational force is less than the buoyant force in magnitude.

Let's denote the mass of the fluid in the sack of water as m_f. In static equilibrium, the buoyant force magnitude F^{buoy} is equal to the gravitational force magnitude $F^{\text{grav}} = m_f g$ experienced by the sack of water so that

$$F^{\text{buoy}} = m_f g. \tag{15-20}$$

In short, the magnitude of the buoyant force is equal to the weight of the water in the sack.

In Fig. 15-18b, we have replaced the sack of water with a stone that exactly fills the hole in Fig. 15-18a. The stone is said to *displace* the water, meaning that it occupies space that would otherwise be occupied by water. The object pushes the water out of the way. We have changed nothing about the shape of the hole, so the forces at the hole's surface must be the same as when the water-filled sack was in place. Thus, the same upward buoyant force that acted on the water-filled sack now acts on the stone. That is, the magnitude F^{buoy} of the buoyant force is still equal to $m_f g$, the weight of the water displaced by the stone.

Unlike the water-filled sack, the stone is not in static equilibrium. Because the stone has a higher density than water, the magnitude of the downward gravitational force F^{grav} on the stone is greater in magnitude than that of the upward buoyant force. This is shown in the free-body diagram to the right of the pool in Fig. 15-18b. The stone thus accelerates downward, sinking to the bottom of the pool. Once on the bottom, the normal force from the floor balances $\vec{F}^{\text{grav}}$ and the stone stops moving.

Let us next exactly fill the hole in Fig. 15-18a with a block of low-density wood, as in Fig. 15-18c. Again, nothing has changed about the forces at the hole's surface, so the magnitude F^{buoy} of the buoyant force is still equal to $m_f g$, the weight of the displaced water. Like the stone, the block is not in static equilibrium. However, this time the magnitude of the gravitational force F^{grav} is less than the buoyant force (as shown to the right of the pool), and so the block accelerates upward, rising to the top surface of the water.

Our results with the sack, stone, and block apply to all fluids and are summarized in **Archimedes' principle:**

> When a body is fully or partially submerged in a fluid, a buoyant force $\vec{F}^{\text{buoy}}$ from the surrounding fluid acts on the body. The buoyant force is directed upward and has a magnitude equal to the weight $m_f g$ of the fluid that has been displaced by the body.

The buoyant force on a body in a fluid has the magnitude

$$F^{\text{buoy}} = m_f g \qquad \text{(buoyant force magnitude)}, \tag{15-21}$$

where m_f is the mass of the fluid that is displaced by the body.

Floating

When we release a block of lightweight wood just above the water in a pool, it moves into the water because the gravitational force on it pulls it downward. As the block displaces more and more water, the magnitude F^{buoy} of the upward buoyant force acting on it increases. Eventually, F^{buoy} is large enough to equal the magnitude F^{grav} of the downward gravitational force on the block, and the block comes to rest. The block is then in static equilibrium and is said to be *floating* in the water. In general,

> When a body floats in a fluid, the magnitude F^{buoy} of the buoyant force on the body is equal to the magnitude F^{grav} of the gravitational force on the body.

In the late evening of August 21, 1986, something (possibly a volcanic tremor) disturbed Cameroon's Lake Nyos, which has a high concentration of dissolved carbon dioxide. The disturbance caused that gas to form bubbles. Being less dense than the surrounding fluid (the water), those bubbles were buoyed to the surface, where they released the carbon dioxide. The gas, being more dense than the surrounding fluid (now the air), rushed down the mountainside like a river, asphyxiating 1700 persons and the scores of animals seen here.

We can write this statement in terms of the magnitude of the forces as

$$F^{\text{buoy}} = F^{\text{grav}}. \tag{15-22}$$

From Eq. 15-21, we know that $F^{\text{buoy}} = m_f g$. Thus,

> When a body floats in a fluid, the magnitude F^{grav} of the gravitational force on the body is equal to the weight $m_f g$ of the fluid that has been displaced by the body.

We can write this statement

$$F^{\text{grav}} = m_f g \qquad \text{(floating condition)}. \tag{15-23}$$

In thinking about whether an object will sink or float, density is the key consideration, *not* the total mass of the object. If the density is less than that of water, the object can displace a mass of water equal to its own weight by being only partially submerged. If the density is greater than that of water, even completely immersing the object produces a buoyant force that is less than the object's weight. In that case, part of the object's weight will be unbalanced; there will be a net downward force and the object will sink to the bottom.

Apparent Weight in a Fluid

If we place a stone on a scale that is calibrated to measure weight, then the reading on the scale is the stone's weight. However, if we do this underwater, the upward buoyant force on the stone from the water decreases the reading. That reading is then an apparent weight. In general, an apparent weight or net downward force on the body is related to the actual weight of a body and the buoyant force on the body by

(apparent weight) = (actual weight) − (magnitude of buoyant force),

which we can write as

$$F^{\text{app}} = F^{\text{grav}} - F^{\text{buoy}} \qquad \text{(apparent weight)}. \tag{15-24}$$

If, in some strange test of strength, you had to lift a heavy stone, you could do it more easily with the stone underwater. Then your applied force would need to exceed only the stone's apparent weight, not its larger actual weight, because the upward buoyant force would help you lift the stone.

The magnitude of the buoyant force on a floating body is equal to the body's weight. Equation 15-24 thus tells us that a floating body has an apparent weight of zero—the body would produce a reading of zero on a scale. (When astronauts prepare to perform a complex task in space, they practice the task floating underwater, where their apparent weight is zero as it is in space.)

READING EXERCISE 15-8: A penguin floats first in fluid A of density ρ_A, then in fluid B of density $\rho_B = 0.95\rho_A$, and then in fluid C of density $\rho_C = 1.10\rho_A$. (a) Rank the fluids according to the magnitude of the buoyant force on the penguin, greatest first. (b) Rank the fluids according to the amount of fluid displaced by the penguin, greatest first. ■

READING EXERCISE 15-9: Each year student teams and colleges and universities design, build, and race full-size canoes made entirely of reinforced concrete. In the figure, a group of University of Maryland boat designers show off their award-winning concrete canoe. How can you get a concrete canoe to float?

■

TOUCHSTONE EXAMPLE 15-5: Iceberg

What fraction of the volume of an iceberg floating in seawater is visible?

SOLUTION ■ Let $V_{iceberg}$ be the total volume of the iceberg. The nonvisible portion is below water and thus is equal to the volume V_{fluid} of the fluid (the seawater) displaced by the iceberg. We seek the fraction (call it frac)

$$frac = \frac{V_{iceberg} - V_{fluid}}{V_{iceberg}} = 1 - \frac{V_{fluid}}{V_{iceberg}}, \qquad (15\text{-}25)$$

but we know neither volume. A **Key Idea** here is that, because the iceberg is floating, Eq. 15-23 ($F^{grav} = m_{fluid}g$) applies. We can write that equation as

$$m_{iceberg}g = m_{fluid}g,$$

from which we see that $m_{iceberg} = m_{fluid}$. Thus, the mass of the iceberg is equal to the mass of the displaced fluid (seawater).

Although we know neither mass, we can relate them to the densities of ice and seawater given in Table 15-2 by using Eq. 15-1 ($\rho = m/V$). Because $m_{iceberg} = m_{fluid}$, we can write

$$\rho_{iceberg}V_{iceberg} = \rho_{fluid}V_{fluid},$$

or

$$\frac{V_{fluid}}{V_{iceberg}} = \frac{\rho_{iceberg}}{\rho_{fluid}}.$$

Substituting this into Eq. 15-25 and then using the known densities, we find

$$frac = 1 - \frac{\rho_{iceberg}}{\rho_{fluid}} = 1 - \frac{917 \text{ kg/m}^3}{1024 \text{ kg/m}^3}$$

$$= 0.10 \text{ or } 10\%. \qquad \text{(Answer)}$$

The fact that only 10% of an iceberg can be seen above water is the source of a common expression: "That's only the tip of the iceberg."

TOUCHSTONE EXAMPLE 15-6: Water-Filled Sack

Consider the thin-walled sack full of water suspended in a swimming pool shown in Fig. 15-17. The atmospheric pressure at the surface of the pool is the mean sea level pressure. Assume that the bottom of the sack is 60 cm below the top of the sack.

(a) If the top of the sack is 40 cm below the surface of the pool, what is the pressure at the center of the sack?

SOLUTION ■ To determine the pressure at the center of the sack we need to find how far it is below the water surface. The middle of a sack with a top to bottom distance of 60 cm is 60 cm/2 = 30 cm below the top of the sack. If the top of the sack is 40 cm below the surface, then the middle of the sack is at a depth given by $\Delta y = -40 \text{ cm} - 30 \text{ cm} = -70 \text{ cm}$. Assuming that the water is in-

compressible, we can use Eq. 15-10 to find the pressure, P_2, at the center of the sack,

$$P_2 = P_1 - \rho_{water}g\,\Delta y = P^{atm} - \rho_{water}g\,\Delta y$$

$$= 101\,325 \text{ Pa} - (998 \text{ kg/m}^3)(9.8 \text{ N/kg})(-0.70\text{m})$$

$$= 108.2 \text{ Pa}. \qquad \text{(Answer)}$$

(b) If the water inside the sack has a mass of 100 kg, what is the buoyant force on the sack?

SOLUTION ■ The **Key Idea** here is that according to Archimedes' principle, the buoyant force has the same magnitude as the weight of the pool water *displaced* by the sack of water. Since the plastic sack is thin with essentially no mass, the weight of the sack with the water in it is the same as the weight of the water the sack displaced. We can use Eq. 6-5 to find the weight of the displaced water,

$$\text{Weight} = |\vec{F}^{\text{ grav}}| = mg = (100\,\text{kg})(9.80\,\text{m/s}^2) = 980\,\text{N}.$$

Since the buoyant force acts upward, using Archimedes' principle (Eq. 15-21),

$$\vec{F}^{\text{ buoy}} = +|\vec{F}^{\text{grav}}|\hat{j} = (980\,\text{N})\hat{j}. \qquad \text{(Answer)}$$

(c) If the swimmer gently pulls the sack down deeper into the pool, what happens to the pressure at the center of the sack?

SOLUTION ■ The **Key Idea** here is that according to Eq. 15-10, the pressure at the center of the sack (and in fact at all locations in the sack) increases as the depth of the sack increases.

(d) What happens to the buoyant force on the sack as it is pulled down further?

SOLUTION ■ The **Key Idea** here is that according to Archimedes' principle, the buoyant force must remain the same as the weight of the water displaced. But the displaced weight doesn't change with depth (to the extent that the local gravitational field strength *g* does not change with depth).

(e) If the buoyant force is caused by the pressure of water on the sack and the pressure increases with depth, why doesn't the buoyant force on the water in the sack *increase*?

SOLUTION ■ The **Key Idea** here is that the buoyant force on the water in the sack is a *net force* that depends on the vector sum of the upward forces on the bottom surface elements of the sack due to the pool's water pressure, and the downward forces on the top surface elements of the sack also due to the pool's water pressure. Ultimately the net force is related to *pressure differences* between the bottom and top sack elements. Since pressure increases linearly with depth, the *pressure differences do not* depend on depth.

FIGURE 15-19 ■ At a certain point, the rising flow of smoke and heated gas changes from steady to turbulent.

15-8 Ideal Fluids in Motion

The motion of *real fluids* is very complicated and not yet fully understood. Instead, we shall discuss the motion of an **ideal fluid,** which is simpler to handle mathematically and yet provides useful results. Here are four assumptions that we make about our ideal fluid, all concerned with flow:

1. **Steady flow** In *steady* (or *laminar*) *flow,* the velocity of the moving fluid at any fixed point does not change with time, either in magnitude or in direction. The gentle flow of water near the center of a quiet stream is steady; that in a chain of rapids is not. Figure 15-19 shows a transition from steady flow to *nonsteady* (or *turbulent*) flow for a rising stream of smoke. The speed of the smoke particles increases as they rise and, at a certain critical speed, the flow changes from steady to nonsteady (that is, from laminar to *nonlaminar* flow).

2. **Incompressible flow** We assume, as we have already done for fluids at rest, that our ideal fluid is incompressible; that is, its density has a constant, uniform value.

3. **Nonviscous flow** Roughly speaking, the viscosity of a fluid is a measure of how resistive the fluid is to flow. For example, thick honey is more resistive to flow than water, and so honey is said to be more viscous than water. Viscosity is the fluid analog of friction between solids; both are mechanisms by which the kinetic energy of moving objects can be transferred to thermal energy. In the absence of friction, a block could glide at constant speed along a horizontal surface. In the same way, an object moving through a nonviscous fluid would experience no *viscous drag force*—that is, no resistive force due to viscosity; it could move at constant speed through the fluid. The British scientist Lord Rayleigh noted that in an ideal fluid a ship's propeller would not work but, on the other hand, a ship (once set into motion) would not need a propeller!

We can make the flow of a fluid visible by adding a *tracer.* This might be a dye injected into many points across a liquid stream (Fig. 15-20) or smoke particles added to

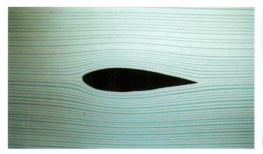

FIGURE 15-20 ■ Streamlined flow around an airfoil.

FIGURE 15-21 ■ Smoke reveals streamlines in airflow past a car in a wind-tunnel test.

a gas flow (Figs. 15-19 and 15-21). Each bit of a tracer follows a *streamline,* which is the path that a tiny element of the fluid would take as the fluid flows. Recall from Chapter 2 that the velocity of a particle is always tangent to the path taken by the particle. Here the particle is the fluid element, and its velocity $\vec{v}$ is always tangent to a streamline (Fig. 15-22). For this reason, two streamlines cannot intersect with a finite fluid velocity, since a fluid element cannot flow in two directions simultaneously. However, at a point of zero velocity, a stagnation point, there is no direction defined and two or more streamlines may intersect at a stagnation point.

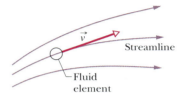

FIGURE 15-22 ■ A fluid element traces out a streamline as it moves. The velocity vector of the element is tangent to the streamline at every point.

15-9 The Equation of Continuity

You may have noticed that you can increase the speed of the water emerging from a garden hose by partially closing the hose opening with your thumb. Apparently the speed $|\vec{v}|$ (often denoted simply as v) of the water depends on the cross-sectional area A through which the water flows. This makes sense when we realize that the faucet is putting water out at a certain rate. If that much water has to get through a smaller hole, it has to go faster.

Here we wish to derive an expression that relates the speed of a fluid v to the cross-sectional area A of the pipe through which it flows. Let's consider the steady flow of an ideal fluid through a pipe with varying cross section, like that in Fig. 15-23. The flow is toward the right, and the pipe segment shown (part of a longer pipe) has length L. The fluid has speeds v_1 at the left end of the segment and v_2 at the right end. The pipe has cross-sectional areas A_1 at the left end and A_2 at the right end. Suppose that in a time interval Δt a volume ΔV of fluid enters the pipe segment at its left end (that volume is colored purple in Fig. 15-23a). Then, because the fluid is incompressible, an identical volume ΔV must emerge from the right end of the segment (it is colored green in Fig. 15-23b).

We can use this common volume ΔV to relate the speeds and areas. To do so, we first consider Fig. 15-24, which shows a side view of a pipe of *uniform* cross-sectional area A. In Fig. 15-24a, a fluid element e is about to pass through the dashed line drawn across the pipe width. The element's speed is v, so during a time interval Δt, the element moves along the pipe a distance $\Delta x = v \Delta t$. The volume ΔV of fluid that has passed through the dashed line in that time interval Δt is

$$\Delta V = A \Delta x = Av \Delta t. \tag{15-26}$$

Applying these ideas to both the left and right ends of the pipe segment in Fig. 15-23, we have

$$\Delta V = A_1 v_1 \Delta t = A_2 v_2 \Delta t$$

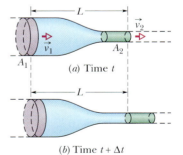

(a) Time t

(b) Time $t + \Delta t$

FIGURE 15-23 ■ Fluid flows from left to right at a steady rate through a tube segment of length L. The fluid's velocity is $\vec{v}_1$ at the left side and $\vec{v}_2$ at the right side. The tube's cross-sectional area is A_1 at the left side and A_2 at the right side. From time t in (a) to time $t + \Delta t$ in (b), the amount of fluid in purple enters at the left side and the equal amount of fluid shown in green emerges at the right side.

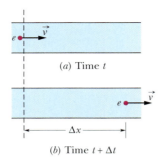

(a) Time t

(b) Time t + Δt

FIGURE 15-24 ■ Fluid flows at a constant velocity $\vec{v}$ through a tube. (*a*) At time *t*, fluid element *e* is about to pass the dashed line. (*b*) At time *t* + Δ*t*, element *e* is a distance Δ*x* = v_x Δ*t* from the dashed line where v_x is the *x*-component of $\vec{v}$.

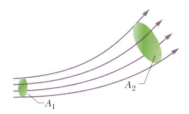

FIGURE 15-25 ■ A tube of flow is defined by the streamlines that form the boundary. The volume flow rate must be the same for all cross sections of the tube of flow.

or
$$A_1v_1 = A_2v_2 \quad \text{(equation of continuity).} \quad (15\text{-}27)$$

This relation between speed and cross-sectional area is called the **equation of continuity** for the flow of an ideal fluid. It tells us that the flow speed increases when we decrease the cross-sectional area through which the fluid flows (as when we partially close off a garden hose with a thumb).

The equation of continuity (Eq. 15-27) applies not only to an actual pipe but also to any so-called *tube of flow*, or imaginary tube whose boundary consists of streamlines. Such a tube acts like a real pipe because no fluid element can cross a streamline. Thus, all the fluid within a tube of flow must remain within its boundary. Figure 15-25 shows a tube of flow in which the cross-sectional area increases from area A_1 to area A_2 along the flow direction. From Eq. 15-27 we know that, with the increase in area, the speed must decrease, as is indicated by the greater spacing between streamlines at the right in Fig. 15-25. Similarly, you can see that in Fig. 15-20 the speed of the flow is greatest just above and just below the cylinder.

We can rewrite the equation of continuity (Eq. 15-27) as
$$R_V = Av = \text{a constant} \quad \text{(volume flow rate, equation of continuity),} \quad (15\text{-}28)$$

in which R_V is the **volume flow rate** of the fluid (volume per unit time). Its SI unit is the cubic meter per second (m³/s). If the density ρ of the fluid is uniform, we can multiply this expression (Eq. 15-28) by that density to get the **mass flow rate** R_m (mass per unit time):
$$R_m = \rho R_V = \rho Av = \text{a constant} \quad \text{(mass flow rate, the equation of continuity).} \quad (15\text{-}29)$$

The SI unit of mass flow rate is the kilogram per second (kg/s). Equation 15-29 says that the mass that flows into the pipe segment of Fig. 15-23 each second must be equal to the mass that flows out of that segment each second.

Engineers can use velocity measurements and the equation of continuity as a tool for determining volume flow rates of incompressible fluids such as oil flowing in a pipe of variable cross section or water flowing in a stream of variable cross section. But the equation of continuity (Eqs. 15-27 to 15-29) can only be used under certain conditions. In deriving it we assumed that fluid isn't being added to or subtracted from the system. For example, we assume that our oil pipe doesn't leak or that new water is not being added to a stream by a tributary as it flows along. In cases where fluid is added or subtracted from a tube of flow, we would find that $v_1A_1 \neq v_2A_2$.

Even when the continuity condition holds, whenever we use the equation of continuity we assume that we have a uniform flow of fluid over a cross-sectional area. In other words, we assume that the velocity of each of the elements of the fluid is moving at the same speed in a direction that is perpendicular to a cross-sectional area. In real pipes, fluid flows more slowly near the surface of the pipe than in the middle and it is not obvious how to calculate the product *vA* at a given location along our pipe.

In the next section we will introduce the mathematical concept of *flux* to help us deal with more realistic situations in which the simplified application of the equation of continuity cannot be used either because we do not have true continuity or because we do not have a nice uniform flow of fluid over a cross-sectional area.

READING EXERCISE 15-10: The figure shows a pipe and gives the volume flow rate (in cm³/s) and the direction of flow for all but one section. What are the volume flow rate and the direction of flow for that section?

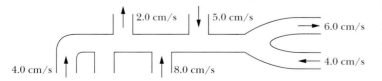

TOUCHSTONE EXAMPLE 15-7: Blood Flow

The cross-sectional area A_1 of the aorta (the major blood vessel emerging from the heart) of a normal resting person is 3 cm², and the average speed v_1 of the blood is 30 cm/s. A typical capillary (diameter ≈ 6 μm) has a cross-sectional area A_2 of 3×10^{-7} cm² and an average flow speed v_2 of 0.05 cm/s. How many capillaries does such a person have?

SOLUTION ■ The **Key Idea** here is that all the blood that passes through the capillaries must have passed through the aorta. Therefore, the volume flow rate through the aorta must equal the total volume flow rate through the capillaries. Let us assume that the capillaries are identical, with the given cross-sectional area A_2 and average flow speed v_2. Then, from Eq. 15-27 we have

$$A_1 v_1 = n A_2 v_2,$$

where n is the number of capillaries. Solving for n yields

$$n = \frac{A_1 v_1}{A_2 v_2} = \frac{(3\ \text{cm}^2)(30\ \text{cm/s})}{(3 \times 10^{-7}\ \text{cm}^2)(0.05\ \text{cm/s})}$$

$$= 6 \times 10^9 \text{ or 6 billion.} \qquad \text{(Answer)}$$

You can easily show that the combined cross-sectional area of the capillaries is about 600 times the cross-sectional area of the aorta.

TOUCHSTONE EXAMPLE 15-8: Necking Down

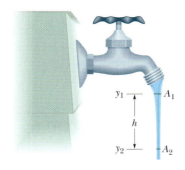

y_1 ─┬─ A_1

h

y_2 ─┴─ A_2

FIGURE 15-26 ■ As water falls from a tap, its speed increases. Because the flow rate must be the same at all cross-sections, the stream must "neck down."

Figure 15-26 shows how the stream of water emerging from a faucet "necks down" as it falls. The indicated cross-sectional areas are $A_1 = 1.2$ cm² and $A_2 = 0.35$ cm². The two levels are separated by a vertical distance $h = 45$ mm. What is the volume flow rate from the tap?

SOLUTION ■ The **Key Idea** here is simply that the volume flow rate through the higher cross section must be the same as that through the lower cross section. Thus, from Eq. 15-27, we have

$$A_1 v_1 = A_2 v_2, \qquad (15\text{-}30)$$

where v_1 and v_2 are the water speeds at the levels corresponding to A_1 and A_2. Adapting the last equation on page 43 to this situation we get

$$v_{2y}^2 = v_{1y}^2 + 2(-g)(y_2 - y_1), \qquad (15\text{-}31)$$

where $h = y_1 - y_2$. Eliminating v_2 between Eqs. 15-30 and 15-31 and solving for v_1, we obtain

$$v_1 = \sqrt{\frac{2ghA_2^2}{A_1^2 - A_2^2}}$$

$$= \sqrt{\frac{(2)(9.8\ \text{m/s}^2)(0.045\ \text{m})(0.35\ \text{cm}^2)^2}{(1.2\ \text{cm}^2)^2 - (0.35\ \text{cm}^2)^2}}$$

$$= 0.286\ \text{m/s} = 28.6\ \text{cm/s}.$$

From Eq. 15-28, the volume flow rate R_V is then

$$R_V = A_1 v_1 = (1.2\ \text{cm}^2)(28.6\ \text{cm/s})$$

$$= 34\ \text{cm}^3/\text{s}. \qquad \text{(Answer)}$$

15-10 Volume Flux

The term *volume flux* is a synonym for volume flow rate, R_V. In this section we consider a more careful mathematical definition of volume flux that will allow us to apply the equation of continuity to more realistic situations. The word "flux" comes from the Latin word meaning "to flow." Often the word "flux" is used in science and engineering to describe the rate of flow or penetration of matter or energy through a surface. However, later in this text we will introduce definitions of electric and magnetic flux. Even though the mathematical definitions of electric and magnetic flux are analogous to that of volume flow rate, it is surprising to find that nothing at all is flowing.

Volume Flux in a Stream

In this section we develop the concept of flux to calculate the rate at which water flows though a cross-sectional area of a stream. Suppose we have a stream of water undergoing steady laminar flow (as described in Sections 15-8 and 15-9). Assume all the water is traveling in the same direction through a wide shallow channel as shown in Fig. 15-27 but that the elements of water closer to the stream bottom and sides are moving more slowly than the water in the top central part of the stream. We cannot simply find a single velocity, $\vec{v}$, at all points along our cross-sectional area.

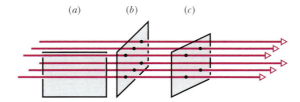

FIGURE 15-27 ■ Water moving in laminar flow through a wide shallow channel. The darker regions in the diagram with longer velocity vectors indicate areas farther from the sides and bottom of the stream where the water encounters less drag and hence flows faster. The flow can be described by a vector field consisting of the velocity vectors at each location in the stream.

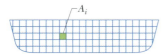

FIGURE 15-28 ■ The ith of many, many small area elements that make up the cross-sectional area of a stream channel. The size of the elements is chosen to be so small that the stream velocity vector is constant over an element of area.

Let us consider stream water passing through many small elements. We can denote the ith area element as ΔA_i, where we use the delta notation (Δ) to signify that the area is part of a larger area, as shown in Fig. 15-28. In fact, we typically choose the area elements ΔA_i to be small enough that the velocity of the part of the stream flowing through it is essentially constant. Suppose that the volume flux, Φ_i, represents the rate at which water flows through the ith element of area.

> **VOLUME FLUX** is defined as the rate at which something passes through an area.

Obviously if the water velocity $\vec{v}$ is parallel to the plane of the area element ΔA_i, water passes by the area without going through it. In that case the flux element, Φ_i, is zero (Fig. 15-29a). On the other hand, if $\vec{v}$ is perpendicular to the plane of the area, the flux element, Φ_i, is a maximum (Fig. 15-29b). If the area is oriented so it is between perpendicular and parallel the volume flux is in between zero and its maximum value (Fig. 15-29c). Thus, the volume flux depends on the *angle* between the velocity vector $\vec{v}$ representing the flow of water and the orientation of the areas shown in Fig. 15-29.

FIGURE 15-29 ■ The velocity vector field of a stream is represented here as imaginary streamlines. The amount of water that passes through a small imaginary area depends on the area's orientation. Three orientations are shown for the same area element A_i. (a) No water passes through the area when it is parallel to the velocity vector. (b) The largest volume of water passes through an area that is perpendicular to the velocity vector. (c) Less water passes through when the orientation of the plane of the area is between perpendicular and parallel.

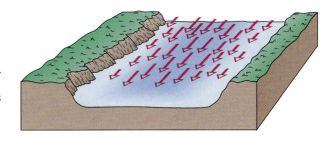

As shown in Eq. 15-26, the volume of water, ΔV_i, that passes in a time Δt through a small area element, A_i, perpendicular to the stream's direction of flow is given by $\Delta V_i = \Delta x \ A_i = (v_{ix}\Delta t) \ A_i$ (Fig. 15-29). Thus, when an area is perpendicular to the x-component of the stream velocity v_x, the flux element, $\Phi_\perp$, is given by the volume rate of flow through the area

$$\Phi_{i\perp} = \frac{\Delta V_i}{\Delta t} = \frac{\Delta x \ A_i}{\Delta t} = v_{ix}A_i. \tag{15-32}$$

Defining the Normal Vector for an Area

In order to find the flux at intermediate angles, it will be convenient to represent the orientation of our small area mathematically. An area uses two of the three dimensions of space, but a direction perpendicular to the plane of the area lies along a single line. Thus it is easier to define the orientation of the ith area element mathematically by a vector having a magnitude of A_i that points in a direction *perpendicular* to the plane of the area. Since the term normal is a synonym for perpendicular, a vector used to describe the direction and magnitude of an area is known as a **normal vector** and can be denoted as $\vec{A}_i$. But in what direction should our normal vector point? If the element of area is part of an imaginary closed container, it is conventional to choose our normal vector for a flat surface pointing *out* of a surface rather than into a surface. The box in Fig. 15-30 has an inside and outside even if it has no front or back. Thus, the normal vectors representing each element of area point outward. If we put a front and back on the box, we would define it as a *closed surface* because if it were a real box, we could trap or enclose something inside of it.

How does the angle between the normal to an area and a stream velocity vector affect the flux? Suppose water in a straight stream channel is moving from left to right as it flows through a small imaginary box shown in Fig. 15-31. What will the flux be through each face of the box as a function of the angle θ between the normal vector of a box face and the stream velocity vector $\vec{v}_i$? Since the flux is a measure of the volume of water that flows through a surface per unit time, it will depend on the component of $\vec{v}_i$ normal to the plane of the area element and the magnitude of the area $|\Delta\vec{A}_i|$ (Fig. 15-30). But as seen in Fig. 15-31, the component of $\vec{v}_i$ along the direction of the normal vector is $|\vec{v}_i|\cos\theta_i$, and so the flux element Φ_i is given by

$$\Phi_i = (|\vec{v}_i|\cos\theta_i)|\vec{A}_i| = \vec{v}_i\cdot\vec{A}_i$$

where A_i represents the ith area element.

To make our definition of flux more generally useful, we will develop an equation for flux in terms of the velocity and area vectors $\vec{v}$ and $\vec{A}$. We define the flux of water through the ith element of area mathematically using the scalar product (introduced in Section 9-4) as

$$\Phi_i \equiv \vec{v}_i\cdot\vec{A}_i \qquad \text{(volume flux definition for a small area element).} \qquad (15\text{-}33)$$

Let us consider how to calculate the net flow of water through an imaginary surface placed in a stream such as the box shown in Fig. 15-31. We can assign a velocity vector to each point in the stream of water. If our area elements are small enough, the velocity vector is the same everywhere on the surface of a given area element. Thus, if we know the area and orientation of each face and the magnitude and direction of the stream velocity vectors, we can use Eq. 15-33 to calculate the net flux through the cross-sectional area of the stream by adding up the products of the normal velocity vector components and the area elements through which water flows. Mathematically this is given by

$$\Phi^{\text{net}} = \Phi_1 + \Phi_2 + \cdots + \Phi_N = \vec{v}_1\cdot\vec{A}_1 + \vec{v}_2\cdot\vec{A}_2 + \cdots + \vec{v}_N\cdot\vec{A}_N = \sum_{n=1}^{N}\vec{v}_n\cdot\vec{A}_n$$

$$(15\text{-}34)$$

where $\vec{v}_1, \vec{v}_2, \vec{v}_N$, and so on represent the velocity vectors at the location of each of the N area elements.

Net Fluid Volume Flux through a Closed Container

Let us consider how fluid flows into face 3 and out of faces 1 and 2 of a small imaginary box we have placed in our stream (Fig. 15-31). If our fluid, in this case water, is

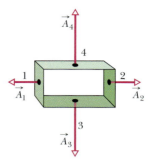

FIGURE 15-30 ■ The directions of the normal vectors are shown for four of the six faces that make up a container. Each normal vector points out of the container. Since area elements 3 and 4 have twice the area as 1 and 2, their normal vectors have twice the length.

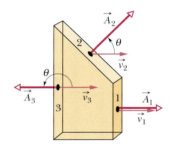

FIGURE 15-31 ■ A small imaginary box with six faces (or surfaces) is placed in an area of the stream where the velocity of the water is uniform over all the faces of the box. The angle between the velocity vector and the normal vectors for three of the six faces of the box are shown.

incompressible and is not created or destroyed inside our imaginary closed surface, we expect that the rate at which it flows into the box should be the same as the rate at which it flows out of the box. The total flux through all the faces of a closed container is known as the **net flux.** Thus, we expect the net flux of an incompressible fluid through the box to be zero. Does the mathematics of our method for finding net flux tell us this?

We have chosen our area elements to be small enough that the flow velocity vector at the location of any particular area element is uniform. However, that doesn't mean that magnitudes of the velocity vectors and relative angle between the velocity vectors and the normal vectors are necessarily the same at the location of each flat surface area. Thus, in general the net flux through a surface is the sum of the flux through each surface area and is still given by the application of Eq. 15-34 to our new situation. Our definition of net flux allows us to deal with a general case for which we have different velocity vectors and normal vector orientations at each face. This might be the case if the stream is very turbulent.

When the velocity and area vectors point in the same general direction as they do in faces 1 and 2, the scalar product of $\vec{v}_i$ and $\vec{A}_i$ is positive. This tells us that water *flowing out of a surface* is defined as a *positive* flux. But the flux at face 3 is different. The velocity and area vectors are in opposite directions and their scalar product is negative. Thus, when water flows *into a surface* the flux is *negative*. Another feature of our box is that the areas associated with the bottom face and the front and back faces all have normal vectors that are perpendicular to the stream velocity vectors. The scalar product rules give us no flux or volume flow through these additional faces. It can be shown that the sum of the negative flux of water into face 3, the positive flux of water out of face 1, and the positive flux out of face 2 add up to zero.

If we refine the equation of continuity to take the direction of flow of fluid into a closed surface as negative volume flux and the flow out of a closed surface as positive volume flux, we indeed expect the net flux through our imaginary box to be zero. This makes sense physically as long as water is incompressible and as long as we can't spontaneously create new water inside the box or remove water from the box. Later when we define electric and magnetic flux, we will see that it may be possible to have a net flux at the boundaries of a closed surface.

READING EXERCISE 15-11: Consider the imaginary box in Fig. 15-31 and assume that it has a width of 4.0 cm, a depth of 1.0 cm, a height on the left side of 8.0 cm, and a height on the right side of 4.0 cm. (a) Find the magnitude of the area of faces 1, 2, and 3, respectively. Report your answer in square meters. (b) What is the total surface area of the box? [Be sure to include the areas of all six area elements (faces) in your calculation.] ■

READING EXERCISE 15-12: Consider the imaginary box in Fig. 15-31 and assume that it is placed in a stream that moves from left to right through the box with a uniform velocity. In other words, we make the simplifying assumption that the velocity vector is the same at every point and on the surface of the small box. Suppose the box has a width of 4.0 cm, a depth of 1.0 cm, a height on the left side of 8.0 cm, and a height on the right side of 4.0 cm. If the magnitude of the stream velocity at the location of the box is 0.50 m/s, (a) find the flux through faces 1, 2, and 3 respectively in m^3/s; (b) explain why only faces 1, 2, and 3 have nonzero flux, and (c) show that the net flux through the closed surface defined by the box is zero. ■

15-11 Bernoulli's Equation

As we discussed in Section 15-8, ideal fluids are incompressible and flow in a streamlined fashion without experiencing friction forces. Water that flows in slow-moving rivers or large pipes acts like an ideal fluid. In this section we present a very useful equation that relates the pressure, velocity, and vertical location of an ideal fluid as it flows.

Consider an ideal fluid flowing through a pipe like that shown in Fig. 15-32. In a time interval Δt, a parcel of fluid of volume ΔV (colored purple in Fig. 15-32a) enters the pipe at the left input end and an identical volume (colored green in Fig. 15-32b) emerges at the right output end. Since we are assuming that the fluid is incompressible, the emerging volume must be the same as the entering volume. In addition we assume that the fluid has a constant density ρ.

Let y_1, v_1, and P_1 be the elevation, speed, and pressure of the fluid entering at the left, and y_2, v_2, and P_2 be the corresponding quantities for the fluid emerging at the right. By applying the principle of conservation of energy to the fluid, we can show that these quantities are related by

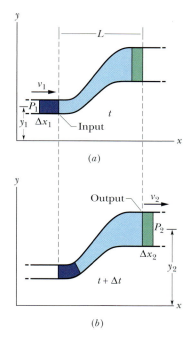

(a)

$$P_1 + \tfrac{1}{2}\rho v_1^2 + \rho g y_1 = P_2 + \tfrac{1}{2}\rho v_2^2 + \rho g y_2. \tag{15-35}$$

We can also write this equation as

$$P + \tfrac{1}{2}\rho v^2 + \rho g y = \text{a constant} \qquad \text{(ideal fluid-flow equation)}. \tag{15-36}$$

(b)

Equations 15-35 and 15-36 are equivalent forms of **Bernoulli's equation,** named after Daniel Bernoulli, who studied fluid flow in the 1700s.* Like the equation of continuity (Eq. 15-28), Bernoulli's equation is not a new principle but simply the reformulation of a familiar principle in a form more suitable to fluid mechanics. As a check let us consider Bernoulli's equation for two special cases.

FIGURE 15-32 ■ Fluid flows at a steady rate through a length L of a tube from the input end at the left to the output end at the right. From time t in (a) to time $t + \Delta t$ in (b), the parcel of fluid of mass m shown in purple enters the input end and the equal amount shown in green emerges from the output end.

Case 1, No flow: Here we set the speeds at the two ends of the pipe equal to zero in Eq. 15-35 so that $v_1 = v_2$. This gives us

$$P_2 = P_1 - \rho g(y_2 - y_1) \qquad \text{(elevation change but no flow).}$$

In this case, Bernoulli's equation simply reduces to Eq. 15-7.

Case 2, No Elevation Change: A major prediction of Bernoulli's equation emerges if we assume the fluid does not change elevation as it flows so that $y_1 = y_2$. Equation 15-35 then becomes

$$P_1 + \tfrac{1}{2}\rho v_1^2 = P_2 + \tfrac{1}{2}\rho v_2^2, \tag{15-37}$$

which tells us that:

> If the speed of a fluid element increases as it travels along a horizontal streamline, the pressure of the fluid must decrease.

Put another way, where the speed of the fluid is relatively high, its pressure is relatively low. Conversely, where the speed of the fluid is relatively low, its pressure is relatively high.

The link between a change in pressure and a change in speed makes sense if you consider what happens to the parcel of fluid as it crosses a boundary between high and low pressure. As the parcel of fluid that is moving from left to right nears a narrow region, the higher pressure behind it exerts a force on it toward the right of

*For irrotational flow (which we assume), the constant in Eq. 15-36 has the same value for all points within the tube of flow; the points do not have to lie along the same streamline. Similarly, the points 1 and 2 in Eq. 15-35 can lie anywhere within the tube of flow.

magnitude $F_1 = P_1 \Delta A$ where ΔA is the cross-sectional area of the parcel. Likewise the lower pressure in front of it exerts a force on it toward the left of magnitude $F_2 = P_2 \Delta A$. But since $P_1 > P_2$, then $F_1 > F_2$ and the parcel experiences a net force to the right. It then accelerates over a short distance and speeds up. In passing from a region of lower pressure to one of higher pressure $F_2 > F_1$, so the parcel slows down.

Bernoulli's equation is strictly valid only to the extent that the fluid is ideal. If viscous forces are present, thermal energy will be involved. We take no account of this in the derivation that follows.

Proof of Bernoulli's Equation

Since our ideal fluid system experiences no friction, we can assume that mechanical energy is conserved by any parcel of fluid lying between the two vertical planes separated by a distance L in Fig. 15-32 as it moves from its initial state (Fig. 15-32a) to its final state (Fig. 15-32b). We also assume that the fluid does not change its properties during this process and that the flow is steady. Thus, we need be concerned only with changes that take place at the input and output ends of the pipe.

The work-kinetic energy theorem tells us that if energy is conserved,

$$W^{\text{net}} = \Delta K. \tag{15-38}$$

In other words, the change in the kinetic energy of our parcel must equal the net work done on it. But the change in kinetic energy results from the change in speed of the parcel of fluid at the ends of the pipe. Thus, if a parcel of fluid flows into end 1 and out end 2 of the pipe, then

$$\begin{aligned} \Delta K &= \tfrac{1}{2} m v_2^2 - \tfrac{1}{2} m v_1^2 \\ &= \tfrac{1}{2} \rho V (v_2^2 - v_1^2), \end{aligned} \tag{15-39}$$

where $m = \rho V$ is the mass of the parcel of fluid of volume V.

The work done on the parcel arises from two sources. One is the negative gravitational work done on the parcel of fluid during the vertical lift of the mass from the input to the output level. A second source of work done on the parcel of fluid results from the forces exerted on it by the fluid behind it and the fluid in front of it as it flows due to pressure differences at various locations in the pipe.

Gravitational Work: The work W^{grav} done by the gravitational force $(m g \hat{\jmath})$ on the parcel of fluid of mass m during the vertical lift of the mass from the input to the output level is given by Eq. 9-14

$$\begin{aligned} W^{\text{grav}} &= -mg(y_2 - y_1) \\ &= -\rho g V (y_2 - y_1). \end{aligned} \tag{15-40}$$

This work is negative because the upward displacement and the downward gravitational force have opposite directions. (Note that in this context the notation W^{grav} has nothing to do with the weight of the parcel.)

Pressure Difference Work: Work is done on the parcel of moving fluid as a result of the pressure difference between its input and output ends. We start by finding the equation for the work needed to move a parcel of fluid through a distance Δx in a pipe of cross section A. If the parcel of fluid is under pressure P, it is given by

$$W = F_x \Delta x = (PA)\Delta x = P(A\Delta x) = PV,$$

where $V = A\, \Delta x$ is the volume of the parcel.

We now consider the parcel of fluid shown at the input end in Fig. 15-32. It is at pressure P_1 and has a volume V. Since the displacement and the force due to the pressure at the input end are in the *same* direction, the work done on it by the fluid behind it is positive and given by $W_1 = +P_1V$. Now consider the work done on the parcel of fluid shown at the output end in Fig. 15-32. It is at pressure P_2 but also has a volume V. Since the displacement and the force exerted due to the pressure at the output end are in *opposite* directions, the work done on it by the fluid behind it is negative and given by $W_2 = +P_2V$. Thus, the total work W^{pd} done on the incoming and emerging parcels due to pressure difference is given by

$$W^{pd} = W_1 - W_2 = P_1V - P_2V$$
$$= -(P_2 - P_1)V. \qquad (15\text{-}41)$$

The work-kinetic energy theorem of Eq. 15-38 now becomes

$$W^{net} = W^{grav} + W^{pd} = \Delta K.$$

Substituting from Eqs. 15-39, 15-40, and 15-41 yields

$$-\rho gV(y_2 - y_1) - V(P_2 - P_1) = \tfrac{1}{2}\rho V(v_2^2 - v_1^2).$$

This, after a slight rearrangement, matches Eq. 15-35, which we set out to prove.

READING EXERCISE 15-13: Water flows smoothly through the pipe shown in the figure, descending in the process. Rank the four numbered sections of pipe according to (a) the volume flow rate R_V through them, (b) the flow speed v through them, and (c) the water pressure P within them, greatest first.

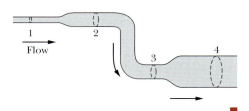

TOUCHSTONE EXAMPLE 15-9: Ethanol Flow

Ethanol of density $\rho = 791$ kg/m^3 flows smoothly through a horizontal pipe that tapers in cross-sectional area from $A_1 = 1.20 \times 10^{-3}$ m^2 to $A_2 = A_1/2$. The pressure difference between the wide and narrow sections of pipe is 4120 Pa. What is the volume flow rate R_V of the ethanol?

SOLUTION ■ One **Key Idea** here is that, because the fluid flowing through the wide section of pipe must entirely pass through the narrow section, the volume flow rate R_V must be the same in the two sections. Thus, from Eq. 15-28,

$$R_V = v_1A_1 = v_2A_2. \qquad (15\text{-}42)$$

However, with two unknown speeds, we cannot evaluate this equation for R_V.

A second **Key Idea** is that, because the flow is smooth, we can apply Bernoulli's equation. From Eq. 15-35, we can write

$$P_1 + \tfrac{1}{2}\rho v_1^2 + \rho gy = P_2 + \tfrac{1}{2}\rho v_2^2 + \rho gy, \qquad (15\text{-}43)$$

where subscripts 1 and 2 refer to the wide and narrow sections of pipe, respectively, and y is their common elevation. This equation hardly seems to help because it does not contain the desired volume flow R_V and it contains the unknown speeds v_1 and v_2.

However, there is a neat way to make it work for us: First, we can use Eq. 15-42 and the fact that $A_2 = A_1/2$ to write

$$v_1 = \frac{R_V}{A_1} \quad \text{and} \quad v_2 = \frac{R_V}{A_2} = \frac{2R_V}{A_1}. \qquad (15\text{-}44)$$

Then we can substitute these expressions into Eq. 15-43 to eliminate the unknown speeds and introduce the desired volume flow rate. Doing this and solving for R_V yield

$$R_V = A_1\sqrt{\frac{2(P_1 - P_2)}{3\rho}}. \qquad (15\text{-}45)$$

We still have a decision to make: We know that the pressure difference between the two sections is 4120 Pa, but does that mean that $P_1 - P_2$ is 4120 Pa or -4120 Pa? We could guess the former is true, or otherwise the square root in Eq. 15-45 would give us an imaginary number. Instead of guessing, however, let's try some reasoning. From Eq. 15-42 we see that speed v_2 in the narrow section (small A_2) must be greater than speed v_1 in the wider section (larger A_1). Recall that if the speed of a fluid increases as it travels along a horizontal path (as here), the pressure of the fluid must de-

crease. Thus, P_1 is greater than P_2, and $P_1 - P_2 = +4120$ Pa. Inserting this and known data into Eq. 15-45 gives

$$R_V = 1.20 \times 10^{-3} \text{ m}^2 \sqrt{\frac{(2)(4120 \text{ Pa})}{(3)(791 \text{ kg/m}^3)}}$$

$$= 2.24 \times 10^{-3} \text{ m}^3/\text{s}. \qquad \text{(Answer)}$$

Problems

SEC. 15-3 ■ PRESSURE AND DENSITY

1. Syringe Find the pressure increase in the fluid in a syringe when a nurse applies a force of 42 N to the syringe's circular piston, which has a radius of 1.1 cm.

2. Three Liquids Three liquids that will not mix are poured into a cylindrical container. The volumes and densities of the liquids are 0.50 L, 2.6 g/cm³; 0.25 L, 1.0 g/cm³; and 0.40 L, 0.80 g/cm³. What is the force on the bottom of the container due to these liquids? One liter $= 1 \text{ L} = 1000 \text{ cm}^3$. (Ignore the contribution due to the atmosphere.)

3. Office Window An office window has dimensions 3.4 m by 2.1 m. As a result of the passage of a storm, the outside air pressure drops to 0.96 atm, but inside the pressure is held at 1.0 atm. What net force pushes out on the window?

4. Front Tires You inflate the front tires on your car to 28 psi. Later, you measure your blood pressure, obtaining a reading of 120/80, the readings being in mm Hg. In countries using the metric system (which is to say, most of the world), these pressures are customarily reported in kilopascals (kPa). In kilopascals, what are (a) your tire pressure and (b) your blood pressure?

5. Fish A fish maintains its depth in fresh water by adjusting the air content of porous bone or air sacs to make its average density the same as that of the water. Suppose that with its air sacs collapsed, a fish has a density of 1.08 g/cm³. To what fraction of its expanded body volume must the fish inflate the air sacs to reduce its density to that of water?

6. Airtight Container An airtight container having a lid with negligible mass and an area of 77 cm² is partially evacuated. If a 480 N force is required to pull the lid off the container and the atmospheric pressure is 1.0×10^5 Pa, what is the air pressure in the container before it is opened?

7. Otto Von Guericke In 1654 Otto von Guericke, inventor of the air pump, gave a demonstration before the noblemen of the Holy Roman Empire in which two teams of eight horses could not pull apart two evacuated brass hemispheres. (a) Assuming that the hemispheres

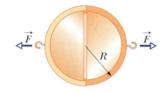

FIGURE 15-33 ■ Problem 7.

have thin walls, so that R in Fig. 15-33 may be considered both the inside and outside radius, show that the force $\vec{F}$ required to pull apart the hemispheres has magnitude $|\vec{F}| = \pi R^2 \Delta P$, where ΔP is the difference between the pressures outside and inside the sphere. (b) Taking R as 30 cm, the inside pressure as 0.10 atm, and the outside pressure as 1.00 atm, find the force mag-

nitude the teams of horses would have had to exert to pull apart the hemispheres. (c) Explain why one team of horses could have proved the point just as well if the hemispheres were attached to a sturdy wall.

SEC. 15-4 ■ GRAVITATIONAL FORCES AND FLUIDS AT REST

8. Hydrostatic Difference Calculate the hydrostatic difference in blood pressure between the brain and the foot in a person of height 1.83 m. The density of blood is $1.06 \times 10^3 \text{ kg/m}^3$.

9. Sewage Outlet The sewage outlet of a house constructed on a slope is 8.2 m below street level. If the sewer is 2.1 m below street level, find the minimum pressure difference that must be created by the sewage pump to transfer waste of average density 900 kg/m³ from outlet to sewer.

10. Phase Diagram Figure 15-34 displays the *phase diagram* of carbon, showing the ranges of temperature and pressure in which carbon will crystallize either as diamond or graphite. What is the minimum depth at which diamonds can form if the temperature at that depth is 1000°C and the rocks there have density 3.1 g/cm³? Assume that, as in a fluid, the pressure at any level is due to the gravitational force on the material lying above that level, and neglect variation of g with depth.

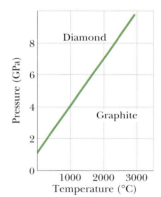

FIGURE 15-34 ■
Problem 10.

11. Swimming Pool A swimming pool has the dimensions 24 m × 9.0 m × 2.5 m. When it is filled with water, what is the force (resulting from the water alone) on (a) the bottom, (b) each short side, and (c) each long side? (d) If you are concerned with the possibility that the concrete walls and floor will collapse, is it appropriate to take the atmospheric pressure into account? Why?

12. Seawater (a) Assuming the density of seawater is 1.03 g/cm³, find the total weight of water on top of a nuclear submarine at a depth of 200 m if its (horizontal cross-sectional) hull area is 3000 m². (b) In atmospheres, what water pressure would a diver experience at this depth? Do you think that occupants of a damaged submarine at this depth could escape without special equipment?

13. Crew Members Crew members attempt to escape from a damaged submarine 100 m below the surface. What force must be applied to a pop-out hatch, which is 1.2 m by 0.60 m, to push it out at that depth? Assume that the density of the ocean water is 1025 kg/m³.

14. Barrel A cylindrical barrel has a narrow tube fixed to the top, as shown (with dimensions) in Fig. 15-35. The vessel is filled with water to the top of the tube. Calculate the ratio of the hydrostatic force on the bottom of the barrel to the gravitational force on the water contained inside the barrel. Why is that ratio not equal to one? (You need not consider the atmospheric pressure.)

15. Cylindrical Vessels Two identical cylindrical vessels with their bases at the same level each contain a liquid of density ρ. The area of each base is A, but in one vessel the liquid height is h_A, and in the other it is h_B. Find the work done by the gravitational force in equalizing the levels when the two vessels are connected.

16. Geological Features in analyzing certain geological features, it is often appropriate to assume that the pressure at some horizontal *level of compensation*, deep inside Earth, is the same over a large region and is equal to the pressure due to the gravitational force on the overlying material. Thus, the pressure on the level of compensation is given by the fluid pressure formula. This model requires, for one thing, that mountains have roots of continental rock extending into the denser mantle (Fig. 15-36). Consider a mountain 6.0 km high. The continental rocks have a density of 2.9 g/cm³, and beneath the continent the mantle has a density of 3.3 g/cm³. Calculate the depth D of the root. (*Hint:* Set the pressure at points a and b equal; the depth y of the level of compensation will cancel out.)

17. Ocean Figure 15-37 shows the juncture of ocean and continent. Find the depth h of the ocean using the level-of-compensation technique presented in Problem 16.

18. L-shaped Tank The L-shaped tank shown in Fig. 15-38 is filled with water and is open at the top. If $d = 5.0$ m, what are (a) the force on face A and (b) the force on face B due to the water?

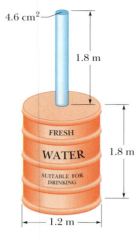

FIGURE 15-35 ■ Problem 14.

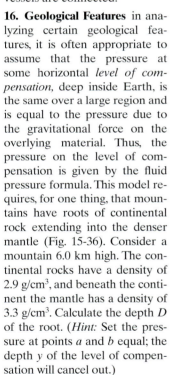

FIGURE 15-36 ■ Problem 16.

FIGURE 15-37 ■ Problem 17.

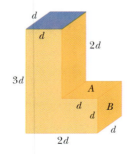

FIGURE 15-38 ■ Problem 18.

19. Water Stands Water stands at a depth D behind the vertical upstream face of a dam, as shown in Fig. I5-39. Let W be the width of the dam. Find (a) the net horizontal force on the dam from the gauge pressure of the water and (b) the net torque due to that force (and thus gauge pressure) about a line through O parallel to the width of the dam. (c) Find the moment arm of the net horizontal force about the line through O.

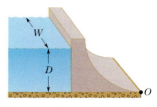

FIGURE 15-39 ■ Problem 19.

SEC. 15-5 ■ MEASURING PRESSURE

20. Lemonade To suck lemonade of density 1000 kg/m³ up a straw to a maximum height of 4.0 cm, what minimum gauge pressure (in atmospheres) must you produce in your mouth?

21. Atmosphere What would be the height of the atmosphere if the air density (a) were uniform and (b) decreased linearly to zero with height? Assume that at sea level the air pressure is 1.0 atm and the air density is 1.3 kg/m³.

SEC. 15-6 ■ PASCAL'S PRINCIPLE

22. Piston A piston of small cross-sectional area a is used in a hydraulic press to exert a small force $\vec{f}$ on the enclosed liquid. A connecting pipe leads to a larger piston of cross-sectional area A (Fig. 15-40). (a) What force magnitude $|\vec{F}|$ will the larger piston sustain without moving? (b) If the small piston has a diameter of 3.80 cm and the large piston one of 53.0 cm, what force magnitude on the small piston will balance a 20.0 kN force on the large piston?

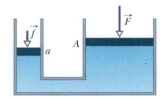

FIGURE 15-40 ■ Problems 22 and 23.

23. Hydraulic Press In the hydraulic press of Problem 22, through what distance must the large piston be moved to raise the small piston a distance of 0.85 m?

SEC. 15-7 ■ ARCHIMEDES' PRINCIPLE

24. A Boat Floats A boat floating in fresh water displaces water weighing 35.6 kN. (a) What is the weight of the water that this boat would displace if it were floating in salt water with a density of 1.10×10^3 kg/m³? (b) Would the volume of the displaced water change? If so, by how much?

25. Iron Anchor An iron anchor of density 7870 kg/m³ appears 200 N lighter in water than in air. (a) What is the volume of the anchor? (b) How much does it weigh in air?

26. Cubical Object In Fig. 15-41 a cubical object of dimensions $L = 0.600$ m on a side and with a mass of 450 kg is suspended by a rope in an open tank of liquid of

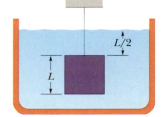

FIGURE 15-41 ■ Problem 26.

density 1030 kg/m³. (a) Find the magnitude of the total downward force on the top of the object from the liquid and the atmosphere, assuming that atmospheric pressure is 1.00 atm. (b) Find the magnitude of the total upward force on the bottom of the object. (c) Find the tension in the rope. (d) Calculate the magnitude of the buoyant force on the object using Archimedes' principle. What relation exists among all these quantities?

27. Block of Wood A block of wood floats in fresh water with two-thirds of its volume submerged. In oil the block floats with 0.90 of its volume submerged. Find the density of (a) the wood and (b) the oil.

28. Blimp A blimp is cruising slowly at low altitude, filled as usual with helium gas. Its maximum useful payload, including crew and cargo, is 1280 kg. The volume of the helium-filled interior space is 5000 m³. The density of helium gas is 0.16 kg/m³, and the density of hydrogen is 0.081 kg/m³. How much more payload could the blimp carry if you replaced the helium with hydrogen? (Why not do it?)

29. Hollow Sphere A hollow sphere of inner radius 8.0 cm and outer radius 9.0 cm floats half-submerged in a liquid of density 800 kg/m³. (a) What is the mass of the sphere? (b) Calculate the density of the material of which the sphere is made.

30. Dead Sea About one-third of the body of a person floating in the Dead Sea will be above the water line. Assuming that the human body density is 0.98 g/cm³, find the density of the water in the Dead Sea. (Why is it so much greater than 1.0 g/cm³?)

31. Iron Shell A hollow spherical iron shell floats almost completely submerged in water. The outer diameter is 60.0 cm, and the density of iron is 7.87 g/cm³. Find the inner diameter.

32. Wood with Lead A block of wood has a mass of 3.67 kg and a density of 600 kg/m³. It is to be loaded with lead so that it will float in water with 0.90 of its volume submerged. What mass of lead is needed (a) if the lead is attached to the top of the wood and (b) if the lead is attached to the bottom of the wood? The density of lead is 1.13×10^4 kg/m³.

33. Iron Casting An iron casting containing a number of cavities weighs 6000 N in air and 4000 N in water. What is the total volume of all the cavities in the casting? The density of iron (that is, a sample with no cavities) is 7.87 g/cm³.

34. Density of Brass Assume the density of brass weights to be 8.0 g/cm³ and that of air to be 0.0012 g/cm³. What percent error arises from neglecting the buoyancy of air in weighing an object of mass m and density ρ on a beam balance?

35. Slab of Ice (a) What is the minimum area of the top surface of a slab of ice 0.30 m thick floating on fresh water that will hold up an automobile of mass 1100 kg? (b) Does it matter where the car is placed on the block of ice?

36. Three Children Three children, each of weight 356 N, make a log raft by lashing together logs of diameter 0.30 m and length 1.80 m. How many logs will be needed to keep them afloat in fresh water? Take the density of the logs to be 800 kg/m³.

37. Metal Rod A metal rod of length 80 cm and mass 1.6 kg has a uniform cross-sectional area of 6.0 cm². Due to a nonuniform density, the center of mass of the rod is 20

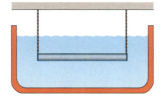

FIGURE 15-42 ■
Problem 37.

cm from one end of the rod. The rod is suspended in a horizontal position in water by ropes attached to both ends (Fig. 15-42). (a) What is the tension in the rope closer to the center of mass? (b) What is the tension in the rope farther from the center of mass? (*Hint:* The buoyancy force on the rod effectively acts at the rod's geometric center.)

38. Floating Car A car has a total mass of 1800 kg. The volume of air space in the passenger compartment is 5.00 m³. The volume of the motor and front wheels is 0.750 m³, and the volume of the rear wheels, gas tank, and trunk is 0.800 m³; water cannot enter these areas. The car is parked on a hill; the handbrake cable snaps and the car rolls down the hill into a lake (Fig. 15-43). (a) At first, no water enters the passenger compartment. How much of the car, in cubic meters, is below the water surface with the car floating as shown? (b) As water slowly enters, the car sinks. How many cubic meters of water are in the car as it disappears below the water surface? (The car, with a heavy load in the trunk, remains horizontal.)

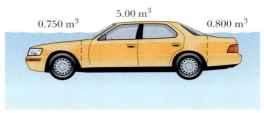

0.750 m³   5.00 m³   0.800 m³

FIGURE 15-43 ■ Problem 38.

SEC. 15-9 ■ THE EQUATION OF CONTINUITY

39. Garden Hose A garden hose with an internal diameter of 1.9 cm is connected to a (stationary) lawn sprinkler that consists merely of an enclosure with 24 holes, each 0.13 cm in diameter. If the water in the hose has a speed of 0.91 m/s, at what speed does it leave the sprinkler holes?

40. Two Streams Two streams merge to form a river. One stream has a width of 8.2 m, depth of 3.4 m, and current speed of 2.3 m/s. The other stream is 6.8 m wide and 3.2 m deep, and flows at 2.6 m/s. The width of the river is 10.5 m, and the current speed is 2.9 m/s. What is its depth?

41. Flooded Basement Water is pumped steadily out of a flooded basement at a speed of 5.0 m/s through a uniform hose of radius 1.0 cm. The hose passes out through a window 3.0 m above the waterline. What is the power of the pump?

42. Water Pipe The water flowing through a 1.9 cm (inside diameter) pipe flows out through three 1.3 cm pipes. (a) If the flow rates in the three smaller pipes are 26, 19, and 11 L/min, what is the flow rate in the 1.9 cm pipe? (b) What is the ratio of the speed of water in the 1.9 cm pipe to that in the pipe carrying 26 L/min?

SEC. 15-11 ■ BERNOULLI'S EQUATION

43. Pipe Increases in Area Water is moving with a speed of 5.0 m/s through a pipe with a cross-sectional area of 4.0 cm². The water gradually descends 10 m as the pipe increases in area to 8.0 cm². (a) What is the speed at the lower level? (b) If the pressure at the upper level is 1.5×10^5 Pa, what is the pressure at the lower level?

44. Torpedoes Models of torpedoes are sometimes tested in a horizontal pipe of flowing water, much as a wind tunnel is used to test model airplanes. Consider a circular pipe of internal diameter 25.0 cm and a torpedo model, aligned along the axis of the pipe, with a diameter of 5.00 cm. The model is to be tested with water flowing past it at 2.50 m/s. (a) With what speed must the water flow in the part of the pipe that is unconstricted by the model? (b) What will the pressure difference be between the constricted and unconstricted parts of the pipe?

45. Basement Pipe A water pipe having a 2.5 cm inside diameter carries water into the basement of a house at a speed of 0.90 m/s and a pressure of 170 kPa. If the pipe tapers to 1.2 cm and rises to the second floor 7.6 m above the input point, what are (a) the speed and (b) the water pressure at the second floor?

46. Water Intake A water intake at a pump storage reservoir (Fig. 15-44) has a cross-sectional area of 0.74 m². The water flows in at a speed of 0.40 m/s. At the generator building 180 m below the intake point, the cross-sectional area is smaller than at the intake and

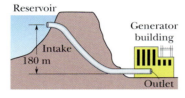

FIGURE 15-44 ■ Problem 46.

the water flows out at 9.5 m/s. What is the difference in pressure, in megapascals, between inlet and outlet?

47. Large Area A tank of large area is filled with water to a depth $D = 0.30$ m. A hole of cross-sectional area $A = 6.5$ cm² in the bottom of the tank allows water to drain out. (a) What is the rate at which water flows out, in cubic meters per second? (b) At what distance below the bottom of the tank is the cross-sectional area of the stream equal to one-half the area of the hole?

48. Air Flows Air flows over the top of an airplane wing of area A with speed $|\vec{v}_{top}|$ and past the underside of the wing (also of area A) with speed $|\vec{v}_{under}|$. Show that in this simplified situation Bernoulli's equation predicts that the magnitude $|\vec{L}|$ of the upward lift force on the wing will be

$$|\vec{L}| = \tfrac{1}{2}\rho A(v_{top}^2 - v_{under}^2),$$

where ρ is the density of the air.

49. Airplane Wing If the speed of flow past the lower surface of an airplane wing is 110 m/s, what speed of flow over the upper surface will give a pressure difference of 900 Pa between upper and lower surfaces? Take the density of air to be 1.30×10^{-3} g/cm³, and see Problem 48.

50. Two Tanks Suppose that two tanks, A and B, each with a large opening at the top, contain different liquids. A small hole is made in the side of each tank at the same depth d below the liquid surface, but the hole in tank A has half the cross-sectional area of the hole in tank B. (a) What is the ratio ρ_A/ρ_B of the densities of the liquids if the mass flow rate is the same for the two holes? (b) What is the ratio of the volume flow rates from the two tanks? (c) To what height above the hole in tank B should liquid be added or drained to equalize the volume flow rates?

51. Water in the Horizontal Pipe In Fig, 15-45, water flows through a horizontal pipe, and then out into the atmosphere at a speed of 15 m/s. The diameters of the left and right sections of the pipe are

FIGURE 15-45 ■
Problem 51.

5.0 cm and 3.0 cm, respectively. (a) What volume of water flows into the atmosphere during a 10 min period? In the left section of the pipe, what are (b) the speed v_B, and (c) the gauge pressure?

52. Beverage Keg An opening of area 0.25 cm² in an otherwise closed beverage keg is 50 cm below the level of the liquid (of density 1.0 g/cm³) in the keg. What is the speed of the liquid flowing through the opening if the gauge pressure in the air space above the liquid is (a) zero and (b) 0.40 atm?

53. Dam The fresh water behind a reservoir dam is 15 m deep. A horizontal pipe 4.0 cm in diameter passes through the dam 6.0 m below the water surface, as shown in Fig. 15-46. A plug secures the pipe opening. (a) Find the magnitude of the frictional force between plug and pipe wall. (b) The plug is removed. What volume of water flows out of the pipe in 3.0 h?

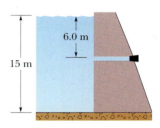

FIGURE 15-46 ■
Problem 53.

54. Filled Tank A tank is filled with water to a height H. A hole is punched in one of the walls at a depth h below the water surface (Fig. 15-47). (a) Show that the distance x from the base of the tank to the point at which the resulting stream strikes the floor is given by $x = 2\sqrt{h(H - h)}$. (b) Could a hole be punched at another depth to produce a second stream that would have the same range? If so, at what depth? (c) At what depth should the hole be placed to make the emerging stream strike the ground at the maximum distance from the base of the tank?

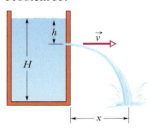

FIGURE 15-47 ■
Problem 54.

55. Venturi Meter A *venturi meter* is used to measure the flow speed of a fluid in a pipe. The meter is connected between two sections of the pipe (Fig. 15-48); the cross-sectional area A of the entrance and exit of the meter matches the pipe's cross-sectional area. At the entrance and exit, the fluid flows through the pipe with speed $v_A = |\vec{v}_A|$. But it flows through a narrow "throat" of cross-sectional area B with speed $v_B = |\vec{v}_B|$. A manometer connects the wider portion of the meter to the narrower portion. The change in the fluid's speed is accompanied by a change ΔP in the fluid's pressure, which

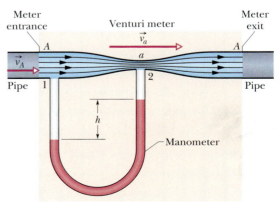

FIGURE 15-48 ■ Problems 55 and 56.

causes a height difference h of the liquid in the two arms of the manometer. (Here ΔP means pressure in the throat minus pressure in the pipe.) (a) By applying Bernoulli's equation and the equation of continuity to points 1 and 2 in Fig. 15-48, show that

$$\vec{v}_A = \sqrt{\frac{2B^2 \Delta P}{\rho(B^2 - A^2)}},$$

where ρ is the density of the fluid. (b) Suppose that the fluid is fresh water, that the cross-sectional areas are 64 cm² in the pipe and 32 cm² in the throat, and that the pressure is 55 kPa in the pipe and 41 kPa in the throat. What is the rate of water flow in cubic meters per second?

56. Venturi Tube Consider the venturi tube of Problem 55 and Fig. 15-48 without the manometer. Let A equal $5a$. Suppose that the pressure P_1 at A is 2.0 atm. Compute the values of (a) $|\vec{V}_A|$ at A and (b) $|\vec{v}_a|$ at a that would make the pressure P_2 at a equal to zero. (c) Compute the corresponding volume flow rate if the diameter at A is 5.0 cm. The phenomenon that occurs at a when P_2 falls to nearly zero is known as cavitation. The water vaporizes into small bubbles.

57. Pitot Tube A pitot tube (Fig. 15-49) is used to determine the airspeed of an airplane. It consists of an outer tube with a number of small holes B (four are shown) that allow air into the tube; that tube is connected to one arm of a U-tube. The other arm of the U-tube is connected to hole A at the front end of the device, which points in the direction the plane is headed. At A the air becomes stagnant so that $v_A = 0$. At B, however, the speed of the air presum-

ably equals the airspeed v of the aircraft. (a) Use Bernoulii's equation to show that

$$v = \sqrt{\frac{2\rho g h}{\rho_{air}}},$$

where ρ is the density of the liquid in the U-tube and h is the difference in the fluid levels in that tube. (b) Suppose that the tube contains alcohol and indicates a level difference h of 26.0 cm. What is the plane's speed relative to the air? The density of the air is 1.03 kg/m³ and that of alcohol is 810 kg/m³.

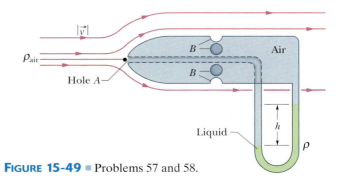

FIGURE 15-49 ■ Problems 57 and 58.

58. High-Altitude Aircraft A pitot tube (see Problem 57) on a high-altitude aircraft measures a differential pressure of 180 Pa. What is the airspeed if the density of the air is 0.031 kg/m³?

Additional Problems

59. Pool Filling You have been asked to review plans for a swimming pool in a new hotel. The water is to be supplied to the hotel by a horizontal main pipe of radius $R_1 = 6.00$ cm, with water under pressure of 2.00 atm. A vertical pipe of radius $R_2 = 1.00$ cm is to carry the water to a height of 9.40 m, where the water is to pour out freely into a square pool of width 10.0 m and (proposed) water depth of 2.00 m. (a) How much time will be required to fill the pool? (b) If more than a few days is considered unacceptable and less than a few hours is considered dangerous, is the filling time acceptable and safe?

60. Hydraulic Engineers Figure 15-50 shows two sections of an old pipe system that runs through a hill. On each side of the hill, the pipe radius is 2.00 cm. However,

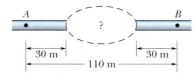

FIGURE 15-50 ■ Problem 60.

the radius of the pipe inside the hill is no longer known. To determine it, hydraulic engineers first establish that water flows through the left-hand and right-hand sections at 2.50 m/s. Then they release a dye in the water at point A and find that it takes 88.8 s to reach point B. What is the radius (or average radius) of the pipe within the hill?

61. Floating and Sinking Suppose you have the following collection of objects: a pencil, a coin, an empty plastic box for CDs with its edges taped shut, the same box opened up, a needle, an un-

opened can of soda pop, and an empty can of soda pop. Which of these objects do you expect will float on water and which will sink? Will it make a difference if you carefully place the object with its largest surface on the surface of the water? In which cases? Discuss the criteria you come up with, explaining carefully why you decided on each one and why it plays a role. After you have written your answer, perform the experiments and compare your results with your predictions.

62. Balloon in a Car Explain why a helium balloon in a closed automobile moves to the front of the car when the car accelerates, whereas the passengers feel pushed backwards. Discuss this in terms of the physics you have learned.

63. At the Pool If an inflated beach ball is placed beneath the surface of a pool and released, it shoots upward, out of the water. Explain why.

64. The Meteor and the Dolphin The curator of a science museum is transporting a chunk of meteor iron (i.e., a piece of iron that fell from the sky—see Fig. 15-51a) from one part of the museum to another. Since the chunk of iron weighs 250 lb and is too big for her to lift by herself, she is using a handtruck (see Fig. 15-51b). While passing through the marine mammals section of the museum, she accidentally hits a bump and the meteorite tips off the handtruck and falls into the dolphin pool. Fortunately, the iron doesn't hit a dolphin, but it quickly sinks to the bottom. "Rats!" she cries. Unfortunately, the meteorite has many sharp edges and she is worried that the dolphins, curious creatures that they are, will come to inspect it and be cut when they rub against it. She wants to get it up out of the pool as quickly as possible. Fortunately, the meteorite has lots of holes in it and there are ropes with hooks on one end lying around. If she could get a hook into one of the holes, she might be able to pull it up to the top, tie the rope around a post, and lever it out with the handtruck. Unfortunately, she remembers that the meteorite is too heavy for her to lift. (a) Will the fact that the meteorite is in the pool under water make it harder or easier for her to lift with the rope? Explain. (b) The meteorite is sitting on the concrete bottom of the pool. Is the force the meteorite exerts on the bottom bigger or smaller than the force it would exert if the pool had no water in it? Explain. (c) Can she lift the meteorite? Calculate how much force she would have to exert on a rope hooked to the meteorite to pull it up from the bottom of the pool. She can lift about 100 pounds, the pool is 12 feet deep, and the density of iron is about 8000 kg/m³.

FIGURE 15-51(*a*) ▪ Problem 64..

FIGURE 15-51(*b*) ▪ Problem 64.

65. Pushing Iron For each of the following partial sentences, indicate whether they are correctly completed by the symbol corresponding to the phrase *greater than* (>), *less than* (<), or *the same as* (=). (a) A chunk of iron is sitting on a table. It is then moved from the table into a bucket of water sitting on the table. The iron now rests on the bottom of the bucket. The force the bucket exerts on the block when the block is sitting on the bottom of the bucket is _____ the force that the table exerted on the block when the block was sitting on the table. (b) A chunk of iron is sitting on a table. It is then moved from the table into a bucket of water sitting on the table. The iron now rests on the bottom of the bucket. The total force on the block when it is sitting on the bottom of the bucket is _____ it was on the table. (c) A chunk of iron is sitting on a table. It is then covered by a bell jar, which has a nozzle connected to a vacuum pump. The air is extracted from the bell jar. The force the table exerts on the block when the block is sitting in a vacuum is _____ the force that the table exerted on the block when the block was sitting in the air. (d) A chunk of iron is sitting on a scale. The iron and the scale are then both immersed in a large vat of water. After being immersed in the water, the scale reading will be _____ the scale reading when they were simply sitting in the air. (Assume the scale would read zero if nothing were sitting on it, even when it is underwater.)

66. The Three-Vase Puzzle* Water is poured to the same level in each of the three vessels shown in Fig. 15-52. Each vessel has the same base area. Since the water is to the same depth in each vessel, each will have the same pressure at the bottom. Since the area and pressure are the same, each liquid should exert the same force on the base of the vessel. Yet, if the vessels are weighed, three different

values are obtained. (The one in the center clearly holds less liquid than the one at the left, so it must weigh less.) How can you justify this apparent contradiction?

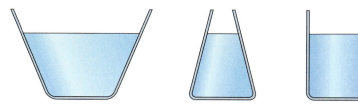

FIGURE 15-52 ▪ Problem 66.

67. Hanging Blocks† Three cubical blocks of equal volume are suspended from strings. Blocks *A* and *B* have the same mass and block *C* has less mass. Each block is lowered into a fish tank and they hang at rest as shown in Fig. 15-53. (a) Is the force exerted by the water on the top surface of block *A* greater than, less than, or equal to the force exerted by the water on the top surface of block *B*? Explain. (b) Is the force exerted by the water on the top surface of block *A* greater than, less than, or equal to the force exerted by the water on the top surface of block *C*? Explain. (c) Is the force exerted on the water by block *C* greater than, less than or equal to the force exerted on the water by block *A*? Explain. (d) Rank the buoyant forces acting on the three blocks from largest to smallest. If any buoyant forces are equal, indicate that explicitly. Explain.

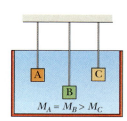

FIGURE 15-53 ▪ Problem 67.

68. Floating Blocks‡ Figure 15-54 shows five blocks increasing in mass from block *A* to block *E* as indicated. The blocks have equal volumes but different masses. The blocks are placed in an aquarium tank filled with water and blocks *B* and *E* come to rest as shown in Fig. 15-54. Sketch on the figure where you would expect blocks *A*, *C*, and *D* to come to rest. (The differences in mass between successive blocks is significant—not just a tiny amount.)

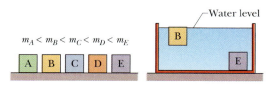

FIGURE 15-54 ▪ Problem 68.

*From A. Arons, *A Guide to Introductory Physics Teaching* (New York: John Wiley, 1990).

†From M. E. Loverude, "Investigation of Student Understanding of Hydrostatics and Thermal Physics and the Underlying Concepts from Mechanics," Ph.D. thesis, University of Washington, 1999.

‡From M. E. Loverude, "Investigation of Student Understanding of Hydrostatics and Thermal Physics and the Underlying Concepts from Mechanics," Ph.D. thesis, University of Washington, 1999.

16 | Oscillations

On September 19, 1985, seismic waves from an earthquake that originated along the west coast of Mexico caused terrible and widespread damage in Mexico City, about 400 km from the origin.

Why did the seismic waves cause such extensive damage in Mexico City but almost none on the way there?

The answer is in this chapter.

16-1 Periodic Motion: An Overview

Any measurable quantity that repeats itself at regular time intervals is defined as undergoing **periodic** behavior. We are surrounded by systems with quantities that vary periodically. The systems with periodic behavior that are most familiar involve obvious mechanical oscillations or motions. There are swinging chandeliers, boats bobbing at anchor, and the surging pistons in the engines of cars.

The motions associated with some periodic behavior are not obvious. For example, we cannot see the oscillations of the air molecules that transmit the sensation of sound, the oscillations of the atoms in a solid that convey the sensation of temperature, and the oscillations of the electrons in the antennas of radio and TV transmitters that convey information. Some examples of periodic changes are shown in Figs. 16-1 and 16-2. They include electrical signals associated with human heart beats and the air pressure changes that occur when musical instruments are played.

It is obvious from looking at Figs. 16-1 and 16-2 that the variations of electrical signals from the heart and the sound pressure from the trumpet are both periodic but quite complex. On the other hand, the oscillation of air pressure caused by the flute is much simpler. In fact the flute pattern looks like the graph of a sine or cosine function. If the periodic variation of a physical quantity over time has the shape of a sine (or cosine) function, we call it a **sinusoidal oscillation.**

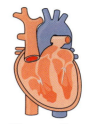

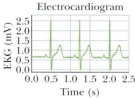

FIGURE 16-1 ■ An electrocardiogram showing the periodic pattern of electrical signals that drive human heart beats. Data recorded with a computer data acquisiton EKG sensor. (Courtesy of Vernier Software and Technology.)

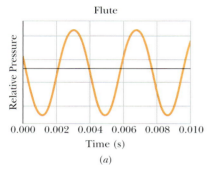

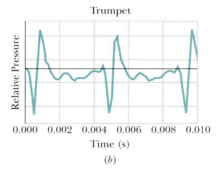

FIGURE 16-2 ■ The disturbance of air molecules causes variations in air pressure near musical instruments. The pattern of these pressure variations repeats itself at regular time intervals so that the sound is periodic. The pressure variations that are proportional to the voltage output of a small microphone can be recorded with a computer data acquisition system. (a) A sustained note from a flute and (b) from a trumpet. (Data courtesy of Vernier Software and Technology.)

Sinusoidal oscillations are surprisingly common and learning about them helps us understand more complex oscillations. For this reason we begin this chapter by exploring the mathematics of sinusoidal oscillations and how oscillations can be related to the uniform circular motion we studied in Chapters 5, 11, and 12. Mastering the mathematical description of sinusoidal motion is critical to acquiring a full understanding of periodic physical systems. It is also vital to obtaining a full appreciation of the transmission of both mechanical and sound waves treated in Chapters 17 and 18.

As you will see, physicists and engineers refer to the sinusoidal motions of particles in mechanical systems as **simple harmonic motion (or SHM).** In fact, most of the chapter is devoted to understanding how certain forces found in our everyday surroundings cause the sinusoidal oscillations that we call SHM.

Although your study of simple harmonic motion will enhance your understanding of mechanical systems, it is also vital to understanding the topics in waves, electricity, magnetism, and light encountered in Chapters 30–37. Finally, a knowledge of SHM provides a basis for understanding modern physics, including the wave nature of the light and how atoms and nuclei absorb and emit energy.

16-2 The Mathematics of Sinusoidal Oscillations

Sinusoidal Oscillations and Uniform Circular Motion

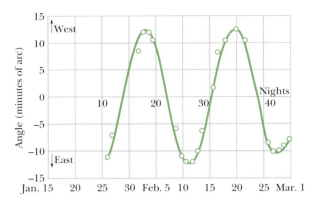

FIGURE 16-3 ■ The angle between Jupiter and its moon Callisto as seen from Earth. The circles are based on Galileo's 1610 measurements. The curve is a best fit, strongly suggesting sinusoidal motion. At Jupiter's mean distance, 10 minutes of arc corresponds to about 2×10^6 km. (Adapted from A. P. French, *Newtonian Mechanics*, W. W. Norton, New York, 1971, p. 288.)

In 1610, Galileo used his newly constructed telescope to discover the four principal moons of Jupiter. Over weeks of observation, each moon seemed to him to be moving back and forth relative to the planet in what today we would call sinusoidal motion; the disk of the planet was the midpoint of the motion. The record of Galileo's observations, written in his own hand, is still available. A. P. French of MIT used Galileo's data to work out the position of the moon Callisto relative to Jupiter. In the results shown in Fig. 16-3, the circles are Galileo's data which looks sinusoidal. The curve shows the best fit of a sinusoidal function to the data. A full oscillation takes about 16.8 days, as can be seen on the plot.

Actually, Callisto moves with essentially constant speed in an essentially circular orbit around Jupiter. The moon's true motion—far from being sinusoidal—is uniform circular motion. What Galileo saw—and what you can see with a good pair of binoculars and a little patience—is the projection of this uniform circular motion on a line in the plane of the motion. We are led by Galileo's remarkable observations to the conclusion that the sinusoidal motion he observed is actually uniform circular motion viewed edge-on. In more formal language:

> The projection of uniform circular motion on a diameter of the circle in which this motion occurs is sinusoidal.

We can explore the relationship between the sinusoidal oscillations and uniform circular motion that we studied in Chapter 5 more carefully using an everyday object instead of a moon that orbits around a distant planet. Consider a spot on a disk that is rotating about an axis at a constant rotational velocity shown in Fig. 16-4. A graph of the spot's vertical displacement x versus time is shown in Fig. 16-5. For this case we have chosen to point the x axis up and the y axis to the right. If we take a series of side views of the disk (so you only see one of the dimensions it is moving in), the projection of the spot on the x axis as a function of time gives us a sinusoidal graph as shown in Fig. 16-4.

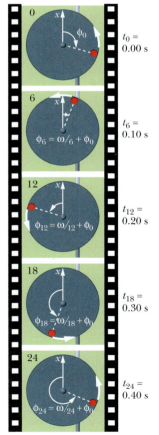

FIGURE 16-4 ■ This selection of frames shows the positions of a spot on a rotating disk every 1/10th of a second. They represent every 6th frame of a video sequence recorded at 60 frames/second. The angle the spot makes with respect to the chosen x axis starts out negative and then increases at a constant rate. *Note*: In order to tie in with Fig. 16-4 we have chosen to orient the x axis vertically.

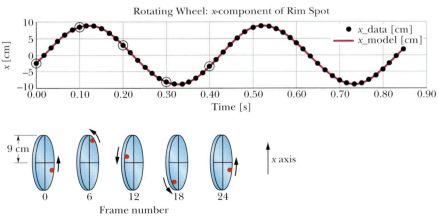

FIGURE 16-5 ■ The motion of a disk with a constant rotational velocity was recorded at 60 frames a second. A plot of the x-component of a spot on the disk has a sinusoidal shape. Thus it oscillates with SHM. The location of the spot on our sideways depictions of the disk is shown below the graph for every 6th frame. *Note:* In order to use conventional polar coordinates to relate the spot on the disk with the x-component of its position, we have chosen the x axis to be vertical.

Period and Frequency

As part of our study of uniform circular motion in Section 5-7 we introduced the idea of a **period** as the time it takes an object rotating about an axis at a regular rate to complete a single revolution. For example, the video frames of a rotating disk in Fig. 16-5 show that it has a period of 0.40 s. In Fig. 16-4, we see that the projection of the spot on our chosen x axis appears to oscillate up and down with a period of $T = 0.40$ s. Thus, the time for one oscillation of the spot's x-component is the same as the period of the rotation of the disk.

Whereas the period tells us the time for one rotation or oscillation, **frequency** is a related quantity that tells us how many oscillations or cycles there are in a given time. For example, we can see from the Fig. 16-4 graph that the spot has a frequency of 2.5 oscillations each second because that's the number of complete oscillations that would occur if we had taken data for 1 s rather than for only 0.85 s. The symbol for frequency is f, and its SI unit is the hertz (abbreviated Hz). Alternative names and symbols for the hertz include

$$1 \text{ hertz} = 1 \text{ Hz} = 1 \text{ cycle/second} = 1 \text{ oscillation/second}. \qquad (16\text{-}1)$$

Clearly, when the period of oscillation is very short there are many more oscillations in a second so the frequency goes up. The converse is true also; when the period is long, the frequency goes down. In fact, we see that the period and frequency of the oscillations shown in Fig. 16-4 are inversely related to each other. This inverse relationship holds in general, so that

$$T = \frac{1}{f}. \qquad (16\text{-}2)$$

The Equation Describing Sinusoidal Motion

We have rather glibly described the graphs shown in Figs. 16-2a and 16-4 as representing sinusoidal functions. We do this because the graphs look like that of a sine or cosine as a function of angle. Recall that the cosine of an angle, ϕ, is defined as the ratio of the distance of a point of interest from the y axis, denoted as x, and the magnitude of the distance of the point from the origin denoted as r as shown in Fig. 16-6. The sine function is similarly defined in terms of a ratio involving the distance from the x axis:

$$\cos \phi \equiv \frac{x}{r} \qquad (16\text{-}3)$$

and

$$\sin \phi \equiv \frac{y}{r}. \qquad (16\text{-}4)$$

In considering rotational positions, we continue the convention of describing angles in radians (or rads) used in Chapters 11 and 12. Although it would be possible to use degrees, radian measure is required if we want to take derivatives of mathematical functions that involve angles. Recall that in Eq. 11-1 the magnitude of the radian is defined as the magnitude of the ratio of the arc length s of a rotating object and the perpendicular distance from its axis of rotation ($|\phi| = s/r$) to the object (or spot). As the spot moves through a complete cycle, it sweeps out an arc length of $s = 2\pi r$ and thus an angle of $(2\pi r)/r = 2\pi$ radians. Since 2π radians $= 360°$, we can convert from radians to degrees by multiplying by the factor

$$\phi \text{ (deg)} = \left(\frac{180°}{\pi \text{ rad}}\right) \phi \text{ (rad)} \qquad \text{(conversion from radians to degrees)}.$$

Obviously we divide by the factor to convert to radians from degrees.

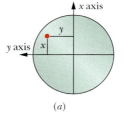

(a)

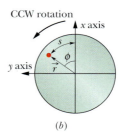

(b)

FIGURE 16-6 ■ The location of any spot on a disk can be described in either (a) Cartesian or (b) polar coordinates. Rotational position is denoted as ϕ. A counterclockwise arc from the axis is defined as a positive ϕ. *Note*: In order to coordinate with Fig. 16-4, we have also chosen the x axis to be vertical here.

How do our definitions of the sine and cosine functions in Eq. 16-3 and Eq. 16-4 lead to the graph shapes shown in Figs. 16-2a and 16-4? Let's use the cosine function as an example. In Fig. 16-5, a spot on a disk is turning with a constant rotational velocity ω. If we denote the initial angular position of the disk spot when $t = 0$ s as ϕ_0, then the angular position increases at a constant rate according to the equation $\phi(t) = \omega t + \phi_0$. This is shown is Fig. 16-5, in which the initial angular position ϕ_0 is taken to be negative since it is marked as being in a clockwise direction relative to the x axis. As time goes on the angular position $\phi(t)$ increases, passes through zero, and becomes positive.

Let us think about what happens in Cartesian coordinates. How does the x-component of the vector, $\vec{r}$, shown in Fig 16-6 vary over time for a counterclockwise rotation? In the time period where $\phi(t)$ is near zero the value of the x-component, denoted as x, does not change very rapidly but after a quarter-turn near $\phi(t) = \pi/2$ the value of x is changing very rapidly. The rate of change of x slows down again near $\phi(t) = \pi$. Between π and 2π, x is negative but its rate of change speeds up and slows down again. This clearly leads to a sinusoidal graph shape that goes on and on as the disk spot turns round and round.

In general for a sinusoidal motion, the value of the x-component of the spot as a function of time *can be described by either a sine or cosine function,* depending on which axis the angle is measured from. The values of x taken relative to the origin are typically called the displacement. Using the cosine function we find that the sinusoidal variation over time of the **displacement,** x, can be represented by the equation

$$x(t) = X \cos(\omega t + \phi_0) \quad \text{(sinusoidal displacement)}, \quad (16\text{-}5)$$

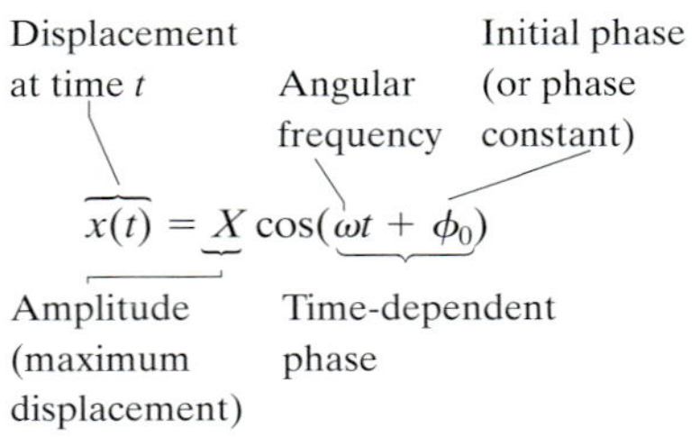

where X, ω, and ϕ_0 are constants.

When a sine function is shifted left by 90° or $+\frac{\pi}{2}$rad it looks exactly like a cosine function. So an alternate, equally viable equation for displacement $x(t)$ would be $x(t) = X \sin(\omega t + \phi_0')$, where $\phi_0' = (\phi_0 + \frac{\pi}{2})$.

For convenience the quantities that determine the shape of the graph of Eq. 16-5 are named and displayed in Fig. 16-7. X is defined as the **amplitude** of the x-component motion. It is a positive constant whose value represents the magnitude of the maximum displacement of the particle in either direction from its so-called equilibrium value. The cosine function in Eq. 16-5 varies between the limits ± 1, so the displacement $x(t)$ varies between the limits $\pm X$. For example, in Fig. 16-4 the amplitude of the cosine curve is obviously 9 cm. This is also the maximum distance from the axis of rotation of the disk to the spot.

The constant ϕ_0 is called the **initial phase.** It is also sometimes called the *phase constant* or *phase angle.* The value of the initial phase ϕ_0 allows us to calculate the magnitude of the displacement of the x-component, denoted as $x(0)$, at the starting time $t = 0$ s. Since the expression $\omega t = 0$ rad when $t = 0$ s, we get $x(0) = X \cos(\phi_0)$ or $\phi_0 = \pm \cos^{-1}(x(0)/X)$. Knowing the initial phase and displacement allows us to determine the initial angular velocity. The initial phase plays the same role as the x_1 or y_1 terms in the kinematic equations because it determines the initial value of the function $x(t)$.

$\phi(t) = \omega t + \phi_0$ is defined as the **time-dependent phase** of $x(t)$. Some time-dependent phases for the rotating disk are shown in Fig. 16-4. Each "frame" shown consists of every 6th frame of a more complete set of video images. The phases in the selection of frames shown in the figure are denoted as ϕ_0, $\omega t_7 + \phi_0$, $\omega t_{13} + \phi_0$, and so on, where the corresponding times are $t_1 = 0.00$ s, $t_6 = 0.10$ s, $t_{12} = 0.30$ s.

We find that if X is a maximum when $t = 0$ s the initial phase is either $\phi_0 = 0$ rad, $\pm 2\pi$ rad, $\pm 4\pi$ rad, . . ., and so on. This is because in this case $x(0) = X \cos(0) = X$. In other words, whenever the initial phase is a multiple of 2π radians, the displacement is a maximum at $t = 0$ s. For simplicity, in the $x(t)$ plots in Fig. 16-8a the initial phase (or phase constant) ϕ_0 has been set to zero radians.

Displacement   Initial phase
at time t Angular   (or phase
        frequency constant)

$$\overline{x(t)} = \underline{X} \cos(\underline{\omega t + \phi_0})$$

Amplitude   Time-dependent
(maximum   phase
displacement)

FIGURE 16-7 ■ A handy reference to the quantities in Eq. 16-5 for simple harmonic motion.

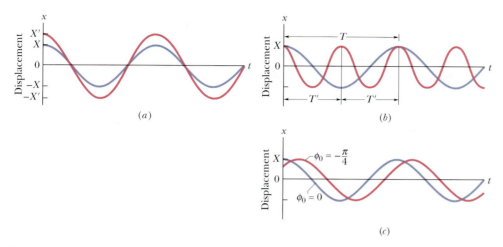

FIGURE 16-8 ■ Graphs of two sinusoidal motions. In each case, the blue curve is obtained from Eq. 16-5 with $\phi_0 = 0$ rad. (a) The red curve differs from the blue curve *only* in that its amplitude X' is greater (the red curve extremes of displacement are higher and lower). (b) The red curve differs from the blue curve *only* in that its period is $T' = T/2$ (the red curve is compressed horizontally). (c) The red curve differs from the blue curve *only* in that $\phi_0 = -\pi/4$ rad rather than zero (the negative value of ϕ_0 shifts the red curve to the right).

In describing sinusoidal motion, the constant ω is known as the **angular frequency** of the motion. When the sinusoidal equation represents the projection of a spot on a steadily rotating object along an axis, *the rotational velocity of the object and the angular frequency of the projection are identical in value.* This is certainly the case in Fig. 16-4.

The SI unit of angular frequency is the radian per second. Figure 16-8 compares $x(t)$ for two sinusoidal motions that differ either in amplitude, in period (and thus in frequency and angular frequency), or in initial phase.

We can use the fact that the rotational velocity of a spinning object and the angular frequency of a corresponding oscillation are identical to derive an important relationship. In particular, we can relate the angular frequency, ω, of an oscillating object to its oscillation frequency f. The derivation goes as follows: A rotating object undergoing uniform circular motion sweeps through an angle of 2π radians in a single period T so that its rotational velocity is given by $\omega = 2\pi/T$. But since the frequency is inversely proportional to the period (that is, since $f = 1/T$), it is obvious that

$$\omega = \frac{2\pi}{T} = 2\pi f. \tag{16-6}$$

READING EXERCISE 16-1: Although it is not conventional to do so, the equation $x(t) = X' \sin[\omega' t + \phi'_0]$ can also be used to describe sinusoidal motion. Suppose the same motion has been described by both the cosine function in Eq. 16-5 and by the sine function shown here. Consider the amplitude, angular frequency, and initial phase associated with the motion. Which factors stay the same? Which will change? Explain. *Hint*: You may want to think about the spot on a disk as it undergoes uniform circular motion. ■

READING EXERCISE 16-2: A particle undergoes sinusoidal oscillations of period T (such as the curve with a value of $+X$ at $t = 0$ s in Fig. 16-8a). Assume the particle is at $-X$ at time $t = 0$. (a) When $t = 2.00T$, where is the particle? At $-X$? $+X$? Zero? Between $-X$ and 0 m? Or between 0 m and $+X$? Answer the same questions for the following times: (b) $t = 3.50T$, and (c) $t = 5.25T$. ■

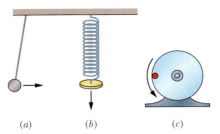

(a)   *(b)*   *(c)*

Figure 16-9 ■ We know that the (*a*) pendulum oscillating at small angles, the (*b*) mass on the spring, and the (*c*) Cartesian components of a spot on a disk undergo sinusoidal motion. These can be made to move with the same period and phase. The spring and the spot on the rotating disk can also have the same amplitude. What are the mathematical characteristics of forces needed to induce sinusoidal motion in a mass–spring system? In a pendulum? Are they the same?

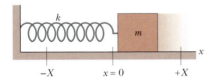

Figure 16-10 ■ We can imagine a *horizontal oscillator* that consists of a mass–spring system. In theory, the mass can oscillate back and forth on a frictionless surface when it is displaced from equilibrium. In practice, this system is very difficult to set up.

16-3 Simple Harmonic Motion: The Mass–Spring System

So far we have described sinusoidal motion mathematically as in Eq. 16-5. We considered a particle on a rotating disk undergoing uniform circular motion, as shown in Fig. 16-9(*c*), and we observed something special about the component along an axis in the plane of motion passing through the center of rotation: It varies sinusoidally. In this section we explore the behavior of particles that move back and forth sinusoidally along a straight line. We will use a mass–spring system as a model, as shown in the middle of Fig. 16-9. In particular, we are interested in: (1) determining the behavior of a particular spring force experimentally and showing that it causes sinusoidal motion of a mass attached to that spring; (2) using Newton's Second Law to predict theoretically that a spring should cause a mass to oscillate with one-dimensional sinusoidal motion; and (3) discussing how our theory predicts some surprising characteristics of that motion that leads us to define it as simple harmonic motion.

In Section 9-5 we explored Hooke's law for an ideal spring. We imagined a horizontally oriented spring attached to a wall at one end and to a mass at the other end (like that shown in Fig. 16-10). We noted that in its relaxed state, the spring exerts no forces. However, if the spring is displaced from its relaxed state so it is stretched or compressed, it exerts a force on anything attached to its ends. The direction of the spring force always acts in a direction *opposite* to its displacement. The force tries to bring the spring to its relaxed state. For this reason, we describe a spring force as a **restoring force.** If we are careful to put the origin of our chosen *x* axis at the relaxed position, then Hooke's law can be expressed using Eq. 9-17,

$$F_x^{\text{spring}} = -kx,$$

where F_x^{spring} is the *x*-component of the spring force, *x* is the displacement from its equilibrium position, and *k* is the spring constant (or stiffness factor).

Experimental Findings: Forces, Displacement, and Time

In physics we usually analyze a simplified model system before considering more complex "real-world" systems. Unfortunately, in picking a model mass–spring system we are presented with a dilemma. The simplest system to model mathematically is a *horizontal oscillator* with a partially extended spring attached to a block that slides on a perfectly frictionless surface as shown in Fig. 16-10. However, it is not easy without special equipment (such as an air track) to set up a friction-free experiment using such a system. The simplest system to set up experimentally is a *vertical oscillator* in which a mass descends as it stretches until there is no net force on the mass. This vertical location of a mass hanging on a spring is called its **equilibrium position.** However, both theory and experiments show that the corresponding components of net forces caused by displacements from equilibrium on masses hanging vertically and horizontal masses are the *same.** Thus, for simplicity, we will consider our model system to be a mass hanging down vertically from a spring.

We start by presenting the results of measurements made on our vertical mass–spring system. Next we show how our experimental knowledge of forces on the system can be used in conjunction with Newton's Second Law to derive its position vs. time equation theoretically. At the same time we can determine theoretically how factors such as mass, spring stiffness, and amplitude influence the period of oscillation. In a later section we will show how the forces experienced by a pendulum bob have the same mathematical characteristics as those that drive a mass on a spring.

---

*The theoretical equivalency of the horizontal and vertical mass-spring system is developed in Touchstone Example 16-3.

A 100 g mass is attached to a light 10.1 g spring and suspended from an electronic force sensor, shown in Fig. 16-11. An ultrasonic motion detector is placed underneath it to record its displacement from equilibrium. The mass is pulled down to a maximum displacement of about 4.0 cm and released. About a half second later, a computer data acquisition system begins to measure the forces exerted by the spring on the mass as a function of the displacement of the mass (Fig. 16-12). At the same time, the displacement of the mass is tracked with the motion detector. The experimental results for the net force on the mass as a function of time are shown in Fig. 16-13a. Similarly, the experimental results for the displacement from equilibrium as a function of time are shown in Fig. 16-13b.

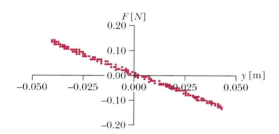

FIGURE 16-12 ■ A graph of the data showing net force on the mass as a function of its displacement from equilibrium. It is obvious that Hooke's law holds with $F_y^{net} = -ky$. A fit to the data shows that the magnitude of the slope of the graph (which is the value of k) is given by $k = 3.23$ N/m.

FIGURE 16-11 ■ A common way to measure forces and displacements for a vertical mass–spring oscillator is to hang a light spring with an attached mass from an electronic force sensor. An ultrasonic motion detector can be placed underneath the mass to measure displacements. If we displace the mass vertically from its equilibrium point, it will oscillate between a maximum value of displacement from equilibrium $+Y$ and a minimum value $-Y$.

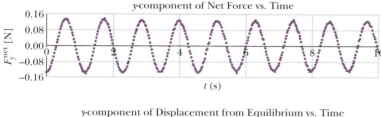

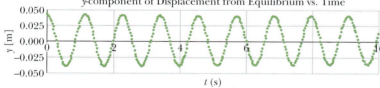

FIGURE 16-13 ■ These graphs of experimental data show that the net force exerted on a 100 g mass by a spring causes it to oscillate with what appears to be sinusoidal motion. Examination of the two graphs shows that on a moment-by-moment basis the net force is proportional to the displacement but oppositely directed.

Careful examination of the graph in Fig. 16-13 gives a period of oscillation for our 100 g mass that is $T^{exp} = 1.1$ s. A model of the plotted data shows that the equation that describes the displacement versus time has exactly the same form as Eq. 16-5. In particular,

$$x(t) = X \cos[\omega t + \phi_0] = (0.040 \text{ m}) \cos[(5.5 \text{ rad/s}) \, t + 0.0 \text{ rad}]. \qquad (16\text{-}7)$$

The value of the maximum displacement or amplitude X is 0.040 m, or only 4.0 cm. The angular frequency, ω, is 5.5 rad/s, and the initial phase (or phase constant), ϕ_0, is 0.0 rad. Equation 16-7 provides experimental verification that our vertical spring–mass system oscillates *sinusoidally*.

We see from Fig. 16-13a that the restoring force component in the y direction is proportional to the displacement component but opposite in sign—the familiar Hooke's law as described earlier in Eq. 9-17 (with y playing the role of x). We can use the experimental fact that the spring force causes our mass to oscillate sinusoidally to develop a definition of simple harmonic motion:

> **Simple harmonic motion (or SHM)** is the sinusoidal motion executed by a particle of mass m subject to a one-dimensional net force that is proportional to the displacement of the particle from equilibrium but opposite in sign.

As we show later in this section, the displacement can be either linear or rotational.

Theoretical Prediction: Spring Forces Cause SHM

We can combine Newton's Second Law with our experimental verification of Eq. 9-17 for a spring–mass system that undergoes SHM to get

$$F^{\text{net}} = -ma_x = -kx, \tag{16-8}$$

where F^{net}, a_x, and x are the respective x-components of force, acceleration, and displacement. This equation can be used to explain why Eq. 16-5 does indeed describe the motion of our mass–spring system. To do this, we need to express the acceleration of the mass as the second derivative of the displacement. By doing this we can rewrite Eq. 16-8 as

$$F^{\text{net}} = ma_x = m\frac{d^2x}{dt^2} = -kx, \quad \text{so that} \quad \frac{d^2x}{dt^2} = -\frac{k}{m}x. \tag{16-9}$$

What happens when we actually take the first and then the second derivative of Eq. 16-5? Do we get a minus sign and a positive constant that can be associated with the ratio k/m? The first derivative is

$$\frac{dx}{dt} = \frac{d[X\cos(\omega t + \phi_0)]}{dt} = -\omega X \sin(\omega t + \phi_0),$$

and the second derivative is

$$\frac{d^2x}{dt^2} = \frac{d[-\omega X \sin(\omega t + \phi_0)]}{dt} = -\omega^2 X \cos(\omega t + \phi_0) = -\omega^2 x. \tag{16-10}$$

We do indeed get a negative sign times a positive constant in front of the x term, but for the theoretical equation derived from Newton's Second Law to match our experimentally determined equation for displacement versus time, we must have the angular frequency be equal to

$$\omega = \sqrt{\frac{k}{m}} \quad \text{(angular frequency)}. \tag{16-11}$$

By combining Eqs. 16-6 and 16-11, we can write, for the **period** of the linear oscillator shown in Fig. 16-11,

$$T = 2\pi\sqrt{\frac{m}{k}} \quad \text{(period)}. \tag{16-12}$$

Equations 16-11 and 16-12 tell us that a large angular frequency (and thus a small period) goes with a stiff spring (large k) and a low-mass object (small m). This seems like a very reasonable pair of relationships. Let us verify whether our data for the mass–spring system satisfies Eq. 16-12. In order to simplify our theoretical model *we have chosen to ignore the relatively small 10.1 g mass of the spring* and assume that the mass of the system can be adequately represented by only the hanging mass. Using this assumption our theoretical model tells us that a spring with k of 3.23 N/m and hanging mass, m, of 0.10 kg will have a period of

$$T^{\text{theory}} = 2\pi\sqrt{\frac{m}{k}} = 2\pi\sqrt{\frac{0.10 \text{ kg}}{3.23 \text{ N/m}}} = 1.1 \text{ s}.$$

This predicted period is the same as our experimental value of $T^{\text{exp}} = 1.1$ s to 2 significant figures.

Describing How Spring Forces Cause Oscillations

One quite surprising result of our theoretical predictions is that Eq. 16-12 tells us that *the period of the oscillations do not depend on amplitude.* You might guess that if you stretch or compress the spring more, the distance the mass travels in an oscillation will be greater. But additional experiments show that period does not depend on amplitude as long as we don't distort the spring so that Hooke's law fails. This is because if we start with a larger initial displacement (amplitude) the forces and accelerations are greater and our mass moves faster.

We can now see why an object that experiences a spring force oscillates. If we hold the object out at some displacement and release it, the net force acts opposite to the displacement so the object will start to accelerate toward its equilibrium position. It moves faster and faster toward the equilibrium position but as it gets there, the force is becoming smaller, so when it gets to its equilibrium position it is moving with some velocity. By Newton's First Law the object keeps going and overshoots. The force now acts in the opposite direction to slow it down, but by the time the force has brought the object to its turn-around point the object has another displacement, this time on the other side. The process repeats, and if there is no friction or damping, the oscillations will go on forever.

Not All Sinusoidal Motions Are SHM

Every oscillating system such as a diving board or a violin string has an element of "springiness" and an element of "inertia" or mass, and thus behaves like the linear oscillator of Fig. 16-11. As long as the forces that act on objects in an oscillating system are linear restoring forces, we will get simple harmonic motion that is sinusoidal. Almost all sinusoidal motions qualify as simple harmonic motion. One notable exception is the sinusoidal motion of the moon Callisto about Jupiter that Galileo observed (Fig. 16-3). The motion Galileo observed is the projection of a nearly circular orbit seen edge-on. The gravitational force law that keeps the moon in orbit is not proportional to its displacement from the center of the orbit. Therefore, Callisto is not undergoing simple harmonic motion. One consequence of this is that the period of the moon's motion is not amplitude-independent—it depends of the orbital radius. Although the motion Galileo observed is sinusoidal, it does not qualify as SHM.

A Rotational Simple Harmonic Oscillator

Figure 16-14 shows a rotational version of a simple harmonic oscillator; the element of springiness or elasticity is associated with the restoring torque the suspension wire can exert when it's twisted. This is rather like the forces a spring can exert when it is stretched or compressed. A device that can exert a restoring torque on an object is called a **torsion oscillator** (or torsion pendulum), with *torsion* referring to the twisting.

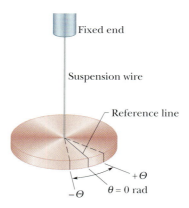

FIGURE 16-14 ■ An angular simple harmonic oscillator, or torsion oscillator, is an angular version of the linear simple harmonic oscillator of Fig. 16-10. The disk oscillates in a horizontal plane; the reference line oscillates with angular amplitude Θ. The twist in the suspension wire stores potential energy as a spring does and provides the restoring torque.

If we rotate the disk in Fig. 16-14 by some rotational displacement of magnitude $|\Theta|$ from its equilibrium position (where the reference line is at $\theta = 0$) and release it, it will oscillate with a rotational amplitude of Θ about that position in **rotational simple harmonic motion.** Displacing the disk through any angle θ in either direction from its equilibrium orientation results in the suspension wire of a restoring torque given by

$$\tau^{\text{net}} = I\alpha = -\kappa\theta. \tag{16-13}$$

Here κ (Greek *kappa*) is a constant, called the **torsion constant,** that depends on the length, diameter, and the suspension wire's shear modulus (as defined in Section 13-5).

Comparison of Eq. 16-13 with Eq. 16-8 leads us to suspect that Eq. 16-13 is the rotational form of Hooke's law, and that we can transform Eq. 16-12, which gives the period of linear SHM, into an equation for the period of rotational SHM; we replace the spring constant k in Eq. 16-12 with its equivalent, the constant κ of Eq. 16-13, and we replace the mass m in Eq. 16-12 with its equivalent, the rotational inertia I of the oscillating disk. These replacements lead to

$$T = 2\pi\sqrt{\frac{I}{\kappa}} \qquad \text{(torsion oscillator)}, \tag{16-14}$$

which is the correct equation for the period of a rotational simple harmonic oscillator, or torsion pendulum.

READING EXERCISE 16-3: The experimental period of the 100 g vertical mass oscillating on the spring shown in Fig. 16-11 is about 3% larger than the period calculated using a simplified theoretical model in which the spring is assumed to be massless, but we know the spring actually has a mass of 10.1 g. Consider the nature of Eq. 16-12 and explain why the measured period should be a bit longer than the theoretical value we reported. No calculation is needed. ■

READING EXERCISE 16-4: Which of the following relationships between the x-component force F_x on a particle and the particle's position x implies simple harmonic oscillation: (a) $F_x = (-5 \text{ N/m})x$, (b) $F_x = (-400 \text{ N/m}^2)x^2$, (c) $F_x = (+10 \text{ N/m})x$, (d) $F_x = (3 \text{ N/m}^2)x^2$? ■

16-4 Velocity and Acceleration for SHM

The Velocity for Simple Harmonic Motion

Let us imagine how the velocity of the mass on the spring in Fig. 16-11 changes as it moves through a complete oscillation cycle. Is the magnitude of the velocity a maximum, a minimum or zero when the magnitude of its displacement is the greatest? Obviously the mass has a velocity of zero when it is turning around. But the mass turns around when the magnitude of the displacement is a maximum. This means that the velocity of the mass is out of phase with its displacement in the same way that the cosine and sine functions are out of phase with each other. By differentiating Eq. 16-5, we find that the expression for the velocity of a particle moving with simple harmonic motion is indeed a sine function whenever the displacement is a cosine function; that is,

$$v(t) = \frac{dx(t)}{dt} = \frac{d}{dt}[X\cos(\omega t + \phi_0)],$$

or
$$v(t) = -\omega X \sin[\omega t + \phi_0] \quad \text{(velocity).}$$
(16-15)

Figure 16-15a is a plot of the x-component of displacement given by Eq. 16-5 with $\phi_0 = 0$ rad. Figure 16-15b shows Eq. 16-15, also with $\phi_0 = 0$ rad. Analogous to the amplitude X in Eq. 16-5, the positive quantity ωX in Eq. 16-15 is the maximum velocity V and is called the **velocity amplitude.** As you can see in Fig. 16-15b, the velocity of the oscillating particle varies between the limits $\pm \omega X$. Note also in that figure that the curve of $v(t)$ is *shifted* (to the left) from the curve of $x(t)$ by one-quarter period; when the magnitude of the displacement is greatest (that is, $x(t) = X$), then the magnitude of the velocity is least [that is, $v(t) = 0$ m/s]. When the magnitude of the displacement is least (that is, zero), the magnitude of the velocity is greatest (that is, $V = \omega X$).

The Acceleration of SHM

Knowing the velocity $v(t)$ for simple harmonic motion, we can find an expression for the acceleration of the oscillating particle by differentiating once more. Thus, we have, from Eq. 16-15,

$$a(t) = \frac{dx(t)}{dt} = \frac{d}{dt}[-\omega X \sin(\omega t + \phi_0)],$$

or
$$a(t) = -\omega^2 X \cos(\omega t + \phi_0) \quad \text{(acceleration).}$$
(16-16)

Figure 16-15c is a plot of Eq. 16-16 for the case where the initial phase is zero (or $\phi_0 = 0$ rad). The positive quantity $\omega^2 X$ in Eq. 16-16 is equal to the maximum acceleration called the **acceleration amplitude** A; that is, the acceleration of the particle varies between the limits $\pm A = \pm \omega^2 X$, as Fig. 16-15c shows. Note also that the curve of $a(t)$ is shifted (to the left) by one-quarter period relative to the curve of $v(t)$.

We can combine Eqs. 16-5 and 16-16 to yield

$$a(t) = -\omega^2 x(t),$$
(16-17)

which is the hallmark of simple harmonic motion:

> In SHM, the acceleration is proportional to the displacement but opposite in sign, and the two quantities are related by the square of the angular frequency.

Thus, as Fig. 16-15 shows, when the displacement has its greatest positive value, the acceleration has its greatest negative value, and conversely, when the displacement is zero, the acceleration is also zero.

READING EXERCISE 16-5: Consider Fig. 16-15b, which shows the velocities of a mass on spring as a function of time. Identify at what time or times (t_1, t_2, t_3, and t_4) the vertical component of velocity of the mass has a maximum value, a minimum value, and is zero. ■

READING EXERCISE 16-6: Consider Fig. 16-15c, which shows the acceleration of a mass on spring as a function of time. In which region or regions (1, 2, 3, or 4) is the vertical component of acceleration of the mass increasing? Decreasing? ■

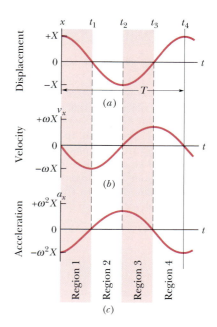

FIGURE 16-15 ■ Assume $t_1 = 0$ s. (a) The displacement $x(t)$ of a particle undergoing SHM with an initial phase of $\phi_0 = 0$ rad. The period T marks one complete oscillation. (b) The velocity $v_x(t)$ of the particle. (c) The acceleration $a_x(t)$ of the particle.

TOUCHSTONE EXAMPLE 16-1: Spider Oscillations

Orb web spiders are commonly found outdoors near buildings. An orb web spider with a mass of about 2 g drops off a tree branch onto the center of her horizontal web (Fig. 16-16). As a result, the web undergoes a vertical displacement from its original location to the equilibrium point of the spider-web system (as determined by the vertical location of the spider when the oscillations damp out).

FIGURE 16-16 ■ Top view of a spider dropping onto her web.

(a) If this vertical displacement of the web and spider from equilibrium is about 0.5 cm and if the vertical restoring force the web exerts on the spider is proportional to her displacement, estimate the frequency and period of the vertical oscillation of the spider-web system.

SOLUTION ■ The **Key Idea** here is that as the spider steps on her web the degree of sag caused by the force of her weight enables us to find the web's spring constant k. Once we know the spring constant we can use that along with her estimated mass to determine the frequency and period of the spider's oscillation.

We choose a y axis that points vertically upward and use Eq. 9-16 $[F_y^{\text{spring}} = -k(y_2 - y_1)]$. By noting that the vertical component of the web's net upward restoring force is $+mg$ and that $y_2 < y_1$, we get

$$k = \frac{F_y^{\text{spring}}}{-(y_2 - y_1)} = \frac{+(2 \times 10^{-3}\,\text{kg})(9.8\,\text{N/kg})}{0.5 \times 10^{-2}\,\text{m}} = 3.9\,\text{N/m}.$$

The angular frequency is then given by Eq. 16-11 as

$$\omega = \sqrt{\frac{k}{m}} = \sqrt{\frac{3.9\,\text{N/m}}{2 \times 10^{-3}\,\text{kg}}} = 44\,\text{rad/s}.$$

But we know that $\omega = 2\pi f$, so

$$f = \frac{\omega}{2\pi} = \frac{44\,\text{rad/s}}{2\pi} = 7.0\,\text{Hz} \quad \text{and} \quad T = \frac{1}{f} = \frac{1}{7\,\text{Hz}} = 0.14\,\text{s}.$$
(Answer)

(b) If we start tracking the spider's oscillation when she is at her lowest vertical position relative to the spider-web system equilibrium point, what is her initial phase?

SOLUTION ■ The **Key Idea** here is that we must take the equilibrium point to be at $y = 0$ m and that $y(0) = -Y$ when $t = 0$ s. Using Eq. 16-5 with y instead of x denoting displacement, we get

$$-Y = Y \cos(\omega t + \phi_0)$$

so that

$$\cos(\phi_0) = -1,$$

which gives

$$\phi_0 = \pm \pi\,\text{rad}. \quad \text{(Answer)}$$

(c) If the maximum displacement of the spider from the equilibrium point of the spider-web system is given by $Y = 1.0$ cm, what is the maximum acceleration of the spider as she oscillates?

SOLUTION ■ The **Key Idea** here is that the magnitude A of the maximum acceleration is the acceleration amplitude $\omega^2 Y$ as shown in Eq. 16-16. Thus,

$$A = \omega^2 Y = (44\,\text{rad/s})^2 (10 \times 10^{-3}\,\text{m}) = 20\,\text{m/s}^2. \quad \text{(Answer)}$$

The maximum magnitude of acceleration occurs when the spider is turning around at the ends of her path. This is when the force on her is a maximum. The graphs in Fig. 16-15 also show the magnitude of displacement and acceleration to be maximum at the same time.

16-5 Gravitational Pendula

We turn now to a class of simple harmonic oscillators in which the restoring force is associated with the gravitational force rather than with the elastic properties of a twisted wire or a spring. Oscillators that depend on gravitational restoring forces or torques hang, and so they are considered to be types of pendulums.

The Simple Pendulum Oscillating at a Small Angle

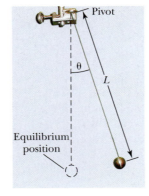

FIGURE 16-17 ■ A pendulum bob swings back and forth at a small angle. Its angular displacement θ from its vertical equilibrium is measured using a rotary motion sensor attached to a computer data acquisition system. The length L of this pendulum measured from the pivot to the center of the bob is 32 cm.

Consider a small particle of mass m (called a *bob*) that hangs from the end of a wire or string. Assume that the wire has a small mass compared to the mass of the particle and that it can't stretch noticeably. If you fix the wire at its upper end, you have constructed a **simple pendulum.** An example of a simple pendulum is shown in Fig. 16-17. The bob is in its equilibrium position when it hangs vertically. But suppose you pull the

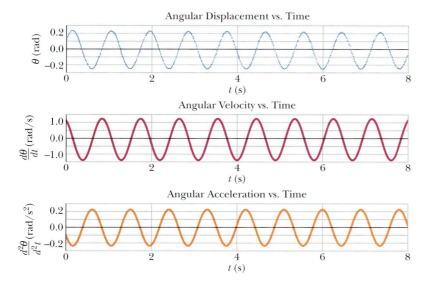

Angular Displacement vs. Time

Angular Velocity vs. Time

Angular Acceleration vs. Time

FIGURE 16-18 ■ Measurements for the *angular displacement* (θ) of the simple pendulum shown in Fig. 16-17 were obtained using a computer data acquisition system outfitted with a rotary motion sensor. A graph of the *angular velocity* ($d\theta/dt$) was also constructed using an algorithm to find a smoothed first derivative of the angular data. This process is repeated using the velocity data to find how the angular acceleration (α) varies with time.

bob up so the initial angular displacement θ between the wire and the vertical is small. What happens to the angle θ between the vertical and the wire as the pendulum bob swings back and forth? You easily see that the bob's motion is periodic. Is it, in fact, simple harmonic motion? If so, what factors does the period T depend on?

We can verify that the simple pendulum shown in Fig. 16-17 undergoes SHM by examining the graphs of the data shown in Fig. 16-18.

Careful examination of the θ vs. t graph in Fig. 16-18 gives a period of oscillation for our pendulum bob of $T^{\exp} = 1.1$ s. A model of the plotted data shows that the equation that describes the angular displacement versus time has exactly the *same form* as Eq. 16-7. In particular,

$$\theta(t) = \Theta \cos(\omega t + \phi_0) = (0.24 \text{ rad}) \cos([7.0 \text{ rad/s}]t - 0.90 \text{ rad}). \quad (16\text{-}18)$$

The value of the maximum angular displacement or amplitude is $\Theta = 0.24$ rad, the angular frequency, ω, is 7.0 rad/s, and the initial phase (or phase constant), ϕ_0, is -0.90 rad. The maximum angular displacement of 0.24 rad can be expressed in degrees as

$$\Theta = 0.24 \text{ rad} = \left(\frac{180°}{\pi \text{ rad}}\right) 0.24 \text{ rad} \approx 15°.$$

If we repeat the experiment for smaller amplitudes, we find that the angular frequency ω stays the same to two significant figures. This fact when combined with the sinusoidal oscillation shown in Eq. 16-18 strongly suggests that the pendulum is undergoing simple harmonic motion. If this is the case then the net horizontal force on the pendulum bob is proportional to its horizontal displacement—at least when angles are small.

Let us see if our experimental results could have been predicted theoretically.

Theoretical Derivation of Simple Pendulum Forces

To find the net horizontal forces on the pendulum, we can set up a free-body diagram using methods introduced in Chapter 6. The forces acting on the bob are the tension force $\vec{F}_{\text{wire} \to \text{bob}}$ that the wire exerts on the bob and the gravitational force $\vec{F}^{\text{grav}}$, as shown in Fig. 16-19b where the string makes an angle θ with the vertical. We resolve $\vec{F}^{\text{grav}}$ into a radial component $|F^{\text{grav}}|\cos \theta$ and a component $|F^{\text{grav}}|\sin \theta$ that is tangent to the path taken by the bob. This tangential component produces a restoring torque about the pendulum's pivot point, because it always acts opposite the displacement of the bob so as to bring the bob back toward its central location. That location is called the *equilibrium position* ($\theta = 0$), because if the pendulum were not swinging it would be at rest at that position.

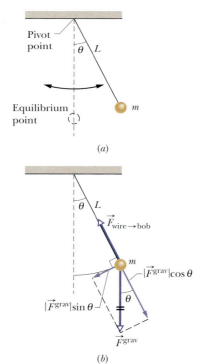

(a)

(b)

FIGURE 16-19 ■ (a) A simple pendulum displaced from its equilibrium position by an angle θ. (b) The forces acting on the bob are the gravitational force $\vec{F}^{\text{grav}}$ and the tension force $\vec{F}_{\text{wire} \to \text{bob}}$ from the wire. The tangential component $|F^{\text{grav}}|\sin \theta$ of the gravitational force is a restoring force that tends to bring the pendulum back to its central position.

From Eq. 11-30 ($|\vec{\tau}| = |\vec{r}||\vec{F}_t|$), we can write the magnitude of the restoring torque as the product of the magnitude of a moment arm ($|\vec{r}| = L$) about the pivot arm and the tangential component of the gravitational force ($|F^{\text{grav}}|\sin\theta$). The tension force in the wire $\vec{F}_{\text{wire}\rightarrow\text{bob}}$ does not contribute to the restoring torque because it always acts parallel to the moment arm. Thus the z-component of the torque can be expressed as

$$\tau_z = -L(|F^{\text{grav}}|\sin\theta). \tag{16-19}$$

The minus sign indicates that the torque acts to reduce θ. Substituting Eq. 16-19 into Eq. 11-32 ($\tau_z^{\text{net}} = I\alpha_z$) and then substituting mg for the magnitude of $\vec{F}^{\text{grav}}$, we obtain

$$-L(mg\sin\theta) = I\alpha_z, \tag{16-20}$$

where I is the pendulum's rotational inertia about the pivot point and α_z is the z-component of its angular acceleration about that point.

We want to focus on the nature of the motion when the maximum (and minimum) angle of displacement is small. Note that whenever an angle θ is small, then the arc length s that the pendulum bob sweeps through (with respect to its equilibrium) and the value of x have essentially the same magnitude as shown in Fig. 16-20, so that

$$\sin\theta = \frac{x}{L} \approx \frac{s}{L} = \theta \quad \text{(approximation for small } \theta\text{)}. \tag{16-21}$$

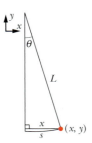

FIGURE 16-20 ■ This diagram illustrates that the arc length s and the value of y are approximately the same when the angle θ is small.

Thus when the angle θ is small, we can replace $\sin\theta$ with θ (expressed in radian measure). (As an example, if $\theta = 15.0° = 0.262$ rad, then $\sin\theta = 0.265$, a difference of only about 1%.) Using this approximation and rearranging terms, we can then express the z-component of angular acceleration as

$$\alpha_z \approx -\frac{mgL}{I}\theta. \tag{16-22}$$

Note that the z axis passes through pivot point and is perpendicular to the plane of oscillation. This equation is the angular equivalent of Eq. 16-5, the hallmark of SHM. It tells us that the angular acceleration $\vec{\alpha} = \alpha_z\hat{k}$ of the pendulum is proportional to the angular displacement $\theta\hat{k}$ but opposite in sign.

Thus, as the pendulum bob moves to, say, the right as in Fig. 16-19a, its acceleration *to the left* increases until the bob turns around and begins moving to the left. Then, when it is on the left, its acceleration to the right tends to return it to the right, and so on, as it swings back and forth in SHM. More precisely, we have verified both experimentally and theoretically that the motion of a *simple pendulum swinging through small angles* is approximately SHM. We can state this restriction to small angles another way: to be correct on predicting the period of a motion to within about 1%, the **angular amplitude** Θ of the motion (the maximum angle of swing) must be about 15° or less.

Comparing Eq. 16-22 and Eq. 16-17, we see that the angular frequency of the pendulum is $\omega = \sqrt{mgL/I}$. Next, if we substitute this expression for ω into Eq. 16-6 ($\omega = 2\pi/T$), we see that the period of the pendulum may be written as

$$T = 2\pi\sqrt{\frac{I}{mgL}}. \tag{16-23}$$

All the mass of a simple pendulum is concentrated in the mass m of the particle-like bob, which is at radius L from the pivot point. Thus, we can use Eq. 11-23 ($I = mr^2$) to write ($I = mL^2$) for the rotational inertia of the pendulum. Substituting this into Eq. 16-23 and simplifying yields

$$T = 2\pi\sqrt{\frac{L}{g}} \quad \text{(simple pendulum, small amplitude)}, \tag{16-24}$$

as a simpler expression for the period of a simple pendulum swinging through only small angles. (We also assume small-angle swinging in the problems of this chapter.) Of course we can use Eq. 16-24 to predict the period of the pendulum of length $L = 32$ cm described in Figs. 16-17 and 16-18. We get

$$T^{\text{theory}} = 2\pi\sqrt{\frac{L}{g}} = 2\pi\sqrt{\frac{0.32 \text{ m}}{9.80 \text{ N/kg}}} = 1.1 \text{ s}.$$

This result matches our experimental result to at least two significant figures. This period is also identical to that of our spring-mass system described by Eq. 16-7. This is because just for fun we chose the pendulum length L so that we would get the same period for the two systems!

A very surprising outcome of this theoretical derivation is that, *for small displacement angles, the period of the simple pendulum does not depend on its bob mass.* If you reflect on this, you should be able to see that the motion of the simple pendulum bob is mass-independent for the same reason that the motion of a falling object does not depend on its mass.

Measuring *g* with a Simple Pendulum

Geologists often use a pendulum to determine the local gravitational strength, *g*, at a particular location on Earth's surface. If a simple pendulum oscillating at small angles is used, we can solve Eq. 16-24 for *g* to get

$$g = \frac{4\pi^2 L}{T^2}. \tag{16-25}$$

Thus, by measuring L and the period T, we can find the value of g. In order to make more precise measurements, a number of refinements are needed. Geophysicists often use a physical pendulum consisting of a solid rod in conjunction with a more sophisticated equation than Eq. 16-25. They can also place the pendulum in an evacuated chamber.

The Physical Pendulum

A "real" pendulum that isn't just a point mass suspended from a massless string is usually called a **physical pendulum,** and it can have a complicated distribution of mass, much different from that of a simple pendulum. Does a physical pendulum also undergo SHM? If so, what is its period?

Figure 16-21 shows an arbitrary physical pendulum displaced to one side by angle θ. The gravitational force $\vec{F}^{\text{grav}}$ acts at its center of mass C, at a distance h from the pivot point O. In spite of their shapes, comparison of Figs. 16-21 and 16-19b reveals only one important difference between an arbitrary physical pendulum and a simple pendulum. For a physical pendulum, the restoring component $|\vec{F}^{\text{grav}}|\sin\theta$ of the gravitational force has a moment arm of distance h about the pivot point rather than of wire length L. In all other respects, an analysis of the physical pendulum would duplicate our analysis of the simple pendulum up through Eq. 16-23. Again, for a small angular amplitude Θ, we would find that the motion is approximately SHM.

If we replace L with h in Eq. 16-23, we can write the period of a physical pendulum as

$$T = 2\pi\sqrt{\frac{I}{mgh}} \qquad \text{(physical pendulum, small amplitude).} \tag{16-26}$$

As with the simple pendulum, I is the rotational inertia of the pendulum about O. However, now I is not simply mL^2. It depends both on the shape of the physical pendulum

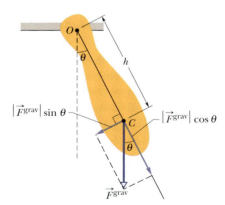

FIGURE 16-21 ■ A physical pendulum. The magnitude of the restoring torque is $h|F^{\text{grav}}|\sin\theta$. When $\theta = 0$, the center of mass C hangs directly below pivot point O.

and on the axis about which it rotates. Table 11-2 shows the rotational inertia equations for common shapes, but these equations describe the rotational inertia about the center of mass (I_{com}). In essentially all physical pendula, the axis of rotation is parallel to an axis through the center of mass but offset by a distance h. In these cases, the parallel axis theorem $I = I_{com} + mh^2$ (Eq. 11-28) can be used to find the required equation for I.

A physical pendulum will not swing if it pivots at its center of mass. Formally, this corresponds to putting $h = 0$ in Eq. 16-26. That equation then predicts $T \rightarrow \infty$, which implies that such a pendulum will never complete one swing.

Corresponding to any physical pendulum that oscillates about a given pivot point O with period T is a simple pendulum of length L_0 with the same period T. We can find L_0 with Eq. 16-24. The point along the physical pendulum at distance L_0 from point O is defined as the *center of oscillation* of the physical pendulum for the given suspension point.

READING EXERCISE 16-7: The vertical acceleration of a falling object is independent of its mass. Likewise the period of a simple pendulum oscillating at small angles is independent of bob mass. Can you explain why in each case? What is similar about the two situations? ■

READING EXERCISE 16-8: Three physical pendula, of masses m, $2m$, and $3m$, have the same shape and size and are suspended at the same point. Rank the masses according to the periods of the pendulum, greatest period first. ■

TOUCHSTONE EXAMPLE 16-2: T-Shaped Pendulum

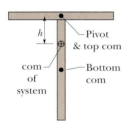

FIGURE 16-22 ■ A T-shaped physical pendulum pivoted at the top of the "T."

A physics student has devised a physical pendulum from two meter sticks of negligible width that are joined together as shown in Fig. 16-22. Assuming the oscillations about the designated pivot occur at small angles, what is the frequency of oscillation?

SOLUTION ■ We need to use Eq. 16-26, which relates the period of a physical pendulum to its rotational inertia (I), mass (m), and moment arm distance (h),

$$T = 2\pi\sqrt{\frac{I}{mgh}}.$$

The **Key Idea** here is to figure out what the rotational inertia of this oddly shaped pendulum is about the chosen pivot and to find the moment arm distance.

The *moment arm h* is the distance from the pivot to the center of mass of the pendulum. To find the com of the T-pendulum we need to clarify our notation. If the total mass of the meter stick system is denoted as m, then the top stick has a mass of $m_{top} = m/2$ and the bottom stick also has a mass of $m_{bottom} = m/2$. If we set $y = 0$ m at the pivot point, then the center of mass of the bottom stick is given by $y_{bottom} = -L/2$. We can now use Eq. 8-11 to find the center of mass of the pendulum (and hence h),

$$Y_{com} = \frac{1}{M_{sys}}(m_{top}y_{top} + m_{bottom}y_{bottom})$$

$$= \frac{1}{m}\left[\left(\frac{m}{2}\right)(0) + \left(\frac{m}{2}\right)\left(-\frac{L}{2}\right)\right] = -\frac{L}{4}.$$

Thus
$$h = |Y_{com}| = \frac{L}{4}. \qquad (16\text{-}27)$$

The *rotational inertia* of the T-shaped system is the sum of the rotational inertias of the top and bottom sticks about the pivot point. The top rod is rotating about its center of mass. The rotational inertia of a stick or rod rotating about its center of mass is shown in Table 11-2 as $I_{com} = \frac{1}{12}ML^2$ so the rotational inertia of the top rod is given by $I_{top} = \frac{1}{12}(m/2)L^2$. The bottom rod is rotating about its end, and we must use the parallel axis theorem (Eq. 11-28) to find its rotational inertia. Since the distance between the axis of rotation of the bottom rod and the pivot point is $L/2$, the rotational inertia of the bottom rod turns out to be $I_{end} = \frac{1}{12}ML^2 + M(L/2)^2 = \frac{1}{3}ML^2$. Thus, the pendulum's total rotational inertia I is given by

$$I = I_{top} + I_{bottom} = \frac{1}{12}(m/2)L^2 + \frac{1}{3}(m/2)L^2 = \frac{5}{24}(mL^2). \qquad (16\text{-}28)$$

Finally we can substitute Eqs. 16-27 and 16-28 into Eq. 16-26 to get

$$T = 2\pi\sqrt{\frac{I}{mgh}} = 2\pi\sqrt{\frac{\frac{5}{24}mL^2}{mg(L/4)}} = 2\pi\sqrt{\frac{20L}{24g}}.$$

Since $L = 1.00$ m,

$$T = 2\pi\sqrt{\frac{20L}{24g}} = 2\pi\sqrt{\frac{20(1.00\text{ m})}{24(9.8\text{ m/s}^2)}} = 1.83\text{ s}. \qquad \text{(Answer)}$$

16-6 Energy in Simple Harmonic Motion

In Chapter 10 we saw that a simple pendulum swinging at a small angle transfers energy back and forth between kinetic energy and potential energy, while the sum of the two—the mechanical energy E of the oscillating Earth–pendulum system—remains constant. What about the linear oscillator made up of the mass–spring system that was considered in Section 16-3? Does it trade energy back and forth as it oscillates? Let us try to answer this question for the linear oscillator using theoretical considerations.

The potential energy of a horizontal linear oscillator like that of Fig. 16-10 is associated entirely with the mass–spring system. Its value depends on how much the spring is stretched or compressed—that is, on $x(t)$. We can use Eqs. 10-14 and 16-5 to find

$$U(t) = \tfrac{1}{2}kx^2 = \tfrac{1}{2}kX^2\cos^2(\omega t + \phi). \qquad (16\text{-}29)$$

Note carefully that for any angle α, a function written in the form $\cos^2\alpha$ (as here) means $(\cos\alpha)^2$. This is *not* the same as a function written as $\cos\alpha^2$, which means $\cos(\alpha^2)$.

The kinetic energy of the system of Fig. 16-11 is associated entirely with the hanging mass. Its value depends on how fast the mass is moving—that is, on $v(t)$. We can use Eq. 16-15 to find

$$K(t) = \tfrac{1}{2}mv^2 = \tfrac{1}{2}m\omega^2 X^2\sin^2(\omega t + \phi_0). \qquad (16\text{-}30)$$

If we use Eq. 16-11 to substitute k/m for ω^2, we can write Eq. 16-30 as

$$K(t) = \tfrac{1}{2}mv^2 = \tfrac{1}{2}kX^2\sin^2(\omega t + \phi_0). \qquad (16\text{-}31)$$

The mechanical energy follows from Eqs. 16-29 and 16-31 and is

$$E = U + K$$
$$= \tfrac{1}{2}kX^2\cos^2(\omega t + \phi_0) + \tfrac{1}{2}kX^2\sin^2(\omega t + \phi_0)$$
$$= \tfrac{1}{2}kX^2[\cos^2(\omega t + \phi_0) + \sin^2(\omega t + \phi_0)].$$

For any angle α,

$$\cos^2\alpha + \sin^2\alpha = 1.$$

Thus, the quantity in the square brackets above is unity and we have

$$E = U + K = \tfrac{1}{2}kX^2. \qquad (16\text{-}32)$$

The mechanical energy of a horizontal mass–spring system oscillator is indeed constant and independent of time. The potential energy and kinetic energy of this oscillator are shown as functions of time t in Fig. 16-23a, and as functions of displacement x in Fig. 16-23b.

Since a linear oscillation trades energy back and forth in a symmetric fashion, it turns out that the average kinetic energy is the same as the average potential energy. Each average energy is 1/2 of the total. In equation form this can be expressed as

$$\langle K \rangle = \langle U \rangle = \frac{E^{\text{tot}}}{2}. \qquad (16\text{-}33)$$

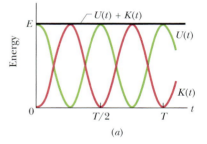

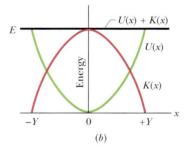

FIGURE 16-23 ▪ (*a*) Potential energy $U(t)$, kinetic energy $K(t)$, and mechanical energy E as functions of time t for a linear harmonic oscillator. Note that all energies are positive and that the potential energy and the kinetic energy peak twice during every period. (*b*) Potential energy $U(t)$, kinetic energy $K(x)$, and mechanical energy E as functions of position x for a linear harmonic oscillator with amplitude Y. For $x = 0$ the energy is all kinetic, and for $x = \pm Y$ it is all potential.

We also examined the behavior of a vertical mass–spring–Earth system (Fig. 16.11). To consider the mechanical energy exchange between potential and kinetic energies in that system, we need to take gravitational energy into account. This complication is treated in Touchstone Example 16-3.

We presented actual data in Fig. 10-14 showing very similar trading of kinetic and gravitational potential energy for a simple pendulum.

READING EXERCISE 16-9: Assume that the spring–block system shown in Fig. 16-10 has a kinetic energy of 3 J and an elastic potential energy of 2 J when the block is at $x = +20$ cm. (a) What is the kinetic energy when the block is at $x = 0$ cm? What is the elastic potential energy when the block is at (b) $x = -X$? ■

TOUCHSTONE EXAMPLE 16-3: Oscillation Energy

Consider the 100 g mass hanging vertically from the spring as shown in Fig. 16-11. It oscillates with a period of $T = 1.1$ s trading between kinetic and potential energy. Its displacement from equilibrium as a function of time is given by Eq. 16-7 and the motion of the mass is shown graphically in Fig. 16-13. Since the 100 g mass is much greater than the mass of the spring, we will ignore the mass of the spring as we analyze the energy transformations during a single oscillation.

(a) What is the total energy of the oscillating system?

SOLUTION ■ Equation 16-32 can be used to express total energy of a SHM oscillator in terms of the spring constant k and the amplitude of the displacement Y. Since the angular frequency is related to k and the oscillating mass m by Eq. 16-11 ($\omega = \sqrt{k/m}$), we can write Eq. 16-32 as

$$E = U + K = \tfrac{1}{2}kY^2 = \tfrac{1}{2}m\omega^2 Y^2. \qquad \text{(Eq. 16-32)}$$

We know that $m = 0.10$ kg. By examining Eq. 16-7 we see that $Y = 0.040$ m and $\omega = 5.5$ rad/s. This gives us a total system energy of

$$E = \tfrac{1}{2}m\omega^2 Y^2 = \tfrac{1}{2}(0.10 \text{ kg})(5.5 \text{ rad/s})^2(0.040 \text{ m})^2$$

$$= 2.4 \times 10^{-3} \text{ J}. \qquad \text{(Answer)}$$

(b) At what time or times during the first oscillation is the total mechanical energy of the mass–spring system equal to its potential energy?

SOLUTION ■ When the system's total and potential energy are the same, its kinetic energy must be zero. That occurs whenever the magnitude of velocity of the mass is zero, which happens when the magnitude of the displacement is a maximum and the mass is turning around. Since the period $T = 1.1$ s, an examination of Fig. 16-13 shows this occurring at $t = 0$ s and 1.1 s. It is also turning around halfway between these times, so

$$E = U \quad \text{at } 0.0 \text{ s}, 0.55 \text{ s, and } 1.1 \text{ s.} \qquad \text{(Answer)}$$

An alternate way to arrive at the same answer is to examine Fig. 16-23 and note that the potential energy peaks at 0, $T/2$, and T.

(c) At what time or times during the first oscillation is the total mechanical energy of the mass–spring system equal to its kinetic energy?

SOLUTION ■ When the system's total and kinetic energy are the same, the speed of the mass is a maximum. That occurs whenever the mass passes through its equilibrium point. According to Eq. 16-7, if we set $t_1 = 0$ then this happens when

$$\cos(\omega t) = \cos([5.5 \text{ rad}]t) = 0,$$

which occurs when

$\omega t = (5.5 \text{ rad})t = \pi/2$ or $3\pi/2$ at about 0.28 s and 0.83 s. (Answer)

An alternate way to arrive at the same answer is to examine Fig. 16-23 and note that the kinetic energy peaks at $T/4$ and at $3T/4$.

(d) At what time or times are the potential and kinetic energy the same so $K = U = \tfrac{1}{2}E$?

SOLUTION ■ An examination of Fig. 16-23 shows that the system's potential and kinetic energy are the same at

$$\tfrac{1}{8}T, \ \tfrac{3}{8}T, \ \tfrac{5}{8}T, \text{ and } \tfrac{7}{8}T, \quad \text{or} \quad 0.138 \text{ s}, 0.413 \text{ s}, 0.688 \text{ s, and } 0.963 \text{ s.}$$

(Answer)

(e) What types of potential energy is stored in the system as it oscillates? Is it really legitimate to use Eqs. 16-29 and 16-32, which measure spring potential energy, when both gravitational and spring potential energy are present?

SOLUTION ■ If our mass–spring system were oscillating horizontally the potential energy would consist entirely of the elastic energy stored in the spring. The situation for the hanging mass is not so simple since we have both spring energy and gravitational potential energy involved. In this case $U = U^{\text{grav}} + U^{\text{spring}}$. Actually the energy relationships presented in Section 16-6 are still valid.

To explain why the energy equations in Section 16-6 are valid, let us choose our origin so that $y = 0$ m at the hanging equilibrium point of the system shown in Fig. 16-24. At any point during an

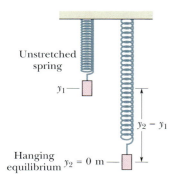

Unstretched spring

y_1 —

$y_2 - y_1$

Hanging equilibrium $y_2 = 0$ m

FIGURE 16-24 ▪ A hanging mass-spring system.

oscillation, the net force on the hanging mass is the sum of an upward (positive) spring force and a downward (negative) gravitational force). This can be expressed in terms of the vertical components as

$$F_y^{\text{net}} = F_y^{\text{spring}} + F_y^{\text{grav}}.$$

When the mass is placed on the spring, the weight of the mass causes the spring to descend from its natural equilibrium point at y_1 to its hanging equilibrium point at y_2. But we chose y_2 = 0 m. At this hanging equilibrium point, the net force on the mass is zero, so that

$$F_y^{\text{net}} = F_y^{\text{spring}} + F_y^{\text{grav}} = -k(y_2 - y_1) - mg$$
$$= -k(0 - y_1) - mg = 0\,\text{N}.$$

A **Key Idea** here is that at the hanging equilibrium location $ky_1 = mg$.

Since we are free to set an arbitrary reference level for gravita-

tional potential, let's set the gravitational potential energy at $y = 0$ m to $U^{\text{grav}}(0) = -\frac{1}{2}mgy_1$ where y_1 is the location of the center of the mass when the spring is unstretched (as shown in Fig. 16-24). Next let's calculate the total potential energy of the spring–mass system when the mass has a displacement of y from its hanging equilibrium position,

$$U = U^{\text{grav}} + U^{\text{spring}} = (mgy - \tfrac{1}{2}mgy_1) + \tfrac{1}{2}k(y - y_1)^2,$$

but $ky_1 = mg$, so that

$$U = (ky_1 y - \tfrac{1}{2}ky_1^2) + \tfrac{1}{2}k(y - y_1)^2$$
$$= (ky_1 y - \tfrac{1}{2}ky_1^2) + (\tfrac{1}{2}ky^2 - ky_1 y + \tfrac{1}{2}ky_1^2)$$
$$= \tfrac{1}{2}ky^2.$$

This is a very important result because it tells us that with regard to energy considerations, the potential energy of a hanging mass when oscillating about its equilibrium position can be treated as if the spring were unstretched at the hanging equilibrium. This depends on choosing our gravitational potential energy reference point carefully (which we are free to do). This result provides theoretical verification for the claim made in Section 16-3, that a mass oscillating from a hanging spring and one oscillating horizontally have the same type of mathematical behavior.

16-7 Damped Simple Harmonic Motion

The equations we have developed to describe harmonic motion predict that it goes on forever with the same amplitude. Whatever starting value of t you put into Eq. 16-5, it will oscillate from then on with the same amplitude it started with. Oscillations in the real world usually die out gradually, transferring mechanical energy to thermal energy by the action of frictional forces. A pendulum will swing only briefly under water, because the water exerts a drag force on the pendulum that quickly eliminates the motion. A pendulum swinging in air does better, but still the motion dies out eventually, because the air exerts a drag force on the pendulum and friction acts at its support. These forces reduce the mechanical energy of the pendulum's motion.

When the motion of an oscillator is reduced by external friction or drag forces, the oscillator and its motion are said to be **damped.** An idealized example of a damped oscillator is shown in Fig. 16-25, where a block with mass m oscillates vertically on a spring with spring constant k. From the block, a rod extends to a vane (both assumed to have negligible mass) that is submerged in a liquid. As the vane moves up and down, the liquid exerts an inhibiting drag force on it and thus on the entire oscillating system. With time, the mechanical energy of the block–spring system decreases, as energy is transferred to thermal energy of the liquid and vane. If the liquid is alcohol that is not very viscous, the drag forces will be small. But a liquid like honey or molasses could exert much larger drag forces.

Theoretical Analysis

Let us assume the liquid in the system shown in Fig. 16-25 exerts a **damping force** $\vec{F}^{\text{drag}}$ that is *proportional in magnitude to the velocity* $\vec{v}$ *of the vane and block* (an

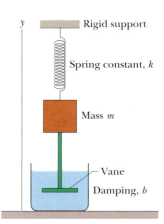

y Rigid support

Spring constant, k

Mass m

Vane

Damping, b

FIGURE 16-25 ▪ An idealized damped simple harmonic oscillator. A vane immersed in a liquid exerts a damping force on the block as the block oscillates parallel to the x axis.

assumption that is accurate if the vane moves slowly). Then, for forces along the y axis in Fig. 16-25, we have

$$\vec{F}^{\,\mathrm{drag}} = -b\vec{v}, \tag{16-34}$$

where b is a **damping constant** that depends on the characteristics of both the vane and the liquid and has the SI unit of kilogram per second. The minus sign indicates that $\vec{F}^{\,\mathrm{drag}}$ opposes the motion.

The y-component of the force on the block from the spring is $F_y^{\mathrm{spring}} = -ky$. Let us assume that the gravitational force on the block is negligible compared to $\vec{F}^{\,\mathrm{drag}}$ and $\vec{F}^{\,\mathrm{spring}}$. Then we can write Newton's Second Law for components along the y axis ($F_y^{\mathrm{net}} = ma_y$) as

$$-bv_y - ky = ma_y. \tag{16-35}$$

Substituting dy/dt for v_y and d^2x/dt^2 for a_y and rearranging give us the differential equation

$$m\frac{d^2y}{dt^2} + b\frac{dy}{dt} + ky = 0. \tag{16-36}$$

The solution of this equation is

$$y(t) = Ye^{-bt/2m}\cos(\omega' t + \phi_0), \tag{16-37}$$

where Y is the initial amplitude, ϕ_0 is the initial phase, and ω' is the angular frequency of the damped oscillator. This angular frequency is given by

$$\omega' = \sqrt{\frac{k}{m} - \frac{b^2}{4m^2}}. \tag{16-38}$$

If $b = 0$ (there is no damping), then Eq. 16-38 reduces to Eq. 16-11 ($\omega = \sqrt{k/m}$) for the angular frequency of an undamped oscillator, and Eq. 16-37 reduces to Eq. 16-5 for the displacement of an undamped oscillator. If the damping constant is small but not zero (so that $b \ll \sqrt{km}$), then $\omega' \approx \omega$. We define the pendulum as **underdamped** whenever $k/m > b^2/4m^2$ so that $\omega' < \omega$.

We can regard Eq. 16-37 as a cosine function whose amplitude, which is $Ye^{-bt/2m}$, gradually decreases with time, as Fig. 16-26 suggests. For an undamped oscillator, the mechanical energy is constant and is given by Eq. 16-32 ($E = \frac{1}{2}kY^2$). If the oscillator is damped, the mechanical energy is not constant but decreases with time. If the damping is small, we can find $E(t)$ by replacing the amplitude, Y, in Eq. 16-32 with $Ye^{-bt/2m}$, the amplitude of the damped oscillations. By doing so, we find that

$$E(t) \approx \tfrac{1}{2}kY^2\,e^{-bt/m}, \tag{16-39}$$

which tells us that, like the amplitude, the mechanical energy decreases exponentially with time.

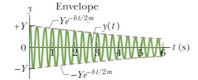

FIGURE 16-26 ■ The displacement function $y(t)$ for the damped oscillator of Fig. 16-25, with $m = 250$ g, $k = 85$ N/m, and $b = 70$ g/s. The amplitude, which is $Ye^{-bt/2m}$, decreases exponentially, though this exponential decrease does not show up well since the damping coefficient b is not large. The exponential drop-off is more pronounced in Fig. 16-28.

Experimental Results

Here we present actual data for another damped oscillator that has identical mathematical behavior to that of the block oscillating in a viscous liquid. It consists of an

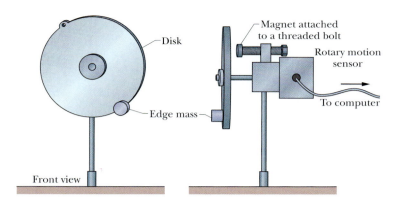

FIGURE 16-27 ■ A physical pendulum consisting of an aluminum disk with an edge mass attached to it. Magnetic damping is used to provide a drag torque that reduces the amplitude of the pendulum over time.

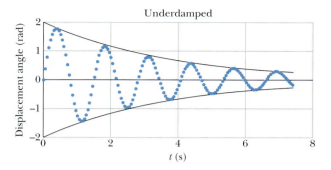

FIGURE 16-28 ■ Actual data for the angular displacement from the vertical equilibrium of the damped physical pendulum (shown in Fig. 16-27). Of most interest here is that the magnetic damping force causes the amplitude of oscillation to decrease exponentially in time. This is shown by the shape of the "envelope function" that best conforms to the displacement maxima or minima, found to be $\Theta e^{-(b/2I)t} = (2 \text{ rad})e^{-(0.27s^{-1})t}$.

edge mass attached to an aluminum disk to create the physical pendulum shown in Fig. 16-27. A strong magnet placed in the vicinity of the disk exerts a drag force on the disk that is proportional to the angular velocity of the disk $d\theta/dt$. (This kind of damping is a result of eddy currents induced in the disk. You can refer to Section 31-5 for more details.)

Both the pendulum and the mass–spring systems have linear restoring forces and velocity-dependent drag forces. Thus, we can modify Eqs. 16-37, 16-38, and 16-39 to describe the pendulum motion. We need to replace the linear displacement y with an angular displacement θ, the spring mass m with I (where I is the total rotational inertia of the disk and its edge mass), and the spring constant k with a torque strength mgL (where in our case L is the radius of the disk and m is the edge mass). This gives us a new motion equation for small angles of pendulum oscillation

$$\theta(t) = \Theta e^{-bt/2I}\cos(\omega' t + \phi_0), \qquad (16\text{-}40)$$

where the angular amplitude (or "envelope" function) $\Theta e^{-bt/2I}$ varies in time.

The feature of the data that we are most interested in here is the way the angular amplitude of the pendulum decreases with time when the pendulum is underdamped. This is shown in Fig. 16-28. The exponential behavior is obvious for these data.

Critical Damping and Overdamping

In Figs. 16-27 and 16-28 we have shown underdamped motions for two different but mathematically similar systems. When the damping coefficient in our systems are so large that

$$\omega' = \sqrt{\frac{mgL}{I} - \frac{b^2}{4I^2}} = 0 \qquad \text{(physical pendulum)}$$

$$\text{or} \quad \omega' = \sqrt{\frac{k}{m} - \frac{b^2}{4m^2}} = 0 \qquad \text{(mass–spring)},$$

the system undergoes **critical damping.** Under this circumstance, a system that is displaced will settle back exponentially to its equilibrium point in the *minimum possible time* without oscillating. We have not presented the equations that describe what happens when the damping coefficient b in Eq. 16-38 gets so large that the square root becomes negative, a condition known as **overdamping.** An overdamped system also settles back to its equilibrium exponentially, but it takes more time (Fig. 16-29).

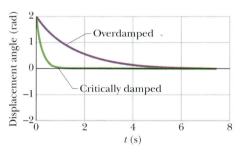

FIGURE 16-29 ■ The theoretically calculated angular displacement vs. time values for the damped pendulum shown in Fig. 16-27. The bottom exponential curve shows what happens when the pendulum edge mass is critically damped and settles back to equilibrium without oscillating in a minimum time of about 1 s. The top exponential curve shows an overdamped condition when the edge mass takes about 6 s to reach equilibrium.

Engineers make use of their understanding of critical damping and overdamping in the design of automobile shock absorbers. A shock absorber is designed so that when a car hits a typical bump, the springs that connect the chassis to the wheels return to equilibrium slowly without oscillating too much. Too little damping causes the passengers to bounce up and down, and too much damping gives a rough ride because the car cannot respond quickly to a bump. Note what happens when you push the front of a car down suddenly and let go. If it returns to equilibrium without oscillating, the car is either critically damped or overdamped. If is oscillates a bit, it is underdamped.

READING EXERCISE 16-10: Here are three sets of values for the damping constant and mass for the damped oscillator of Fig. 16-25. Using Eq. 16-39, rank the sets according to the time required for the mechanical energy to decrease to one-fourth of its initial value, greatest first. No calculations are needed.

Set 1:	b_0	m_0
Set 2:	$6b_0$	$4m_0$
Set 3:	$3b_0$	m_0

■

TOUCHSTONE EXAMPLE 16-4: Damped Mass–Spring

For the damped oscillator of Fig. 16-25, $m = 250$ g, $k = 85$ N/m, and $b = 70$ g/s.

(a) What is the period of the motion?

SOLUTION ■ The **Key Idea** here is that because $b \ll \sqrt{km} = 4.6$ kg/s, the period is approximately that of the undamped oscillator. From Eq. 16-12, we then have

$$T = 2\pi\sqrt{\frac{m}{k}} = 2\pi\sqrt{\frac{0.25 \text{ kg}}{85 \text{ N/m}}} = 0.34 \text{ s.} \quad \text{(Answer)}$$

(b) How long does it take for the amplitude of the damped oscillations to drop to half its initial value?

SOLUTION ■ Now the **Key Idea** is that the amplitude at time t is displayed in Eq. 16-37 as $Ye^{-bt/2m}$. It has the value Y at $t = 0$. Thus, we must find the value of t for which

$$Ye^{-bt/2m} = \tfrac{1}{2}Y.$$

Canceling Y and taking the natural logarithm of the equation that remains, we have $\ln \tfrac{1}{2}$ on the right side and

$$\ln(e^{-bt/2m}) = -bt/2m$$

on the left side. Thus,

$$t = \frac{-2m \ln\tfrac{1}{2}}{b} = \frac{-(2)(0.25 \text{ kg})(\ln\tfrac{1}{2})}{0.070 \text{ kg/s}}$$

$$= 5.0 \text{ s.} \quad \text{(Answer)}$$

Because $T = 0.34$ s, this is about 15 periods of oscillation.

(c) How long does it take for the mechanical energy to drop to one-half its initial value?

SOLUTION ■ Here the **Key Idea** is that, from Eq. 16-39, the mechanical energy at time t is $\tfrac{1}{2}kY^2e^{-bt/m}$. It has the value $\tfrac{1}{2}kY^2$ at $t = 0$. Thus, we must find the value of t for which

$$\tfrac{1}{2}kY^2e^{-bt/m} = \tfrac{1}{2}(\tfrac{1}{2}kY^2).$$

If we divide both sides of this equation by $\tfrac{1}{2}kY^2$ and solve for t as we did above, we find

$$t = \frac{-m \ln\tfrac{1}{2}}{b} = \frac{-(0.25 \text{ kg})(\ln\tfrac{1}{2})}{0.070 \text{ kg/s}} = 2.5 \text{ s.} \quad \text{(Answer)}$$

This is exactly half the time we calculated in (b), or about 7.5 periods of oscillation. Figure 16-30 was drawn to illustrate this touchstone example. Since the system's mechanical energy depends on the square of the amplitude, it decreases more rapidly than the amplitude does.

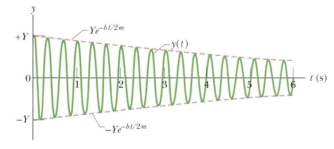

FIGURE 16-30 ■ (a) This diagram shows a period of about 0.34 s. (b) Since amplitude has dropped to about $0.7Y$ after 2.5 s, the diagram shows a system mechanical energy of $(0.7Y)^2 = 0.5Y^2$ at that time.

16-8 Forced Oscillations and Resonance

Sometimes we would like to maintain oscillations longer than they would naturally continue because of the damping forces. For example, if you were on a swing that was given only one big push, you would go up and back a few times before the mechanical energy was completely lost and you came to a stop. Although we cannot totally eliminate such loss of mechanical energy, we can replenish the energy from some source. As an example, you know that by swinging your legs or torso you can "pump" a swing to maintain or enhance the oscillations. In doing this, you transfer biochemical energy to mechanical energy of the oscillating system.

A person swinging without pumping their legs or without anyone pushing undergoes *free oscillation*. However, if someone pushes the swing periodically, the swing has *forced*, or *driven*, *oscillations*. Two angular frequencies are associated with a system undergoing driven oscillations: (1) the *natural* angular frequency ω of the system while oscillating freely with the inevitable damping forces present, and (2) the angular frequency ω_d of the external driving force.

We can use Fig. 16-25 to represent an idealized forced simple harmonic oscillator if we cause the structure marked "rigid support" to move up and down at an angular frequency ω_d that we can adjust. This forced oscillator will settle down to oscillate at our chosen angular frequency ω_d of the driving force, and its displacement $y(t)$ is given by

$$y(t) = Y\cos(\omega_d t + \phi_1), \tag{16-41}$$

where Y is the amplitude of the oscillations.

How large the displacement amplitude Y is depends on a complicated function of ω_d and ω. The maximum velocity $V = \omega' Y$ of the oscillations is easier to describe: it is greatest when

$$\omega_d = \omega' \quad \text{(resonance)}, \tag{16-42}$$

which is a condition called **resonance.** Equation 16-42 is also *approximately* the condition at which the displacement amplitude Y of the oscillations is greatest. Thus, if you push a swing at its natural angular frequency, the displacement and maximum velocities will increase to large values, a fact that children learn quickly by trial and error. If you push at other angular frequencies, either higher or lower, the displacement and maximum velocities will be smaller.

Figure 16-31 shows how the displacement amplitude of a damped, driven oscillator depends on the angular frequency ω_d of the driving force, for three values of the damping coefficient b. Note that for all three the amplitude is approximately greatest when $\omega_d/\omega' = 1$—that is, when the resonance condition of Eq. 16-42 is satisfied. The curves of Fig. 16-31 show that less damping gives a taller and narrower *resonance peak*. A system with a tall, narrow resonance curve is referred to as having a "high-Q."

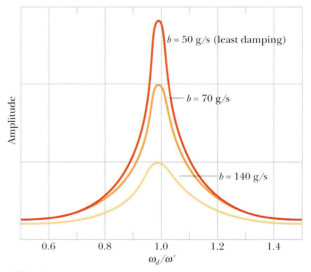

FIGURE 16-31 ■ The displacement amplitude of a forced oscillator varies as the angular frequency ω_d of the driving force is varied. The amplitude is greatest *approximately* at $\omega_d/\omega' = 1$, the resonance condition. The curves here correspond to three values of the damping constant b.

Resonance and Earthquake Damage

All mechanical structures have one or more natural angular frequencies, and if a structure is subjected to a strong external driving force that matches one of these angular frequencies, the resulting oscillations of the structure may rupture it. Thus, for example, aircraft designers must make sure that none of the natural angular frequencies at which a wing can oscillate matches the angular frequency of the engines in flight. A wing that flaps violently at certain engine speeds would obviously be dangerous.

Mexico's earthquake in September 1985 was a major earthquake (8.1 on the Richter scale), but the seismic waves from it should have been too weak to cause extensive damage when they reached Mexico City about 400 km away. However, Mexico City is largely built on an ancient lake bed, where the soil is still soft with water. Although the amplitude of the seismic waves was weak in the firmer ground en route to Mexico City, their amplitude substantially increased in the loose soil of the city. Maximum accelerations of the waves were as much as $0.20g$, and the angular frequency was (surprisingly) concentrated around 3 rad/s. Not only did the ground begin oscillating with large amplitudes, but also many of the buildings with intermediate height had resonant angular frequencies of about 3 rad/s. Most of those buildings collapsed during the violent shaking, while shorter buildings (higher resonant angular frequencies) and taller buildings (with lower resonant angular frequencies) remained standing.

Problems

SEC. 16-4 ■ VELOCITY AND ACCELERATION FOR SHM

1. Object Undergoing SHM An object undergoing simple harmonic motion takes 0.25 s to travel from one point of zero velocity to the next such point. The distance between those points is 36 cm. Calculate the object's (a) period, (b) frequency, and (c) amplitude.

2. Oscillating Block An oscillating block–spring system takes 0.75 s to begin repeating its motion. Find its (a) period, (b) frequency in hertz, and (c) angular frequency in radians per second.

3. Oscillator An oscillator consists of a block of mass 0.500 kg connected to a spring. When set into oscillation with amplitude 35.0 cm, the oscillator repeats its motion every 0.500 s. Find (a) the period, (b) the frequency, (c) the angular frequency, (d) the spring constant, (e) the maximum speed, and (f) the magnitude of the maximum force on the block from the spring.

4. Maximum Acceleration What is the maximum acceleration of a platform that oscillates with an amplitude of 2.20 cm at a frequency of 6.60 Hz?

5. Loudspeaker A loudspeaker produces a musical sound by means of the oscillation of a diaphragm. If the amplitude of oscillation is limited to 1.0×10^{-3} mm, what frequencies will result in the magnitude of the diaphragm's acceleration exceeding g?

6. Spring Balance The scale of a spring balance that reads from 0 to 15.0 kg is 12.0 cm long. A package suspended from the balance is found to oscillate vertically with a frequency of 2.00 Hz. (a) What is the spring constant? (b) How much does the package weigh?

7. A Particle of Mass A particle with a mass of 1.00×10^{-20} kg is oscillating with simple harmonic motion with a period of 1.00×10^{-5} s and a maximum speed of 1.00×10^{3} m/s. Calculate (a) the angular frequency and (b) the maximum displacement of the particle.

8. A Small Body A small body of mass 0.12 kg is undergoing simple harmonic motion of amplitude 8.5 cm and period 0.20 s. (a) What is the magnitude of the maximum force acting on it? (b) If the oscillations are produced by a spring, what is the spring constant?

9. Electric Shaver In an electric shaver, the blade moves back and forth over a distance of 2.0 mm in simple harmonic motion, with frequency 120 Hz. Find (a) the amplitude, (b) the maximum blade speed, and (c) the magnitude of the maximum blade acceleration.

10. Speaker Diaphragm A loudspeaker diaphragm is oscillating in simple harmonic motion with a frequency of 440 Hz and a maximum displacement of 0.75 mm. What are (a) the angular frequency, (b) the maximum speed and (c) the magnitude of the maximum acceleration?

11. Automobile Spring An automobile can be considered to be mounted on four identical springs as far as vertical oscillations are concerned. The springs of a certain car are adjusted so that the oscillations have a frequency of 3.00 Hz. (a) What is the spring constant of each spring if the mass of the car is 1450 kg and the mass is evenly distributed over the springs? (b) What will be the oscillation frequency if five passengers, averaging 73.0 kg each, ride in the car? (Again, consider an even distribution of mass.)

12. A Body Oscillates A body oscillates with simple harmonic motion according to the equation

$$x = (6.0 \text{ m}) \cos[(3\pi \text{ rad/s})t + \pi/3 \text{ rad}].$$

At $t = 2.0$ s, what are (a) the displacement, (b) the velocity, (c) the acceleration, and (d) the phase of the motion? Also, what are (e) the frequency and (f) the period of the motion?

13. Piston in Cylinder The piston in the cylinder head of a locomotive has a stroke (twice the amplitude) of 0.76 m. If the piston moves with simple harmonic motion with a frequency of 180 rev/min, what is its maximum speed?

14. BMMD Astronauts sometimes use a device called a body-mass measuring device (BMMD). Designed for use on orbiting space vehicles, its purpose is to allow astronauts to measure their mass in the "weightless" conditions in Earth orbit. The BMMD is a spring-mounted chair; an astronaut measures his or her period of oscillation in the chair; the mass follows from the formula for the period of an oscillating block–spring system. (a) If M is the mass of the astronaut and m the effective mass of that part of the BMMD that also oscillates, show that

$$M = (k/4\pi^2)T^2 - m,$$

where T is the period of oscillation and k is the spring constant. (b) The spring constant was $k = 605.6$ N/m for the BMMD on Skylab Mission Two; the period of oscillation of the empty chair was 0.90149 s. Calculate the effective mass of the chair. (c) With an astronaut in the chair, the period of oscillation became 2.08832 s. Calculate the mass of the astronaut.

15. Harbor At a certain harbor, the tides cause the ocean surface to rise and fall a distance d (from highest level to lowest level) in simple harmonic motion, with a period of 12.5 h. How long does it take for the water to fall a distance $d/4$ from its highest level?

16. Two Blocks In Fig. 16-32 two blocks ($m = 1.0$ kg and $M = 10$ kg) and a spring ($k = 200$ N/m) are arranged on a horizontal, frictionless surface. The coefficient of static friction between the two blocks is 0.40. What amplitude of simple harmonic motion of the spring–blocks system puts the smaller block on the verge of slipping over the larger block?

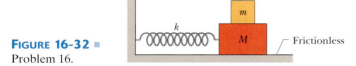

FIGURE 16-32 ■
Problem 16.

17. Shake Table A block is on a horizontal surface (a shake table) that is moving back and forth horizontally with simple harmonic motion of frequency 2.0 Hz. The coefficient of static friction between block and surface is 0.50. How great can the amplitude of the SHM be if the block is not to slip along the surface?

18. Block and Piston A block rides on a piston that is moving vertically with simple harmonic motion. (a) If the SHM has period 1.0 s, at what amplitude of motion will the block and piston separate? (b) If the piston has an amplitude of 5.0 cm, what is the maximum frequency for which the block and piston will be in contact continuously?

19. Oscillator An oscillator consists of a block attached to a spring ($k = 400$ N/m). At some time t, the position (measured from the system's equilibrium location), velocity, and acceleration of the block are $x = 0.100$ m, $v = -13.6$ m/s, and $a = -123$ m/s². Calculate (a) the frequency of oscillation, (b) the mass of the block, and (c) the amplitude of the motion.

20. Simple Harmonic Oscillator A simple harmonic oscillator consists of a block of mass 2.00 kg attached to a spring of spring constant 100 N/m. When $t = 1.00$ s, the position and velocity of the block are $x = 0.129$ m and $v = 3.415$ m/s. (a) What is the amplitude of the oscillations? What were the (b) position and (c) velocity of the block at $t = 0$ s?

21. Massless Spring A massless spring hangs from the ceiling with a small object attached to its lower end. The object is initially held at rest in a position y_1 such that the spring is at its rest length. The object is then released from y_1 and oscillates up and down, with its lowest position being 10 cm below y_1. (a) What is the frequency of the oscillation? (b) What is the speed of the object when it is 8.0 cm below the initial position? (c) An object of mass 300 g is attached to the first object, after which the system oscillates with half the original frequency. What is the mass of the first object? (d) Relative to y_1 where is the new equilibrium (rest) position with both objects attached to the spring?

22. Two Particles Two particles execute simple harmonic motion of the same amplitude and frequency along close parallel lines. They pass each other moving in opposite directions each time their displacement is half their amplitude. What is their phase difference?

23. Two Particles Oscillate Two particles oscillate in simple harmonic motion along a common straight-line segment of length A. Each particle has a period of 1.5 s, but they differ in phase by $\pi/6$ rad. (a) How far apart are they (in terms of A) 0.50 s after the lagging particle leaves one end of the path? (b) Are they then moving in the same direction, toward each other, or away from each other?

24. Two Identical Springs In Fig. 16-33, two identical springs of spring constant k are attached to a block of mass m and to fixed supports. Show

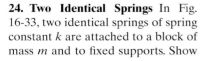

FIGURE 16-33 ■ Problems 24 and 25.

that the block's frequency of oscillation on the frictionless surface is

$$f = \frac{1}{2\pi}\sqrt{\frac{2k}{m}}.$$

25. Block and Two Springs Suppose that the two springs in Fig. 16-33 have different spring constants k_1 and k_2. Show that the frequency f of oscillation of the block is then given by

$$f = \sqrt{f_1^2 + f_2^2},$$

where f_1 and f_2 are the frequencies at which the block would oscillate if connected only to spring 1 or only to spring 2.

26. Tuning Fork The end of one of the prongs of a tuning fork that executes simple harmonic motion of frequency 1000 Hz has an amplitude of 0.40 mm. Find (a) the magnitude of the maximum acceleration and (b) the maximum speed of the end of the prong. Find (c) the magnitude of the acceleration and (d) the speed of the end of the prong when the end has a displacement of 0.20 mm.

27. Two Springs Are Joined In Fig. 16-34, two springs are joined and connected to a block of mass m. The surface is frictionless. If the springs both have spring constant k, show that

$$f = \frac{1}{2\pi}\sqrt{\frac{k}{2m}}$$

gives the block's frequency of oscillation.

28. Block on Incline In Fig. 16-35, a block weighing 14.0 N, which slides without friction on a 40.0° incline, is connected to the top of the incline by a massless spring of unstretched length 0.450 m and spring constant 120 N/m. (a) How far from the top of the incline does the block stop? (b) If the block is pulled slightly down the incline and released, what is the period of the resulting oscillations?

29. Unstretched Length A uniform spring with unstretched length L and spring constant k is cut into two pieces of unstretched lengths L_1 and L_2, with $L_1 = nL_2$. What are the corresponding spring constants (a) k_1 and (b) k_2 in terms of n and k? If a block is attached to the original spring, as in Fig. 16-10, it oscillates with frequency f. If the spring is replaced with the piece L_1 or L_2, the corresponding frequency is f_1 or f_2. Find (c) f_1 and (d) f_2 in terms of f.

30. Ore Cars In Fig. 16-36, three 10 000 kg ore cars are held at rest on a 30° incline on a mine railway using a cable that is parallel to the incline. The cable stretches 15 cm just before the coupling between the two lower cars breaks, detaching the lowest car. Assuming that the cable obeys Hooke's law, find (a) the frequency and (b) the amplitude of the resulting oscillations of the remaining two cars.

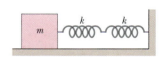

FIGURE 16-34 ■
Problem 27.

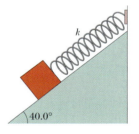

FIGURE 16-35 ■
Problem 28.

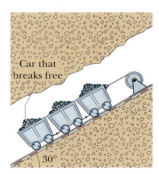

FIGURE 16-36 ■
Problem 30.

31. Balance Wheel The balance wheel of a watch oscillates with a rotational amplitude of π rad and a period of 0.500 s. Find (a) the maximum rotational speed of the wheel, (b) the rotational speed of the wheel when its displacement is $\pi/2$ rad, and (c) the magnitude of the rotational acceleration of the wheel when its displacement is $\pi/4$ rad.

32. Flat Disk A flat uniform circular disk has a mass of 3.00 kg and a radius of 70.0 cm. It is suspended in a horizontal plane by a vertical wire attached to its center. If the disk is rotated 2.50 rad about the wire, a torque of 0.0600 N · m is required to maintain that orientation. Calculate (a) the rotational inertia of the disk about the wire, (b) the torsion constant, and (c) the angular frequency of this torsion pendulum when it is set oscillating.

SEC. 16-5 ■ GRAVITATIONAL PENDULA

33. Simple Pendulum What is the length of a simple pendulum that marks seconds by completing a full swing from left to right and then back again every 2.0 s?

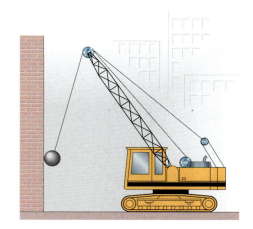

FIGURE 16-37 ■ Problem 34.

34. Demolition Ball In Fig. 16-37, a 2500 kg demolition ball swings from the end of a crane. The length of the swinging segment of cable is 17 m. (a) Find the period of the swinging, assuming that the system can be treated as a simple pendulum. (b) Does the period depend on the ball's mass?

35. Physical Pendulum A physical pendulum consists of a meter stick that is pivoted at a small hole drilled through the stick a distance d from the 50 cm mark. The period of oscillation is 2.5 s. Find d.

36. Trapeze A performer seated on a trapeze is swinging back and forth with a period of 8.85 s. If she stands up, thus raising the center of mass of the *trapeze + performer* system by 35.0 cm, what will be the new period of the system? Treat *trapeze + performer* as a simple pendulum.

37. Pivoting Long Rod A pendulum is formed by pivoting a long thin rod of length L and mass m about a point on the rod that is a distance d above the center of the rod. (a) Find the period of this pendulum in terms of d, L, m, and g, assuming small-amplitude swinging. What happens to the period if (b) d is decreased, (c) L is increased, or (d) m is increased?

FIGURE 16-38 ■ Problem 38.

38. Solid Disk In Fig. 16-38, a physical pendulum consists of a uniform solid disk (of mass M and radius R) supported in a vertical plane by a pivot located a distance d from the center of the disk. The disk is displaced by a small angle and released. Find an expression for the period of the resulting simple harmonic motion.

39. Oscillating Physical Pendulum The pendulum in Fig. 16-39, consists of a uniform disk with radius 10.0 cm and mass 500 g attached to a uniform rod with length 500 mm and mass 270 g. (a) Calculate the rotational inertia of the pendulum about the pivot point. (b) What is the distance between the pivot point and the center of mass of the pendulum? (c) Calculate the period of oscillation.

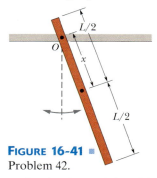

FIGURE 16-39 ■ Problem 39.

40. Pendulum with Disk A uniform circular disk whose radius R is 12.5 cm is suspended as a physical pendulum from a point on its rim. (a) What is its period? (b) At what radial distance $r < R$ is there a pivot point that gives the same period?

41. Long Uniform Rod In the overhead view of Fig. 16-40, a long uniform rod of length L and mass m is free to rotate in a horizontal plane about a vertical axis through its center. A spring with force constant k is connected horizontally between one end of the rod and a fixed wall. When the rod is in equilibrium, it is parallel to the wall. What is the period of the small oscillations that result when the rod is rotated slightly and released?

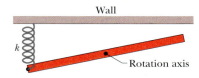

FIGURE 16-40 ■ Problem 41.

42. A Stick A stick with length L oscillates as a physical pendulum, pivoted about point O in Fig. 16-41: (a) Derive an expression for the period of the pendulum in terms of L and x, the distance from the pivot point to the center of mass of the pendulum. (b) For what value of x/L is the period a minimum? (c) Show that if $L = 1.00$ m and $g = 9.80$ m/s^2, this minimum period is 1.53 s.

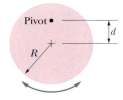

FIGURE 16-41 ■ Problem 42.

43. Frequency What is the frequency of a simple pendulum 2.0 m long (a) in a room, (b) in an elevator accelerating upward at a rate of 2.0 m/s^2, and (c) in free fall?

44. In a Car A simple pendulum of length L and mass m is suspended in a car that is traveling with constant speed $|\vec{v}|$ around a circle of radius R. If the pendulum undergoes small oscillations in a radial direction about its equilibrium position, what will be its frequency of oscillation?

45. The Bob The bob on a simple pendulum of length R moves in an arc of a circle. (a) By considering that the radial acceleration of the bob as it moves through its equilibrium position is that for uniform circular motion (v^2/R), show that the tension in the string at that position is $mg(1 + \Theta^2)$ if the angular amplitude Θ is small. (See "Trigonometric Expansions" in Appendix E.)

(b) Is the tension at other positions of the bob greater, smaller, or the same?

46. Angular Amplitude For a simple pendulum, find the angular amplitude Θ at which the restoring torque required for simple harmonic motion deviates from the actual restoring torque by 1.0%. (See "Trigonometric Expansions" in Appendix E.)

47. Wheel Rotates A wheel is free to rotate about its fixed axle. A spring is attached to one of its spokes a distance r from the axle, as shown in Fig. 16-42. (a) Assuming that the wheel is a hoop of mass m and radius R, obtain the angular frequency of small oscillations of this system in terms of m, R, r, and the spring constant k. How does the result change if (b) $r = R$ and (c) $r = 0$?

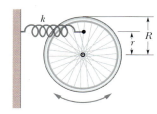

FIGURE 16-42 ■
Problem 47.

SEC. 16-6 ■ ENERGY IN SIMPLE HARMONIC MOTION

48. Large Slingshot A (hypothetical) large slingshot is stretched 1.50 m to launch a 130 g projectile with speed sufficient to escape from Earth (11.2 km/s). Assume the elastic bands of the slingshot obey Hooke's law. (a) What is the spring constant of the device, if all the elastic potential energy is converted to kinetic energy? (b) Assume that an average person can exert a force of 220 N. How many people are required to stretch the elastic bands?

49. Mechanical Energy Find the mechanical energy of a block–spring system having a spring constant of 1.3 N/cm and an oscillation amplitude of 2.4 cm.

50. Block–Spring An oscillating block–spring system has a mechanical energy of 1.00 J, an amplitude of 10.0 cm, and a maximum speed of 1.20 m/s. Find (a) the spring constant, (b) the mass of the block, and (c) the frequency of oscillation.

51. Horizontal Frictionless A 5.00 kg object on a horizontal frictionless surface is attached to a spring with spring constant 1000 N/m. The object is displaced from equilibrium 50.0 cm horizontally and given an initial velocity of 10.0 m/s back toward the equilibrium position. (a) What is the frequency of the motion? What are (b) the initial potential energy of the block–spring system, (c) the initial kinetic energy, and (d) the amplitude of the oscillation?

52. Block of Mass M A block of mass M, at rest on a horizontal frictionless table, is attached to a rigid support by a spring of constant k. A bullet of mass m and velocity $\vec{v}$ strikes the block as shown in Fig. 16-43. The bullet is embedded in the block. Determine (a) the speed of the block immediately after the collision and (b) the amplitude of the resulting simple harmonic motion.

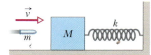

FIGURE 16-43 ■ Problem 52.

53. Displacement in SHM When the displacement in SHM is one-half the amplitude X, what fraction of the total energy is (a) kinetic energy and (b) potential energy? (c) At what displacement, in terms of the amplitude, is the energy of the system half kinetic energy and half potential energy?

54. Particle Undergoing SHM A 10 g particle is undergoing simple harmonic motion with an amplitude of 2.0×10^{-3} m and a maximum

acceleration of magnitude 8.0×10^{-3} m/s². The phase constant is $-\pi/3$ rad. (a) Write an equation for the force on the particle as a function of time. (b) What is the period of the motion? (c) What is the maximum speed of the particle? (d) What is the total mechanical energy of this simple harmonic oscillator?

55. Block Suspended from Spring A 4.0 kg block is suspended from a spring with a spring constant of 500 N/m. A 50 g bullet is fired into the block from directly below with a speed of 150 m/s and becomes embedded in the block. (a) Find the amplitude of the resulting simple harmonic motion. (b) What fraction of the original kinetic energy of the bullet is transferred to mechanical energy of the harmonic oscillator?

56. Vertical Spring A vertical spring stretches 9.6 cm when a 1.3 kg block is hung from its end. (a) Calculate the spring constant. This block is then displaced an additional 5.0 cm downward and released from rest. Find (b) the period, (c) the frequency, (d) the amplitude, and (e) the maximum speed of the resulting SHM.

SEC. 16-7 ■ DAMPED SIMPLE HARMONIC MOTION

57. Amplitude Ratio In Touchstone Example 16-4, what is the ratio of the amplitude of the damped oscillations to the initial amplitude when 20 full oscillations have elapsed?

58. Lightly Damped The amplitude of a lightly damped oscillator decreases by 3.0% during each cycle. What fraction of the mechanical energy of the oscillator is lost in each full oscillation?

59. System Shown For the system shown in Fig. 16-25, the block has a mass of 1.50 kg and the spring constant is 8.00 N/m. The damping force is given by $-b(dy/dt)$, where $b = 230$ g/s. Suppose that the block is initially pulled down a distance 12.0 cm and released. (a) Calculate the time required for the amplitude of the resulting oscillations to fall to one-third of its initial value. (b) How many oscillations are made by the block in this time?

60. Suspension System Assume that you are examining the oscillation characteristics of the suspension system of a 2000 kg automobile. The suspension "sags" 10 cm when the entire automobile is placed on it. Also, the amplitude of oscillation decreases by 50% during one complete oscillation. Estimate the values of (a) the spring constant k and (b) the damping constant b for the spring and shock absorber system of one wheel, assuming each wheel supports 500 kg.

SEC. 16-8 ■ FORCED OSCILLATIONS AND RESONANCE

61. Rigid Support For Eq. 16-41, suppose the amplitude Y is given by

$$Y = \frac{F^{\text{max}}}{[m^2(\omega_d^2 - \omega^2)^2 + b^2\omega_d^2]^{1/2}},$$

where F^{max} is the (constant) amplitude of the external oscillating force exerted on the spring by the rigid support in Fig. 16-25. At resonance, what are (a) the amplitude and (b) the velocity amplitude of the oscillating object?

62. Washboard Road A 1000 kg car carrying four 82 kg people travels over a rough "washboard" dirt road with corrugations 4.0 m apart, which cause the car to bounce on its spring suspension. The car bounces with maximum amplitude when its speed is 16 km/h. The car now stops, and the four people get out. By how much does the car body rise on its suspension due to this decrease in mass?

Additional Problems

63. Where's the Force? A 50 gram mass is hanging from a spring whose unstretched length is 10 cm and whose spring constant is 2.5 N/m, as shown in Fig. 16-44. In the list below are described five situations. In some of the situations, the mass is at rest and remains at rest. In other situations, at the instant described, the mass is in the middle of an oscillation initiated by a person pulling the mass downward 5 cm from its equilibrium position and releasing it. Ignore both air resistance and internal damping in the spring. At the time the situation occurs, indicate whether the force vector requested points up (U), down (D), or has a magnitude of zero newtons (0). (a) The force on the mass exerted by the spring when the mass is at its equilibrium position and is at rest. (b) The force on the mass exerted by the spring when the mass is at its equilibrium position and is moving downward. (c) The net force on the mass when the mass is at its equilibrium position and is moving upward. (d) The force on the mass exerted by the spring when it is at the top of its oscillation. (e) The net force on the mass when it is at the top of its oscillation.

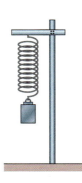

FIGURE 16-44 ■ Problem 63.

64. Swinging Ball A pendulum consisting of a massive ball on a light but nearly rigid rod is shown at two successive times in Fig. 16-45. The maximum angle of displacement of the pendulum from its equilibrium point is 25°. (a) If the length of the rod is *R* and the mass of the ball is *m*, find an expression for the speed of the ball at the point shown in Fig. 16-45*b*. (b) If the ball in Fig 16-45*b* is moving to the left at the instant of the snapshot, in what direction does its acceleration point at this instant in time? (c) In what direction does the acceleration of the ball in Fig. 16-45*a* point?

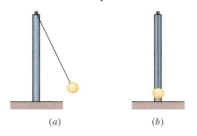

(*a*)   (*b*)

FIGURE 16-45 ■ Problem 64.

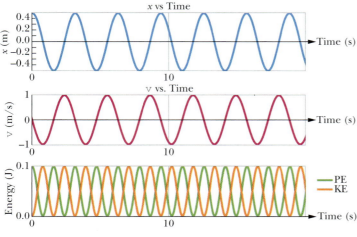

FIGURE 16-46 ■ Problem 65.

65. Oscillating Energies The graphs in Fig. 16-46 represent a computer simulation of a mass on a spring. The kinetic and potential energies are plotted in the bottom graph, in addition to the position and velocity of the oscillator (upper two graphs). The potential and kinetic energy curves oscillate, but not about zero. They also seem to oscillate twice as fast as the position and velocity curves. Is this correct? Explain.

66. Oscillating Graphs A mass is hanging from a spring off the edge of a table. The position of the mass is measured by a sonic ranger sitting on the floor 25 cm below the mass's equilibrium position. At some time, the mass is started oscillating. At a later time, the sonic ranger begins to take data.

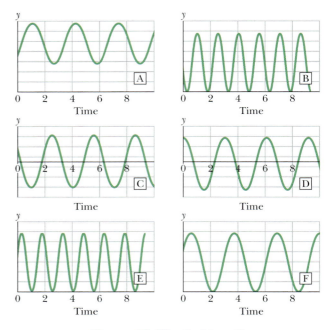

FIGURE 16-47 ■ Problem 66.

Figure 16-47 shows a series of graphs associated with the motion of the mass and a series of physical quantities. The graph labeled (A) is a graph of the mass's position as measured by the ranger. For each physical quantity, identify which graph could represent that quantity for this situation. If none are possible, answer N.

(a) Velocity of the mass
(b) Net force on the mass
(c) Force exerted by the spring on the mass
(d) Kinetic energy of the spring–mass system
(e) Potential energy of the spring–mass–Earth system
(f) Gravitational potential energy of the spring–mass–Earth system

67. Pendulum Graphs Some of the graphs shown in Fig. 16-48 represent the motion of a pendulum—a massive ball attached to a rigid, nearly massless rod, which in turn is attached to a rigid, nearly frictionless pivot. Four graphs of the pendulum's angle as a function of time are shown. Below are a set of four initial conditions and a denial. Match each graph with its most likely initial conditions (or with the denial). Note that the scales on the y axes are not necessarily the same. (There is not necessarily a one-to-one match.)

(1) $\theta_0 = 120°, (d\theta/dt)_0 = 0°/s$
(2) $\theta_0 = 173°, (d\theta/dt)_0 = 30°/s$
(3) $\theta_0 = 6°, (d\theta/dt)_0 = 0°/s$
(4) $\theta_0 = 173°, (d\theta/dt)_0 = 0°/s$
(5) Not a possible pendulum graph.

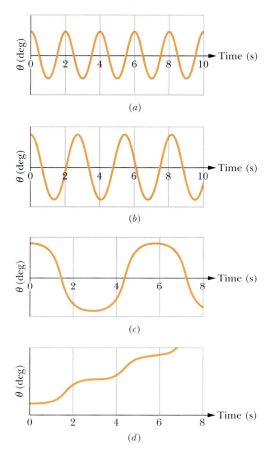

(a)

(b)

(c)

(d)

FIGURE 16-48 ■ Problem 67.

68. Swingin' in the Rain There is a forest in Italy that has many waterfalls. At one of the waterfalls, a long rope hangs down from the top of the cliff near the waterfall and has a seat on the bottom. Adventurous visitors could hop onto the seat and swing down into the waterfall. Their starting angle seems to be about 20°. It takes one of these adventurous people 8 seconds to swing out and back. Estimate the length of the rope and the speed with which they pass through the waterfall.

69. To What Angle? A small metal ball of mass m hangs from a pivot by a rigid, light metal rod of length R. The ball is swinging back and forth with an amplitude that remains small throughout its motion, $\theta^{max} \leq 5°$. Ignore all damping. (a) The equation of motion of this ideal pendulum can be derived in a variety of ways and is

$$\frac{d^2\theta}{dt^2} = -\frac{g}{R}\sin\theta.$$

For small angles, show how this can be replaced by an approximate equation of motion that can be solved more easily than the

one given. (b) Write a general solution for the approximate equation of motion you obtained in (a) that works for any starting angle and angular velocity (as long as the angles stay in the range where the approximation is OK). Demonstrate that what you have written *is* a solution and show that at a time $t = 0$ your solution can have any given starting position and velocity. (c) If the length of the rod is 0.3 m, the mass of the ball is 0.2 kg, and the clock is started at a time when the ball is passing through the center ($\theta = 0$) and is moving with an angular speed of 0.1 rad/s, find the maximum angle your solution says the ball will reach. Can you use the approximate equation of motion for this motion? If the starting angle is not small, you cannot easily solve the equation of motion without a computer. But there are still things you can do. (d) Derive the energy conservation equation for the motion of the pendulum. (Do *not* use the small-amplitude approximation.) (e) If the pendulum is released from a starting angle of θ_1, what will be the maximum speed it travels at any point on its swing?

70. What's Wrong with cos? Observation of the oscillation of a mass on the end of a spring reveals that the detailed structure of the position as a function of time is fit very well by a function of the form

$$x(t) = X\cos(\omega t + \phi_0).$$

Yet subsequent observations give convincing evidence that this cannot be a good representation of the motion for long time periods. Explain what observation leads to this conclusion and resolve the apparent contradiction.

71. Where Is the Energy? A block of mass m is attached to a spring of spring constant k that is attached to a wall as shown in Fig. 16-49. (a) If the block starts at time $t = 0$ with the spring being at its rest length but the block having a velocity v_1, find a solution for the mass's position at all subsequent times. Make any assumptions you like in order to have a plausible but solvable model, but state your assumptions explicitly. (b) Are the energies at the times

FIGURE 16-49 ■ Problem 71.

$$t = 0, \quad t = \frac{\pi}{2}\sqrt{\frac{m}{k}}, \quad t = 2\pi\sqrt{\frac{m}{k}}$$

kinetic, potential, or a mixture of the two? (This is *not* a short-answer question. Show how you know.)

72. Damped Oscillator A class looked at the oscillation of a mass on a spring. They observed for 10 seconds and found its oscillation was well fit by assuming that the mass's motion was governed by Newton's Second Law with the spring force $F_x^{spring} = -k\Delta x$, where Δx represents the stretch or squeeze of the spring and k is the spring constant. However, it was also clear that this was not an adequate representation for times on the order of 10 minutes, since by that time the mass had stopped oscillating. (a) Suppose the mass was started at an initial position x_1 with a velocity 0. Write down the solution, $x(t)$ and $v_x(t)$, for the equations of motion of the mass using only the spring force. What is the total energy of the oscillating mass? (b) Assume that there is also a velocity-dependent force (the *damping* force) that the spring exerts on an object when it is moving. Let's make the simplest assumption that the dynamic-spring force is linear in the velocity, $F_x^{spring-dyn} = -\gamma v_x$. Assume further that over one period of oscillation the velocity-dependent piece is small. Therefore, take $x(t)$ and $v_x(t)$ to be given by the oscillation without damping. Calculate the work done by the damping force over one period in terms of the parameters k, m, and γ and the fraction of the energy lost in one period. (c) The mass used was 1 kg. From the description of the experiment, estimate the spring constant and the number of periods it took to lose half the energy. Use this and your result from (a) to estimate the approximate size of the damping constant γ.

73. Catching a Pellet and Oscillating

The following problem is a standard problem found in this text (and in many others). A block of mass M is at rest on a horizontal frictionless table. It is attached to a rigid support by a spring of constant k. A clay pellet having mass m and velocity v strikes the

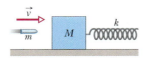

FIGURE 16-50
Problem 73.

block as shown in the figure and sticks to it. See Fig. 16-50. (a) Determine the velocity of the block immediately after the collision. (b) Determine the amplitude of the resulting simple harmonic motion. In order to solve this problem you must make a number of simplifying assumption; some are stated in the problem and some are not. First, solve the problem as stated. (c) Discuss the approximations you had to make in order to solve the problem. (There are at least five.)

74. Bungee Jump
As part of an open house, a physics department sets up a bungee jump from the top of a crane. See Fig. 16-51. Assume that one end of an elastic band will be firmly attached to the top of the crane and the other to the waist of a courageous participant. The participant will step off the edge of the crane house to be slowed and brought back up by the elastic band before hitting the ground. Assume the elastic band behaves like a Hooke's law spring. Estimate the length and spring constant of the elastic you would recommend be used.

FIGURE 16-51 ■ Problem 74.

17 | Transverse Mechanical Waves

When a beetle moves along the sand within a few tens of centimeters of this sand scorpion, the scorpion immediately turns toward the beetle and dashes to it (for lunch). The scorpion can do this without seeing (it is nocturnal) or hearing the beetle.

How can the scorpion so precisely locate its prey?

The answer is in this chapter.

17-1 Waves and Particles

Two ways to get in touch with a friend in a distant city are to write a letter and to use the telephone.

The first choice (the letter) involves the concept of "particles": A material object moves from one point to another, carrying information with it. Most of the preceding chapters deal with particles or with systems of particles.

The second choice (the telephone) involves the concept of "waves," the subject of this chapter and the next. In your telephone call, a sound wave carries your message from your vocal cords to the telephone. There, an electromagnetic wave takes over, passing along a copper wire or an optical fiber or through the atmosphere, possibly by way of a communications satellite. At the receiving end there is another sound wave, from a telephone to your friend's ear. Although the message is passed, nothing that you have touched reaches your friend. For example, if you have the flu, she can't catch it by talking to you on the phone since no matter (and therefore no virus) is passed between the two of you. In a wave, momentum and energy move from one point to another but no material object makes that journey.

Leonardo da Vinci understood waves when he wrote of water waves: "It often happens that the wave flees the place of its creation, while the water does not; like the waves made in a field of grain by the wind, where we see the waves running across the field while the grain remains in place."

Particles and *waves* are the two important entities described in classical physics. Both entities have positions and velocities associated with them. But there are ways in which these two entities are very different. The word *particle* suggests a tiny concentration of matter capable of transmitting energy and momentum through its movement from one place to another. For example, a baseball is a particle that can transmit the energy and momentum imparted to it by a pitcher to the catcher by moving from one to the other. Alternatively the energy and momentum associated with a *wave* is spread out in space and is transmitted without any matter moving from one place to another. For example, when the pitcher shouts at a catcher, she can disturb the air particles in her vicinity, but the sound wave that transmits her voice depends on a chain of temporary disturbances of the air and not on molecules moving from the pitcher to the catcher.

In this chapter and the next we will put particles aside for a while and learn about mechanical waves and how they travel. Then in Chapters 34, 36, and 37, we will explore the behavior of electromagnetic waves including radio and light waves.

17-2 Types of Waves

Waves are of three main types:

1. *Mechanical waves.* These waves are most familiar because we encounter them almost constantly; common examples include water waves, sound waves, and seismic waves. All these waves have certain central features: They are governed by Newton's laws, and they can only move through a material medium, such as water, air, and rock.

2. *Electromagnetic waves.* These waves are less familiar, but you use them constantly; common examples include visible and ultraviolet light, radio and television waves, microwaves, x-rays, and radar waves. These waves can transmit energy and momentum without a material medium. Light waves from stars, for example, travel through the vacuum of space to reach us. Electromagnetic waves are treated in Chapters 32, 34, 36 and 37.

3. *Matter waves.* Although these waves are commonly used in modern technology, they are probably very unfamiliar to you. These waves are associated with electrons, protons, and other fundamental particles, and even atoms and molecules. Because we commonly think of these things as constituting matter, such waves are called matter waves. Their behavior is described by the laws of quantum mechanics.

Much of what we discuss in this chapter applies to waves of all kinds. However, for specific examples we shall refer to mechanical waves because they are the simplest and most familiar. In this chapter we consider how to best describe ideal mechanical waves mathematically. An **ideal** mechanical wave does not lose mechanical energy or change its shape as it travels through a medium.

17-3 Pulses and Waves

Consider Figure 17-1, which shows a "Slinky wave demonstrator" consisting of a very long Slinky that hangs from long, evenly spaced strings. If we give a quick pull outward (to the left) on the left end of the Slinky and then give it a quick push back inward (to the right), a compression-expansion disturbance like that shown in Fig. 17-1 will propagate along the length of the Slinky. (You need to look carefully at Fig. 17-1, rather than Fig. 17-2, to see the effect). Since the back-and-forth oscillations of the Slinky coils are parallel to the direction in which the disturbance travels, the motion is said to be **longitudinal.**

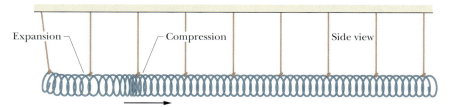

FIGURE 17-1 ■ A large Slinky demonstrator hanging from long, evenly spaced strings. Note the compression disturbance moving from left to right near the second and third strings with an expansion just behind it.

On the other hand, if we give the end of the Slinky a quick jerk back and forth (into and out of the page) at right angles to the line of the Slinky, a disturbance like that shown in Fig. 17-2 results and moves down the length of the Slinky. In this case, the displacement of the coils in the Slinky is *perpendicular* to the direction in which the disturbance travels (i.e., along the length of the Slinky). Such a disturbance is called **transverse.**

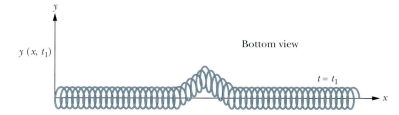

FIGURE 17-2 ■ A transverse or sideways disturbance or pulse moving from left to right along a Slinky wave demonstrator. In this case the viewer is lying on the floor underneath the Slinky looking up at time $t = t_1$.

In both cases discussed above, the singular disturbance is referred to as a **wave pulse** or **short wave.** If we repeat the motion that causes the disturbance (either the back-forth jerk or the push-pull motion) at regular time intervals, the result is a traveling and repeating disturbance that is referred to as a **continuous wave.** Just as with single pulses, waves can be longitudinal or transverse.

For example, if you give one end of a stretched string a single up-and-down jerk, a single *pulse* travels along the string as in Fig. 17-3a. This pulse and its motion can occur because the string is under tension. During the upward part of the jerk, when you pull your end of the string upward, it begins to pull upward on the

FIGURE 17-3 ■ (*a*) A single pulse travels along a stretched string. A typical string piece (marked with a dot) moves up and back only once as the pulse passes. Since the string piece's displacement is perpendicular to the wave direction, the pulse is called a *transverse wave*. (*b*) A continuous sinusoidal displacement at the left end also causes the piece of string to move up and down in a transverse direction except now the wave is continuous.

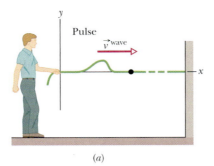

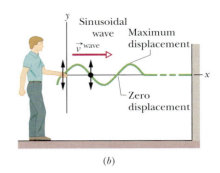

(*a*) (*b*)

adjacent section of the string via tension between the two sections. As the adjacent section moves upward, it begins to pull the next section upward, and so on. Meanwhile, suppose that you have now pulled down on your end of the string, completing the up-down stroke. As each section is moving upward in turn, it begins to be pulled back downward by neighboring sections that are already on the way down. The net result is that a distortion in the string's shape (the pulse) moves along the string at some velocity $\vec{v}^{\,wave}$. Note that although the disturbance moves down the string to the right, the parts of the string itself just move up and down. This is an extremely important, but subtle, point regarding pulses and waves. The wave or pulse travels, but the particles that make up the medium (string or otherwise) do not.

Sinusoidal Transverse Waves

If, instead of a single up-down stroke, you move your hand up and down repeatedly with simple harmonic motion as in Fig. 17-3*b*, a continuous *wave* travels along the string with a wave velocity denoted by $\vec{v}^{\,wave} = v_x^{wave}\hat{i}$. The vertical displacement of the string is perpendicular to the direction that the wave propagates (along the string), and so this wave is called a **transverse wave.** The larger the vertical displacement of your hand during the up-down stroke, the higher the peak and the lower the valley of the wave will be. This characteristic, measured as the magnitude of the maximum displacement of a small bit of string from its equilibrium position as the wave passes through it, is called the **amplitude** Y of the wave. This definition of amplitude is very similar to the one developed in the last chapter in regard to oscillations.

If we take a photograph of a transverse wave (perhaps the wave in the string of Fig. 17-3*b*) at some time $t = t_1$, we can see the wave shape and try to find a mathematical function that describes that shape or **wave form.** For example, if the motion of your hand is a sinusoidal function of time, the wave has a sinusoidal shape at any given instant, as in Fig. 17-3*b*. That is, the wave has the shape of a continuously repeating sine curve or cosine curve. (We consider here only an "undamped" string, in which no friction-like forces within the string cause the wave to die out as it travels along. In addition, we assume that the string is so long that we need not consider a wave rebounding from the far end and that the wave amplitude is small.)

The wave in Fig. 17-3*b* travels along a single line, so we call it a one-dimensional wave. For example, sound waves traveling in a pipe can only move in one dimension (but they can travel in two or three dimensions in other circumstances). For the wave in Fig. 17-3*b*, or any other one-dimensional wave or pulse, the convention is to define the line of travel as the *x* axis. As we can see from Fig. 17-3*b*, the vertical displacement of a bit of string at a given time depends on how far down along the string we make the measurement. Hence, we say that the vertical displacement of the string is a function of the horizontal position, *x*. The distance (in this example the horizontal distance) the wave travels in one cycle (for example, the peak-to-peak or valley-to-valley distance) is defined as the **wavelength** λ.

How a Sand Scorpion Catches Its Prey

The sand scorpion shown in the photograph opening this chapter uses waves of both transverse and longitudinal motion to locate its prey. When a beetle even slightly disturbs the sand, it sends pulses along the sand's surface (Fig. 17-4). One set of pulses is longitudinal, traveling with speed $|\vec{v}^{\,\text{L-wave}}| = v^{\text{L-wave}} = 150$ m/s. A second set is transverse, traveling with speed $|\vec{v}^{\,\text{T-wave}}| = v^{\text{T-wave}} = 50$ m/s.

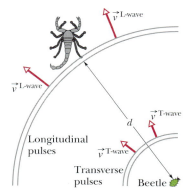

FIGURE 17-4 ■ A beetle's motion sends fast longitudinal pulses and slower transverse pulses along the sand's surface. The sand scorpion first intercepts the longitudinal pulses; here, it is the rear-most right leg that senses the pulses earliest.

The scorpion, with its eight legs spread roughly in a circle about 5 cm in diameter, intercepts the faster longitudinal pulses first and learns the direction of the beetle; it is in the direction of whichever leg is disturbed earliest by the pulses. The scorpion then senses the time interval Δt between that first interception and the interception of the slower transverse waves and uses it to determine the distance d to the beetle. The time interval is given by

$$\Delta t = \frac{d}{v^{\text{T-wave}}} - \frac{d}{v^{\text{L-wave}}}.$$

Solving for the distance gives us

$$d = (75 \text{ m/s})\, \Delta t.$$

For example, if $\Delta t = 4.0$ ms, then $d = 30$ cm, which gives the scorpion a perfect fix on the beetle.

Of course, the scorpion no more does these calculations in its head as it hunts a beetle than you do a conscious recalculation of the location of your center of mass as you decide in what way to throw out your arms when you slip. Instead, instinct and habits based on successful personal experiences rule the scorpion (and you) in such situations. But there is physics behind such "instincts"—they are not arbitrarily successful.

Waves and Oscillations

As we discussed above, if we take a picture of a wave, we can freeze the wave in time and investigate the spatial characteristics of the wave, like the wave's amplitude and its length. Alternatively, we could pick a special place, x_P, along the wave and peek at it through a slit, as in Fig. 17-5. As we peek though the slit, we would see the bits of string or Slinky rise and fall as the wave travels through space. If we made measurements while peeking, we could plot the displacement as a function of time for this location $x = x_P$. Doing so allows us to focus in on the characteristics of the wave that are associated with the passage of time. When we do this, we see several aspects of a wave that remind us of the oscillations we learned about in the last chapter.

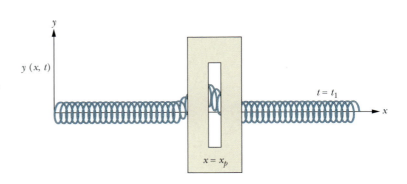

FIGURE 17-5 ■ A transverse wave traveling along a hanging Slinky as seen from the perspective of someone looking up through a slit while lying on the floor. The x axis is horizontal and chosen to be along the axis of the Slinky. The y axis is also in the horizontal plane but it lies along the direction of transverse displacement of Slinky coils. The viewer watches displacement y from the Slinky's equilibrium point as a function of time at a fixed position x_P.

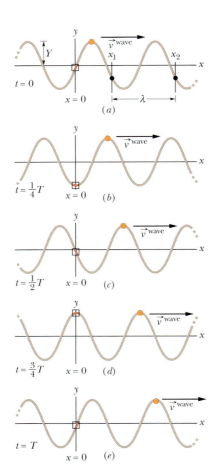

FIGURE 17-6 ■ Five "snapshots" of a traveling string wave taken at times $0, \frac{1}{4}T, \frac{1}{2}T, \frac{3}{4}T$, and T, where T is the period of oscillation of a fixed piece of string. A particular bit of string (the piece around $x = 0$) has been boxed so you can see how it moves back and forth as the continuous wave disturbance travels horizontally to the right along the string. The amplitude Y is indicated. A wavelength λ is also indicated. The circle in each snapshot shows the location of the wave crest as it moves from left to right.

For example, let us examine the small piece of string boxed in Fig. 17-6. This figure shows five "snapshots" of a sinusoidal wave traveling in the positive direction of an x axis. The time between snapshots is constant. The movement of the wave is indicated by the rightward progress of the short arrow pointing to a high point of the wave. From snapshot to snapshot, the short arrow moves to the right with the wave shape, but the pieces of the string move *only* parallel to the y axis. Let us follow the motion of the boxed string segment at $x = 0$. In the first snapshot (Fig. 17-6a), it is at displacement $y = 0$. In the next snapshot, it is at its extreme downward displacement because a *valley* (or extreme low point) of the wave is passing through it. It then moves back up through $y = 0$. In the fourth snapshot, it is at its extreme upward displacement because a *peak* (or extreme high point) of the wave is passing through it. In the fifth snapshot, it is again at $y = 0$, having completed one full oscillation. Notice that the wave in Fig. 17-6 moves to the right by $\frac{1}{4}\lambda$ from one snapshot to the next. Thus, by the fifth snapshot, it has moved to the right by 1λ. The time required for the cycle of rising and falling to repeat is called the **period** T of the wave.

READING EXERCISE 17-1: The figure just below shows an asymmetric pulse traveling from left to right along a stretched string. If this snapshot was taken at time $t = 0$ s: (a) which figure represents the graph of the displacement y versus position x at $t = 0$ s? (b) Which figure represents the graph of the displacement y versus time t at $x = x_1$ (as marked with a dot on the snapshot)?

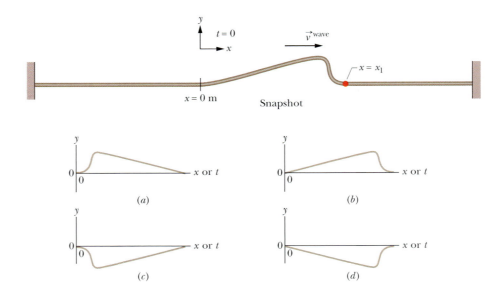

■

TOUCHSTONE EXAMPLE 17-1: Propagating Pulse

Consider a pulse propagating along a long stretched string in the x direction as shown in Fig. 17-7. Note that the scale in the y direction has been exaggerated for visibility.

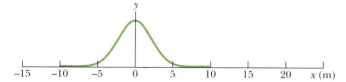

FIGURE 17-7 ■ A pulse propagates along a string in the x direction.

(a) Sketch the shape of the string after the wave pulse has traveled a distance of 10 m to the right. Explain why you sketched the shape you did.

SOLUTION ■ The **Key Idea** here is that the crest of the wave, which was at $x = 0$ m, has moved so that it is now at $x = 10$ m, but the wave shape does not change. This is like the continuous sinusoidal wave shown in Fig. 17-6 that is depicted as moving without shape change. Figure 17-8 shows the sketch.

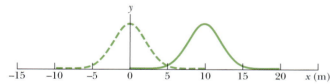

FIGURE 17-8 ■ The pulse after it has moved a distance of 10 m to the right. The original position of the pulse is the dashed line.

(b) If the wave takes 2.0 s to move the 10 m to the right, what are its velocity and wave speed?

SOLUTION ■ The **Key Idea** is that we assume that the wave shape does not change and that it moves at a constant rate determined by the motion of its crest. If the wave moves to the right 10 m in 2 s, then

$$\vec{v}^{\text{ wave}} = v_x^{\text{wave}}\,\hat{i} = \left(\frac{10\text{ m}}{2\text{ s}}\right)\hat{i} = (5\text{ m/s})\,\hat{i}$$

(wave velocity L → R movement),

and
$$v^{\text{wave}} = |\vec{v}^{\text{ wave}}| = 5\text{ m/s}$$

(wave speed—or velocity magnitude).

(c) Imagine that a small piece of essentially massless tape is placed on the string at $x = 0$ m when $t = 0$ s. Describe the motion of the tape for the next 2 s.

SOLUTION ■ The **Key Idea** here is that as the wave pulse moves along the piece of string with the tape on it, the piece of tape does not move in the horizontal direction. Only its displacement in the y direction changes. At first the tape is at the location of the wave crest and so is at its maximum value of y. It then moves down toward a zero value of y, slowly at first and then more rapidly. In a little more than a second, the tape comes to rest at $y = 0$. The wave pulse has passed and so the tape remains there.

(d) Suppose the pulse shown in Fig. 17-7 is moving to the left instead. Sketch the shape of the string after the wave pulse has traveled a distance of 5 m to the left. Explain why you sketched the shape you did.

SOLUTION ■ The **Key Idea** here is that the crest of the wave which was at $x = 0$ m, has moved without changing shape so that it is now at $x = -5$ m. Figure 17-9 shows the sketch.

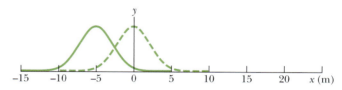

FIGURE 17-9 ■ The pulse after it has moved a distance of 5 m to the left. The original position of the pulse is the dashed line.

(e) If the wave takes 1 s to move the 5 m to the left, what are its velocity and wave speed?

SOLUTION ■ Once again we assume that the wave shape does not change and that it moves at a constant rate determined by the motion of its crest. If the wave moves 5 m to the left in 1 s, then

$$\vec{v}^{\text{ wave}} = v_x^{\text{wave}}\,\hat{i} = \left(\frac{-5\text{ m}}{1\text{ s}}\right)\hat{i} = (-5\text{m/s})\,\hat{i}$$

(wave velocity R → L movement),

and
$$v^{\text{wave}} = |\vec{v}^{\text{ wave}}| = 5\text{ m/s}$$

(wave speed—or velocity magnitude).

17-4 The Mathematical Expression for a Sinusoidal Wave

We found that the disturbance (whether pulse or wave, transverse or longitudinal) depends on both position x and time t. If we call the displacement y, we can write $y = f(x,t)$ or $y(x,t)$ to represent this functional dependence on time and position. In the example of the transverse pulse traveling along a Slinky as pictured in Fig. 17-2,

$y(x,t)$ represents the transverse (vertical) displacement of the Slinky rings from their equilibrium position at given position x and time t. (Alternatively, in the longitudinal wave on the Slinky shown in Fig. 17-1, $y(x,t)$ could represent the number of Slinky coils per centimeter at a given x and t.)

We can completely describe any wave or pulse that does not change shape over time and travels at a constant velocity using the relation $y = f(x,t)$, in which y is the displacement as a function f of the time t and the position x. In general, a wave can have any shape so long as it is not too sharp. The trick then is to find the correct expression for the function, $f(x,t)$.

Fortunately, it turns out that any shape pulse or wave can be constructed by adding up different sinusoidal oscillations. This makes the description of sinusoidal waves especially useful. So, for the rest of this section we'll discuss the properties and descriptions of continuous waves produced by displacing a stretched string using a sinusoidal motion like that shown in Fig. 17-3b. We will start by using the equation we developed in Chapter 16 to describe for sinusoidal motion at the location of a single piece of string. As we did in looking through the slit in Fig. 17-5, we will only let time vary. Next we can consider how to describe a snapshot that records the displacement of many pieces of the string at a single time. Finally, we can combine our snapshot with the results of peeking through a slit to get a single equation that ought to describe the propagation of a single sinusoidal wave. Basically we are trying to describe the displacement y of every piece of the string from its equilibrium point at every time. We are looking for $y(x,t)$.

Looking Through a Slit: Sinusoidal Wave Displacement at $x = 0$

If we choose a coordinate system so that $x = 0$ m at the left end of the string in Fig. 17-3b, then the motion at the left end of the string can be described using Eq. 16-5 with the string displacement from equilibrium represented by $y(x,t) = y(0,t)$ rather than by simply $y(t)$. To simplify our consideration we assume that the initial phase of the string oscillation at $x = 0$ m and $t = 0$ s is zero. This gives us

$$y(0,t) = Y\cos(\omega t), \qquad \text{(Eq. 16-5)}$$

where the angular frequency can be related to the period of oscillation by $\omega = 2\pi/T$. Although we use the cosine function in Chapter 16 to describe simple harmonic motion, *it is customary to use the sine function to describe wave motion.* As we mentioned in Chapter 16, when a sine function is shifted to the left by $\pi/2$ it looks like a cosine function. So we can also describe the same string displacement as a function of time at $x = 0$ m as

$$y(0,t) = Y\cos(\omega t) = Y\sin(\omega t + \pi/2). \qquad (17\text{-}1)$$

Note that using the sine function requires a different, nonzero initial phase angle given by $\pi/2$. If we locate our slit at another nonzero value of x as shown in Fig. 17-5, then the initial phase (at $t = 0$ s) will often turn out to be different from $\pi/2$. In fact this initial phase is a function of the location x of the piece of string we are considering.

A Snapshot: Sinusoidal Wave Displacement at $t = 0$

Imagine that the man has been moving the end of the string up and down as shown in Fig. 17-3b for a long time using a sinusoidal motion. Instead of looking through a slit as time varies, we take a snapshot of the string at a time $t = 0$s similar to that shown in Fig. 17-3b. Then we expect our snapshot to be described by the equation

$$y(x,0) = Y\sin(kx + \pi/2) \qquad (17\text{-}2)$$

where k is a constant and the "initial" phase when x is zero must also be $\pi/2$. Note that if the snapshot of the string were taken at another time, the initial phase would probably be different.

Combining Expressions for *x* and *t*

Equation 17-1 describes the displacement at all times for just the piece of string located at $x = 0$ m. Equation 17-2 describes the displacement of all the pieces of string at $t = 0$s. We can make an intelligent guess that the equation describing $y(x,t)$ is some combination of these two expressions given by

$$y(x,t) = Y\sin[(kx \pm \omega t) + \pi/2)],\qquad(17\text{-}3)$$

where $\pi/2$ represents the initial phase when $x = 0$ m and $t = 0$ s for the special case we considered. In general we can describe the motion of our sinusoidal wave with an arbitrary initial phase by modifying Eq. 17-3 to get

$$y(x,t) = Y\sin[(kx \pm \omega t) + \phi_0)]\qquad\text{(sinusoidal wave motion, arbitrary initial phase),}\qquad(17\text{-}4)$$

where ϕ_0 is the initial phase (or phase constant) when both $x = 0$m and $t = 0$s. The $\pm$ sign refers to the direction of motion of the wave as we shall see in Section 17-5. In cases where the initial phase is not important, we can simplify Eq. 17-3 by choosing an initial time and origin of the x axis that lies along the line of motion of the wave so that $\phi_0 = 0$ rad.

Amplitude and Phase

The argument $\phi(x,t) = (kx \pm \omega t) + \phi_0$ of the sine function is called the **time- and space-dependent phase** of the wave where k and ω are constants that we can determine for a particular wave. Although the phase $\phi(x,t)$ is a function of both time and position, it is *neither* a time *nor* a position. Rather, the phase $\phi(x,t)$ *must* be an angle because we can only take the sine of angles.

As the wave sweeps through a string segment at a particular position x, the phase changes linearly with time t. This means that the sine also changes, oscillating between $+1$ and -1. Its extreme positive value $(+1)$ corresponds to a peak of the wave moving through the segment; then, the value of y at position x is Y. Its extreme negative value (-1) corresponds to a valley of the wave moving through the segment; then, the value of y at position x is $-Y$. Thus, the sine function and the time-dependent phase of a wave correspond to the oscillation of a string segment, and the amplitude of the wave determines the extremes of the segment's displacement.

Wavelength and Wave Number

In order to come up with a more useful mathematical expression for the wave, we must more carefully investigate the nature of the phase of the wave. We know that the sine function repeats itself every 2π radians. We also know that the wavelength λ of a wave is the distance (parallel to the direction of the wave's travel) between repetitions of the shape of the wave (or *wave form*). A typical wavelength is marked in Fig. 17-6a, which is a snapshot of the wave at time $t = 0$ s. If we take the simple case for which the initial phase is zero, at that time Eq. 17-4, $y(x,t) = Y\sin[(kx \pm \omega t) + \phi_0)]$ becomes

$$y(x,0) = Y\sin(kx).\qquad(17\text{-}5)$$

By definition, the displacement y is the same at both ends of one wavelength—that is, y is the same at $x = x_1$ and at $x = x_1 + \lambda$. (Again, x does not represent a displacement here but just the horizontal position of a string element. By the equation above,

$$Y\sin kx_1 = Y\sin k(x_1 + \lambda)$$
$$= Y\sin(kx_1 + k\lambda). \tag{17-6}$$

A sine function begins to repeat itself when its angle (or argument) is increased by 2π rad, so in the equation above, we must have $k\lambda = 2\pi$, or

$$k = \frac{2\pi}{\lambda} \qquad \text{(wave number).} \tag{17-7}$$

We call k the **wave number.** The wave number is inversely proportional to the wavelength. Its SI unit is the radian per meter, or the inverse meter. (Note that the symbol k here does *not* represent a spring constant as previously.)

Period, Angular Frequency, and Frequency

In contrast to the graphs in Fig. 17-6, which represent pictures of the string at a particular instant, we will now focus on a particular bit of string and consider how it moves as a function of time. As we noted earlier, if you were to peek at the string through a slit, you would see that the single segment of the string at that position moves up and down in simple harmonic motion. Figure 17-10 shows a graph of the displacement y of a bit of string versus time t at a certain position along the string, taken to be $x = 0$. This motion is described by $y(x,t) = Y \sin(kx \pm \omega t + \phi_0)$. Setting $x = 0$ and arbitrarily choosing a zero phase constant and a negative time-dependent term, we get

$$y(0,t) = Y\sin(-\omega t) = -Y\sin(\omega t). \tag{17-8}$$

Here we have made use of the fact that $\sin(-\alpha) = -\sin\alpha$. Figure 17-10 is a graph of this equation. Be careful though: This figure *does not* show the shape of the wave. It shows a graph of the time-dependent variation in the displacement of a small bit of the string.

Since we have defined the period of oscillation T of a wave to be the time any string segment takes to move through one full oscillation, we can apply this equation to both ends of this time interval. Equating the results yields

$$Y \sin \omega t_1 = Y \sin \omega(t_1 + T)$$
$$Y \sin \omega t_1 = Y \sin(\omega t_1 + \omega T) \tag{17-9}$$

which can be true only if $\omega T = 2\pi$. In other words,

$$\omega = \frac{2\pi}{T} \qquad \text{(angular frequency).} \tag{17-10}$$

We call ω the **angular frequency** of the wave; its SI unit is the radian per second.

Recall from Chapter 16 that the **frequency** f of an oscillation is defined as $1/T$. Hence, its relationship to angular frequency is

$$f = \frac{1}{T} = \frac{\omega}{2\pi} \qquad \text{(frequency).} \tag{17-11}$$

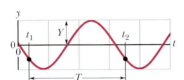

FIGURE 17-10 ■ A graph with unspecified units of the displacement of the string segment at $x = 0$ as a function of time, as the sinusoidal wave of Fig. 17-6 passes through it. The amplitude Y is indicated. A typical period T, measured from an arbitrary time t_1, is also indicated.

Like the frequency of simple harmonic motion in Chapter 16, this frequency f is a number of oscillations per unit time—here, the number made by a string segment as the wave moves through it. As in Chapter 16, f is usually measured in hertz or its multiples, such as kilohertz.

To summarize our discussion of Eq. 17-4 given by $y(x,t) = Y \sin[(kx \pm \omega t) + \phi_0]$, the names of the quantities are displayed in Fig. 17-11 for your reference.

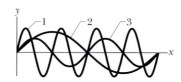

FIGURE 17-11 ■ The names of the quantities in Eq. 17-4, for a transverse sinusoidal wave.

READING EXERCISE 17-2: The figure is a composite of three snapshots, each of a wave traveling along a particular string. The time-dependent phases for the waves are given by (a) $(2 \text{ rad/m})x - (4 \text{ rad/s})t$, (b) $(4 \text{ rad/m})x - (8 \text{ rad/s})t$, and (c) $(8 \text{ rad/m})x - (16 \text{ rad/s})t$. Which phase corresponds to each snapshot? ■

TOUCHSTONE EXAMPLE 17-2: Traveling Wave Equations

A traveling sinusoidal wave train can be described in terms of a wave number k and an angular frequency ω using Eq. 17-4. If the wave train is moving to the right and has an initial phase of $\phi_0 = 0$ rad then we can write our wave equation as

$$y(x,t) = Y \sin(kx - \omega t).$$

Use the definition of various terms along with Eqs. 17-7, 17-10, and 17-11 to:

(a) Explain what T and λ mean physically.

SOLUTION ■ The symbols λ and T stand for the wavelength and the period of the wave, respectively.

The wavelength, λ, is the distance one has to move along the string at a fixed instant in time (for instance, when looking at a snapshot of the wave) in order to pass by a full oscillation of the wave.

The period, T, is the time one has to wait at a particular point in space (along an axis that is lined up with the undisturbed string) for a small piece of the string to go through one complete oscillation.

(b) Show that displacement of the string from its undisturbed condition given by $y(x,t)$ can also be described by the equation

$$y(x,t) = Y \sin\left[2\pi\left(\frac{x}{\lambda} - ft\right)\right].$$

SOLUTION ■ The **Key Idea** here is that the most important symbols in the sinusoidal wave equation are x and t. They represent the variables. All the other symbols represent constants that describe a particular wave. As a result, every sinusoidal wave with a zero initial phase will look like $\sin[(\text{mess})x \pm (\text{another mess})t]$. By identifying how the "mess" in each situation is related to the constants in another situation, the transformation of Eq. 17-4 into different forms is straightforward. Also we only need to show that the argument given by terms inside the square brackets are the same as the argument $[kx - \omega t]$. For the situation at hand we can use the relationships in Eqs. 17-7, 17-10, and 17-11 ($k = 2\pi/\lambda$, $\omega = 2\pi/T$, and $f = 1/T = \omega/2\pi$) to verify that

$$2\pi\left(\frac{x}{\lambda} - ft\right) = \frac{2\pi}{\lambda}x - 2\pi ft = kx - \frac{2\pi}{T}t = kx - \omega t. \quad \text{(Answer)}$$

(c) Show that displacement of the string from its undisturbed condition given by $y(x, t)$ can also be described by the equation

$$y(x,t) = Y \sin\left[2\pi\left(\frac{x}{\lambda} - \frac{t}{T}\right)\right].$$

SOLUTION ■ Using the same approach as we used in part (b), we see that

$$2\pi\left(\frac{x}{\lambda} - \frac{t}{T}\right) = 2\pi\left(\frac{x}{\lambda} - ft\right),$$

but we showed in part (a) that

$$2\pi\left(\frac{x}{\lambda} - ft\right) = kx - \omega t. \quad \text{(Answer)}$$

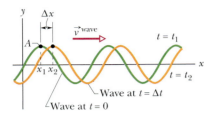

FIGURE 17-12 ■ One end of a string that is under tension is moved up and down continuously in a sinusoidal fashion as shown in Fig. 17-3b. Two snapshots of the sinusoidal wave that results are shown, at time $t = t_1$ and then at time t_2. As the wave moves to the right with an x-component of velocity v_x^{wave}, the entire sine curve shifts by Δx during Δt. The string segment at x_1 moves down during $\Delta t = t_2 - t_1$, while the piece of string at x_2 moves up as the local wave crest moves from left to right.

17-5 Wave Velocity

Figure 17-12 shows two snapshots of a sinusoidal wave taken a small time interval Δt apart. The wave is traveling in the positive x direction (to the right in Fig. 17-12b), and the entire wave pattern is moving a distance Δx in that direction during the interval Δt. The ratio $\Delta x/\Delta t$ (or, in the differential limit, dx/dt) is the **wave velocity** component along the x axis, which we denote as v_x^{wave}. How can we find the wave velocity?

As the wave in Fig. 17-12 moves, each point of the moving wave form, such as point A marked on a peak, retains its displacement y. (Points on the string do not retain their displacement, but points on the wave *form* do.) That is, time passes and the location of point A changes (so x and t are both changing), but the value of $y(x,t)$ associated with point A does not change; it remains the maximum value. If point A retains its displacement as it moves, the phase in $y(x,t) = Y \sin(kx \pm \omega t)$ must remain a constant. So,

$$kx \pm \omega t = \text{a constant.} \tag{17-12}$$

Note that although this argument is constant, in this case, both x and t are increasing. [The time t always increases and x increases because the wave is moving in the (rightward) positive direction]. In order for the phase $kx \pm \omega t$ to remain constant with both t and x increasing, the negative sign must be chosen. In other words,

> The equation for a sinusoidal wave traveling right is $y(x,t) = Y \sin(kx - \omega t)$ and the equation for a sinusoidal wave moving left is $y(x,t) = Y \sin(kx + \omega t)$.

By assuming that the wave moves at a constant velocity $\vec{v}^{wave}$, we can express its velocity in terms of how far an imaginary point on the wave form has moved in a short time interval. Since the wave form only moves parallel to the x axis, we can write

$$\vec{v}^{wave} \equiv \frac{\Delta \vec{x}}{\Delta t} \qquad \text{(definition of wave velocity)},$$

or in component form

$$v_x^{wave} = \frac{\Delta x}{\Delta t}.$$

If we take the time interval under consideration to be the period of the wave T, then we know that the wave travels one wavelength λ. So the magnitude of the wave velocity, or wave speed, is

$$\left| v_x^{wave} \right| = v^{wave} = \frac{\lambda}{T} = \lambda f = \frac{\omega}{k} \qquad \text{(speed related to wave length and period).} \tag{17-13}$$

Are all traveling waves sinusoidal? Not necessarily. Consider a wave of arbitrary shape, given by

$$y(x,t) = h(kx \pm \omega t), \tag{17-14}$$

where h represents *any* function, the sine or cosine function being one possibility. Since the variables x and t enter in the combination $kx \pm \omega t$, it is possible for waves to have other shapes associated with them. For example, a wave pulse traveling from left to right could be represented by

$$y(x,t) = \frac{Y}{(kx - \omega t)^2 + 1} \qquad \text{(a possible traveling wave pulse).}$$

Particle Velocity and Acceleration in Mechanical Waves

In discussing the velocity of a wave as we did above, it is important to remember that the magnitude of this velocity is the speed at which the wave disturbance moves, *not* the speed of the bits or pieces of string through which the wave travels. To repeat the quote of Leonardo da Vinci, consider " . . . waves made in a field of grain by the wind, where we see the waves running across the field while the grain remains in place."

We have already begun to develop an understanding of the velocity of the wave itself. We now consider the velocity of the particles that are displaced as a transverse wave moves through. A piece of string can only move in the y direction as a disturbance passes by. Thus, at a given time t we can denote the velocity of the piece of string that is located at x in terms of its y-component $v_y^{\text{string}} = v_y(x,t)$. Since only the string pieces can move in the y direction, we often shorten this y-component notation to simplify $v_y(x,t)$. To find the velocity component $v_y(x,t)$ as a sinusoidal disturbance passes through, we simply need to differentiate the expression for the displacement of the string segment, $y(x,t) = Y \sin[(kx \pm \omega t) + \phi_0)]$ with respect to time while holding x constant. This is similar to what we did to find the velocity of a mass oscillating at the end of a spring in Section 16-4,

$$v_y = \frac{\partial y(x,t)}{\partial t} = \pm \omega Y \cos[(kx \pm \omega t) + \phi_0], \qquad (17\text{-}15)$$

where the use of the symbol $\partial/\partial t$ signifies that we are taking a partial derivative in which the location along the string, x, is considered to be a constant. If we find a second partial derivative using Eq. 17-15, we have the following expression for the acceleration of the piece of string at location x at a time t,

$$a_y = \frac{\partial v_y}{\partial t} = \frac{\partial^2 y(x,t)}{\partial t^2} = \mp \omega^2 Y \sin[(kx \pm \omega t) + \phi_0]. \qquad (17\text{-}16)$$

TOUCHSTONE EXAMPLE 17-3: Wave on a String

A wave traveling along a string is described by

$$y(x,t) = (0.00327 \text{ m}) \sin[(72.1 \text{ rad/m})x - (2.72 \text{ rad/s})t], \quad (17\text{-}17)$$

where $y(x,t)$ represents the transverse displacement from equilibrium of a small piece of string with a horizontal location of x at time t.

(a) What is the amplitude of this wave?

SOLUTION ■ The **Key Idea** is that Eq. 17-17 is of the same form as Eq. 17-4, with $\phi_0 = 0$ rad,

$$y = Y\sin(kx - \omega t + \phi_0), \qquad (17\text{-}18)$$

so we have a sinusoidal wave. By comparing the two equations, we see that the amplitude is

$$Y = 0.00327 \text{ m} = 3.27 \text{ mm}. \qquad \text{(Answer)}$$

(b) What are the wavelength, period, and frequency of this wave?

SOLUTION ■ By comparing Eqs. 17-17 and 17-18, we see that the wave number and angular frequency are

$$k = 72.1 \text{ rad/m} \quad \text{and} \quad \omega = 2.72 \text{ rad/s}.$$

We then relate wavelength λ to k via Eq. 17-7:

$$\lambda = \frac{2\pi}{k} = \frac{2\pi \text{ rad}}{72.1 \text{ rad/m}} \qquad \text{(Answer)}$$
$$= 0.0871 \text{ m} = 8.71 \text{ cm}.$$

Next, we relate T to ω using Eq. 17-10:

$$T = \frac{2\pi}{\omega} = \frac{2\pi \text{ rad}}{2.72 \text{ rad/s}} = 2.31 \text{ s}, \qquad \text{(Answer)}$$

and from Eq. 17-11 we have

$$f = \frac{1}{T} = \frac{1}{2.31 \text{ s}} = 0.433 \text{ Hz}. \qquad \text{(Answer)}$$

(c) What is the velocity with which the wave moves along the string?

SOLUTION ■ The speed of the wave is given by Eq. 17-13:

$$v^{wave} = \frac{\omega}{k} = \frac{2.72 \text{ rad/s}}{72.1 \text{ rad/m}} = 0.0377 \text{ m/s} \quad \text{(Answer)}$$
$$= 3.77 \text{ cm/s}.$$

Because the phase in Eq. 17-17 contains the variable x, the wave is moving along the x axis. According to Eq. 17-4, the *minus* sign in front of the ωt term indicates that the wave is moving in the *positive* x direction. (Note that the (b) and (c) answers are independent of the wave amplitude.)

(d) What is the displacement y of the small piece of string located at $x = 22.5$ cm when $t = 18.9$ s?

SOLUTION ■ The **Key Idea** here is that Eq. 17-17 gives the displacement as a function of position x and time t. Substituting the given values into the equation yields

$$y = (0.00327 \text{ m}) \sin[(72.1 \text{ rad/m}) \times (0.225 \text{ m}) - (2.72 \text{ rad/s})$$
$$\times (18.9 \text{ s})]$$
$$= (0.00327 \text{ m}) \sin[-35.1855 \text{ rad}] = (0.00327 \text{ m})(0.588)$$
$$= 0.00192 \text{ m} = 1.92 \text{ mm}. \quad \text{(Answer)}$$

Thus, the transverse displacement from the string's equilibrium position is positive.

TOUCHSTONE EXAMPLE 17-4: Transverse String Motion

In Touchstone Example 17-3d, we showed that at $t = 18.9$ s the transverse displacement component y of the piece of string at $x = 0.255$ m due to the wave of Eq. 17-17 is 1.92 mm.

(a) What is v_y, the transverse velocity component of the same element of the string, at that time? (This speed, which is associated with the transverse oscillation of an element of the string, is in the y direction. Do not confuse it with $\vec{v}^{\,wave}$, the constant velocity at which the *wave form* travels along the x axis.)

SOLUTION ■ The **Key Idea** here is that as the wave travels, the displacement of each piece of string is transverse, where x represents the horizontal position of a piece of string. Another **Key Idea** is that the horizontal position, x, of a piece of string never changes. In general, the displacement of a piece of string is given by

$$y(x,t) = Y \sin(kx - \omega t + \phi_0). \quad (17\text{-}19)$$

For a piece of string at a certain location (x, y) we find the rate of change of y by taking the partial derivative of Eq. 17-19 with respect to t while treating x as a constant. Here we have $\phi_0 = 0$ rad, so according to Eq. 17-15,

$$v_y = \frac{\partial y}{\partial t} = -\omega Y \cos(kx - \omega t), \quad (17\text{-}20)$$

where the amplitude Y is always taken as positive. Next, substituting numerical values from Touchstone Example 17-3 we obtain

$$v_y = (-2.72 \text{ rad/s})(3.27 \text{ mm}) \cos(-35.1855 \text{ rad})$$
$$= 7.20 \text{ mm/s}. \quad \text{(Answer)}$$

Thus, at $t = 18.9$ s, the piece of string at the horizontal location $x = 22.5$ cm is moving in the y direction, with a velocity of 7.20 mm/s.

(b) What is the transverse acceleration component a_y of the same element at that time?

SOLUTION ■ The **Key Idea** here is that the transverse acceleration component a_y is the rate at which the transverse velocity of a given piece of string is changing. From Eq. 17-20, again treating x as constant but allowing t to vary, we find

$$a_y = \frac{\partial v_y}{\partial t} = -\omega^2 Y \sin(kx - \omega t).$$

Comparison with Eq. 17-19 shows that we can write this as

$$a_y = -\omega^2 y.$$

We see that the transverse acceleration component of a piece of string that is oscillating is proportional to its transverse component of displacement but opposite in sign. This is completely consistent with the action of the string piece—namely, that it is moving transversely in simple harmonic motion. Substituting numerical values yields

$$a_y = -(2.72 \text{ rad/s})^2(1.92 \text{ mm})$$
$$= -14.2 \text{ mm/s}^2. \quad \text{(Answer)}$$

Thus, at $t = 18.9$ s, the piece of string at $x = 22.5$ cm is displaced from its equilibrium position by 1.92 mm in the positive direction and has an acceleration of magnitude 14.2 mm/s^2 in the negative y direction.

17-6 Wave Speed on a Stretched String

If we distort a string that is under tension, we observe that this distortion (or wave) will travel rapidly along the string (Fig. 17-1). This is true whether the distortion is a single pulse (Figs. 17-2 or 17-3a) or continuous sinusoidal pattern (Fig. 17-3b). In Section 17-3 we asserted that a net force on a piece of string could result from forces exerted on its ends by the pieces of string just adjacent to it. If these end forces act in different directions on a string segment, the string piece can move in a vertical direction. As it moves vertically, it can exert a force on adjacent string pieces. This could cause a set of disturbances to propagate along the string as a wave.

In general, the speed of a wave is determined by the properties of the medium through which it travels. For instance, how do the properties of the stretched string affect the speed of a traveling wave? Intuitively we expect that the wave will travel faster when the string is stretched to a higher tension and that the wave will travel more slowly on a thicker string where each segment of string has a greater mass. In this section we use the impulse-momentum theorem—a form of Newton's Second Law—to derive an equation that relates the speed of the traveling wave to the tension of the string and its massiveness. In our derivation we imagine that the string consists of many tiny segments or pieces that are connected together.

Consider a horizontal stretched string as shown in Fig. 17-13. Suppose the string is under tension—a nonvector quantity we denote as F^{tension}*. What happens if the tiny segment at one end of the string is jerked abruptly upward and then returned to its equilibrium position? The pulse that results propagates along the pieces of string as a wave like the one shown in Fig. 17-13. Although the amplitude in Fig. 17-13 is exaggerated for clarity, we assume the wave amplitude is small enough that: (1) the tension in the string does not change significantly except at the boundary between a displaced and undisplaced segment of string (as shown in Fig. 17-13a) and (2) the angle θ between the horizontal and any segment (or small piece) of string that is displaced is so small that $\sin\theta \approx \tan\theta \approx \theta$. We also assume that the pulse propagates along the string at a *constant* wave velocity $\vec{v}^{\text{wave}} = v_x^{\text{wave}}\hat{i}$. As is often the case in physics, we can test the validity of our assumptions by seeing whether the expression we derive for the wave speed is consistent with observation.

Let us consider how a segment of string moves as a wave pulse travels by it in a short time interval. If we define the mass per unit length of the string as its linear mass density, μ, then the mass of the segment of string that the wave pulse affects in a time interval Δt is given by

$$m = \mu v^{\text{wave}}\Delta t, \qquad (17\text{-}21)$$

where v^{wave} is the speed of propagation of the wave pulse. We start by examining the tension forces that act at each end of our small string segment. For example, in Fig. 17-13a the segment at the crest of the wave experiences a net downward force, but the segment at the leading edge of the wave pulse experiences a net upward force. Thus a short time Δt later, the front segment has moved up while the segment at the crest has moved down. If we take all the string segments into account, we see that the entire pulse shape "moves" to the right even though each of the string segments has only moved vertically—either up or down. This is shown in Fig. 17-13b.

What happens when the leading edge of the pulse encounters a segment of the string? A vertical net force $\vec{F}^{\text{net}} = F_y^{\text{net}}\hat{j}$ due to the unbalanced tension forces at

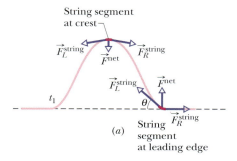

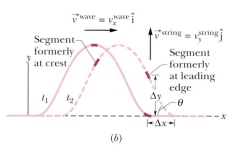

(b)

FIGURE 17-13 ■ Depictions of a traveling wave pulse. The amplitude is exaggerated for clarity. (a) A single pulse travels to the right along a string under constant tension. The piece of string at the wave crest experiences a net force downward while the string segment at the leading edge experiences a net upward force. (b) Two snapshots of the wave at times t_1 and t_2 show the wave form traveling with velocity $\vec{v}^{\text{wave}}$, so it shifts right a distance Δx during a time interval $\Delta t = t_2 - t_1$. As a result of the net forces on the piece of strings, the string segment at the wave crest moves down during Δt while the piece of string at the leading edge moves up.

*To avoid confusing wave period, denoted as T, with tension, we use the notation F^{tension} for tension here and in the next chapter.

the string segment's ends shown in Fig. 17-13a causes the segment to accelerate upward. In fact, the piece of string just to the left of the leading edge is stretched so its tension is greater than in the undisturbed string piece just to the right. As a result, the string piece at the leading edge undergoes a change in vertical position of Δy and a momentum change, $\Delta \vec{p}_y$ during the time Δt. We can use the impulse-momentum theorem to describe the motion of the piece of string in terms of the vertical components of net force and momentum as

$$F_y^{\text{net}} \Delta t = \Delta p_y. \tag{17-22}$$

The segment of string has a mass m, and we assume the segment has no vertical velocity at time t_1. If we denote the vertical velocity change imparted to the segment of string in the short time interval Δt as $\Delta \vec{v}_y^{\text{string}}$, then

$$F_y^{\text{net}} \Delta t = m \Delta v_y^{\text{string}} = m(v_y^{\text{string}} - 0) = m v_y^{\text{string}}.$$

But we can see from Fig. 17-13a that the magnitude of the net force on the string segment is approximately $F^{\text{tension}} \sin \theta$, where F^{tension} is a scalar quantity denoting the tension in the string. The impulse delivered to the string segment by the traveling wave can be related to the newly acquired vertical velocity of the string segment by

$$(F^{\text{tension}} \sin \theta) \Delta t \approx m |v_y^{\text{string}}| = m |\vec{v}^{\text{string}}|.$$

From Fig. 17-13b we also see that the ratio of the magnitude of the vertical velocity of the leading edge mass segment and the magnitude of the horizontal wave velocity of the temporary disturbance is given by

$$\tan \theta = \frac{\Delta y}{\Delta x} = \frac{|\vec{v}^{\text{string}}|}{|\vec{v}^{\text{wave}}|} = \frac{v^{\text{string}}}{v^{\text{wave}}}. \tag{17-23}$$

Combining the last two expressions we derived gives us

$$(F^{\text{tension}} \sin \theta) \Delta t = m v^{\text{wave}} \tan \theta \quad \text{or} \quad F^{\text{tension}} = \frac{m}{\Delta t} \left(\frac{v^{\text{wave}}}{\cos \theta} \right). \tag{17-24}$$

At this point we need to express the mass of the segment in terms of the length of the string segment and the linear density of the string. Thus the mass of the segment in the string can be determined as the density multiplied by the length of the segment so that $m = \mu \Delta x$. However, during the time interval Δt the waveform has moved a distance $\Delta x = |\vec{v}^{\text{wave}}| \Delta t$. We can now rewrite Eq. 17-24 as

$$F^{\text{tension}} = \mu (v^{\text{wave}})^2 \frac{1}{\cos \theta},$$

but for small-amplitude waves the angle of string displacement relative to the horizontal, $\cos \theta$ is approximately 1, so

$$(v^{\text{wave}})^2 = \frac{F^{\text{tension}}}{\mu}.$$

Therefore,
$$v^{\text{wave}} = \sqrt{\frac{F^{\text{tension}}}{\mu}} \qquad \text{(speed).} \tag{17-25}$$

Equation 17-25 gives the speed of the pulse in Fig. 17-13 and the speed of *any* other wave pulse or continuous wave on the same string under the same tension. If we take the square root of the ratio of F^{tension} (dimension MLT^{-2}) and μ (dimension ML^{-1}) we get a dimension of velocity. Thus Eq. 17-25 is dimensionally correct.

Equation 17-25 tells us that

> The speed of a small-amplitude pulse traveling along a stretched string depends only on the tension and linear density of the string and not on the amplitude or frequency of the disturbance.

The amplitude independence is much like that for systems oscillating with simple harmonic motion. It indicates a proportionality between the net forces exerted on a piece of string and its displacement from equilibrium.

If the simplifying assumptions we used in the derivation of Eq. 17-25 are reasonable, then we expect that measurements of wave speed, tension, and linear mass density can be used to verify Eq. 17-25. Indeed, this theoretical prediction has been experimentally confirmed many times. In fact, it turns out that Eq. 17-25 is just as valid for continuous waves as it is for a single wave pulse, and so the speed of a continuous wave is not a function of frequency. The frequency of the wave is fixed entirely by whatever generates the wave (for example, the person jerking the string up and down in Fig. 17-3*b*). However, once the frequency of a continuous wave is set, the wave speed determines the relationship between frequency and wavelength since $\lambda = v^{\text{wave}}/f$ (Eq. 17-13).

READING EXERCISE 17-3: In the derivation of the wave speed in a string shown above, there are references to two different velocities; one is referred to as $\vec{v}_y^{\text{string}}$ and one is referred to as $\vec{v}_x^{\text{wave}}$. As completely as possible, describe the differences between these two velocities. Explain which velocity is used in deriving the expression for the mass, m, of a string piece being displaced and why. Explain which velocity is used to find the y-component of momentum change, Δp_y, of a piece of string and why. ■

READING EXERCISE 17-4: You send a continuous traveling wave along a string by moving one end up and down sinusoidally. If you increase the frequency of the oscillations, do (a) the speed of the wave and (b) the wavelength of the wave increase, decrease, or remain the same? If, instead, you increase the tension in the string, do (c) the speed of the wave and (d) the wavelength of the wave increase, decrease, or remain the same? ■

17-7 Energy and Power Transported by a Traveling Wave in a String

When we start up a wave or pulse in a stretched string, *we* provide the energy for the motion of the string. We impart a momentum to a segment of the string. As the wave or pulse moves away, the momentum is transferred from one segment of the string to the next. In addition, the wave transports energy as both kinetic energy and elastic potential energy. Let us consider each of these forms of energy in turn.

Kinetic Energy

A segment of the string of mass dm, oscillating transversely in simple harmonic motion as the wave passes through it, has kinetic energy associated with its transverse velocity v_y. When the segment is rushing through its $y = 0$ position (segment b in Fig. 17-14), its transverse velocity — and thus its kinetic energy — is a maximum. When

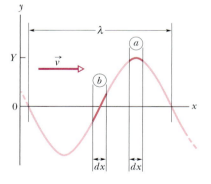

FIGURE 17-14 ■ A snapshot of a traveling wave on a string at time $t = 0$. String segment a is at displacement $y = Y$, and string segment b is at displacement $y = 0$. The kinetic energy of the string segment at each position depends on the transverse velocity of the segment. The potential energy depends on the amount by which the string is displaced from equilibrium as the wave passes through it.

the segment is at its extreme position $y = Y$ (as is segment a), its transverse velocity—and thus its kinetic energy—is zero.

Elastic Potential Energy

To send a sinusoidal wave along a previously straight string, the wave must necessarily stretch the string. As a string segment of length dx oscillates transversely, its length must increase and decrease in a periodic way if the string segment is to fit the sinusoidal wave form. Elastic potential energy is associated with these length changes, just as for a spring.

When the string segment is at its $y = Y$ position (segment a in Fig. 17-14), its length has its normal undisturbed value dx, so its elastic potential energy is zero. However, when the segment is rushing through its $y = 0$ position, it is stretched to its maximum extent, and its elastic potential energy then is a maximum.

Energy Transport

The oscillating string segment thus has both its maximum kinetic energy and its maximum elastic potential energy at $y = 0$. In the snapshot of Fig. 17-14, the regions of the string at maximum displacement have no energy, and the regions at zero displacement have maximum energy. As the wave travels along the string, forces due to the tension in the string continuously do work to transfer energy from regions with energy to regions with no energy.

Suppose we set up a wave on a string stretched along a horizontal x axis so that Eq. 17-4 describes the string's displacement. We might send a wave along the string by oscillating one end of the string, as in Fig. 17-3b. In doing so, we provide energy for the motion and stretching of the string—as the string sections oscillate perpendicularly to the x axis, they have kinetic energy and elastic potential energy. As the wave moves into sections that were previously at rest, energy is transferred into those new sections. Thus, we say that the wave *transports* the energy along the string.

The Rate of Energy Transmission

The kinetic energy dK associated with a string segment of mass dm is given by

$$dK = \tfrac{1}{2} dm \, (v_y^{\text{string}})^2, \tag{17-26}$$

where v_y^{string} is the transverse component of velocity of the oscillating string segment. If we assume the initial phase ϕ_0 is zero, then the y-component of the string element velocity is given by Eq. 17-15 as

$$v_y^{\text{string}} = -\omega Y \cos(kx - \omega t).$$

Using this relation and putting $dm = \mu \, dx$, we rewrite Eq. 17-26 as

$$dK = \tfrac{1}{2}(\mu \, dx)[(-\omega Y)\cos^2(kx - \omega t)]^2. \tag{17-27}$$

Dividing both sides of Eq. 17-27 by dt, we get the rate at which the kinetic energy of a string segment changes and thus the rate at which kinetic energy is carried along by the wave. The result is given by

$$\frac{dK}{dt} = \tfrac{1}{2}\mu\frac{dx}{dt}\omega^2 Y^2\cos^2(kx - \omega t) = \tfrac{1}{2}\mu v^{\text{wave}}\omega^2 Y^2\cos^2(kx - \omega t), \tag{17-28}$$

where the ratio dx/dt, which is positive for a wave moving from left to right, has been replaced by the wave speed v^{wave}. The *average* rate at which kinetic energy is transported is

$$\frac{dK}{dt} = \langle \tfrac{1}{2}\mu v^{\text{wave}}\omega^2 Y^2 [\cos^2(kx - \omega t)] \rangle \tag{17-29}$$

$$= \tfrac{1}{4}\mu v^{\text{wave}}\omega^2 Y^2.$$

Here we have taken the average over an integer number of wavelengths and have used the fact that the average value of the square of a cosine function over an integer number of periods is $\tfrac{1}{2}$.

Elastic potential energy is also carried along with the wave, and at the same average rate given by Eq. 17-29. Although we shall not examine the proof, you should recall that, in an oscillating system such as a pendulum or a spring–block system, the average kinetic energy and the average potential energy are indeed equal.

The **average power,** which is the average rate at which energy of both kinds is transmitted by the wave, is then

$$\langle P \rangle = 2\langle \frac{dK}{dt} \rangle, \tag{17-30}$$

or, from Eq. 17-29,

$$\langle P \rangle = \tfrac{1}{2}\mu v^{\text{wave}}\omega^2 Y^2 \qquad \text{(average power)}. \tag{17-31}$$

The factors μ and v^{wave} in this equation depend on the material and tension of the string. The factors ω and Y depend on the process that generates the wave.

> The dependence of the average power of a wave on the square of its amplitude and also on the square of its angular frequency is a general result, true for sinusoidal waves of all types.

17-8 The Principle of Superposition for Waves

It often happens that two or more waves pass simultaneously through the same region. When we listen to a concert, for example, sound waves from many instruments fall simultaneously on our eardrums. The electrons in the antennas of our radio and television receivers are set in motion by the net effect of many electromagnetic waves from many different broadcasting centers. The water of a lake or harbor may be churned up by waves in the wakes of many boats.

In many real-world cases, we find the interaction between two overlapping waves to be quite complex. However, we also observe that when wave disturbances move along the same straight line and their amplitudes are small, like those shown in Fig. 17-15, their interaction is well behaved. We observe that when well behaved waves interact, they produce a resultant wave with an amplitude equal to the sum of the amplitudes of the two original waves. This effect is seen in Fig. 17-15, where a sequence of snapshots of two pulses traveling in opposite directions on the same stretched string is shown.

> When two well behaved linear waves overlap, the displacement of each point on the string is the sum of the two displacements it would have had from each wave independently.

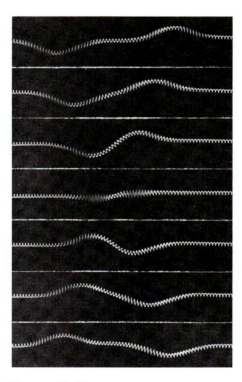

FIGURE 17-15 ■ An experimental demonstration of wave pulse superposition. In this sequence of movie frames two pulses are set in motion in opposite directions along a taut spring. The pulses have almost the same shape except that the one starting on the right has positive displacements while the other starting on the left has negative displacements. In the fourth frame the two pulses almost cancel each other.

Moreover, we observe that each pulse moves through the other, as if the other were not present:

> When well behaved linear waves overlap, one wave does not in any way alter the travel of the other.

This superposition is one of the many ways that waves and particles differ. When particles overlap by colliding they alter each other's motions.

In order to make these two important observations quantitative, let $y_1(x,t)$ and $y_2(x,t)$ be the displacements associated with two waves traveling simultaneously along the same stretched string. Since the result of the overlapping waves is a resultant wave that is the sum of the two, the displacement of the string when the waves overlap is given by the algebraic sum

$$y'(x,t) = y_1(x,t) + y_2(x,t). \qquad (17\text{-}32)$$

This is an example of the **principle of superposition,** which says that when several effects occur simultaneously, their net effect is the sum of the individual effects.

We can look in more detail at how the sum of two waves passing through each other can look rather odd while the waves overlap, but after they pass each other they look more normal again. This situation is shown in the visualization of the superposition of two waves having different amplitudes in Figs. 17-16 and 17-17.

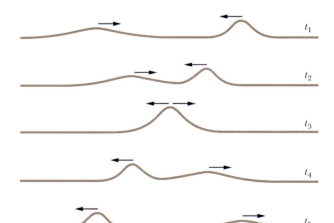

FIGURE 17-16 ▪ A visualization of two pulses on a taut string passing through each other. A short broad pulse moves from left to right while a sharp tall pulse moves from right to left. When they overlap at times t_2, t_3, and t_4, the wave form is the sum or superposition of the displacement of each pulse at each location along the string. A close-up of this type of superposition is shown in Fig. 17-17.

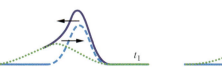

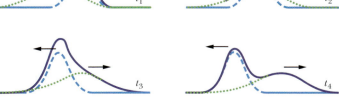

FIGURE 17-17 ▪ Closeup view of a short, broad pulse and a tall sharp pulse passing through each other. The vertical scale relative to the horizontal scale has been enlarged to show more details. *Note that the resultant wave is the sum of the displacements contributed by each of the two pulses at each location along the string.*

TOUCHSTONE EXAMPLE 17-5: Overlapping Pulses

At the time $t = 0$ s, the string has the shape shown in Fig. 17-18. The pulse on the left is moving toward the right and the pulse on the right is moving toward the left. Assume that each box in Fig. 17-18 is 1 cm square.

(a) The leading edges of the pulses will just touch in a time of 0.05 s. What is the speed with which each pulse is traveling?

SOLUTION ■ The **Key Ideas** here are that (1) since these waves travel on the same string they must have the same speed so their leading edges will meet in the middle, and (2) if the wave pulses don't change shape, the motion of any location on the wave shape tells us about the velocity and speed of the wave as a whole. At first the leading edges of the waves are 4.0 cm apart, so each wave has time to travel 2.0 cm in the 0.05 s. This gives us a wave speed of

$$v^{\text{wave}} = |\vec{v}^{\text{wave}}| = \left(\frac{2.0 \text{ cm}}{0.05 \text{ s}}\right) = 40 \text{ cm/s} \qquad \text{(speed of each wave).}$$
(Answer)

(b) Two points on the string are marked with red dots and with the letters A and B. At the instant shown, what are the velocities of the dots? Give magnitude and direction (up, down, left, right, or some combination of them).

SOLUTION ■ There are three **Key Ideas** here. (1) The piece of string located at a given point can only move up or down but NOT along the string. (2) If the displacement of the string is less just behind a point on a wave (relative to its direction of motion),

then the velocity of the piece of string is downward in the negative y direction. The opposite is true if the displacement behind the string is greater, then the velocity is upward. (3) The change Δy of a point on a string as a wave disturbance passes depends on both the wave velocity component $v_x^{\text{wave}} = \Delta x/\Delta t$ and the current slope of the wave shape $\Delta y/\Delta x$ at the point of interest along the string.

At point A: Wave slope is $\Delta y/\Delta x = -2$ so that $\Delta y = -2\Delta x$. But the x-component of the wave speed is given by $v_x^{\text{wave}} = \Delta x/\Delta t = 40$ cm/s, so

$$v_{Ay} = \frac{\Delta y}{\Delta t} = -\frac{2\Delta x}{\Delta t} = -80 \text{ cm/s} \qquad \text{(downward motion).} \quad \text{(Answer)}$$

At point B: Wave slope is $\Delta y/\Delta x = +\frac{2}{3}$ so that $\Delta y = +(\frac{2}{3})\Delta x$. Therefore,

$$v_{By} = \frac{\Delta y}{\Delta t} = \frac{+(\frac{2}{3})\Delta x}{\Delta t} = 27 \text{ cm} \qquad \text{(upward motion).} \quad \text{(Answer)}$$

(c) In Fig. 17-19 are dashed lines indicating where the pulses would be at a time $t = 0.075$ s. Draw a heavy line to show what the shape of the string would look like at this instant. Explain why you think it would look like your sketch.

SOLUTION ■ The **Key Idea** here is that we can use the principle of superposition to sum the contribution of each wave. Right in the middle where the waves overlap, they add up to a constant because the two overlapping slopes are equal in magnitude and opposite in sign (Fig. 17-20).

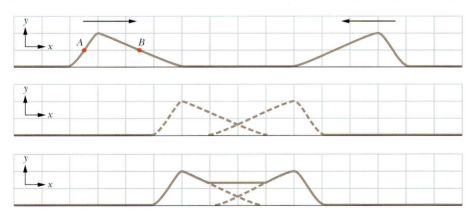

FIGURE 17-18 ■ Two pulses move toward each other on a string.

FIGURE 17-19 ■ Overlapping pulses at time $t = 0.075$ s.

FIGURE 17-20 ■ Answer to the question posed in part (c).

17-9 Interference of Waves

Suppose we send two sinusoidal waves of the same wavelength and amplitude in the same direction along a stretched string. The superposition principle applies. What resultant wave does it predict for the string?

The resultant wave depends on the extent to which the waves are *in phase* (in step) with respect to each other—that is, how much one wave form is shifted from the other wave form. If the waves are exactly in phase (so that the peaks and valleys of one are exactly aligned with those of the other), they combine to double the displacement of either wave acting alone. If they are exactly out of phase (the peaks of one are exactly aligned with the valleys of the other), they combine to cancel everywhere, and the string remains straight. We call this phenomenon of combining waves **interference,** and the waves are said to **interfere.** (These terms refer only to the displacements of the waves; the travel of the waves is unaffected.)

Let one wave traveling along a stretched string be given by

$$y_1(x,t) = Y\sin(kx - \omega t), \tag{17-33}$$

and another, shifted from the first, by an initial phase of ϕ_0 so that

$$y_2(x,t) = Y\sin(kx - \omega t + \phi_0). \tag{17-34}$$

The waves in question have the same angular frequency ω (that is, the same frequency f), the same wave number k (that is, the same wavelength λ), and the same amplitude Y. They both travel in the positive direction of the x axis, with the same speed, given by $|v^{\text{wave}}| = \sqrt{F^{\text{tension}}/\mu}$ (Eq. 17-25). They differ only by the nonzero initial phase ϕ_0 of wave 2, which we call the **phase difference.** These waves are said to be *out of phase* by ϕ_0 or to have a *phase difference* of $\Delta\phi = (\phi_0)_2 - (\phi_0)_1 = \phi_0$, or one wave is said to be *phase-shifted* from the other by ϕ_0.

From the principle of superposition, the resultant wave is the algebraic sum of the two interfering waves and has displacement

$$\begin{aligned} y'(x,t) &= y_1(x,t) + y_2(x,t) \\ &= Y\sin(kx - \omega t) + Y\sin(kx - \omega t + \phi_0). \end{aligned} \tag{17-35}$$

In Appendix E we see that we can write the sum of the sines of two angles α and β as

$$\sin\alpha + \sin\beta = 2\sin\tfrac{1}{2}(\alpha + \beta)\cos\tfrac{1}{2}(\alpha - \beta). \tag{17-36}$$

Applying this relation to Eq. 17-35 leads to

$$y'(x,t) = [2Y\cos\tfrac{1}{2}\phi_0]\sin(kx - \omega t + \tfrac{1}{2}\phi_0), \tag{17-37}$$

where ϕ_0 represents the phase difference $\Delta\phi$ between the two waves. As Fig. 17-21 shows, the resultant wave is also a sinusoidal wave traveling in the direction of increasing x. It is the only wave you would actually see on the string (you would *not* see the two interfering waves of Eqs. 17-33 and 17-34).

> If two sinusoidal waves of the same amplitude and wavelength travel in the *same* direction along a stretched string, they interfere to produce a resultant sinusoidal wave traveling in that direction.

The resultant wave differs from the interfering waves in two respects: (1) its initial *phase* is $\tfrac{1}{2}\phi_0$, and (2) its amplitude Y' is the quantity in the brackets in Eq. 17-37:

$$Y' = 2Y\cos\tfrac{1}{2}\phi_0 \quad \text{(amplitude).} \tag{17-38}$$

The resultant wave of Eq. 17-37, due to the interference of two sinusoidal transverse waves, is also a sinusoidal transverse wave, with an amplitude and an oscillating term.

Let's consider a couple of special and very important cases. If $\phi_0 = 0$ rad (or $0°$), the two interfering waves are exactly in phase, as in Fig. 17-22a. Then Eq. 17-37 reduces to

$$y'(x,t) = 2Y\sin(kx - \omega t) \quad (\phi_0 = 0 \text{ rad}). \tag{17-39}$$

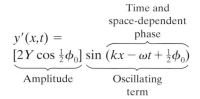

Time and space-dependent phase

$$y'(x,t) = \underbrace{[2Y\cos\tfrac{1}{2}\phi_0]}_{\text{Amplitude}}\underbrace{\sin(kx - \omega t + \tfrac{1}{2}\phi_0)}_{\substack{\text{Oscillating} \\ \text{term}}}$$

FIGURE 17-21 ■ The resultant wave of Eq. 17-37, due to the interference of two sinusoidal transverse waves, is also a sinusoidal wave, with an amplitude and an oscillating term.

This resultant wave is plotted in Fig. 17-22d. Note from both that figure and $y'(x,t) = 2Y\sin(kx - \omega t)$ that the amplitude of the resultant wave is twice the amplitude of either interfering wave. That is the greatest amplitude the resultant wave can have, because the cosine term in $y'(x,t) = [2Y\cos\frac{1}{2}\phi_0]\sin(kx - \omega t + \frac{1}{2}\phi_0)$ has its greatest value (unity) when $\phi_0 = 0$. Interference that produces the greatest possible amplitude is called *fully constructive interference*.

If $\Delta\phi = \phi_0 \pi$ rad (or 180°), the interfering waves are exactly out of phase as in Fig. 17-22b. Then $\cos\frac{1}{2}\phi_0$ becomes $\cos \pi/2 = 0$, and the amplitude of the resultant wave as given by Eq. 17-38 is zero. We then have, for all values of x and t,

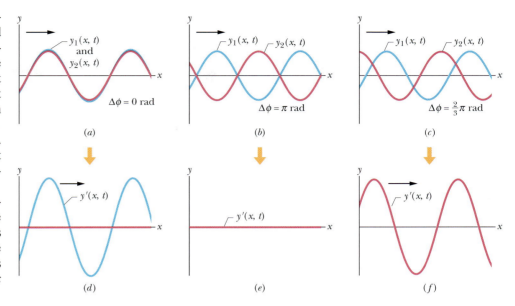

(a)   (b)   (c)

(d)   (e)   (f)

FIGURE 17-22 ■ Two sinusoidal waves with the same k, ω, and y, $y_1(x,t)$ and $y_2(x,t)$, travel along a string in the positive x direction. Here the units for ϕ_0 and t are unspecified. They interfere to give a resultant wave $y'(x,t)$. The resultant wave is what is actually seen on the string. The phase difference $\Delta\phi$ between the two interfering waves is (a) 0 rad or 0°, (b) π rad or 180°, and (c) $\frac{2}{3}\pi$ rad or 120°. The corresponding resultant waves are shown in (d), (e), and (f).

$$y'(x,t) = 0 \qquad (\phi = \pi \text{ rad}). \qquad (17\text{-}40)$$

The resultant wave is plotted in Fig. 17-22e. Although we sent two waves along the string, we see no motion of the string. This type of interference is called *fully destructive interference*.

Because a sinusoidal wave repeats its shape every 2π rad, a phase difference $\Delta\phi_0 = 2\pi$ rad (or 360°) corresponds to a shift of one wave relative to the other wave by a distance equivalent to one wavelength. Thus, phase differences can be described in terms of wavelengths as well as angles. For example, in Fig. 17-22b the waves may be said to be 0.50 wavelength out of phase. Table 17-1 shows some other examples of phase differences and the interference they produce. Note that when interference is neither fully constructive nor fully destructive, it is called *intermediate interference*. The amplitude of the resultant wave is then intermediate between 0 and $2Y$. For example, from Table 17-1, if the interfering waves have a phase difference of 120° ($\Delta\phi = \frac{2}{3}\pi$ rad = 0.33 wavelength), then the resultant wave has an amplitude of Y, the same as the interfering waves (see Figs. 17-22c and f).

TABLE 17-1

Phase Differences and Resulting Interference Types[a]

Phase Difference, ($\Delta\phi$), in			Amplitude of Resultant Wave	Type of Interference
Degrees	Radians	Wavelengths		
0	0	0	$2Y$	Fully constructive
120	$\frac{2}{3}\pi$	0.33	Y	Intermediate
180	π	0.50	0	Fully destructive
240	$\frac{4}{3}\pi$	0.67	Y	Intermediate
360	2π	1.00	$2Y$	Fully constructive
865	15.1	2.40	$0.60Y$	Intermediate

[a] The phase difference is between two otherwise identical waves, with amplitude Y, moving in the same direction.

TOUCHSTONE EXAMPLE 17-6: Two Sine Waves

Two identical sinusoidal waves, moving in the same direction along a stretched string, interfere with each other. The amplitude Y of each wave is 9.8 mm, and the phase difference $\Delta\phi = \phi_0$ between them is 100°.

(a) What is the amplitude Y of the resultant wave due to the interference of these two waves, and what type of interference occurs?

SOLUTION ■ The **Key Idea** here is that these are identical sinusoidal waves traveling in the *same direction* along a string, so they interfere to produce a sinusoidal traveling wave. Because they are identical except for their initial phase, they have the *same amplitude*. Thus, the amplitude Y' of the resultant wave is given by Eq. 17-38:

$$Y' = 2Y' \cos\tfrac{1}{2}\phi_0 = (2)(9.8 \text{ mm}) \cos(100°/2)$$

$$= 12.6 \text{ mm}. \qquad \text{(Answer)}$$

Here we have assumed that wave 2 has an initial phase of 100° relative to wave 1. We can tell that the interference is *intermediate* in two ways. The phase difference is between 0 and 180° and, correspondingly, amplitude Y' is between 0 and $2Y$ (= 19.6 mm).

(b) What phase difference, in radians and wavelengths, will give the resultant wave an amplitude of 4.9 mm?

SOLUTION ■ The same **Key Idea** applies here as in part (a), but now we are given Y' and seek ϕ_0. From Eq. 17-38,

$$Y' = 2Y \cos\tfrac{1}{2}\phi_0.$$

We now have

$$4.9 \text{ mm} = (2)(9.8 \text{ mm}) \cos\tfrac{1}{2}\phi_0,$$

which gives us (with a calculator in the radian mode)

$$\phi_0 = 2 \cos^{-1}\frac{4.9 \text{ mm}}{(2)(9.8 \text{ mm})}$$

$$= \pm 2.636 \text{ rad} \approx \pm 2.6 \text{ rad}. \qquad \text{(Answer)}$$

There are two solutions because we can obtain the same resultant wave by letting the first wave *lead* (travel ahead of) or lag (travel behind) the second wave by 2.6 rad. In wavelengths, the phase difference is

$$\frac{\phi_0}{2\pi \text{ rad/wavelength}} = \frac{\pm 2.636 \text{ rad}}{2\pi \text{ rad/wavelength}}$$

$$= \pm 0.42 \text{ wavelength.} \qquad \text{(Answer)}$$

17-10 Reflections at a Boundary and Standing Waves

In the preceding two sections, we discussed two sinusoidal waves of the same wavelength and amplitude traveling *in the same direction* along a stretched string. What if they travel in opposite directions? We can again find the resultant wave by applying the superposition principle. Figure 17-23 suggests the situation graphically. It shows the two combining waves, one traveling to the left in Fig. 17-23a, the other to the right

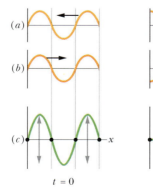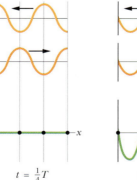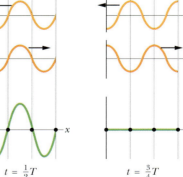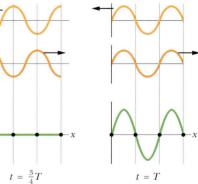

$$t = 0 \qquad t = \tfrac{1}{4}T \qquad t = \tfrac{1}{2}T \qquad t = \tfrac{3}{4}T \qquad t = T$$

FIGURE 17-23 ■ (a) Five snapshots of a wave traveling to the left, at the times t indicated below part (c) (T is the period of oscillation). (b) Five snapshots of a wave identical to that in (a) but traveling to the right, at the same times t. (c) Corresponding snapshots for the superposition of the two waves on the same string. At $t = 0, \tfrac{1}{2}T$, and T, fully constructive interference occurs because of the alignment of peaks with peaks and valleys with valleys. At $t = \tfrac{1}{4}T$ and $\tfrac{3}{4}T$, fully destructive interference occurs because of the alignment of peaks with valleys. Some points (the nodes, marked with dots) never oscillate; some points (the antinodes) oscillate the most.

in Fig. 17-23*b*. Figure 17-23*c* shows their sum, obtained by applying the superposition principle graphically.

The outstanding feature of the resultant wave is that there are places along the string, called **nodes,** where the string never moves. Four such nodes are marked by dots in Fig. 17-23*c*. Halfway between adjacent nodes are **antinodes,** where the amplitude of the resultant wave is a maximum. Wave patterns such as that of Fig. 17-23*c* are called **standing waves** because the wave patterns do not move left or right; the locations of the maxima and minima do not change.

> If two sinusoidal waves of the same amplitude and wavelength travel in *opposite* directions along a stretched string, their interference with each other produces a standing wave.

To analyze a standing wave, we represent the two combining waves with the equations

$$y_1(x,t) = Y\sin(kx - \omega t) \tag{17-41}$$

and
$$y_2(x,t) = Y\sin(kx + \omega t). \tag{17-42}$$

The principle of superposition gives, for the combined wave,

$$y'(x, t) = y_1(x, t) + y_2(x, t) = Y\sin(kx - \omega t) + Y\sin(kx + \omega t).$$

Applying the trigonometric relation of $\sin\alpha + \sin\beta = 2\sin\frac{1}{2}(\alpha + \beta)\cos\frac{1}{2}(\alpha - \beta)$ leads to

$$y'(x,t) = [2Y\sin kx]\cos\omega t, \tag{17-43}$$

which is displayed in Fig. 17-24. This equation does not describe a traveling wave because it is not of the form of $y(x,t) = h(kx \pm \omega t)$. Instead, it describes a standing wave.

The quantity $2Y\sin kx$ in the brackets of $y'(x,t) = [2Y\sin kx]\cos\omega t$ can be viewed as the amplitude of oscillation of the string segment that is located at position x. However, since an amplitude is always positive and $\sin kx$ can be negative, we take the absolute value of the quantity $2Y\sin kx$ to be the amplitude at x.

In a traveling sinusoidal wave, the amplitude of the wave is the same for all string segments. That is not true for a standing wave, in which the amplitude *varies with position.* In the standing wave of $y'(x,t) = [2Y\sin kx]\cos\omega t$, for example, the amplitude is zero for values of kx that give $\sin kx = 0$. Those values are

$$kx = n\pi, \qquad \text{for } n = 0, 1, 2, \ldots. \tag{17-44}$$

Substituting $k = 2\pi/\lambda$ in this equation and rearranging, we get

$$x = n\frac{\lambda}{2}, \qquad \text{for } n = 0, 1, 2, \ldots \quad \text{(nodes)}, \tag{17-45}$$

as the positions of zero amplitude—the nodes—for the standing wave of Eq. 17-43. Note that adjacent nodes are separated by $\lambda/2$, half a wavelength.

The amplitude of the standing wave of $y'(x,t) = [2Y\sin kx]\cos\omega t$ has a maximum value of $2Y$, which occurs for values of kx that give $|\sin kx| = 1$. Those values are

$$kx = \tfrac{1}{2}\pi, \tfrac{3}{2}\pi, \tfrac{5}{2}\pi, \ldots$$
$$= (n + \tfrac{1}{2})\pi, \qquad \text{for } n = 0, 1, 2, \ldots. \tag{17-46}$$

Displacement
$$\overbrace{y'(x,t)} = \underbrace{[2Y\sin kx]}\underbrace{\cos\omega t}$$
Amplitude   Oscillating
at position x term

FIGURE 17-24 ■ The resultant wave of Eq. 17-43 is a standing wave and is due to the interference of two sinusoidal waves of the same amplitude and wavelength that travel in opposite directions with the same initial phase.

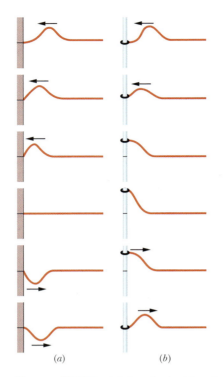

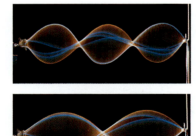

(a)   (b)

FIGURE 17-25 ▪ (a) A pulse incident from the right is reflected at the left end of the string, which is tied to a wall. Note that the reflected pulse is inverted from the incident pulse. (b) Here the left end of the string is tied to a ring that can slide without friction up and down the rod. Now the pulse is not inverted by the reflection.

FIGURE 17-26 ▪ Stroboscopic photographs reveal (imperfect) standing wave patterns on a string being made to oscillate by a vibrator at the left end. The patterns occur at certain frequencies of oscillation.

Substituting $k = 2\pi/\lambda$ in Eq. 17-43 and rearranging, we get

$$x = (n + \tfrac{1}{2})\frac{\lambda}{2} \qquad \text{for } n = 0, 1, 2, \ldots \qquad \text{(antinodes)}, \qquad (17\text{-}47)$$

as the positions of maximum amplitude—the antinodes—of the standing wave of Eq. 17-43. The antinodes are separated by $\lambda/2$ and are located halfway between pairs of nodes.

Reflections at a Boundary

We can set up a standing wave in a stretched string by allowing a traveling wave to be reflected from the far end of the string so that it travels back through itself. The incident (original) wave and the reflected wave can then be described by Eqs. 17-41 and 17-42, respectively, and they can combine to form a pattern of standing waves.

In Fig. 17-25, we use a single pulse to show how such reflections take place. In Fig. 17-25a, the string is fixed at its left end. When the pulse arrives at that end, it exerts an upward force on the support (the wall). By Newton's Third Law, the support exerts an opposite force of equal magnitude on the string. This force generates a pulse at the support. This pulse travels back along the string in the direction opposite that of the incident pulse causing transverse string displacements that are inverted. In a "hard" reflection of this kind, there must be a node at the support because the string is fixed there. The reflected and incident pulses must have opposite signs, so as to cancel each other at that point.

In Fig. 17-25b, the left end of the string is fastened to a light ring that is free to slide without friction along a rod. When the incident pulse arrives, the ring moves up the rod. As the ring moves, it pulls on the string, stretching the string and producing a reflected pulse with the same sign and amplitude as the incident pulse. This reflected pulse is not inverted. Thus, in such a "soft" reflection, the incident and reflected pulses reinforce each other, creating an antinode at the end of the string; the maximum displacement of the ring is twice the amplitude of either of these pulses. The same types of reflections occur for continuous sinusoidal waves.

READING EXERCISE 17-5: Two waves with the same amplitude and wavelength interfere in three different situations to produce resultant waves with the following equations:
(1) $y'(x,t) = (0.004 \text{ m})\sin((5 \text{ rad/m})x - (4 \times 10^3 \text{ rad/s})t)$,
(2) $y'(x,t) = [(0.004 \text{ m})\sin((5 \text{ rad/m})x)]\cos((4 \times 10^3 \text{ rad/s})t)$, and
(3) $y'(x,t) = (0.004 \text{ m})\sin((5 \text{ rad/m})x + (4 \times 10^3 \text{ rad/s})t)$.
In which situation are the two combining waves traveling (a) toward positive x, (b) toward negative x, and (c) in opposite directions? ▪

17-11 Standing Waves and Resonance

Consider a string, such as a guitar string, that is stretched between two clamps. Suppose we send a sinusoidal wave of a certain frequency along the string, say, toward the right. When the wave reaches the right end, it reflects and begins to travel back to the left. That left-going wave then overlaps the wave that is still traveling to the right. When the left-going wave reaches the left end, it reflects again and the newly reflected wave begins to travel to the right, overlapping the left-going and right-going waves. In short, we very soon have many overlapping traveling waves, which interfere with one another.

For certain frequencies, the interference produces a standing wave pattern (or normal **oscillation mode**) with nodes and large antinodes like those in Fig. 17-26. Such a standing wave is said to be produced at **resonance,** and the string is said to *resonate* at these certain frequencies, called **resonant frequencies.** This standing wave resonance is not unlike the harmonic oscillator resonance discussed in Section 16-8. If the string is oscillated at some frequency other than a resonant frequency, a standing wave is not set up. Then the interference of the right-going and left-going traveling waves results in only small (perhaps imperceptible) oscillations of the string.

Let a string be stretched between two clamps separated by a fixed distance L. To find expressions for the resonant frequencies of the string, we note that a node must exist at each of its ends, because each end is fixed and cannot oscillate. The simplest pattern that meets this key requirement is that in Fig. 17-27a, which shows the string at both its extreme displacements (one solid and one dashed, together forming a single "loop"). There is only one antinode, which is at the center of the string. Note that half a wavelength spans the length L, which we take to be the string's length. Thus, for this pattern, $\lambda/2 = L$. This condition tells us that if the left-going and right-going traveling waves are to set up this pattern by their interference, they must have the wavelength $\lambda = 2L$.

A second simple pattern meeting the requirement of nodes at the fixed ends is shown in Fig. 17-27b. This pattern has three nodes and two antinodes and is said to be a two-loop pattern. For the left-going and right-going waves to set it up, they must have a wavelength $\lambda = L$. A third pattern is shown in Fig. 17-27c. It has four nodes, three antinodes, and three loops, and the wavelength is $\lambda = \frac{2}{3}L$. We could continue this progression by drawing increasingly more complicated patterns. In each step of the progression, the pattern would have one more node and one more antinode than the preceding step, and an additional $\lambda/2$ would be fitted into the distance L.

Thus, a standing wave can be set up on a string of length L by a wave with a wavelength equal to one of the values

$$\lambda = \frac{2L}{n}, \qquad \text{for } n = 1, 2, 3. \dots \tag{17-48}$$

The resonant frequencies that correspond to these wavelengths follow from Eq. 17-16:

$$f = \frac{v^{\text{wave}}}{\lambda} = n\frac{v^{\text{wave}}}{2L}, \qquad \text{for } n = 1, 2, 3, \dots \quad \text{(string fixed at both ends)}, \tag{17-49}$$

where v^{wave} is the speed of traveling waves on the string.

Equation 17-49 tells us that the resonant frequencies are integer multiples of the lowest resonant frequency, $f = v^{\text{wave}}/2L$, which corresponds to $n = 1$. The oscillation mode with that lowest frequency is called the *fundamental mode* or the *first harmonic*. The *second harmonic* is the oscillation mode with $n = 2$, the *third harmonic* is that with $n = 3$, and so on. The frequencies associated with these modes are often labeled f_1, f_2, f_3 and so on. The collection of all possible oscillation modes is called the **harmonic series,** and n is called the **harmonic number** of the nth harmonic.

The phenomenon of resonance is common to all oscillating systems and can occur in two and three dimensions. For example, Fig. 17-28 shows a two-dimensional standing wave pattern on the oscillating head of a kettledrum.

READING EXERCISE 17-6: In the following series of resonant frequencies, one frequency (lower than 400 Hz) is missing: 150, 225, 300, 375 Hz. (a) What is the missing frequency? (b) What is the frequency of the seventh harmonic? ■

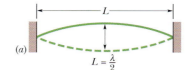
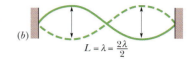
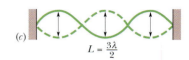

FIGURE 17-27 ■ A string, stretched between two clamps, is made to oscillate in standing-wave patterns. (a) The simplest possible pattern consists of one *loop*, which refers to the composite shape formed by the string in its extreme displacements (the solid and dashed lines). (b) The next simplest pattern has two loops. (c) The next has three loops.

FIGURE 17-28 ■ One of many possible standing wave patterns for a kettledrum head, made visible by dark powder sprinkled on the drumhead. As the head is set into oscillation at a single frequency by a mechanical vibrator at the upper left of the photograph, the powder collects at the nodes, which are circles and straight lines (rather than points) in this two-dimensional example.

TOUCHSTONE EXAMPLE 17-7: String Harmonics

In Fig. 17-29, a string, tied to a sinusoidal vibrator at P and running over a fixed pulley at Q, is stretched by a block of mass m. The separation L between P and Q is 1.2 m, the linear density of the string is 1.6 g/m, and the frequency f of the vibrator is fixed at 120 Hz. The amplitude of the motion at P is small enough for that point to be considered a node. A node also exists at Q.

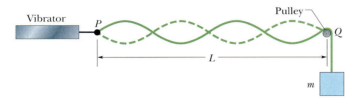

FIGURE 17-29 ■ A string under tension connected to a vibrator. For a fixed vibrator frequency, standing wave patterns will occur for certain values of the string tension.

(a) What mass m allows the vibrator to set up the fourth harmonic on the string?

SOLUTION ■ One **Key Idea** here is that the string will resonate at only certain frequencies, determined by the wave speed $|\vec{v}^{\text{ wave}}|$ on the string and the length L of the string. From Eq. 17-49, these resonant frequencies are

$$f = n\frac{v^{\text{ wave}}}{2L}, \quad \text{for } n = 1, 2, 3, \ldots. \quad (17\text{-}50)$$

To set up the fourth harmonic (for which $n = 4$), we need to adjust the right side of this equation, with $n = 4$, so that the left side equals the frequency of the vibrator (120 Hz).

We cannot adjust L in Eq. 17-50; it is set. However, a second **Key Idea** is that we *can* adjust v^{wave} because it depends on how much mass m we hang on the string. According to Eq. 17-25, wave speed $v^{\text{wave}} = \sqrt{F^{\text{tension}}/\mu}$. Here the tension F^{tension} in the string is equal to the magnitude of the weight mg of the block. Thus,

$$v^{\text{wave}} = \sqrt{\frac{F^{\text{tension}}}{\mu}} = \sqrt{\frac{mg}{\mu}}. \quad (17\text{-}51)$$

Substituting v^{wave} from Eq. 17-51 into Eq. 17-50, setting $n = 4$ for the fourth harmonic, and solving for m give us

$$m = \frac{4L^2 f^2 \mu}{n^2 g} = \frac{(4)(1.2\text{ m})^2(120\text{ Hz})^2(0.0016\text{ kg/m})}{(4)^2(9.8\text{ m/s}^2)}$$

$$= 0.846\text{ kg} \approx 0.85\text{ kg}. \quad \text{(Answer)} \quad (17\text{-}52)$$

(b) What standing wave mode is set up if $m = 1.00$ kg?

SOLUTION ■ If we insert this value of m into Eq. 17-52 and solve for n, we find that $n = 3.7$. A **Key Idea** here is that n must be an integer, so $n = 3.7$ is impossible. Thus, with $m = 1.00$ kg, the vibrator cannot set up a standing wave on the string, and any oscillation of the string will be small, perhaps even imperceptible.

17-12 Phasors

We can represent a wave on a string (or any other type of wave) with a **phasor.** In essence, a phasor is an arrow that has a magnitude equal to the amplitude of the wave and that rotates around an origin; the angular speed of the phasor is equal to the angular frequency ω of the wave. For example, the wave

$$y_1(x,t) = Y_1 \sin(kx - \omega t), \quad (17\text{-}53)$$

which has an initial phase of $(\phi_0)_1 = 0$ rad. It is represented at a time $t > 0$ s, and location x along a string by the phasor shown in Fig. 17-30a. The magnitude of the phasor is the amplitude Y_1 of the wave. As the phasor rotates around the origin at angular speed ω, its projection y_1 on the vertical axis varies sinusoidally, from a maximum of Y_1 through zero to a minimum of $-Y_1$ and then back to Y_1. This variation corresponds to the sinusoidal variation in the displacement $y_1(x,t)$ of any chosen point x along the string as the wave passes through it.

When two waves travel along the same string in the same direction, we can represent them and their resultant wave in a *phasor diagram*. The phasors in Fig. 17-30b represent the wave of Eq. 17-53 and a second wave given by

$$y_2(x,t) = Y_2 \sin(kx - \omega t + \phi_0), \quad (17\text{-}54)$$

and has an initial phase of $(\phi_0)_2$. So, this second wave is phase-shifted from the first wave with a phase difference of $\Delta\phi = (\phi_0)_2 - (\phi_0)_1 = \phi_0 - 0 = \phi_0$. Because

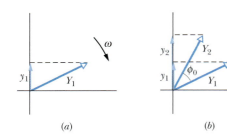 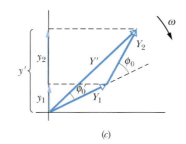

(a)   (b)   (c)

FIGURE 17-30 ■ (a) A phasor of magnitude Y_1 rotating about an origin at angular speed ω represents a sinusoidal wave with an initial phase of zero. The phasor's projection y_1 on the vertical axis represents the displacement at a point X at a time t through which the wave passes. (b) A second phasor, also of angular speed ω but of magnitude Y_2 and rotating at a constant angle ϕ_0 from the first phasor, represents a second wave, with an initial phase of ϕ_0. (c) The resultant wave of the two waves is represented by the vector-like phasor sum Y' of the two phasors. The projection y' on the vertical axis represents the displacement of the some point x and time t as that resultant wave passes through it.

the phasors rotate at the same angular speed ω, the angle between the two phasors is always $\Delta\phi = \phi_0$. If $\Delta\phi$ is a *positive* quantity, then the phasor for wave 2 *lags* the phasor for wave 1 as they rotate, as drawn in Fig. 17-30b. If $\Delta\phi$ is a negative quantity, then the phasor for wave 2 *leads* the phasor for wave 1.

Since waves y_1 and y_2 have the same wave number k and angular frequency ω, we know from Eq. 17-37 that if the two waves have the same amplitude ($Y = Y_1 = Y_2$) then their resultant is of the form

$$y'(x,t) = Y'\sin(kx - \omega t + \phi'_0), \qquad (17\text{-}55)$$

where $Y' = 2Y\cos(\frac{1}{2}\phi_0)$ is the amplitude of the resultant wave and $\phi'_0 = \frac{1}{2}\phi_0$ is its phase. We used Eq. 17-37 to determine the equations for Y' and ϕ'_0 by superimposing the two combining waves.

Finding the superimposed wave is much easier using a phasor diagram. This is because even though phasors are not really vectors that have defined vector dot and cross products, they add like vectors. So even if the amplitudes Y_1 and Y_2 and frequencies ω_1 and ω_2 of two waves are not the same, we can use a vector-like phasor sum to find an equation for the resultant wave at any instant during their rotation. For example, we simply use the rules of vector addition to sum of the two phasors at any instant during their rotation. Fig. 17-30c shows how phasor Y_2 can be shifted to the head of phasor Y_1. The magnitude of the phasor sum equals the amplitude Y' in Eq. 17-55. The angle between the phasor sum and the phasor for y_1 equals the initial phase of the combined wave given in Eq. 17-55 as $\phi'_0 = \frac{1}{2}\phi_0$.

Although we have shown how phasors can be combined for a situation for which the amplitudes and frequencies are the same, it is important to note that:

> We can use phasors to combine waves *even if their amplitudes and frequencies are different.*

TOUCHSTONE EXAMPLE 17-8: Combining Two Waves

Two sinusoidal waves $y_1(x, t)$ and $y_2(x, t)$ have the same wavelength and travel together in the same direction along a string. Their amplitudes are $Y_1 = 4.0$ mm and $Y_2 = 3.0$ mm, and their initial phases are 0 and $\pi/3$ rad, respectively. What are the amplitude Y' and initial phase ϕ'_0 of the resultant wave (in the form of Eq. 17-55)?

SOLUTION ■ One **Key Idea** here is that the two waves have a number of properties in common: they travel along the same string and so have the same speed v^{wave}, angular wave number $k(= 2\pi/\lambda)$, and the same angular frequency $\omega(= kv^{\text{wave}})$.

A second **Key Idea** is that the waves can be represented by phasors rotating at the same angular speed ω about an origin. Because the initial phase for wave 2 is *greater* than that for wave 1 by

$\pi/3$, phasor 2 must *lag* phasor 1 by $\pi/3$ rad in their clockwise rotation, as shown in Fig. 17-31. The resultant wave due to the interference of waves 1 and 2 can then be represented by a phasor that is a vector–like sum of phasors 1 and 2.

To simplify the phasor summation, we drew phasors 1 and 2 in Fig. 17-31a at the instant when phasor 1 lies along the horizontal axis and the lagging phasor 2 at positive angle $\pi/3$ rad. In Fig. 17-31b we shifted phasor 2 so its tail is at the head of phasor 1. Then we draw the phasor Y' of the resultant wave from the tail of phasor 1 to the head of phasor 2. The initial phase ϕ'_0 of the combined is the angle phasor 2 makes with respect to phasor 1.

Though phasors are not really vectors, they can be added using vector addition rules. To find values for Y' and ϕ'_0 we can sum phasors 1 and 2 directly on a vector-capable calculator, by

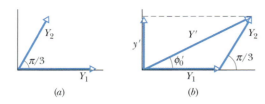

(a)   (b)

FIGURE 17-31 ▪ (a) Two phasors of magnitudes Y_1 and Y_2 with phase difference $\pi/3$. (b) Phasor addition at any instant during their rotation gives the magnitude Y' of the phasor for the resultant wave.

adding a vector of magnitude 4.0 mm and angle 0 rad to a vector of magnitude 3.0 mm and angle $\pi/3$ rad, or we can add the phasors using vector components. For the horizontal components we have

$$Y_h' = Y_1 \cos 0 + Y_2 \cos \pi/3$$

$$= 4.0 \text{ mm} + (3.0 \text{ mm}) \cos \pi/3 = 5.50 \text{ mm}.$$

For the vertical components we have

$$Y_v' = Y_1 \sin 0 + Y_2 \sin \pi/3$$

$$= 0 + (3.0 \text{ mm}) \sin \pi/3 = 2.60 \text{ mm}.$$

Thus, the resultant wave has an amplitude of

$$Y' = \sqrt{(5.50 \text{ mm})^2 + (2.60 \text{ mm})^2}$$

$$= 6.1 \text{ mm} \qquad \text{(Answer)}$$

and an initial resultant wave phase of

$$\phi_0' = \tan^{-1}\frac{2.60 \text{ mm}}{5.50 \text{ mm}} = 0.44 \text{ rad}. \qquad \text{(Answer)}$$

From Fig. 17-31b, the initial phase for wave 2, given by ϕ_0, is a *positive* angle relative to the initial phase of wave 1. Thus, the resultant wave *lags* wave 1 by an initial phase of $\phi'_0 = 0.44$ rad. From Eq. 17-55, we can write the resultant wave as

$$y'(x,t) = (6.1 \text{ mm}) \sin(kx - \omega t + 0.44 \text{ rad}). \quad \text{(Answer)}$$

Problems

SEC. 17-5 ▪ WAVE VELOCITY

1. Angular Frequency A wave has an angular frequency of 110 rad/s and a wavelength of 1.80 m. Calculate (a) the angular wave number and (b) the speed of the wave.

2. Electromagnetic Waves The speed of electromagnetic waves (which include visible light, radio, and x-rays) in vacuum is 3.0×10^8 m/s. (a) Wavelengths of visible light waves range from about 400 nm in the violet to about 700 nm in the red. What is the range of frequencies of these waves? (b) The range of frequencies for shortwave radio (for example, FM radio and VHF television) is 1.5 to 300 MHz. What is the corresponding wavelength range? (c) X-ray wavelengths range from about 5.0 nm to about 1.0×10^{-2} nm. What is the frequency range for x-rays?

3. Sinusoidal Wave A sinusoidal wave travels along a string. The time for a particular point to move from maximum displacement to zero is 0.170 s. What are the (a) period and (b) frequency? (c) The wave length is 1.40 m; what is the wave speed?

4. Write the Equation Write the equation for a sinusoidal wave traveling in the negative direction along an x axis and having an amplitude of 0.010 m, a frequency of 550 Hz, and a speed of 330 m/s.

5. Show That Show that

$$y(x,t) = Y \sin k(x - vt), \qquad y(x,t) = Y \sin 2\pi\left(\frac{x}{\lambda} - ft\right),$$

$$y(x,t) = Y \sin \omega\left(\frac{x}{v} - t\right), \qquad y(x,t) = Y \sin 2\pi\left(\frac{x}{\lambda} - \frac{t}{T}\right)$$

are all equivalent to $y(x,t) = Y \sin (kx - \omega t)$.

6. Equation of a Transverse The equation of a transverse wave traveling along a very long string is $y(x,t) = (6.0 \text{ cm}) \sin \{(0.020 \pi \text{ rad/cm}) x + (4.0 \pi \text{ rad/s})t\}$ where x and y are expressed in centimeters and t is in seconds. Determine (a) the amplitude, (b) the wavelength, (c) the frequency, (d) the speed, (e) the direction of propagation of the wave, and (f) the maximum transverse speed of a particle in the string. (g) What is the transverse displacement at $x = 3.5$ cm when $t = 0.26$ s?

7. Write an Equation (a) Write an equation describing a sinusoidal transverse wave traveling on a cord in the $+x$ direction with a wavelength of 10 cm, a frequency of 400 Hz, and an amplitude of 2.0 cm. (b) What is the maximum speed of a point on the cord? (c) What is the speed of the wave?

8. Transverse Sinusoidal A transverse sinusoidal wave of wavelength 20 cm is moving along a string in the positive x direction. The transverse displacement of the string particle at $x = 0$ cm as a function of time is shown in Fig. 17-32. (a) Make a rough sketch of one wavelength of the wave (the portion between $x = 0$ cm and $x = 20$ cm) at time $t = 0$ s. (b) What is the speed of the wave? (c) Write the equation for the wave with all the constants evaluated. (d) What is the transverse velocity of the particle at $x = 0$ m at $t = 5.0$ s?

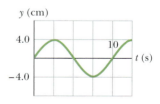

FIGURE 17-32 ▪ Problem 8.

9. Sinusoidal Wave Two A sinusoidal wave of frequency 500 Hz has a speed of 350 m/s. (a) How far apart are two points that differ in phase by $\pi/3$ rad? (b) What is the phase difference between two displacements at a certain point at times 1.00 ms apart?

SEC. 17-6 ■ WAVE SPEED ON A STRETCHED STRING

10. Violin Strings The heaviest and lightest strings on a certain violin have linear densities of 3.0 and 0.29 g/m. (a) What is the ratio of the diameter of the heaviest string to that of the lightest string, assuming that the strings are of the same material? (b) What is the ratio of speeds if the strings have the same tension?

11. What Is the Speed What is the speed of a transverse wave in a rope of length 2.00 m and mass 60.0 g under a tension of 500 N?

12. Wire Clamped The tension in a wire clamped at both ends is doubled without appreciably changing the wire's length between the clamps. What is the ratio of the new to the old wave speed for transverse waves traveling along this wire?

13. Linear Density The linear density of a string is 1.6×10^{-4} kg/m. A transverse wave on the string is described by the equation

$$y(x,t) = (0.021 \text{ m}) \sin[(2.0 \text{ rad/m})x + (30 \text{ rad/s})t].$$

What is (a) the wave speed and (b) the tension in the string?

14. The Equation of The equation of a transverse wave on a string is

$$y(x,t) = (2.0 \text{ mm}) \sin[(20 \text{ rad/m})x - (600 \text{ rad/s})t].$$

The tension in the string is 15 N. (a) What is the wave speed? (b) Find the linear density of the string in grams per meter.

15. Stretched String A stretched string has a mass per unit length of 5.0 g/cm and a tension of 10 N. A sinusoidal wave on this string has an amplitude of 0.12 mm and a frequency of 100 Hz and is traveling in the negative direction of x. Write an equation for this wave.

16. The Fastest Wave What is the fastest transverse wave that can be sent along a steel wire? For safety reasons, the maximum tensile stress to which steel wires should be subjected is 7.0×10^8 N/m². The density of steel is 7800 kg/m³. Show that your answer does not depend on the diameter of the wire.

17. Single Particle A sinusoidal transverse wave of amplitude Y and wavelength λ travels on a stretched cord. (a) Find the ratio of the maximum particle speed (the speed with which a single particle in the cord moves transverse to the wave) to the wave speed. (b) If a wave having a certain wavelength and amplitude is sent along a cord, would this speed ratio depend on the material of which the cord is made such as wire or nylon?

18. Displacement of Particles A sinusoidal wave is traveling on a string with speed 40 cm/s. The displacement of the particles of the string at $x = 10$ cm is found to vary with time according to the equation $y(x,t) = (5.0 \text{ cm}) \sin [1.0 \text{ rad/cm} - (4.0 \text{ rad/s})t]$. The linear density of the string is 4.0 g/cm. What are the (a) frequency and (b) wavelength of the wave? (c) Write the general equation giving the transverse displacement of the particles of the string as a function of position and time. (d) Calculate the tension in the string.

19. As a Function of Position A sinusoidal transverse wave is traveling along a string in the negative direction of an x axis. Figure 17-33 shows a plot of the displacement as

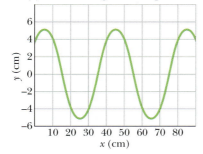

FIGURE 17-33 ■ Problem 19.

a function of position at time $t = 0$ s; the y intercept is 4.0 cm. The string tension is 3.6 N, and its linear density is 25 g/m. Find the (a) amplitude, (b) wavelength, (c) wave speed, and (d) period of the wave. (e) Find the maximum transverse speed of a particle in the string. (f) Write an equation describing the traveling wave.

20. Three Pulleys, One Mass In Fig. 17-34a string 1 has a linear density of 3.00 g/m, and string 2 has a linear density of 5.00 g/m. They are under tension owing to the hanging block of mass $M = 500$ g. Calculate the wave speed on (a) string 1 and (b) string 2. (*Hint*: When a string loops halfway around a pulley, it pulls on the pulley with a net force that is twice the tension in the string.) Next the block is divided into two blocks (with $M_1 + M_2 = M$) and the apparatus is rearranged as shown in Fig. 17-34b. Find (c) M_1 and (d) M_2 such that the wave speeds in the two strings are equal.

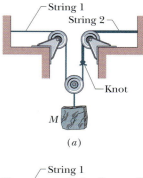

(a)

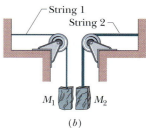

(b)

FIGURE 17-34 ■ Problem 20.

21. Two Pulses Two A wire 10.0 m long and having a mass of 100 g is stretched under a tension of 250 N. If two pulses, separated in time by 30.0 ms, are generated, one at each end of the wire, where will the pulses first meet?

22. Baseball Rubber Band The type of rubber band used inside some baseballs and golf balls obeys Hooke's law over a wide range of elongation of the band. A segment of this material has an unstretched length ℓ and a mass m. When a force $\vec{F}$ is applied, the band stretches an additional length $\Delta\ell$. (a) What is the speed (in terms of m, $\Delta\ell$, and the spring constant k) of transverse waves on this stretched rubber band? (b) Using your answer to (a), show that the time required for a transverse pulse to travel the length of the rubber band is proportional to $1/\sqrt{\Delta\ell}$ if $\Delta\ell \ll \ell$ and is constant if $\Delta\ell \gg \ell$.

23. Uniform Rope A uniform rope of mass m and length L hangs from a ceiling. (a) Show that the speed of a transverse wave on the rope is a function of y, the distance from the lower end, and is given by $v^{\text{wave}} = \sqrt{gy}$. (b) Show that the time a transverse wave takes to travel the length of the rope is given by $t = 2\sqrt{L/g}$.

SEC. 17-7 ■ ENERGY AND POWER TRANSPORTED BY A TRAVELING WAVE IN A STRING

24. Average Power A string along which waves can travel is 2.70 m long and has a mass of 260 g. The tension in the string is 36.0 N. What must be the frequency of traveling waves of amplitude 7.70 mm for the average power to be 85.0 W?

SEC. 17-9 ■ INTERFERENCE OF WAVES

25. Two Identical Traveling Waves Two identical traveling waves, moving in the same direction, are out of phase by $\pi/2$ rad. What is the amplitude of the resultant wave in terms of the common amplitude Y of the two combining waves?

26. What Phase Difference What phase difference between two otherwise identical traveling waves, moving in the same direction along a stretched string, will result in the combined wave having an amplitude 1.50 times that of the common amplitude of the two combining waves? Express your answer in (a) degrees, (b) radians, and (c) wavelengths.

27. Identical Except for Phase Two sinusoidal waves, identical except for phase, travel in the same direction along a string and interfere to produce a resultant wave given by $y'(x,t) = (3.0$ mm$)$ $\sin [(20$ rad/m$)x - (4.0$ rad/s$)t + 0.820$ rad$]$, with x in meters and t in seconds. What are (a) the wavelength λ of the two waves, (b) the phase difference between them, and (c) their amplitude Y?

SEC. 17-11 ■ STANDING WAVES AND RESONANCE

28. A String Under Tension A string under tension F^{tension} oscillates in the third harmonic at frequency f_3, and the waves on the string have wavelength λ_3. If the tension is increased to $\tau_f = 4\tau_i$ and the string is again made to oscillate in the third harmonic, what then are (a) the frequency of oscillation in terms of f_3 and (b) the wavelength of the waves in terms of λ_3?

29. Nylon Guitar String A nylon guitar string has a linear density of 7.2 g/m and is under a tension of 150 N. The fixed supports are 90 cm apart. The string is oscillating in the standing wave pattern shown in Fig. 17-35. Calculate the (a) speed, (b) wavelength, and (c) frequency of the traveling waves whose superposition gives this standing wave.

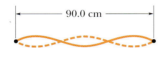

FIGURE 17-35 ■ Problem 29.

90.0 cm

30. Two Sinusoidal Waves Two sinusoidal waves with identical wavelengths and amplitudes travel in opposite directions along a string with a speed of 10 cm/s. If the time interval between instants when the string is flat is 0.50 s, what is the wavelength of the waves?

31. String Fixed at Both Ends A string fixed at both ends is 8.40 m long and has a mass of 0.120 kg. It is subjected to a tension of 96.0 N and set oscillating. (a) What is the speed of the waves on the string? (b) What is the longest possible wavelength for standing wave? (c) Give the frequency of that wave.

32. Between Fixed Supports A 125 cm length of string has a mass of 2.00 g. It is stretched with a tension of 7.00 N between fixed supports. (a) What is the wave speed for this string? (b) What is the lowest resonant frequency of this string?

33. Three Lowest Frequencies What are the three lowest frequencies for standing waves on a wire 10.0 m long having a mass of 100 g, which is stretched under a tension of 250 N?

34. String A, String B String A is stretched between two clamps separated by distance L. String B, with the same linear density and under the same tension as string A, is stretched between two clamps separated by distance $4L$. Consider the first eight harmonics of string B. Which, if any, has a resonant frequency that matches a resonant frequency of string A?

35. Resonant Frequencies A string that is stretched between fixed supports separated by 75.0 cm has resonant frequencies of 420 and 315 Hz, with no intermediate resonant frequencies. What are (a) the lowest resonant frequency and (b) the wave speed?

36. Two Pulses In Fig. 17-36, two pulses travel along a string in opposite directions. The wave speed v^{wave} is 2.0 m/s and the pulses are 6.0 cm apart at $t = 0$. (a) Sketch the wave patterns when t is equal to 5.0, 10, 15, 20, and 25 ms. (b) In what form (or type) is the energy of the pulses at $t = 15$ ms?

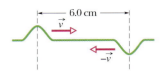

FIGURE 17-36 ■ Problem 36.

37. A String Oscillates A string oscillates according to the equation

$$y'(x,t) = (0.50 \text{ cm}) \sin \left[\left(\frac{\pi}{3} \text{ rad/cm}\right)x\right] \cos[(40\pi \text{ rad/s})t].$$

What are the (a) amplitude and (b) speed of the two waves (identical except for direction of travel) whose superposition gives this oscillation? (c) What is the distance between nodes? (d) What is the speed of a particle of the string at the position $x = 1.5$ cm when $t = \frac{9}{8}$ s?

38. Standing Wave A standing wave results from the sum of two transverse waves traveling in opposite directions given by

$$y_1 = (0.05 \text{ m }) \cos ((\pi \text{ rad/m})x - (4\pi \text{ rad/s})t)$$

and

$$y_2 = (0.05 \text{ m}) \cos ((\pi \text{ rad/m})x + (4\pi \text{ rad/s})t).$$

(a) What is the smallest positive value of x that corresponds to a node? (b) At what times during the interval $0 \le t \le 0.50$ s will the particle at $x = 0.00$ m have zero velocity?

39. Three-Loop Standing Wave A string 3.0 m long is oscillating as a three-loop standing wave with an amplitude of 1.0 cm. The wave speed is 100 m/s. (a) What is the frequency? (b) Write equations for two waves that, when combined, will result in this standing wave.

40. In an Experiment In an experiment on standing waves, a string 90 cm long is attached to the prong of an electrically driven tuning fork that oscillates perpendicular to the length of the string at a frequency of 60 Hz. The mass of the string is 0.044 kg. What tension must the string be under (weights are attached to the other end) if it is to oscillate in four loops?

41. Tuning Fork Oscillation of a 600 Hz tuning fork sets up standing waves in a string clamped at both ends. The wave speed for the string is 400 m/s. The standing wave has four loops and an amplitude of 2.0 mm. (a) What is the length of the string? (b) Write an equation for the displacement of the string as a function of position and time.

42. Second Harmonic A rope, under a tension of 200 N and fixed at both ends, oscillates in a second-harmonic standing wave pattern. The displacement of the rope is given by

$$y(x,t) = (0.10 \text{ m})(\sin \left(\left(\frac{\pi}{2} \text{ rad/m}\right)x\right) \sin ((12\pi \text{ rad/s})t),$$

where $x = 0.00$ m at one end of the rope, x is in meters, and t is in seconds. What are (a) the length of the rope, (b) the speed of the waves on the rope, and (c) the mass of the rope? (d) If the rope oscillates in a third-harmonic standing wave pattern, what will be the period of oscillation?

43. A Generator A generator at one end of a very long string creates a wave given by

$$y(x,t) = (6.0 \text{ cm}) \cos \frac{\pi}{2} [(2.0 \text{ rad/m})x + (8.0 \text{ rad/s})t],$$

and one at the other end creates the wave

$$y(x,t) = (6.0 \text{ cm}) \cos \frac{\pi}{2} [(2.0 \text{ rad/m})x - (8.0 \text{ rad/s})t].$$

Calculate the (a) frequency, (b) wavelength, and (c) speed of each wave. At what x values are the (d) nodes and (e) antinodes?

44. Standing Wave Pattern A standing wave pattern on a string is described by

$$y(x,t) = (0.040 \text{ m}) \sin((5\pi \text{ rad/m})x) \cos((40\pi \text{ rad/s})t).$$

(a) Determine the location of all nodes for $0 \le x \le 0.40$ m. (b) What is the period of the oscillatory motion of any (nonnode) point on the string? What are (c) the speed and (d) the amplitude of the two traveling waves that interfere to produce this wave? (e) At what times for $0.000 \text{ s} \le t \le 0.050$ s will all the points on the string have zero transverse velocity?

45. Maximum Kinetic Energy Show that the maximum kinetic energy in each loop of a standing wave produced by two traveling waves of identical amplitudes is $2\pi^2 \mu Y^2 f v^{\text{wave}}$.

46. Antinode For a certain transverse standing wave on a long string, an antinode is at $x = 0.00$ m and a node is at $x = 0.10$ m. The displacement $y(t)$ of the string particle at $x = 0.00$ m is shown in Fig. 17-37. When $t = 0.50$ s, what are the displacements of the string particles at (a) $x = 0.20$ m and (b) $x = 0.30$ m? At $x = 0.20$ m, what are the transverse velocities of the string particles at (c) $t = 0.50$ s and (d) $t = 1.0$ s? (e) Sketch the standing wave at $t = 0.50$ s for the range $x = 0.00$ m to $x = 0.40$ m.

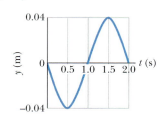

FIGURE 17-37 ■ Problem 46.

47. Aluminum Wire In Fig. 17-38, an aluminum wire, of length $L_1 = 60.0$ cm, cross-sectional area 1.00×10^{-2} cm², and density 2.60 g/cm³, is joined to a steel wire, of density 7.80 g/cm³ and the same cross-sectional area. The compound wire, loaded with a block of mass $m = 10.0$ kg, is arranged so that the distance L_2 from the joint to the supporting pulley is 86.6 cm. Transverse waves are set up in the wire by using an external source of variable frequency; a node is located at the pulley. (a) Find the lowest frequency of excitation for which standing waves are observed such that the joint in the wire is one of the nodes. (b) How many nodes are observed at this frequency?

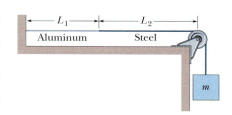

FIGURE 17-38 ■ Problem 47.

SEC. 17-12 ■ PHASORS

48. Of the Same Period Two sinusoidal waves of the same period, with amplitudes of 5.0 and 7.0 mm, travel in the same direction along a stretched string; they produce a resultant wave with an amplitude of 9.0 mm. The initial phase of the 5.0 mm wave is 0.0 rad. What is the initial phase of the 7.0 mm wave?

49. Amplitude of the Resultant Determine the amplitude of the resultant wave when two sinusoidal string waves having the same frequency and traveling in the same direction on the same string are combined, if their amplitudes are 3.0 cm and 4.0 cm and they have initial phases of 0.0 and $\pi/2$ rad, respectively.

50. Three Sinusoidal Waves Three sinusoidal waves of the same frequency travel along a string in the positive direction of an x axis. Their amplitudes are y_1, $y_1/2$, and $y_1/3$, and their initial phases are 0, $\pi/2$, and π rad, respectively. What are (a) the amplitude and (b) the phase constant of the resultant wave? (c) Plot the wave form of the resultant wave at $t = 0.00$ s, and discuss its behavior as t increases.

Additional Problems

51. Graphing a Pulse on a String Consider the motion of a pulse on a long taut string. We will choose our coordinate system so that when the string is at rest, the string lies along the x axis of the coordinate system. We will take the positive direction of the x axis to be to the right on this page and the positive direction of the y axis to be up. Ignore gravity. A pulse is started on the string moving to the right. At a time $t_1 = 0$ s a photograph of the string would look like Fig. 17-39A. A point on the string to the right of the pulse is marked by a spot of red paint.

For each of the items that follow, identify which figure (B-F) would look most like the graph of the indicated quantity. (Take the positive axis as up.) If none of the figures look like you expect the graph to look, write N. Assume that the axes would be labeled with correct units.

(a) The graph of the y displacement of the spot of red paint as a function of time.

(b) The graph of the x velocity of the spot of red paint as a function of time.

(c) The graph of the y velocity of the spot of red paint as a function of time.

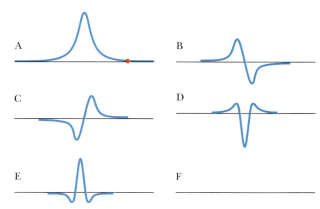

FIGURE 17-39 ■ Problem 51.

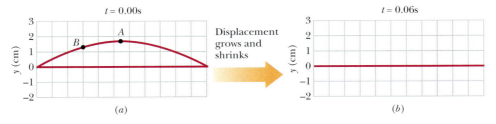

(a)

(b)

FIGURE 17-40 ■ Problem 52.

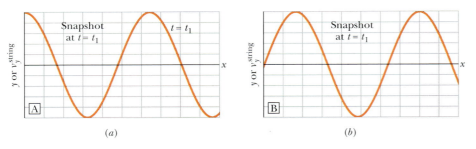

(a)

(b)

FIGURE 17-41 ■ Problem 53.

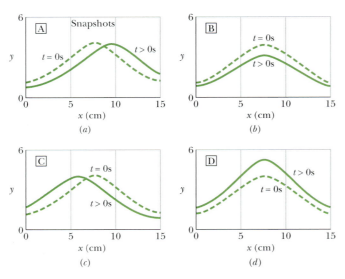

(a)

(b)

(c)

(d)

FIGURE 17-42(a) ■ Problem 55.

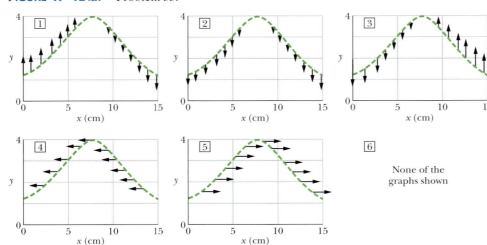

None of the
graphs shown

FIGURE 17-42(b) ■ Problem 55.

(d) The graph of the y-component of the force on the piece of string marked by the red spot as a function of time.

52. Motion in a Standing Wave A string is connected at one end to a vibrating reed and on the other to a weight pulling the string tight over a pulley. At the time $t = 0.00$ s a photograph of the string looks like Fig. 17-40a. The displacement of the wave is then observed to grow, reaching a maximum of 3 cm at $t = 0.02$ s. The displacement of the wave then decreases with time and at time $t = 0.06$ s appears flat, as shown in Fig. 17-40b.

Each box in the grid has a side of 1 cm. (a) Two points on the string are marked with heavy black dots and with the letters A and B. At $t = 0.00$ s, in what direction is each of them moving and which one (if any) is moving faster? (b) At $t = 0.06$ s, mark the position of the two points on the string in the figure (b). In what direction is each of them moving and which one (if any) is moving faster? (c) What is the period of the oscillation? (d) Explain what could be changed to make the velocity of point A zero throughout the oscillation. Give two different ways to do this, and explain why such a change would work.

53. Displacement and Velocity Patterns in Waves Each graph shown in Fig. 17-41 may represent either a picture of the shape of a wave on a string at a particular instant in time, t_1, or the transverse (up and down) velocity of the mass points of that string at that time. Depending on which it is, the wave on the string may be:

(a) A right-moving traveling wave, (b) A left-moving traveling wave, (c) A part of a standing wave with a displacement that increases in time and is shown at time t_1, (d) A part of a standing wave with a displacement that decreases in time and is shown at time t_1. For each of the following cases, decide what the wave is doing and choose one of the four letters (a)–(d).

i. Graph A is a graph of the string's shape and graph B is a graph of the string's velocity.

ii. Graph A is a graph of both the string's shape and the string's velocity.

iii. Graph B is a graph of the string's shape and graph A is a graph of the string's velocity.

54. Interpreting an Oscillatory Equation Consider the equation

$$y(x,t) = Y \cos(at) \sin(bx)$$

Let x represent some position and t represent time. (a) Describe a physical situation represented by this equation. As part of your description include a sketch and a written description. Indicate what y and x correspond to in the situation you describe. (b) How, if at all, would the physical situation you described in part (a) be different if a were twice as large? Explain how you determined your answer. (c) How, if at all, would the physical situation you described in part (a) be different if b were twice as large? Explain how you determined your answer.

55. Waves and Velocities The four graphs labeled (A)–(D) in Fig. 17-42a show snapshots of waves on a long, taut string. The dashed line shows a picture of a part of the string at the time $t = 0$ s. The solid line is a picture at a time a little bit later. Each of these pictures looks the same at $t = 0$ s, but the results differ because the parts of the string have different velocities at $t = 0$ in each case.

For each of the four cases, select one of the six patterns shown in Fig. 17-42b as the correct velocity pattern for the motion of adjacent string elements to lead to the solid line in Fig. 17-42a, showing the displacement y. (Note the arrows indicate mainly direction. Their lengths are scaled somewhat in proportion to their magnitude, but not strictly so.)

56. Making a Pulse Move A transverse wave pulse is traveling in the $+x$ direction along a long stretched string. The origin is taken at a point on the string that is far from the ends. The speed of the wave is v (a positive number). At time $t = 0$ s, the displacement of the string is described by the function

$$y = f(x,0) = \begin{cases} A\left(1 - \left|\dfrac{x}{\ell}\right|\right) & |x| \le \ell \\ 0 & |x| \ge \ell \end{cases}$$

(a) Construct a graph that portrays the actual shape of the string at $t = 0$ s for the case $A = \ell/2$. (b) Sketch a graph of the wave form at the following times: $1. t = \ell/v, 2. t = 2\ell/v, 3. t = -3\ell/2v$.

57. Comparing Waves Figure 17-43 shows a snapshot of a piece of a wave at a time $t = 0$ s. Make four sketches of this picture and use a dotted line to sketch what each pulse would look like at a slightly later time (a time that is small compared to the time it would take the pulse to move a distance equal to

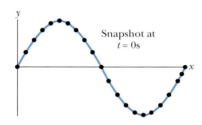

FIGURE 17-43 ■ Problem 57.

its own width but large enough to see a change in the shape of the string) for the following four cases:

A. The pulse is a traveling wave moving to the right.
B. The pulse is a traveling wave moving to the left.
C. The pulse is a standing wave with a displacement that increases in time.
D. The pulse is a standing wave with a displacement that decreases in time.

On each picture, draw arrows to show the velocity of the marked points at time $t = 0$ s.

58. Combining Pulses Figure 17-44 shows graphs that could represent properties of pulses on a stretched string. For the situation and the properties (a)–(e), select which graph provides the best representation of the given property. If none of the graphs are correct, write "none."

Two pulses are started on a stretched string. At time $t = t_1$, an upward pulse is started on the right that moves to the left. At the same instant, a downward pulse is started on the left that moves to the right. At t_1 their peaks are separated by a distance 2s. The distance between the pulses is much larger than their individual widths. The pulses move on the string with a speed v_1. The scales in

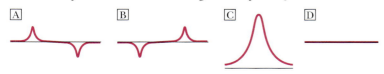

FIGURE 17-44 ■ Problem 58.

the graphs are arbitrary and not necessarily the same. (a) Which graph best represents the appearance of the string at time t_1? (b) Which graph best represents the appearance of the string at a time $t_1 + s/v_1$? (c) Which graph best represents the appearance of the string at a time $t_1 + 2s/v_1$? (d) Which graph best represents the velocity of the string at a time $t_1 + s/v_1$? (e) Which graph best represents the appearance of the string at a time $t_1 + s/v_1 + \varepsilon$ where ε is small compared to s/v_1?

59. Moving a Nonsymmetric Triangular Pulse A long taut spring is started at a time $t = 0$ with a pulse moving in the $+x$ direction in the shape given by the function $f(x)$ with

$$f(x) = \begin{cases} \tfrac{1}{4}x + 1 & -4 < x < 0 \\ -x + 1 & 0 < x < 1 \\ 0 & \text{otherwise} \end{cases}$$

(The units of x and f are in <u>centimeters</u>.) (a) Draw a labeled graph showing the shape of the string at $t = 0$. (b) If the mass density of the string is 50 g/m and it is under a tension of 5 N, draw a labeled graph that shows the shape and position of the string at a time $t = 0.001$ s. (c) Write the solution of the wave equation that explicitly gives the displacement of any piece of the string at any time. (d) What is the speed of a piece of string that is moving up after the pulse has reached it but before it has risen to its maximum displacement? What is its speed while it is returning to its original position after the pulse's peak has passed it?

60. Which Wave Is Which? Figure 17-45a shows a picture of a string at a time t_1. The pieces of the string are each moving with velocities that are indicated by arrows in the picture. (The vertical displacements are small and don't show up in the picture.) Figure 17-45b shows five graphs that could give the shape of the string at the instant for which the velocities are displayed above. (Note: The vertical scale magnifies the displacement by a factor of 100.)

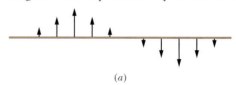

(a)

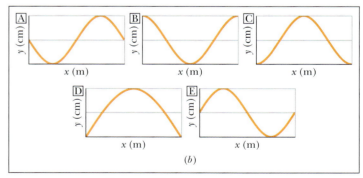

(b)

FIGURE 17-45 ■ Problem 60.

On your paper, place the letters A–E. Next to these letters, indicate for the graphs labeled by those letters, whether the string is

moving as a: (L) left-traveling wave, (R) right-traveling wave, (S+) standing wave with a displacement that increases in time, (S−) standing wave with a displacement that decreases in time, or (N) none of the above.

61. Spring vs. String In this course, we analyzed the motion of a mass on a spring and the oscillations of a taut string. Discuss these two systems, explaining similarities and differences, and give an equation of motion for each.

62. Parsing a Pulse
An instructor is demonstrating the motion of waves on a long, taut spring. He is holding the spring at

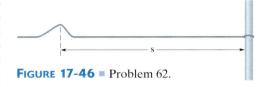

FIGURE 17-46 ■ Problem 62.

one end and will move it so the spring will move back and forth on the floor. The spring is rigidly connected to a metal rod at its other end. The spring is under a tension T and it has a mass density μ. The instructor starts a pulse moving toward the right as shown in Fig. 17-46. The pulse is triangular and is *not* symmetric. The figure is shown at a time t_1. (a) Calculate the time τ it will take the peak of the pulse to reach the wall (to travel a distance s). (b) What will the spring look like at the time $t_1 + \tau$? Draw a carefully constructed and labeled diagram to show what it looks like and how you got your result. (c) What will the spring look like a bit later—say, at a time $t_1 + 2\tau$? What is responsible for this result? (d) The width of the pulse is 0.5 m. If the tension in the spring is 5 N and it has a mass density of 0.1 kg/m, how much time did the professor take to generate the pulse?

63. Modified Harmonics A taut string is tied down at both ends. Assume that the fundamental mode of oscillation of the string has a period T_0. For each of the changes described below give the factor by which the period changes. For example, if the change described resulted in a period twice as long, you would put the number "2." Do not accumulate changes. That is, before each change, assume you are back at the original starting situation. (a) The string is replaced by one of twice the mass but of the same length. (b) The wave length of the starting shape is divided by three. (c) The amplitude of the oscillation is doubled. (d) The tension of the string is halved.

64. Spring vs. String: Graphs Consider two physical systems: System A is a mass hanging from a light spring fixed at one end to a ringstand on a table above the floor. System B is a long spring of uniform density held under tension and able to move transversely in a horizontal plane.

Figure 17-47 shows three graphs labeled #1, #2, and #3, with unmarked axes. They could represent many quantities in the physical systems A and B. For the five physical quantities below, indicate which of the

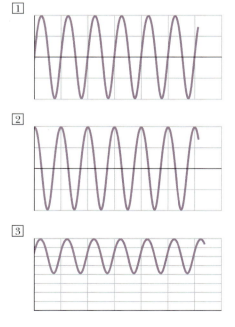

FIGURE 17-47 ■ Problem 64.

graphs could be obtained for the system indicated. If more than one graph applies, give *all* the possible choices. If none applies, write N.

(a) The height of the mass hanging from the spring (system A) above the ground as a function of time for some time interval after the mass has been set into oscillation

(b) The transverse displacement from equilibrium of some portion of the long spring (system B) as a function of position at a particular instant of time while it is carrying a harmonic wave

(c) The velocity of the mass hanging from the spring (system A) as a function of time given that its displacement from equilibrium as a function of time is given by graph #1

(d) The transverse velocity of a small piece of the long spring (system B) as a function of time as a single pulse moves down the spring

(e) The velocity of a small piece of the long spring (system B) as a function of time given that its displacement from equilibrium as a function of time is given by graph #2

65. Explaining the Wave Equation The wave equation

$$\frac{\partial^2 y}{\partial x^2} = \frac{1}{v_1^2}\frac{\partial^2 y}{\partial t^2}$$

is often used to describe the transverse displacement of waves on a stretched spring. Explain the meaning of each of the elements of this equation with reference to the physical spring and discuss under what circumstances you expect it to be a good description.

66. Propagating a Gaussian Pulse Figure 17-48 shows the shape of a pulse on a stretched spring at the time $t = 0$ s. The dispacement of the spring from its equilibrium position at that time is given by

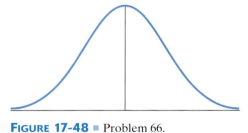

FIGURE 17-48 ■ Problem 66.

$$F(x) = Ae^{-(x/b)^2}$$

The pulse is moving in the positive x direction with a velocity v_1. (a) Sketch a graph showing the shape of the spring at a later time, $t = t_2$. Specify the height and position of the peak in terms of the symbols given. (b) Write an equation for the displacement of any portion of the spring at any time $y(x,t)$. (c) Sketch a graph of the velocity of the piece of the spring at the position $x = 2b$ as a function of time.

67. Varying a Pulse* A long, taut string is attached to a distant wall as shown in Fig. 17-49. A demonstrator moves her hand and creates a very small amplitude pulse that reaches the wall in a time t_1. A small red dot is painted on the string halfway between the demonstrator's hand and the wall. For each situation below, state which of the actions 1–10 listed (taken by itself) will produce the desired result. For each question, *more than one answer may be correct*. If so, give them all.

FIGURE 17-49 ■ Problem 67.

*From the *Wave Test* by M. Wittmann.

Tell how, if at all, the demonstrator can repeat the original experiment to produce: (a) A pulse that takes a longer time to reach the wall (b) A pulse that is wider than the original pulse (c) A pulse that makes the red dot travel a further distance than in the original experiment

1. Move her hand more quickly (but still only up and down once and still by the same amount)
2. Move her hand more slowly (but still only up and down once and still by the same amount)
3. Move her hand a larger distance but up and down in the same amount of time
4. Move her hand a smaller distance but up and down in the same amount of time
5. Use a heavier string of the same length under the same tension
6. Use a lighter string of the same length under the same tension
7. Use a string of the same density, but decrease the tension
8. Use a string of the same density, but increase the tension
9. Put more force into the wave
10. Put less force into the wave

68. Reflecting on a Textbook Error Figure 17-50 is taken from the first edition of a popular standard textbook. Column (a) shows a pulse approaching and reflecting from a fixed end. Column (b) a pulse approaching and reflecting from a free end (ring sliding on a frictionless rod). At least six (6!) of the figures are incorrect. State which and explain why. (*Hint*: The second edition of the book fixed most of the problems by making the pulses symmetric.)

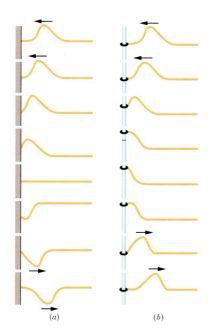

(a) (b)

FIGURE 17-50 ■ Problem 68.

69. Graphing Another Pulse on a String Figure 17-51*a* shows a photograph of a pulse on a taut string moving to the right. The red dot the right of the figure is a small bead of negligble mass attached to the string.

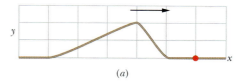

(a)

FIGURE 17-51(a) ■ Problem 69.

For each of the following quantities, select the letter of the graph in Fig. 17-51*b* that could provide a correct graph of the quantity (if the vertical axis were assigned the proper units) and write it on your answer sheet. If none of the graphs could work, write N.

(a) The vertical (up-down) displacement of the bead
(b) The vertical velocity of the bead
(c) The horizontal (left-right) displacement of the bead
(d) The horizontal velocity of the bead

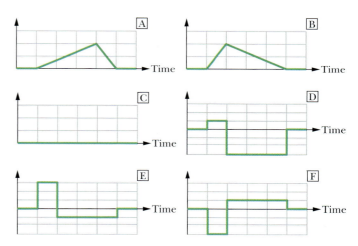

FIGURE 17-51(b) ■ Problem 69.

18 | Sound Waves

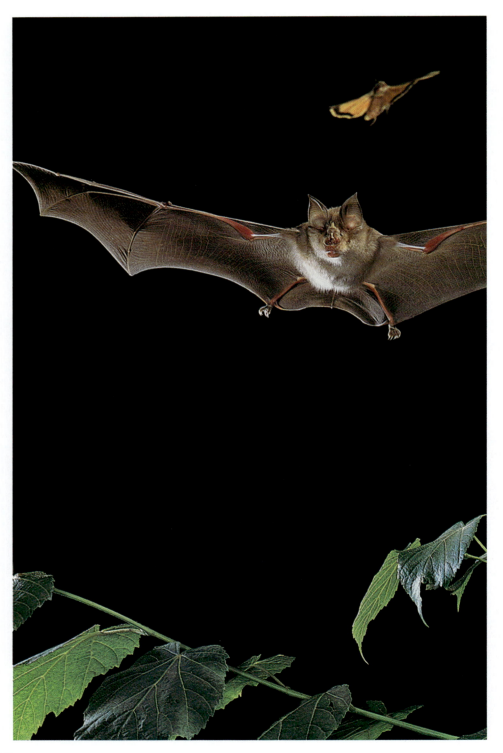

This horseshoe bat not only can locate a moth flying in total darkness, but can also determine the moth's relative speed, to home in on the insect.

How does the bat's detection system work, and how can a moth "jam" the system or otherwise reduce its effectiveness?

The answer is in this chapter.

18-1 Sound Waves

What Is a Sound Wave?

As we saw in Chapter 17, there are two types of mechanical waves that require a material medium to exist: *transverse waves* that involve displacements of a medium perpendicular to the direction in which the wave travels, and *longitudinal waves* that have displacements of a medium parallel to the direction of wave travel. Examples of longitudinal waves include the expansion/compression waves in the Slinky described in the last chapter (Fig. 17-1) and the waves in an air-filled pipe shown in Fig. 18-1.

When we hear sounds, we are detecting longitudinal waves passing through air that have frequencies within the range of human hearing. A **sound wave** can be defined more broadly as a longitudinal wave of any frequency passing through a medium. The medium can be a solid, liquid, or gas. For example, geological prospecting teams use sound waves to probe the Earth's crust for oil. Ships carry sound-emitting gear (sonar) to detect underwater obstacles. Medical personnel use high-frequency sound waves (ultrasound) to create computer-processed images of soft tissues (as shown in Fig. 18-2). Physics students use ultrasound pulses to track the motion of objects in the laboratory.

In this chapter we focus on the characteristics of audible sound waves that travel through air. We start by considering some differences in how waves propagate in one-, two-, and three-dimensional spaces.

Wave Dimensions

Unlike the wave pulses that travel in a straight line along a string, a sound wave from a small "point-like" source is usually not constrained to travel in only one direction. The same is true for the electromagnetic waves that we study in Chapters 34–38. In order to understand how sound waves and electromagnetic waves travel in more than one dimension, we need to consider how the dimensionality of the space through which a wave propagates affects it.

In the previous chapter we observed that the crest of a wave passing along a one-dimensional string moves along a line. If we constrain a sound wave to travel in

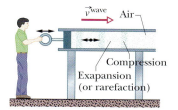

FIGURE 18-1 ■ A sound wave is set up in an air-filled pipe by moving a piston back and forth. Because the oscillations of a segment of the air (represented by the black dot) are parallel to the direction in which the wave travels, the wave is a *longitudinal wave.*

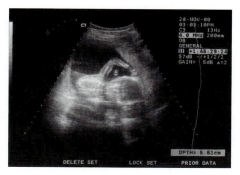

FIGURE 18-2 ■ An image of a fetus flexing an arm. This image is made with ultrasound waves that have a frequency of 4 MHz—two hundred times higher than the threshold of human hearing.

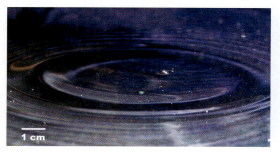

FIGURE 18-3 ■ Water on the surface of a tank of water is disturbed as a steel ball of diameter 0.79 cm falls into the water. The wave crest propagates outward from the source of the disturbance in a widening circle. The wave crest moves at a constant speed of approximately 30 cm/s.

a long tube, its compression wave crests would lie in planes perpendicular to the axis of the tube as shown in Fig. 18-1. Such a one-dimensional wave is known as a **plane wave.** For example, the relatively narrow beam of sound waves that are emitted from an ultrasonic motion detector described in Section 1-8 behave like one-dimensional plane waves.

On the other hand, if you put a small sound source between two large sheets of plywood, the sound wave crests will propagate outward in expanding circles. Although we cannot see sound waves as they propagate in two dimensions, we have all seen two-dimensional surface waves on water. For example, when a pebble is dropped onto the surface of a pond, a wave crest propagates out from the pebble in an expanding circle as shown in Fig. 18-3. This two-dimensional wave is known as a **circular wave.**

If you snap your fingers or ring a tiny sleigh bell, a compression wave is created. At distances that are large relative to the size of the source, compression wave crests travel out in three dimensions as expanding spheres (Fig. 18-4). A three-dimensional wave is defined as a **spherical wave,** provided there is no preferred direction for the propagation of the wave energy.

It is important to understand that the dimensionality of a sound source (such as a one-dimensional wave in a guitar string) and the dimensionality of a sound wave produced by the source are not necessarily the same. Sound waves often travel through a medium with uniform density such as air or water. In this case, we can associate the dimensionality of a wave with the curvature of its wave crest as it spreads, rather than with the dimensionality of the source. In general, we can define the dimensionality of a wave in terms of any point on a propagating waveform. This is because for a wave of a given dimension, the *shape* of the wave crests (maxima) and wave troughs (minima) and points of no deflection (nodes) are the same. For example, in a two-dimensional wave the crests, troughs, and points of zero deflection are all circular or at least circular arcs.

It is useful for us to define a **wavefront** as the collection of all adjacent points on an expanding wave that have the same phase. For example, a wavefront can be the collection of all adjacent crest points. Or it can be taken to be a collection of all adjacent nodes, or all adjacent troughs. **Rays** are defined as lines directed perpendicular to the wavefronts that indicate the direction of travel of the wavefronts. The short double arrows superimposed on the rays of Fig. 18-4 indicate that the longitudinal oscillations of the air, which transmits the wave, are parallel to the rays.

A one-dimensional or plane wave is defined as any wave that has a wavefront that lies in a plane. The wavefront of a two-dimensional or circular wave lies along an expanding circle. The wavefront of a three-dimensional or spherical wave lies along an expanding sphere. These definitions are often still useful in situations where we do not have ideal point sources. For instance, a flat speaker mounted in the door of a car only emits waves in a "forward" direction. In this case we would have half-spherical three-dimensional wavefronts.

As the wavefronts move outward and their radii become larger, their curvatures decrease. Far from the source, these spheres associated with a three-dimensional wave are so large that we lose track of their curved nature all together. For example, when you listen to a singer in a concert hall, your ear is so far away from her that you cannot detect any curvature in the wavefronts she sends out. In such cases we can treat the *local* portion of a wavefront as if it lies in a plane. Thus, there are times in our study of sound when we will treat sound waves as if they are one-dimensional plane waves. Similarly, in Chapter 36 when we study how light waves interfere when passing through slits and traveling parallel to a two-dimensional surface, we can treat light waves as if they were two-dimensional rather than three-dimensional.

At this point let us return to our primary task in this chapter, which is to learn more about the nature of sound waves.

(a)

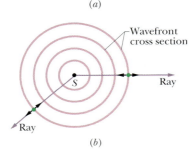

Wavefront cross section

S

Ray

Ray

(b)

FIGURE 18-4 ■ (a) A small sleigh bell that is rung at a steady rate. (b) If the distances from the bell to its listeners are large compared to the size of the bell, then the bell acts as a three-dimensional point source of sound. Each compression wave crest moves out in an expanding sphere. The two-dimensional cross-section of each of four wave crest spheres is shown. The short double arrows signify that the air particles oscillate parallel to the direction of motion of the wave crests. The lines drawn perpendicular to the wave crests are *rays.*

18-2 The Speed of Sound

The ability to calculate the speed of a sound wave as a function of the properties of the medium through which it travels is of great practical importance. For example, we can use our knowledge of the speed of sound in air to estimate how far away a lightning bolt is. Since sound travels approximately 0.2 mi/s in air, it will take about 5 s to travel one mile. Thus a 5 s interval between a lightning flash and a thunderclap tells us that the electrical storm is about a mile away. The ultrasonic motion detectors used in many physics laboratories bounce sound pulses off objects in their surroundings. The time of travel of the sound pulses emitted and then reflected from the motion detector is used to measure the distance between it and the reflecting object.

What properties of a medium does the speed of sound depend on? Let's draw an analogy between the speed of sound and the speed of a wave traveling along a string (Section 17-6). We found that the wave speed increases with the tension, T, in the string. However, the tension determines the magnitude of the restoring force that brings a displaced section of string back toward its equilibrium position. We also found, as expected, that the wave speed decreases with the mass of each disturbed length of the string. This makes sense since the linear mass density of the string, μ, is an inertial property that determines how rapidly, or slowly, the string can respond to the restoring forces acting on it. Thus, we can generalize (Eq. 17-25), which we derived for the wave speed along a stretched string, to

$$v^{wave} = |\vec{v}^{wave}| = \sqrt{\frac{F^{tension}}{\mu}} = \sqrt{\frac{restoring\ property}{inertial\ property}}. \qquad \text{(Eq. 17-25)}$$

If the medium is a fluid (such as air or water) and the wave is longitudinal, we can guess that the inertial property, corresponding to the linear density of the string μ, is the volume density ρ of the fluid. What shall we define as its restoring property?

As a sound wave passes through a fluid, elements of mass in the fluid undergo compressions and expansions due to pressure differences within the fluid. When a mass element in the fluid is compressed it has a higher pressure and pushes on adjacent fluid. This new mass element of fluid becomes compressed into a smaller volume. Then it pushes in turn on another mass element, and so on. This is how the compression wave travels through a fluid.

As shown in Fig. 18-5, the restoring property of the fluid is determined by the extent to which an element of mass changes volume when it experiences a difference in pressure (force per unit area). As a compression wave travels, the density of each mass element increases temporarily. When a fluid mass element becomes more dense its pressure rises, causing a difference in pressure between it and the mass element down the line. Just as the factor k tells us how stiff a spring is, we would like to define a factor B that tells us how "stiff" a three-dimensional medium is. The ratio of the pressure difference to the relative volume change is a property of the fluid we will call the **bulk modulus,** and define as

$$B \equiv -\frac{\Delta P}{\Delta V/V} \qquad \text{(definition of bulk modulus),} \qquad \text{(18-1)}$$

where $\Delta V/V$ is the fractional change in volume produced by a change in pressure ΔP. [As explained in Section 15-3, the SI unit for pressure is the newton per square meter, which is given a special name, the *pascal* (Pa).] The minus sign signifies that a rise in pressure on a medium causes its volume to decrease.

For the pressure changes associated with small amplitude sound waves, the bulk modulus in a given material is approximately constant and is a restoring factor just as

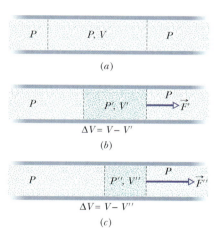

FIGURE 18-5 ■ (a) An undisturbed tube of fluid at pressure P. (b) A compression wave passing through a fluid with a small bulk modulus causes a mass element to be compressed to a smaller volume, V'. (c) The same type of compression wave with a larger bulk modulus. It causes a mass element to undergo a greater volume reduction.

tension is in a string or k is in a spring. So we can substitute B for F^{tension} and ρ for μ in Eq. 17-25 ($v^{\text{wave}} = \sqrt{F^{\text{tension}}/\mu}$) to get

$$v^{\text{wave}} = \left| \vec{v}^{\text{wave}} \right| = \sqrt{\frac{B}{\rho}} \qquad \text{(speed of sound in a fluid).} \qquad (18\text{-}2)$$

Notice that as is the case for waves on a string, we expect the wave velocity to be independent of frequency and amplitude.

It can be easily shown that the dimensions of $\sqrt{B/\rho}$ are those of a velocity. It is possible to derive Eq. 18-2 mathematically using methods similar to those used in Section 17-6 to find the expression for the wave speed along a stretched string. Once again experimental results confirm the validity of Eq. 18-2.

Table 18-1 lists the speed of sound in various media.

TABLE 18-1
The Speed of Sound[a]

Medium	Speed (m/s)	Density (kg/m³)	Bulk Modulus (N/m²)
Gases			
Air (0°C)	331		
Air (20°C)	343	1.21	1.4×10^5
Helium	965		
Hydrogen	1284		
Liquids			
Water (0°C)	1402		
Water (20°C)	1482	990	2.2×10^9
Seawater[b]	1522		
Solids			
Aluminum	6420		
Steel	5941	7850	1.6×10^{11}
Granite	6000		

[a] At 0°C and 1 atm pressure, except where noted.
[b] At 20°C and 3.5% salinity.

Note that the density of water is almost 1000 times greater than the density of air. If this were the only relevant factor, we would expect from Eq. 18-2 that the speed of sound in water would be considerably less than the speed of sound in air. However, Table 18-1 shows us that the reverse is true. We conclude (again from Eq. 18-2) that the bulk modulus of water must be more than 1000 times greater than that of air. This is indeed the case. Water is less compressible than air, which (see Eq. 18-1) is another way of saying that its bulk modulus is much greater.

FIGURE 18-6 ■ Measurement of the speed of sound. The sound pressure variations from a finger snap travel down a 2.4 m tube and back. A microphone sensor connected to a computer data acquisition system is used to measure the time between the initial sound pulses and the reflected pulses.

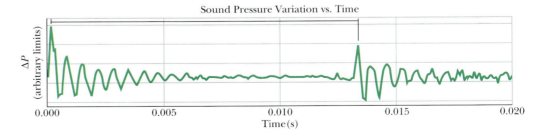

Sound Pressure Variation vs. Time

A graph that shows the outcome of a simple measurement of the speed of sound in air is shown in Fig. 18-6. Analysis of the data shows a speed, which is temperature-dependent, of about 345 m/s.

More About Traveling Sound Waves in Air

If the hanging Slinky discussed in Chapter 17 is pushed back and forth at one end with a continuous sinusoidal motion, we can set up a series of compressions and rarefactions like those shown in Fig. 18-7a. We can do the same thing when pushing a piston back and forth in a column of air as shown in Fig. 18-1. Figure 18-7b displays such a wave traveling rightward through a long air-filled tube. The piston's rightward motion compresses the air next to it; the piston's leftward motion allows the element of air to move back to the left and the pressure to decrease. As each element of air pushes on the next element in turn, the right-left motion of the air and the change in its pressure is passed to the next bit of air along the tube.

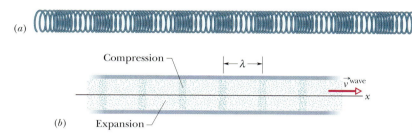

FIGURE 18-7 ■ (a) A continuous longitudinal wave traveling along a hanging Slinky. (b) A sound wave, traveling through a long air-filled tube with speed $\vec{v}^{\text{wave}}$, consists of a moving, periodic pattern of expansions and compressions of the air. The wave is shown at an arbitrary instant. As the wave passes, a fluid element of thickness Δx (not shown) oscillates left and right in simple harmonic motion about its equilibrium position.

The alternating compressions and rarefactions (reductions in pressure) propagate as a sound wave. As a reminder of what we learned in Chapter 17, note that it is the oscillations of pressure that propagate, not the air molecules. However, the air molecules (or molecules of another fluid) are disturbed and oscillate back and forth about their initial positions.

As the wave moves, the air pressure changes at any position x in Fig. 18-7b in a sinusoidal fashion, like the displacement of a string element in a transverse wave, however the pressure variations are not transverse, they are longitudinal. To describe this pressure change from the local atmospheric pressure as a function of position and time for purely sinusoidal oscillations, consider Fig. 18-8. We can model our equation on Eq. 17-4, which describes the propagation of a sinusoidal transverse wave along a string. However, instead of the vertical displacement of a bit of string, we are interested in how pressure varies from an equilibrium value. Modifying Eq. 17-4 gives us

$$\Delta P(x, t) = \Delta P^{\text{max}} \sin[(kx \pm \omega t) + \phi_0], \qquad (18\text{-}3)$$

where ΔP^{max} is called the **pressure amplitude,** which is the maximum change from local atmospheric pressure due to the wave. It turns out that for typical sound waves that

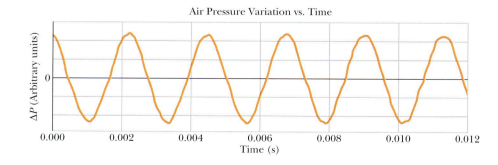

FIGURE 18-8 ■ A computer data acquisition system is used to measure air pressure variations from equilibrium using a small microphone. The sampling rate was 25 000 data points/second. This pressure variation was caused by a speaker cone that oscillated sinusoidally.

we hear, ΔP^{max} is extremely small compared to the equilibrium pressure P^{atm} present when no wave is present.

The wave number k, angular frequency ω, frequency f, wavelength λ, wave speed v^{wave}, period T, and time- and space-dependent phase $\phi(x, t)$ for a longitudinal sound wave are defined and interrelated exactly as for a transverse wave. However, note that the wavelength λ is now the distance (again along the direction of travel) in which the pattern of compression and expansion due to the wave begins to repeat itself (see Fig. 18-7b). Also note that Eq. 17-13 still holds so that the wave speed is given by $v^{\text{wave}} = \lambda f = \omega/k$.

A negative value of ΔP in Fig. 18-8 and Eq. 18-3 corresponds to an expansion of a packet of air, and a positive value corresponds to a compression.

READING EXERCISE 18-1: Verify that Eq. 18-2 is dimensionally correct. In other words, show that the term $\sqrt{B/\rho}$ has the dimensions of velocity in SI units. ■

READING EXERCISE 18-2: Examine Fig. 18-6. The maximum of the finger snap pulse set is at 0.0002 s and the maximum of the reflected pulse is at 0.0133 s. Use these times to calculate the measured value of the wave speed for the set of sound pulses as they travel back and forth through the air inside the 2.26-m-long tube. Is your calculated speed approximately the same as the speed of sound of air at room temperature as reported in Table 18-1? *Note:* The air temperature in the room was not recorded. ■

TOUCHSTONE EXAMPLE 18-1: Sound Arrival Delay

One clue used by your brain to determine the direction of a source of sound is the time delay Δt between the arrival of the sound at the ear closer to the source and the arrival at the farther ear. Assume that the source is distant so that a wavefront from it is approximately planar when it reaches you, and let D represent the separation between your ears.
(a) Find an expression that gives Δt in terms of D and the angle θ between the direction of the source and the forward direction.

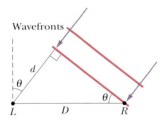

FIGURE 18-9 ■ A wavefront travels a distance $d(= D \sin \theta)$ farther to reach the left ear (L) than to reach the right ear (R).

SOLUTION ■ The situation is shown (from an overhead view) in Fig. 18-9, where wavefronts approach you from a source that is located in front of you and to your right. The **Key Idea** here is that the time delay Δt is due to the distance d that each wavefront must travel to reach your left ear (L) after it reaches your right ear (R). From Fig. 18-9, we find

$$\Delta t = \frac{d}{v^{\text{air}}} = \frac{D \sin \theta}{v^{\text{air}}}, \qquad \text{(Answer)} \quad (18\text{-}4)$$

where v^{air} is the speed of the sound wave in air. Based on a lifetime of experience, your brain correlates each detected value of Δt (from zero to the maximum value) with a value of θ (from zero to 90°) for

the direction of the sound source.

(b) Suppose that you are submerged in water at 20°C when a wavefront arrives from directly to your right ($\theta = 90°$). Based on the time-delay clue, at what angle θ from the forward direction does the source seem to be?

SOLUTION ■ The **Key Idea** here is that the speed is now the speed of the sound wave in water, v^{water}, so in Eq. 18-4 we substitute v^{water} for v^{air} and 90° for θ, finding that

$$\Delta t_w = \frac{D \sin \theta}{v^{\text{water}}} = \frac{D}{v^{\text{water}}}, \qquad (18\text{-}5)$$

Since v^{water} is about four times v^{air}, delay Δt_w is about one-fourth the maximum time delay in air. Based on experience, your brain will process the water time delay as if it occurred in air. Thus, the sound source appears to be at an angle θ smaller than 90°. To find that apparent angle we substitute the time delay D/v^{water} from Eq. 18-5 for Δt in Eq. 18-4, obtaining

$$\frac{D}{v^{\text{water}}} = \frac{D \sin \theta}{v^{\text{air}}}. \qquad (18\text{-}6)$$

Then, to solve for θ we substitute $v^{\text{air}} = 343$ m/s and $v^{\text{water}} = 1482$ m/s (from Table 18-1) into Eq. 18-5, finding

$$\sin \theta = \frac{v^{\text{air}}}{v^{\text{water}}} = \frac{343 \text{ m/s}}{1482 \text{ m/s}} = 0.231,$$

and thus

$$\theta = 13°. \qquad \text{(Answer)}$$

18-3 Interference

Like transverse waves, sound waves undergo interference when two waves pass through the same point at the same time. Let us consider, in particular, the interference between two sound waves with the same frequency, wavelength, amplitude, and phase. The only difference between these waves is that they are traveling in slightly different directions. Figure 18-10 shows how we can set up such a situation: Two point sources S_1 and S_2 emit sound waves that are in phase. Thus, the sources themselves are said to be in phase; that is, as the waves emerge from the sources, their displacements are always identical. We are interested in the waves that then travel through point P in Fig. 18-10. We assume that the distance to P is much greater than the distance between the sources so that the waves are traveling in almost the same direction at P.

If the waves (which both start out at the same point in their pressure oscillation, i.e., *in phase*) traveled along paths with identical lengths to reach point P, they would still be in phase there. In this case, the displacements of the two waves would add. As with transverse waves, this means that they would undergo fully constructive interference there. However, in Fig. 18-10, path L_2 traveled by the wave from S_2 is longer than path L_1 traveled by the wave from S_1. The difference in path lengths means that the waves may not be in phase at point P and so might be at different points in their oscillations.

We can use the definition of phase (from Section 17-4) to determine the phase difference. We specified that the waves had the same phase at locations S_1 and S_2. So we can set $(\phi_0)_1 = (\phi_0)_2$. Thus, the phase difference when the two waves arrive at point P at a common time t is given by

$$\Delta\phi = \phi_2(x, t) - \phi_1(x, t) = [kL_2 - \omega t + (\phi_0)_2] - [kL_1 - \omega t + (\phi_0)_1] = k(L_2 - L_1). \tag{18-7}$$

Indeed the magnitude of the phase difference $\Delta\phi$ at P depends on the **path length difference** $\Delta L = |L_2 - L_1|$ of the two waves. Since the wave number $k = 2\pi/\lambda$, we can rewrite $\Delta\phi = k\,\Delta L$ as

$$|\Delta\phi| = \frac{\Delta L}{\lambda} 2\pi \qquad \text{(path length–phase difference relation).} \tag{18-8}$$

Fully constructive interference occurs when the phase difference, $\Delta\phi$, is zero, 2π, or any integer multiple of 2π.

We can write this condition as

$$|\Delta\phi| = m(2\pi), \qquad \text{for } m = 0, 1, 2, \ldots \qquad \text{(fully constructive interference),} \tag{18-9}$$

or from Eq. 18-8, this also occurs when the ratio $\Delta L/\lambda$ is

$$\frac{\Delta L}{\lambda} = m \qquad \text{(fully constructive interference),} \tag{18-10}$$

where m is zero or any positive integer.

For example, if the magnitude of the path length difference $\Delta L = |L_2 - L_1|$ in Fig. 18-10 is equal to 2λ, then $\Delta L/\lambda = 2$ and the waves undergo fully constructive interference at point P. The interference is fully constructive because the wave from S_2 is phase-shifted relative to the wave from S_1 by 2λ, putting the two waves *exactly in phase* at P.

Fully destructive interference occurs when the magnitude of the phase difference between the two waves $\Delta\phi$ is an odd multiple of π.

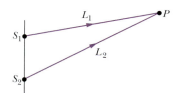

FIGURE 18-10 ■ Two point sources S_1 and S_2 emit spherical sound waves in phase. The rays indicate that the waves pass through a common point P. (Recall that rays are lines that run perpendicular to the wavefront and indicate direction of travel.)

We can write this as

$$|\Delta\phi| = (2m + 1)\pi \qquad \text{(fully destructive interference),} \qquad (18\text{-}11)$$

where once again m is zero or any positive integer. From $|\Delta\phi| = (\Delta L/\lambda)2\pi$ (Eq. 18-8), this we see occurs when the ratio $\Delta L/\lambda$ is

$$\frac{\Delta L}{\lambda} = \frac{2m + 1}{2} = m + \tfrac{1}{2} \qquad \text{(fully destructive interference).} \qquad (18\text{-}12)$$

For example, if the path length difference $\Delta L = |L_2 - L_1|$ in Fig. 18-10 is equal to 2.5λ, then $\Delta L/\lambda = 2.5$ and the waves undergo fully destructive interference at point P. The interference is fully destructive because the wave from S_2 is phase-shifted relative to the wave from S_1 by 2.5 wavelengths, which puts the two waves *exactly out of phase* at P.

Of course, two waves could produce intermediate interference as, say, when $\Delta L/\lambda = 1.2$. This would be closer to fully constructive interference ($\Delta L/\lambda = 1.0$) than to fully destructive interference ($\Delta L/\lambda = 1.5$).

If the initial phase of one of the interfering waves is different than that of the other, we would have to derive a different set of conditions for constructive and destructive interference.

TOUCHSTONE EXAMPLE 18-2: Constructive Interference

In Fig. 18-11a, two point sources S_1 and S_2, which are in phase and separated by distance $D = 1.5\lambda$, emit identical sound waves of wavelength λ.

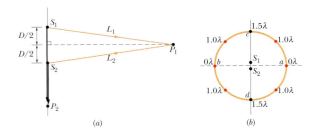

(a)   (b)

FIGURE 18-11 ■ (a) Two point sources S_1 and S_2, separated by distance D, emit spherical sound waves in phase. The waves travel equal distances to reach point P_1. Point P_2 is on the line extending through S_1 and S_2. (b) The path length difference (in terms of wavelength) between the wave from S_1 and S_2, at eight points on a large circle around the sources.

(a) What is the path length difference of the waves from S_1 and S_2 at point P_1, which lies on the perpendicular bisector of distance D, at a distance greater than D from the sources? What type of interference occurs at P_1?

SOLUTION ■ The **Key Idea** here is that, because the waves travel identical distances to reach P_1, their path length difference is

$$\Delta L = 0.0\lambda. \qquad \text{(Answer)}$$

From Eq. 18-10, this means that the waves undergo fully constructive interference at P_1.

(b) What are the path length difference and type of interference at point P_2 in Fig. 18-11a?

SOLUTION ■ Now the **Key Idea** is that the wave from S_1 travels the extra distance D ($= 1.5\lambda$) to reach P_2. Thus, the path length difference is

$$\Delta L = 1.5\lambda. \qquad \text{(Answer)}$$

From Eq. 18-12, this means that the waves are exactly out of phase at P_2 and undergo fully destructive interference there.

(c) Figure 18-11b shows a circle with a radius much greater than D, centered on the midpoint between sources S_1 and S_2. What is the number of points N around this circle at which the interference is fully constructive?

SOLUTION ■ Imagine that, starting at point a, we move clockwise along the circle to point d. One **Key Idea** here is that as we move to point d, the path length difference ΔL increases and so the type of interference changes. From (a), we know that the path length difference is $\Delta L = 0.0\lambda$ at point a. From (b), we know that $\Delta L = 1.5\lambda$ at point d. Thus, there must be one point along the circle between a and d at which $\Delta L = \lambda$, as indicated in Fig. 18-11b. From Eq. 18-10, fully constructive interference occurs at that point. Also, there can be no other point along the way from point a to point d at which fully constructive interference occurs, because there is no other integer than 1 between 0.0 and 1.5.

Another **Key Idea** here is to use symmetry to locate the other points of fully constructive interference along the rest of the circle. Symmetry about line cd gives us point b, at which $\Delta L = 0\lambda$. Also, there are three more points at which $\Delta L = \lambda$. In all we have constructive interference at

$$N = 6 \text{ points.} \qquad \text{(Answer)}$$

18-4 Intensity and Sound Level

In Section 17-7 we derived the fact that the average power in a one-dimensional continuous sinusoidal wave that propagates along a string depends on the squares of both the frequency and amplitude as shown in Eq. 17-31 ($\langle\text{Power}\rangle = \frac{1}{2}\mu v^{\text{wave}}\omega^2 Y^2$).* Let's compare Eq. 17-31 to the equation for the power transmission of a sinusoidal one-dimensional sound wave. Suppose our sound is produced by an oscillating piston in an air-filled pipe of cross-sectional area A shown in Fig. 18-1. The relationship for the average power transmission of the sound waves is given by

$$\langle\text{Power}\rangle = \frac{1}{2}(\rho A)v^{\text{wave}}\omega^2 S^2, \tag{18-13}$$

where S represents a maximum displacement. In this case S denotes the maximum longitudinal displacement about an equilibrium point that particles in the medium carrying the sound undergo. This power relationship can be derived by calculating the average kinetic and potential energy of a given volume of air that undergoes simple harmonic motion. But without using a formal derivation we can see that Eq. 17-31 and Eq. 18-13 are essentially the same. Both S and Y are displacements and the term ρA is equal to the linear density μ of the air in the tube.

In previous sections we have been using maximum pressure difference, $\Delta P^{\max}$, rather than maximum particle displacement, S, as an indicator of how strong a sinusoidal sound disturbance is. It can be shown that these two measures of wave strength are related, by the expression

$$\Delta P^{\max} = v^{\text{wave}}\rho\omega S.$$

If you have ever tried to sleep while someone played loud music nearby, you are well aware that there is more to sound than frequency, wavelength, and speed. Humans also detect how *loud* a sound is. Although the human ear does detect pressure amplitudes, it turns out that we are more sensitive to the *energy fluctuations* in a propagating wave than we are to the pressure alone. Hence, we will define a new energy related quantity associated with waves. The **intensity** I of a sound wave at a surface is the average rate per unit area, A, at which energy is transferred by the wave through or onto the surface. This is what we commonly refer to as loudness. By definition, the intensity of a wave is

$$I \equiv \frac{\langle\text{Power}\rangle}{A}, \tag{18-14}$$

where "Power" is the time rate of energy transfer (the power) of the sound wave, and A is the area of the surface intercepting the sound.

We can use Eq. 18-13 to show the relationship between the energy-related quantity *intensity* and the maximum particle displacement associated with a sound wave,

$$I = \frac{1}{2}\rho v^{\text{wave}}\omega^2 S^2. \tag{18-15}$$

The equation $\Delta P^{\max} = v^{\text{wave}}\rho\omega S$ shown above can be used to express the intensity as a function of pressure change, so that

The intensity of a continuous sinusoidal wave is proportional to both the square of its displacement amplitude and the square of maximum pressure change in the medium caused by the sound waves.

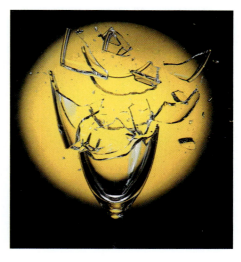

Sound can cause the wall of a drinking glass to oscillate. If the sound produces a standing wave of oscillations and if the intensity of the sound is large enough, the glass will shatter.

---

*In order to avoid confusion between power, usually denoted as P, and pressure, also denoted as P, we choose to spell out the word Power in this section.

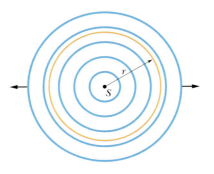

FIGURE 18-12 ■ A point source S emits sound waves uniformly in all directions. The waves pass through an imaginary sphere of radius r that is centered on S.

Variation of Intensity with Distance

How intensity varies with distance from a real sound source is often complex. Some real sources (like loudspeakers) may transmit sound only in particular directions, and objects in the surroundings usually produce echoes (reflected sound waves) that overlap the direct sound waves. In some situations, however, we can ignore echoes and assume that the sound source is a point source that emits the sound *isotropically*—that is, with equal intensity in all directions. The wavefronts spreading from such an isotropic point source S at a particular instant are shown in Fig. 18-12.

Let us assume that the mechanical energy of the sound waves is conserved as they spread from this source. Let us also center an imaginary sphere of radius r on the source, as shown in Fig. 18-12. All the energy emitted by the source must pass through the surface of the sphere. Thus, the time rate at which energy is transferred through the surface by the sound waves must equal the time rate at which energy is emitted by the source (that is, the power, Power_s, of the source). From the fact that intensity is equal to the ratio of power to area (Eq. 18-14), the intensity I at the sphere must then be

$$I = \frac{\text{Power}_s}{4\pi r^2}, \tag{18-16}$$

where $4\pi r^2$ is the area of the sphere. Equation 18-16 tells us that the intensity of sound from an isotropic point source decreases with the square of the distance r from the source.

The Decibel Scale

The displacement amplitude at the human ear ranges from about 10^{-5} m for the loudest tolerable sound to about 10^{-11} m for the faintest detectable sound, a ratio of 10^6. From our discussions above, we know that the intensity of a sound varies as the *square* of its amplitudes, so the ratio of intensities at these two limits of the human auditory system is 10^{12}. Humans can hear over an enormous range of intensities.

We deal with such an enormous range of values by using base 10 logarithms. Consider the relation

$$y = \log x,$$

in which x and y are variables. It is a property of this equation that if we *multiply* x by 10, then y increases by 1. To see this, we write

$$y' = \log(10x) = \log 10 + \log x = 1 + y.$$

Similarly, if we multiply x by 10^{12}, y increases by only 12. An important characteristic of human hearing is that we have logarithmic ears. Within the normal range of hearing when the measured sound intensity increases by a factor of 10, the sound only seems twice as loud to us.

Thus, instead of speaking of the intensity I of a sound wave, it is much more convenient to speak of its **sound level** β, defined as

$$\beta = (10 \text{ dB})\log\frac{I}{I_0}. \tag{18-17}$$

Here dB is the abbreviation for **decibel,** the unit of sound level, a name that was chosen to recognize the work of Alexander Graham Bell. I_0 in Eq. 18-17 is a standard reference intensity ($= 10^{-12}$ W/m^2), chosen because it is near the lower limit of the human range of hearing at a frequency of 1 kHz. For $I = I_0$, Eq. 18-17 gives $\beta = 10 \log 1 = 0$, so our standard reference level, which we can barely hear, corresponds to zero decibels. Then β increases by 10 dB every time the sound intensity increases by an order of magnitude (a factor of 10). But as we mentioned, a 10 dB increase seems to be twice the loudness. And the sound of a typical conversation, which has a β of 60 dB, corresponds to an intensity that is 10^6 times our normal hearing threshold! Table 18-2 lists the sound levels for a variety of environments.

TABLE 18-2
Some Sound Levels (dB)

Hearing threshold	0	Rock concert	110
Rustle of leaves	10	Pain threshold	120
Conversation	60	Jet engine	130

The decibel scale is one of many logarithmic scales used by scientists. Others include the pH scale, star magnitudes, and the Richter scale used to determine the severity of earthquakes.

READING EXERCISE 18-3: The figure indicates the near edge of three small patches 1, 2, and 3 that lie on the surfaces of two imaginary spheres; the spheres are centered on an isotropic point source S of sound. The rates at which energy is transmitted through the three patches by the sound waves are equal. Rank the patches according to (a) the intensity of the sound on them and (b) their area, greatest first. ■

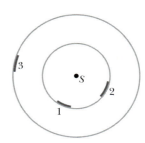

TOUCHSTONE EXAMPLE 18-3: Spark Sound

An electric spark jumps along a straight line of length $L = 10$ m, emitting a pulse of sound that travels radially outward from the spark. (The spark is said to be a line source of sound.) The power of the emission is $Power_s = 1.6 \times 10^4$ W.

(a) What is the intensity I of the sound when it reaches a distance $r = 6$ m from the spark?

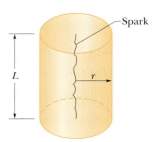

FIGURE 18-13 ■ A spark along a straight line of length L emits sound waves radially outward. The waves pass through an imaginary cylinder of radius r and length L that is centered on the spark.

SOLUTION ■ Let us center an imaginary cylinder of radius $r = 6$ m and length $L = 10$ m (open at both ends) on the spark, as shown in Fig. 18-13. One **Key Idea** here is that the intensity I at the cylindrical surface is the ratio $\langle Power \rangle/A$ of the average rate per unit area at which sound energy passes through a surface of area A on the cylinder. Another **Key Idea** is to assume that the principle of conservation of energy applies to the sound energy. This means that the Power or rate at which energy is transferred through the cylinder must equal the rate at which energy is emitted by the source. Putting these ideas together and noting that the area of a length L of the cylindrical surface is $A = 2\pi rL$, and neglecting the relatively small amount of sound energy

that flows out through the ends of the cylinder, we have, from Eq. 18-14,

$$I = \frac{\langle Power \rangle}{A} = \frac{\langle Power_s \rangle}{2\pi rL}. \quad (18\text{-}18)$$

This tells us that the intensity of the sound from a line source decreases with distance r (rather than with the square of distance r as for the point source described in Eq. 18-16). Substituting the given data, we find

$$I = \frac{1.6 \times 10^4 \text{ W}}{2\pi(6\text{ m})(10\text{ m})} = 42.4 \text{ W/m}^2 \approx 42 \text{ W/m}^2. \quad (Answer)$$

(b) At what average $\langle Power_d \rangle$ is sound energy intercepted by an acoustic detector of area $A_d = 2.0$ cm^2? Assume that the detector is aimed at the spark and located a distance $r = 6$ m from the spark.

SOLUTION ■ Applying the first **Key Idea** of part (a), we know that the intensity of sound at the detector is the ratio of the energy transfer rate $\langle Power_d \rangle$ there to the detector's area A_d:

$$I = \frac{\langle Power_d \rangle}{A_d}. \quad (18\text{-}19)$$

We can imagine that the detector lies on the cylindrical surface of (a). Then the sound intensity at the detector is the intensity I ($= 42.4$ W/m^2) at the cylindrical surface. Solving Eq. 18-19 for $\langle Power_d \rangle$ gives us

$$\langle Power_d \rangle = (42.4 \text{ W/m}^2)(2.0 \times 10^{-4}\text{m}^2) = 8.5 \text{ mW}. \quad (Answer)$$

TOUCHSTONE EXAMPLE 18-4: Concert Sounds

In 1976, the Who set a record for the loudest concert—the sound level 46 m in front of the speaker systems was $\beta_2 = 120$ dB. What is the ratio of the intensity I_2 of the band at that spot to the intensity I_1 of a jackhammer operating at sound level $\beta_1 = 92$ dB?

SOLUTION ■ The **Key Idea** here is that for both the Who and the jackhammer, the sound level β is related to the intensity by the definition of sound level in Eq. 18-17. For the Who, we have

$$\beta_2 = (10 \text{ dB}) \log \frac{I_2}{I_0},$$

and for the jackhammer, we have

$$\beta_1 = (10 \text{ dB}) \log \frac{I_1}{I_0}.$$

The difference in the sound levels is

$$\beta_2 - \beta_1 = (10 \text{ dB}) \left(\log \frac{I_2}{I_0} - \log \frac{I_1}{I_0} \right). \quad (18\text{-}20)$$

Using the identity

$$\log \frac{a}{b} - \log \frac{c}{d} = \log \frac{ad}{bc},$$

we can rewrite Eq. 18-20 as

$$\beta_2 - \beta_1 = (10 \text{ dB}) \log \frac{I_2}{I_1}. \quad (18\text{-}21)$$

Rearranging and substituting the known sound levels now yield

$$\log \frac{I_2}{I_1} = \frac{\beta_2 - \beta_1}{10 \text{ dB}} = \frac{120 \text{ dB} - 92 \text{ dB}}{10 \text{ dB}} = 2.8.$$

Taking the antilog of the far left and far right sides of this equation (the antilog key on your calculator is probably marked as 10^x), we find

$$\frac{I_2}{I_1} = \log^{-1} 2.8 = 630. \quad \text{(Answer)}$$

Thus, the Who was *very* loud.

Temporary exposure to sound intensities as great as those of a jackhammer and the 1976 Who concert results in a temporary reduction of hearing. Repeated or prolonged exposure can result in permanent reduction of hearing (Fig. 18-14). Loss of hearing is a clear risk for any one continually listening to, say, heavy metal at high volume, especially on headphones.

FIGURE 18-14 ■ Pete Townshend of the Who, playing in front of a speaker system. He suffered a permanent reduction in his hearing ability due to his exposure to high-intensity sound, not so much during on-stage performances as from wearing headphones in recording studios and at home.

FIGURE 18-15 ■ The air column within a fujara oscillates when that traditional Slovakian instrument is played.

18-5 Sources of Musical Sound

Musical sounds can be set up by oscillating strings (guitar, piano, violin), membranes (kettledrum, snare drum), air columns (flute, oboe, pipe organ, and the fujara of Fig. 18-15), wooden blocks or steel bars (marimba, xylophone), and many other oscillating bodies. Most instruments involve more than a single oscillating part. In the violin, for example, both the strings and the body of the instrument participate in producing the music.

Standing Waves and Musical Instruments

Recall from Chapter 17 that standing waves can be set up on a stretched string that is fixed at both ends. They arise because waves traveling along the string are reflected back onto the string at each end. If the wavelength of the waves is suitably matched to the length of the string, the superposition of waves traveling in opposite directions produces a standing wave pattern (or oscillation mode). The wavelength required of the waves for such a match is one that corresponds to a *resonant frequency* of the string. The advantage of setting up standing waves is that the string then oscillates with a large, sustained amplitude, pushing back and forth against the surrounding air and thus generating a noticeable sound wave with the same frequency as the oscillations of the string. This production of sound is of obvious importance to, say, a guitarist.

We can set up standing waves of sound in an air-filled pipe in a similar way. As sound waves travel through the air in the pipe, they are reflected at each end and travel back through the pipe. (The reflection occurs even if an end is open, but the reflection is not as complete as when the end is closed.) If the wavelength of the sound waves is suitably matched to the length of the pipe, the superposition of waves traveling in opposite directions through the pipe sets up a standing wave pattern. The wavelength required of the sound waves for such a match is one that corresponds to a resonant frequency of the pipe. The advantage of such a standing wave is that the air in the pipe oscillates with a large, sustained amplitude, emitting at any open end a sound wave that has the same frequency as the oscillations in the pipe. This emission of sound is of obvious importance to, say, an organist.

Many other aspects of standing sound wave patterns are similar to those of string waves: The closed end of a pipe is like the fixed end of a string in that there must be a displacement node (zero displacement in a string, zero motion of molecules in a fluid like air) located there. A zero particle velocity immediately against the closed end of the pipe leads to a large change in pressure (a pressure antinode). The open end of a pipe is like the end of a string attached to a freely moving ring, as in Fig. 17-19b, in that there must be a displacement antinode (maximum displacement in a string, maximum motion of molecules in a fluid like air) located there. This too makes sense, as the molecules of air (or other fluid) are completely unconstrained once they leave the end of the pipe. The large particle velocities here lead to very small changes in pressure (a pressure antinode). To be perfectly precise, the antinode for the open end of a pipe is located slightly beyond the end. We will ignore this difference in our discussions.

So, the simplest standing wave pattern that can be set up in a pipe with two open ends is one with a particle displacement antinode (maximum particle velocities and zero change in pressure) at both ends and no other antinodes between them. Under these conditions, the only way that there can be a standing wave pattern at all is for there to be a displacement node (zero particle velocity and maximum change in pressure) in the middle of the pipe. This is shown in Fig. 18-16a. An alternate representation of the standing wave pattern as a graph of maximum molecular displacement at each position along the length of the page is shown in Fig. 18-16b.

Harmonics

The standing wave pattern of Fig. 18-16a is called the *fundamental mode* or *first harmonic*. For it to be set up, the sound waves in a pipe of length L must have a wavelength given by $L = \lambda/2$, so that $\lambda = 2L$. Several more standing sound wave patterns for a pipe with two open ends are shown in Fig. 18-17a with graphs of air molecule displacements from equilibrium shown next to the pipes. The *second harmonic* requires sound waves of wavelength $\lambda = L$, the *third harmonic* requires wavelength $\lambda = 2L/3$, and so on.

More generally, the resonant frequencies for a pipe of length L with two open ends correspond to the wavelengths

$$\lambda = \frac{2L}{n} \qquad \text{(pipe, two open ends).} \qquad (18\text{-}22)$$

Here n is a positive integer (1, 2, 3, ... etc.) called the *harmonic number*. The resonant frequencies for a pipe with two open ends are then given by

$$f = \frac{v^{\text{air}}}{\lambda} = n\left(\frac{v^{\text{air}}}{2L}\right) \qquad \text{(pipe, two open ends),} \qquad (18\text{-}23)$$

where v^{air} is the speed of the sound wave in air.

Figure 18-17b shows (using pressure variation graphs) some of the standing sound wave patterns that can be set up in a pipe with only one open end. As required, across

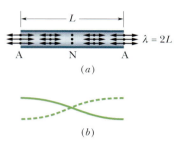

FIGURE 18-16 ■ (a) The simplest standing wave pattern of air molecule displacements from equilibrium for (longitudinal) sound waves in a pipe with both ends open. It shows an antinode (A) across each end and a node (N) across the middle of the pipe. (The longitudinal displacements represented by the double arrows are greatly exaggerated.) (b) A graph of maximum molecular displacements versus position.

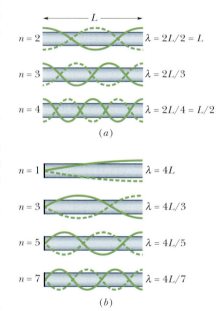

FIGURE 18-17 ■ Standing wave patterns are shown as graphs of the extremes of maximum and minimum air molecule displacements versus location along the pipe. (a) With *both* ends of the pipe open, any harmonic can be set up in the pipe. (b) With only *one* end open, only odd harmonics can be set up.

the open end there is a displacement antinode (maximum particle velocities and zero pressure change) and across the closed end there is a displacement node (zero particle velocities and maximum pressure change). The simplest pattern requires sound waves having a wavelength given by $L = \lambda/4$, so that $\lambda = 4L$. The next simplest pattern requires a wavelength given by $L = 3\lambda/4$, so that $\lambda = 4L/3$, and so on.

More generally, the resonant frequencies for a pipe of length L with only one open end correspond to the wavelengths

$$\lambda = \frac{4L}{2n + 1} \qquad \text{for } n = 0, 1, 2, \ldots, \qquad (18\text{-}24)$$

where n still represents a positive integer with $2n + 1$ giving us odd positive integers. The resonant frequencies are then given by

$$f = \frac{v^{\text{air}}}{\lambda} = (2n + 1)\frac{v^{\text{air}}}{\lambda}n \qquad \text{(pipe, one open end)}. \qquad (18\text{-}25)$$

Note again that only odd harmonics can exist in a pipe with one open end. For example, the second harmonic, with $2n + 1 = 2$, cannot be set up in such a pipe. Note also that for such a pipe the numeric adjective (e.g., first, second, third, . . .) before the word harmonic in a phrase such as "the third harmonic" always refers to the harmonic number n and not to the nth *possible* harmonic.

The length of a musical instrument is related to the range of frequencies over which the instrument is designed to function, and smaller length implies higher frequencies. Figure 18-18, for example, shows the saxophone and violin families, with their frequency ranges suggested by the piano keyboard. Note that, for every instrument, there is overlap with its higher- and lower-frequency neighbors.

What Makes Instruments Distinctive?

In any oscillating system that gives rise to a musical sound, whether it is a violin string or the air in an organ pipe, the fundamental and one or more of the higher harmonics are usually generated simultaneously. Thus, you hear them together—that is, superimposed into a combined wave. When different instruments are playing the same note, they produce the same fundamental frequency but different intensities for the higher harmonics. For example, the fourth harmonic of middle C might be relatively loud on one instrument

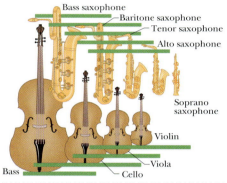

FIGURE 18-18 ■ The saxophone and violin families, showing the relations between instrument length and frequency range. The frequency range of each instrument is indicated by a horizontal bar along a frequency scale suggested by the piano keyboard at the bottom; the frequency increases toward the right.

FIGURE 18-19 ■ The wave forms produced by a violin, a hammer dulcimer struck with a felt and a wood hammer, and a guitar. The fundamental frequency is listed for each wave. These graphs of the relative sound pressure variation versus time are produced using a computer data acquisition system with a microphone sensor attached.

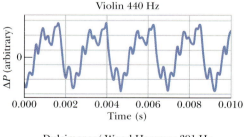

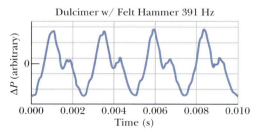

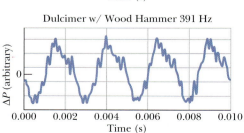

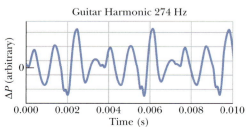

and relatively quiet or even missing on another. Thus, different instruments sound different to you even when they are playing the same note. That would be the case for the combined waves shown for the violin, hammer dulcimer and guitar in Fig. 18-19 even if these three instruments were playing the same note (which they are not).

In Fig. 18-19 we can also see that the pressure variation versus time for the hammer dulcimer hit with a wood hammer has more high frequency harmonics than the corresponding graph for the dulcimer hit with a felt hammer. But, in both cases the dulcimer has the same fundamental frequency of 391 Hz. There is a mathematical technique know as **Fourier analysis** that can be used to determine the amplitudes of each of the harmonic frequencies that combines to make a full sound. A fast Fourier transform (FFT) of the two hammer dulcimer sounds is shown in Fig. 18-20. As you can see, the relative amplitudes for harmonic frequencies between 1500 Hz and 2500 Hz are much higher when the dulcimer is hit with a wood hammer than when it is hit with a felt hammer.

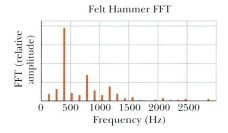

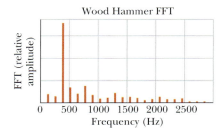

FIGURE 18-20 ■ A fast Fourier transform showing the amplitudes of the harmonics for a hammered dulcimer when it is (*a*) struck with a felt hammer and (*b*) struck with a wood hammer. The wood hammer stimulates more very high harmonics.

READING EXERCISE 18-4: Pipe A, with length L, and pipe B, with length $2L$, both have two open ends. Which harmonic of pipe B has the same frequency as the fundamental of pipe A? ■

TOUCHSTONE EXAMPLE 18-5: Sound in a Tube

Weak background noises from a room set up the fundamental standing wave in a cardboard tube of length $L = 67.0$ cm with two open ends. Assume that the speed of sound in the air within the tube is 343 m/s.

(a) What frequency do you hear from the tube?

SOLUTION ■ The **Key Idea** here is that, with both pipe ends open, we have a symmetric situation in which the standing wave has a displacement antinode at each end of the tube. The standing wave displacement variation pattern is that of Fig. 18-16*b*. The frequency is given by Eq. 18-23 with $n = 1$ for the fundamental mode:

$$f = \frac{nv^{\text{air}}}{2L} = \frac{(1)(343 \text{ m/s})}{(2)(0.670 \text{ m})} = 256 \text{ Hz}. \quad \text{(Answer)}$$

If the background noises set up any higher harmonics, such as the second harmonic, you must also hear frequencies that are *integer* multiples of 256 Hz.

(b) If you jam your ear against one end of the tube, what fundamental frequency do you hear from the tube?

SOLUTION ■ The **Key Idea** now is that, with your ear effectively closing one end of the tube we have an asymmetric situation—a displacement antinode still exists at the open end but a displacement node is now at the other (closed) end. The standing wave pattern is the top one in Fig. 18-17*b*. The frequency is given by Eq. 18-25 with $n = 1$ for the fundamental mode:

$$f = \frac{nv^{\text{air}}}{4L} = \frac{(1)(343 \text{ m/s})}{4(0.670 \text{ m})} = 128 \text{ Hz}. \quad \text{(Answer)}$$

If the background noises set up any higher harmonics, they will be *odd* multiples of 128 Hz. That means that the frequency of 256 Hz (which is an even multiple) cannot now occur.

18-6 Beats

If you listen separately to two sounds whose frequencies are, say, 50 and 52 Hz, most of us cannot tell which one has the higher frequency. However, if the sounds reach our ears simultaneously, what we hear is a sound whose frequency turns out to be 51 Hz, the *average* of the two combining frequencies. We also hear a striking variation in the intensity of this sound—it increases and decreases in slow, wavering **beats** that repeat at a frequency of 2 Hz, the *difference* between the two combining frequencies. Figure 18-21 shows this beat phenomenon (for sounds of frequencies of 8 Hz and 10 Hz).

FIGURE 18-21 ■ (*a*) The pressure variations ΔP of two sound waves as they would be detected separately are plotted as a function of time. The frequencies of the waves are nearly equal. (*b*) If the two waves are detected simultaneously the resultant pressure variation is the superposition of the two waves. Notice how the waves cancel each other at the center of the plot and reinforce each other at the two ends of the plot. (*c*) When plotted over a longer time period the two waves shown in (*a*) and (*b*) show a beat pattern of 2 Hz if the frequencies of the original waves are 10 and 12 Hz.

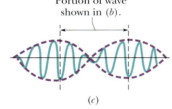

Portion of wave shown in (*b*).

(*c*)

Let the time-dependent variations of pressure due to two sound waves at a particular location be

$$\Delta P_1 = \Delta P^{\max}\cos\omega_1 t \quad \text{and} \quad \Delta P_2 = \Delta P^{\max}\cos\omega_2 t, \tag{18-26}$$

where $\omega_1 > \omega_2$. We have assumed, for simplicity, that the waves have the same amplitude and initial phase. According to the superposition principle, the resultant pressure variation is

$$\Delta P = \Delta P_1 + \Delta P_2 = \Delta P^{\max}(\cos\omega_1 t + \cos\omega_2 t).$$

Using the trigonometric identity (see Appendix E)

$$\cos\alpha + \cos\beta = 2\cos\tfrac{1}{2}(\alpha - \beta)\cos\tfrac{1}{2}(\alpha + \beta)$$

allows us to write the resultant pressure variation as

$$\Delta P = 2\Delta P^{\max}\cos\tfrac{1}{2}(\omega_1 - \omega_2)t \, \cos\tfrac{1}{2}(\omega_1 + \omega_2)t. \tag{18-27}$$

If we write

$$\omega' = \tfrac{1}{2}(\omega_1 - \omega_2) \quad \text{and} \quad \omega = \tfrac{1}{2}(\omega_1 + \omega_2) \tag{18-28}$$

we can then write Eq. 18-27 as

$$\Delta P(t) = [2\Delta P^{\max}\cos\omega' t]\cos\omega t. \tag{18-29}$$

We now assume that the angular frequencies ω_1 and ω_2 of the combining waves are almost equal, which means that $\omega \gg \omega'$ in Eq. 18-28. We can then regard Eq. 18-29 as a cosine function whose angular frequency is ω and whose amplitude (which is not constant but varies with angular frequency ω') is the quantity in the square brackets.

A maximum amplitude will occur whenever $\cos\omega't$ in Eq. 18-29 has the value $+1$ or -1, which happens twice in each repetition of the cosine function. Because $\cos\omega't$ has angular frequency ω', the angular frequency ω_{beat} at which beats occur is $\omega_{beat} = 2\omega'$. Then, with the aid of Eq. 18-28, we can write

$$\omega_{beat} = 2\omega' = (2)(\tfrac{1}{2})(\omega_1 - \omega_2) = \omega_1 - \omega_2.$$

Because $\omega = 2\pi f$, we can recast this as

$$f_{beat} = f_1 - f_2 \qquad \text{(beat frequency)}. \qquad (18\text{-}30)$$

Musicians use the beat phenomenon in tuning their instruments. If an instrument is sounded against a standard frequency (for example, the lead oboe's reference A) and tuned until the beat disappears, then the instrument is in tune with that standard. In musical Vienna, concert A (440 Hz) is available as a telephone service for the benefit of the city's many professional and amateur musicians.

Types of Superposition

So far the three simple cases we have discussed involving the superposition of sound waves are:

1. *Standing waves* created by the interference of two identical waves moving the opposite directions with the same speed

2. *Wave interference* caused by waves that have the same frequency, wavelength, and amplitude that are moving in almost the same direction and overlap at some point

3. *Beats* created by two waves with the same amplitude moving in the same direction with slightly different frequencies

There are many other more complex types of interference of interest to scientists and engineers. The mathematical techniques used to analyze these situations are similar to those we introduced in this section.

18-7 The Doppler Effect

An ambulance is parked by the side of the highway, sounding its 1000 Hz siren. If you are also parked by the highway, you will hear that same frequency. However, if there is relative motion between you and the ambulance, either toward or away from each other, you will hear a different frequency. For example, if you are driving *toward* the ambulance at 120 km/h (about 75 mi/h), you will hear a *higher* frequency (1096 Hz, an *increase* of 96 Hz). If you are driving *away from* the ambulance at that same speed, you will hear a *lower* frequency (904 Hz, a *decrease* of 96 Hz).

These motion-related frequency changes are examples of the **Doppler effect.** The effect was proposed (although not fully worked out) in 1842 by Austrian physicist Johann Christian Doppler. It was tested experimentally in 1845 by Buys Ballot in Holland, "using a locomotive drawing an open car with several trumpeters."

The Doppler effect holds not only for sound waves but also for electromagnetic waves, including microwaves, radio waves, and visible light. Here, we shall consider only sound waves for the special case where no wind is present. This means that we shall measure the speeds of a source S of sound waves and a detector D of those waves *relative to a body of air that is not moving*. We shall assume that S and D move either directly toward or directly away from each other, at speeds less than the speed of sound.

If either the detector or the source is moving, or both are moving, the emitted frequency f and the detected frequency f' are related by

$$f' = f\frac{v^{air} \pm v_D}{v^{air} \pm v_S} \qquad \text{(1D Doppler effect for sound in still air),} \qquad (18\text{-}31)$$

where v^{air} is the speed of sound through the air, v_D is the detector's speed relative to the air, and v_S is the source's speed relative to the air. The choice of plus or minus signs is set by this rule:

When a sound source and a detector are moving toward each other, the sign on the sound's speed and the detector's speed must give upward shifts in frequency. When a sound source and a detector are moving away from each other, the signs on the speeds must give downward shifts in frequency.

In short, toward means shift up, and away means shift down.

Here are some examples of the rule. If the detector moves toward the source, use the plus sign in the numerator of Eq. 18-31 to get a shift up in the frequency. If it moves away, use the minus sign in the numerator to get a shift down. If it is stationary, substitute 0 for v_D. If the source moves toward the detector, use the minus sign in the denominator of Eq. 18-31 to get a shift up in the frequency. If it moves away, use the plus sign in the denominator to get a shift down. If the source is stationary, substitute 0 for v_S.

Next, we derive equations for the Doppler effect for two specific situations and then derive Eq. 18-31 for the general situation.

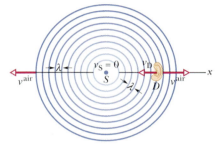

FIGURE 18-22 ■ A stationary source of sound S emits spherical wavefronts, shown one wavelength apart, that expand outward at speed v^{air}. A sound detector D, represented by an ear, moves with velocity $\vec{v}_D$ toward the source. The detector senses a higher frequency because of its motion.

1. When the detector moves relative to the air and the source is stationary relative to the air, the motion changes the frequency at which the detector intercepts wavefronts and thus the detected frequency of the sound wave.

2. When the source moves relative to the air and the detector is stationary relative to the air, the motion changes the wavelength of the sound wave and thus the detected frequency (recall that frequency is related to wavelength).

Detector Moving; Source Stationary

In Fig. 18-22, a detector D (represented by an ear) is moving at speed v_D toward a stationary source S that emits spherical wavefronts, of wavelength λ and frequency f, moving at the speed v^{air} of sound in air. A cross section of the wavefronts are drawn one wavelength apart. The frequency detected by detector D is the rate at which D intercepts wavefronts (or individual wavelengths). If D were stationary, that rate would be f, but since D is moving into the wavefronts, the rate of interception is greater, and thus the detected frequency f' is greater than f.

Let us for the moment consider the situation in which D is stationary (Fig. 18-23). In time t, the wavefronts move to the right a distance $v^{air}t$. The number of wavelengths in that distance $v^{air}t$ is the number of wavelengths intercepted by D in time t, and that number is $v^{air}t/\lambda$. The rate at which D intercepts wavelengths, which is the frequency f detected by D, is

$$f = \frac{v^{air}t/\lambda}{t} = \frac{v^{air}}{\lambda}. \qquad (18\text{-}32)$$

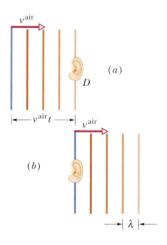

FIGURE 18-23 ■ Wavefronts of Fig. 18-22, assumed planar, (a) reach and (b) pass a stationary detector D; they move a distance $v^{air}t$ to the right in time t.

In this situation, with D stationary, there is no Doppler effect—the frequency detected by D is the frequency emitted by S.

Now let us again consider the situation in which D moves opposite the wavefronts (Fig. 18-24). In time t, the wavefronts move to the right a distance $v^{air}t$ as previously, but now D moves to the left a distance v_Dt. Thus, in this time t, the distance moved by the wavefronts relative to D is $v^{air}t + v_Dt$. The number of wavelengths in this relative distance $v^{air}t + v_Dt$ is the number of wavelengths intercepted by D in time t, and is $(v^{air}t + v_Dt)/\lambda$. The *rate* at which D intercepts wavelengths in this situation is the frequency f', given by

$$f' = \frac{(v^{air}t + v_Dt)/\lambda}{t} = \frac{v^{air} + v_D}{\lambda}. \tag{18-33}$$

From Eq. 18-32, we have $\lambda = v^{air}/f$. Then Eq. 18-33 becomes

$$f' = \frac{v^{air} + v_D}{v^{air}/f} = f\left(\frac{v^{air} + v_D}{v^{air}}\right). \tag{18-34}$$

Note that in Eq. 18-34 f' must be greater than f unless the detector is stationary so that $v_D = 0$.

Similarly, we can find the frequency detected by D if D moves away from the source. In this situation, the wavefronts move a distance $v^{air}t - \vec{v}_Dt$ relative to D in time t, and f' is given by

$$f' = f\left(\frac{v^{air} - v_D}{v^{air}}\right). \tag{18-35}$$

In Eq. 18-35, f' must be less than f unless $v_D = 0$.

We can summarize Eqs. 18-34 and 18-35 with

$$f' = f\left(\frac{v^{air} \pm v_D}{v^{air}}\right) \qquad \text{(detector moving; source stationary).} \tag{18-36}$$

Source Moving; Detector Stationary

Let detector D be stationary with respect to the body of air, and let source S move toward D at speed $|\vec{v}_s|$ (Fig. 18-25). The motion of S changes the wavelength of the sound waves it emits, and thus the frequency detected by D.

To see this change, let $T(= 1/f)$ be the time between the emission of any pair of successive wavefronts W_1 and W_2. During T, wavefront W_1 moves a distance $v^{air}T$ and the source moves a distance v_sT. At the end of T, wavefront W_2 is emitted. In the direction in which S moves, the distance between W_1 and W_2, which is the wavelength λ' of the waves moving in that direction, is $v^{air}T - v_sT$. If D detects those waves, it detects frequency f' given by

$$f' = \frac{v^{air}}{\lambda'} = \frac{v^{air}}{v^{air}T - v_sT} = \frac{v^{air}}{v^{air}/f - v_s/f}$$

$$= f\frac{v^{air}}{v^{air} - v_s}. \tag{18-37}$$

Note that f' must be greater than f unless $|\vec{v}_s| = 0$.

In the direction opposite that taken by S, the wavelength λ' of the waves is $v^{air}T - v_sT$. If D detects those waves, it detects frequency f', given by

$$f' = f\frac{v^{air}}{v^{air} + v_s}. \tag{18-38}$$

Now f' must be less than f unless $v_s = 0$.

We can summarize Eqs. 18-37 and 18-38 with

$$f' = f\frac{v^{air}}{v^{air} \pm v_s} \qquad \text{(source moving; detector stationary).} \tag{18-39}$$

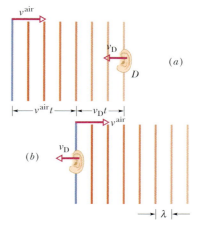

FIGURE 18-24 ■ Wavefronts (*a*) reach and (*b*) pass detector D, which moves opposite the wavefronts. In time t, the wavefronts move a distance $v^{air}t$ to the right and D moves a distance v_Dt to the left.

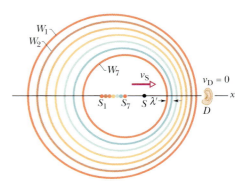

FIGURE 18-25 ■ A detector D is stationary, and a source S is moving toward it at speed v_S. Wavefront W_1 was emitted when the source was at S_1, wavefront W_7 when the source was at S_7. At the moment depicted, the source is at S. The detector perceives a higher frequency because the moving source, chasing its own wavefronts, emits a reduced wavelength λ' in the direction of its motion.

General Doppler Effect Equation for Sound

For the one-dimensional case, we can now derive the general Doppler effect equation by replacing f in Eq. 18-39 (the frequency of the source) with f' of Eq. 18-36 (the frequency associated with motion of the detector). The result is Eq. 18-31 for the general Doppler effect.

That general equation holds not only when both detector and source are moving but also in the two specific situations we just discussed. For the situation in which the detector is moving and the source is stationary, substitution of $v_S = 0$ into Eq. 18-31 gives us Eq. 18-36, which we previously found. For the situation in which the source is moving and the detector is stationary, substitution of $v_D = 0$ into Eq. 18-30 gives us Eq. 18-39, which we previously found. Thus, Eq. 18-31 is the equation to remember.

Similar Doppler frequency shifts occur with electromagnetic waves treated in Chapters 34 and 38, but the equations are slightly different (see Section 38-14).

The Doppler Effect in More Than One Dimension

We can even use Eqs. 18-36 and 18-39 if the source and/or the detector are moving relative to each other at constant velocities that *don't lie along the same line*. Simply draw a line from one object to the other at a particular time when the sound of interest is being emitted and call it the x axis. Start by finding the x-components of the velocities. You can then use the speeds along the x axis in Eq. 18-36 or Eq. 18-37 to determine the frequency shifts. For example, the Doppler shift for light waves is used by astronomers to help them determine how fast a distant object is moving relative to Earth. However, they can only use Doppler shift measurements to find the object's velocity *component* along a line between Earth and the object.

Bat Navigation and Feeding

An example of the use of the three-dimensional Doppler effect in nature is provided by bats. Bats navigate and search out prey by emitting and then detecting reflections of ultrasonic waves. These are sound waves with frequencies greater than can be heard by a human.* For example, a horseshoe bat emits ultrasonic waves at 83 kHz, well above the 20 kHz limit of human hearing.

After the sound is emitted through the bat's nostrils, it might reflect (echo) from a moth, and then return to the bat's ears. The motions of the bat and the moth relative to the air cause the frequency heard by the bat to differ by a few kilohertz from the frequency it emitted. The bat automatically translates this difference into a relative speed between itself and the moth, so it can zero in on the moth.

Some moths evade capture by flying away from the direction in which they hear ultrasonic waves. That choice of flight path reduces the frequency difference between what the bat emits and what it hears, and then the bat may not notice the echo. Some moths avoid capture by clicking to produce their own ultrasonic waves, thus "jamming" the detection system and confusing the bat. (Surprisingly, moths and bats do all this without first studying physics.)

READING EXERCISE 18-5: The figure indicates the directions of motion of a sound source and a detector for six situations in stationary air. In general, $v_0 \neq v_S$. For each situation, is the detected frequency greater than or less than the emitted frequency, or can't we tell without more information about the actual speeds?

	Source	Detector		Source	Detector
(a)	$\longrightarrow$	• 0 speed	(d)	$\longleftarrow$	$\longleftarrow$
(b)	$\longleftarrow$	• 0 speed	(e)	$\longrightarrow$	$\longleftarrow$
(c)	$\longrightarrow$	$\longrightarrow$	(f)	$\longleftarrow$	$\longrightarrow$

*Ultrasonic motion detectors use the same technique for locating objects.

TOUCHSTONE EXAMPLE 18-6: Rocket Sounds

A rocket moves at a speed of 242 m/s directly toward a stationary pole (through stationary air) while emitting sound waves at frequency $f = 1250$ Hz.

(a) What frequency f' is measured by a detector that is attached to the pole?

SOLUTION ■ We can find f' with Eq. 18-31 for the general Doppler effect. The **Key Idea** here is that, because the sound source (the rocket) moves through the air *toward* the stationary detector on the pole, we need to choose the sign on v_S that gives a *shift up* in the frequency of the sound. Thus, in Eq. 18-31 we use the minus sign in the denominator. We then substitute 0 for the detector speed v_D, 242 m/s for the source speed v_S, 343 m/s for the speed of sound v^{air} (from Table 18-1), and 1250 Hz for the emitted frequency f. We find

$$f' = f\frac{v^{\text{air}} \pm v_D}{v^{\text{air}} \pm v_S} = (1250 \text{ Hz})\frac{343 \text{ m/s} \pm 0}{343 \text{ m/s} - 242\text{m/s}},$$

$$= 4245 \text{ Hz} \approx 4250 \text{ Hz}, \qquad \text{(Answer)}$$

which, indeed, is a greater frequency than the emitted frequency.

(b) Some of the sound reaching the pole reflects back to the rocket as an echo. What frequency f'' does a detector on the rocket detect for the echo?

SOLUTION ■ Two **Key Ideas** here are the following:

1. The pole is now the source of sound (because it is the source of the echo), and the rocket's detector is now the detector (because it detects the echo).

2. The frequency of the sound emitted by the source (the pole) is equal to f', the frequency of the sound the pole intercepts and reflects.

We can rewrite Eq. 18-31 in terms of the source frequency f' and the detected frequency f'' as

$$f'' = f'\frac{v^{\text{air}} \pm v_D}{v^{\text{air}} \pm v_S}. \qquad (18\text{-}40)$$

A third **Key Idea** here is that, because the detector (on the rocket) moves through the air *toward* the stationary source, we need to use the *sign* on v_D that gives a *shift up* in the frequency of the sound. Thus, we use the plus sign in the numerator of Eq. 18-40. Also, we substitute $v_D = 242$ m/s, $v_S = 0$, $v^{\text{air}} = 343$ m/s, and $f' = 4245$ Hz. We find

$$f'' = (4245 \text{ Hz})\frac{343 \text{ m/s} + 242 \text{ m/s}}{343 \text{ m/s} \pm 0}$$

$$= 7240 \text{ Hz}, \qquad \text{(Answer)}$$

which, indeed, is greater than the frequency of the sound reflected by the pole.

18-8 Supersonic Speeds; Shock Waves

If a source is moving toward a stationary detector at a speed equal to the speed of sound in a medium—that is, if $v_S = v^{\text{air}}$ or $v_S = v^{\text{water}}$ and so on—Eqs. 18-31 and 18-39 predict that the detected frequency f' will be infinitely great. This means that the

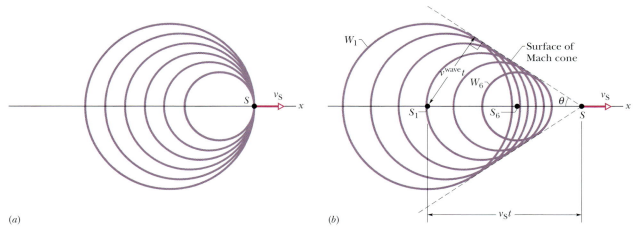

(a)   (b)

FIGURE 18-26 ■ Cross-sectional drawings of: (a) A source of sound S moves at speed v_S equal to the speed of sound and thus as fast as the wavefronts it generates. (b) A source S moves at speed v_S faster than the speed of sound and thus faster than the wavefronts. When the source was at position S_1 it generated wavefront W_1, and at position S_6 it generated W_6. All the spherical wavefronts expand at the speed of sound v^{wave} and bunch along the surface of a cone called the Mach cone, forming a shock wave. The surface of the cone has half-angle θ and is tangent to all the wavefronts.

source is moving so fast that it keeps pace with its own spherical wavefronts, as Fig. 18-26a suggests. What happens when the speed of the source exceeds the speed of sound?

For such supersonic speeds, Eqs. 18-31 and 18-39 no longer apply. Figure 18-26b depicts the spherical wavefronts that originated at various positions of the source. The radius of any wavefront in this figure is $v^{wave}t$, where v^{wave} is the speed of sound in the medium (for example, v^{air} or v^{water}) and t is the time that has elapsed since the source emitted that wavefront. Note that all the wavefronts bunch along a V-shaped envelope in the two-dimensional drawing of Fig. 18-26b. The wavefronts actually extend in three dimensions, and the bunching forms a cone called the *Mach cone*. A *shock wave* is said to exist along the surface of this cone, because the bunching of wavefronts causes an abrupt rise and fall of air pressure as the surface passes by any point. An observer hears the wavefront bunching as a sharp loud sound known as a **sonic boom.** From Fig. 18-26b, we see that the half-angle θ of the cone, called the *Mach cone angle*, is given by

$$\sin \theta = \frac{v^{wave}t}{v_S t} = \frac{v^{wave}}{v_S} \qquad \text{(Mach cone angle).} \qquad (18\text{-}41)$$

The ratio v_S/v^{wave} is called the *Mach number*. When you hear that a particular plane has flown at Mach 2.3, it means that its speed was 2.3 times the speed of sound in the air through which the plane was flying. There is a common misconception that a sonic boom is a single burst of sound that is generated at the moment a plane breaks the sound barrier. However, the shock wave is generated continuously as long as the speed of the plane is greater than the speed of sound. When a shock wavefront generated by a supersonic aircraft (Fig. 18-27) passes by an observer she hears the sonic boom.

Part of the sound that is heard when a rifle is fired is the sonic boom produced by the bullet. A sonic boom can also be heard from a long bullwhip when it is snapped quickly: Near the end of the whip's motion, its tip is moving faster than sound and produces a small sonic boom—the *crack* of the whip.

FIGURE 18-27 ■ A cloud formation is produced off the wings of this Navy jet. One plausible explanation is that the cloud forms because of the sudden decrease in air pressure that results as a supersonic shock wave propagates. This causes water vapor in the air to condense.

Problems

Where needed in the problems, use

$$\text{speed of sound in air} = 343 \text{ m/s} \quad and$$
$$\text{density of air} = 1.21 \text{ kg/m}^3$$

unless otherwise specified.

SEC. 18-2 ■ THE SPEED OF SOUND

1. Devise a Rule Devise a rule for finding your distance in kilometers from a lightning flash by counting the seconds from the time you see the flash until you hear the thunder. Assume that the sound travels to you along a straight line.

2. Outdoor Concert You are at a large outdoor concert, seated 300 m from the speaker system. The concert is also being broadcast live via satellite (at the speed of light, 3.0×10^8 m/s). Consider a listener 5000 km away who receives the broadcast. Who hears the music first, you or the listener and by what time difference?

3. Two Spectators Two spectators at a soccer game in Montjuic Stadium see, and a moment later hear, the ball being kicked on the playing field. The time delay for one spectator is 0.23 s and for the other 0.12 s. Sight lines from the two spectators to the player kicking the ball meet at an angle of 90°. (a) How far is each spectator from the player? (b) How far are the spectators from each other?

4. Column of Soldiers A column of soldiers, marching at 120 paces per minute, keep in step with the beat of a drummer at the head of the column. It is observed that the soldiers in the rear end of the column are striding forward with the left foot when the drummer is advancing with the right. What is the approximate length of the column?

5. Earthquakes Earthquakes generate sound waves inside Earth. Unlike a gas, Earth can experience both transverse (S) and longitudinal (P) sound waves. Typically, the speed of S waves is about 4.5 km/s, and that of P waves 8.0 km/s. A seismograph records P and S waves from an earthquake. The first P waves arrive 3.0 min before the first S waves (Fig. 18-28). Assuming the waves travel in a straight line, how far away does the earthquake occur?

6. Speed of Sound The speed of sound in a certain metal is v^{metal}. One end of a long pipe of that metal of length L is struck a hard blow. A listener at the other end hears two sounds, one from the wave that travels along the pipe and the other from the wave that travels through the air. (a) If v^{air} is the speed of sound in air, what time interval Δt elapses between the arrivals of the two sounds? (b) Suppose that $\Delta t = 1.00$ s and the metal is steel. Find the length L.

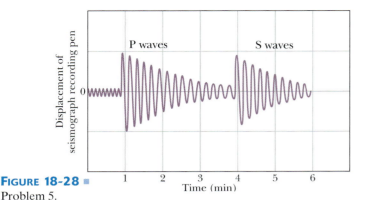

FIGURE 18-28 ■
Problem 5.

7. Stone Is Dropped A stone is dropped into a well. The sound of the splash is heard 3.00 s later. What is the depth of the well?

8. Audible Frequency The audible frequency range for normal hearing is from about 20 Hz to 20 kHz. What are the wavelengths of sound waves at these frequencies?

9. Diagnostic Ultrasound Diagnostic ultrasound of frequency 4.50 MHz is used to examine tumors in soft tissue. (a) What is the wavelength in air of such a sound wave? (b) If the speed of sound in tissue is 1500 m/s, what is the wavelength of this wave in tissue?

10. Pressure in Traveling Wave The pressure in a traveling sound wave is given by the equation

$$\Delta P(x, t) = (1.50 \text{ Pa}) \sin \pi [(0.900 \text{ rad/m})x - (315 \text{ rad/s})t].$$

Find the (a) pressure amplitude, (b) frequency, (c) wavelength, and (d) speed of the wave.

SEC. 18-3 ■ INTERFERENCE

11. Two Loudspeakers In Fig. 18-29, two loudspeakers, separated by a distance of 2.00 m, are in phase. Assume the amplitudes of the sound from the speakers are approximately the same at the position of a listener, who is 3.75 m directly in front of one of the speakers. (a) For what frequencies in the audible range (20 Hz to 20 kHz) does the listener hear a minimum signal? (b) For what frequencies is the signal a maximum?

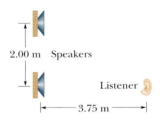

FIGURE 18-29 ■ Problem 11

12. Two Point Sources Two point sources of sound waves of identical wavelength λ and amplitude are separated by distance $D = 2.0\lambda$. The sources are in phase. (a) How many points of maximum signal (that is, maximum constructive interference) lie along a large circle around the sources? (b) How many points of minimum signal (destructive interference) lie around the circle?

13. Loudspeakers on Outdoor Stage Two loudspeakers are located 3.55 m apart on an outdoor stage. A listener is 18.3 m from one and 19.5 m from the other. During the sound check, a signal generator drives the two speakers in phase with the same amplitude and frequency. The transmitted frequency is swept through the audible range (20 Hz to 20 kHz). (a) What are the three lowest frequencies at which the listener will hear a minimum signal because of destruc-

tive interference? (b) What are the three lowest frequencies at which the listener will hear a maximum signal?

14. Two Sound Waves Two sound waves, from two different sources with the same frequency, 540 Hz, travel in the same direction at 330 m/s. The sources are in phase. What is the phase difference of the waves at a point that is 4.40 m from one source and 4.00 m from the other?

15. Half-Circle In Fig. 18-30, sound with a 40.0 cm wavelength travels rightward from a source and through a tube that consists of a straight portion and a half-circle.

FIGURE 18-30 ■ Problem 15.

Part of the sound wave travels through the half-circle and then rejoins the rest of the wave, which goes directly through the straight portion. This rejoining results in interference. What is the smallest radius r that results in an intensity minimum at the detector?

SEC. 18-4 ■ INTENSITY AND SOUND LEVEL

16. Point Source A 1.0 W point source emits sound waves isotropically. Assuming that the energy of the waves is conserved, find the intensity (a) 10 m from the source and (b) 2.5 m from the source.

17. A Source Emits A source emits sound waves isotropically. The intensity of the waves 2.50 m from the source is 1.91×10^{-4} W/m². Assuming that the energy of the waves is conserved, find the power of the source.

18. Differ in Level Two sounds differ in sound level by 1.00 dB. What is the ratio of the greater intensity to the smaller intensity?

19. Increased in Level A certain sound source is increased in sound level by 30 dB. By what multiple is (a) its intensity increased and (b) its pressure amplitude increased?

20. The Source of a Sound The source of a sound wave has a power of 1.00 μW. If it is a point source, (a) what is the intensity 3.00 m away and (b) what is the sound level in decibels at that distance?

21. One in Air, One in Water (a) If two sound waves, one in air and one in (fresh) water, are equal in intensity, what is the ratio of the pressure amplitude of the wave in water to that of the wave in air? Assume the water and the air are at 20°C. (See Table 15-2.) (b) If the pressure amplitudes are equal instead, what is the ratio of the intensities of the waves?

22. Noisy Freight Train Assume that a noisy freight train on a straight track emits a cylindrical, expanding sound wave, and that the air absorbs no energy. How does the amplitude $\Delta P^{\max}$ of the wave depend on the perpendicular distance r from the source?

23. Ratios Find the ratios (greater to smaller) of (a) the intensities, and (b) the pressure amplitudes for two sounds whose sound levels differ by 37 dB.

24. Point Source Two A point source emits 30.0 W of sound isotropically. A small microphone intercepts the sound in an area of 0.750 cm², 200 m from the source. Calculate (a) the sound intensity there and (b) the power intercepted by the microphone.

25. Acoustic Interferometer Figure 18-31 shows an air-filled, acoustic interferometer, used to demonstrate

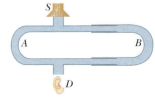

FIGURE 18-31 ■ Problem 25.

the interference of sound waves. Sound source S is an oscillating di-aphragm; D is a sound detector, such as the ear or a microphone. Path SBD can be varied in length, but path SAD is fixed. At D, the sound wave coming along path SBD interferes with that coming along path SAD. In one demonstration, the sound intensity at D has a minimum value of 100 units at one position of the movable arm and continuously climbs to a maximum value of 900 units when that arm is shifted by 1.65 cm. Find (a) the frequency of the sound emitted by the source and (b) the ratio of the amplitude at D of the SAD wave to that of the SBD wave. (c) How can it happen that these waves have different amplitudes, considering that they originate at the same source?

SEC. 18-5 ■ SOURCES OF MUSICAL SOUND

26. A Violin String A violin string 15.0 cm long and fixed at both ends oscillates in its $n = 1$ mode. The speed of waves on the string is 250 m/s, and the speed of sound in air is 348 m/s. What are (a) the frequency and (b) the wavelength of the emitted sound wave?

27. Organ Pipe Organ pipe A, with both ends open, had a fundamental frequency of 300 Hz. The third harmonic of organ pipe B, with one end open, has the same frequency as the second harmonic of pipe A. How long are (a) pipe A and (b) pipe B?

28. Glass Tube The water level in a vertical glass tube 1.00 m long can be adjusted to any position in the tube. A tuning fork vibrating at 686 Hz is held just over the open top end of the tube, to set up a standing wave of sound in the air-filled top portion of the tube. (That air-filled top portion acts as a tube with one end closed and the other end open.) At what positions of the water level is there resonance?

29. Speed of Waves (a) Find the speed of waves on a violin string of mass 800 mg and length 22.0 cm if the fundamental frequency is 920 Hz. (b) What is the tension in the string? For the fundamental, what is the wavelength of (c) the waves on the string and (d) the sound waves emitted by the string?

30. Violin String A certain violin string is 30 cm long between its fixed ends and has a mass of 2.0 g. The "open" string (no applied finger) sounds an A note (440 Hz). (a) To play a C note (523 Hz), how far down the string must one place a finger? (b) What is the ratio of the wavelength of the string waves required for an A note to that required for a C note? (c) What is the ratio of the wavelength of the sound wave for an A note to that for a C note?

31. Small Loudspeaker In Fig. 18-32, S is a small loudspeaker driven by an audio oscillator and amplifier, adjustable in frequency from 1000 to 2000 Hz only. Tube D is a piece of cylindrical sheet-metal pipe 45.7 cm long and open at both ends. (a) If the speed of sound in air is 344 m/s at the existing temperature, at what frequencies will resonance occur in the pipe when the frequency emitted by the speaker is varied from 1000 Hz to 2000 Hz? (b) Sketch the standing wave (using the style of Fig. 18-16b) for each resonant frequency.

FIGURE 18-32 ■
Problem 31.

32. Cello String A string on a cello has length L, for which the fundamental frequency is f. (a) By what length l must the string be shortened by fingering to change the fundamental frequency to rf? (b) What is l if $L = 0.80$ m and $r = 1.2$? (c) For $r = 1.2$, what is the ratio of the wavelength of the new sound wave emitted by the string to that of the wave emitted before fingering?

33. Well A well with vertical sides and water at the bottom resonates at 7.00 Hz and at no lower frequency. (The air-filled portion of the well acts as a tube with one closed end and one open end.) The air in the well has a density of 1.10 kg/m³ and a bulk modulus of 1.33×10^5 Pa. How far down in the well is the water surface?

34. A Tube A tube 1.20 m long is closed at one end. A stretched wire is placed near the open end. The wire is 0.330 m long and has a mass of 9.60 g. It is fixed at both ends and oscillates in its fundamental mode. By resonance, it sets the air column in the tube into oscillation at that column's fundamental frequency. Find (a) that frequency and (b) the tension in the wire.

35. Pulsating Variable Star The period of a pulsating variable star may be estimated by considering the star to be executing *radial* longitudinal pulsations in the fundamental standing wave mode. That is, the star's radius varies periodically with time, with a displacement antinode at the star's surface. (a) Would you expect the center of the star to be a displacement node or antinode? (b) By analogy with a pipe with one open end, show that the period of pulsation T is given by

$$T = \frac{4R}{\langle v \rangle},$$

where R is the equilibrium radius of the star and $\langle v \rangle$ is the average sound speed in the material of the star. (c) Typical white dwarf stars are composed of material with a bulk modulus of 1.33×10^{22} Pa and a density of 10^{10} kg/m³. They have radii equal to 9.0×10^{-3} solar radius. What is the approximate pulsation period of a white dwarf?

36. Pipes A and B Pipe A, which is 1.2 m long and open at both ends, oscillates at its third lowest harmonic frequency. It is filled with air for which the speed of sound is 343 m/s. Pipe B, which is closed at one end, oscillates at its second lowest harmonic frequency. These frequencies of pipes A and B happen to match. (a) If an x axis extends along the interior of pipe A, with $x = 0$ at one end, where along the axis are the displacement nodes? (b) How long is pipe B? (c) What is the lowest harmonic frequency of pipe A?

37. Violin String and Loudspeaker A violin string 30.0 cm long with linear density 0.650 g/m is placed near a loudspeaker that is fed by an audio oscillator of variable frequency. It is found that the string is set into oscillation only at the frequencies 880 and 1320 Hz as the frequency of the oscillator is varied over the range 500–1500 Hz. What is the tension in the string?

SEC. 18-6 ■ BEATS

38. Too Tightly Stretched The A string of a violin is a little too tightly stretched. Four beats per second are heard when the string is sounded together with a tuning fork that is oscillating accurately at concert A (440 Hz). What is the period of the violin string oscillation?

39. Unknown Tuning Fork A tuning fork of unknown frequency makes three beats per second with a standard fork of frequency

384 Hz. The beat frequency decreases when a small piece of wax is put on a prong of the first fork. What is the freauency of this fork?

40. Five Tuning Forks You have five tuning forks that oscillate at close but different frequencies. What are the (a) maximum and (b) minimum number of different beat frequencies you can produce by sounding the forks two at a time depending on how the frequencies differ?

41. Two Piano Wires Two identical piano wires have a fundamental frequency of 600 Hz when kept under the same tension. What fractional increase in the tension of one wire will lead to the occurrence of 6 beats/s when both wires oscillate simultaneously?

SEC. 18-7 ■ THE DOPPLER EFFECT

42. Trooper B Trooper B is chasing speeder A along a straight stretch of road. Both are moving at a speed of 160 km/h. Trooper B, failing to catch up, sounds his siren again. Take the speed of sound in air to be 343 m/s and the frequency of the source to be 500 Hz. What is the Doppler shift in the frequency heard by speeder A?

43. Turbine Whine The 16 000 Hz whine of the turbines in the jet engines of an aircraft moving with speed 200 m/s is heard at what frequency by the pilot of a second craft trying to overtake the first at a speed of 250 m/s?

44. Ambulance Siren An ambulance with a siren emitting a whine at 1600 Hz overtakes and passes a cyclist pedaling a bike at 2.44 m/s. After being passed, the cyclist hears a frequency of 1590 Hz. How fast is the ambulance moving?

45. A Whistle A whistle of frequency 540 Hz moves in a circle of radius 60.0 cm at a rotational speed of 15.0 rad/s. What are (a) the lowest and (b) the highest frequencies heard by a listener a long distance away, at rest with respect to the center of the circle?

46. Motion Detector A stationary motion detector sends sound waves of frequency 0.150 MHz toward a truck approaching at a speed of 45.0 m/s. What is the frequency of the waves reflected back to the detector?

47. French Submarine A French submarine and a U.S. submarine move toward each other during maneuvers in motionless water in the North Atlantic (Fig. 18-33). The French sub moves at 50.0 km/h, and the U.S. sub at 70.0 km/h. The French sub sends out a sonar signal (sound wave in water) at 1000 Hz. Sonar waves travel at 5470 km/h. (a) What is the signal's frequency as detected by the U.S. sub? (b) What frequency is detected by the French sub in the signal reflected back to it by the U.S. sub?

French 50.0 km/h   U.S. 70.0 km/h

FIGURE 18-33 ■ Problem 47.

48. Sound Source A sound source A and a reflecting surface B move directly toward each other. Relative to the air, the speed of source A is 29.9 m/s, the speed of surface B is 65.8 m/s, and the speed of sound is 329 m/s. The source emits waves at frequency 1200 Hz as measured in the source frame. In the reflector frame, what are (a) the frequency and (b) the wavelength of the arriving sound waves? In the source frame, what are (c) the frequency and (d) the wavelength of the sound waves reflected back to the source?

49. Burglar Alarm An acoustic burglar alarm consists of a source emitting waves of frequency 28.0 kHz. What is the beat frequency between the source waves and the waves reflected from an intruder walking at an average speed of 0.950 m/s directly away from the alarm?

50. A Bat A bat is flitting about in a cave, navigating via ultrasonic bleeps. Assume that the sound emission frequency of the bat is 39 000 Hz. During one fast swoop directly toward a flat wall surface, the bat is moving at 0.025 times the speed of sound in air. What frequency does the bat hear reflected off the wall?

51. Girl in Window A girl is sitting near the open window of a train that is moving at a velocity of 10.00 m/s to the east. The girl's uncle stands near the tracks and watches the train move away. The locomotive whistle emits sound at frequency 500.0 Hz. The air is still. (a) What frequency does the uncle hear? (b) What frequency does the girl hear? A wind begins to blow from the east at 10.00 m/s. (c) What frequency does the uncle now hear? (d) What frequency does the girl now hear?

52. Civil Defense Official A 2000 Hz siren and a civil defense official are both at rest with respect to the ground. What frequency does the official hear if the wind is blowing at 12 m/s (a) from source to official and (b) from official to source?

53. Two Trains Two trains are traveling toward each other at 30.5 m/s relative to the ground. One train is blowing a whistle at 500 Hz. (a) What frequency is heard on the other train in still air? (b) What frequency is heard on the other train if the wind is blowing at 30.5 m/s toward the whistle and away from the listener? (c) What frequency is heard if the wind direction is reversed?

SEC. 18-8 ■ SUPERSONIC SPEEDS; SHOCK WAVES

54. Bullet Fired A bullet is fired with a speed of 685 m/s. Find the half angle made by the shock cone with the line of motion of the bullet.

55. Jet Plane A jet plane passes over you at a height of 5000 m and a speed of Mach 1.5. (a) Find the Mach cone half angle. (b) How long after the jet passes directly overhead does the shock wave reach you? Use 331 m/s for the speed of sound.

56. Plane Flies A plane flies at 1.25 times the speed of sound. Its sonic boom reaches a man on the ground 1.00 min after the plane passes directly overhead. What is the altitude of the plane? Assume the speed of sound to be 330 m/s.

Additional Problems

57. Building a Pipe Organ You decide to build a pipe organ in your dormitory room using PVC pipe. Estimate whether you could build an organ that would cover the entire range of human hearing without bending any pipes.

58. Arranging the Patio Speakers You have set up two stereo speakers on your back patio railing as shown in the top view diagram in Fig. 18-34. You are worried that at certain positions you will lose frequencies as a result of interference. The coordinate grid on the edge of the picture has its large tick marks separated by 1 meter. For ease of calculation, make the following assumptions:

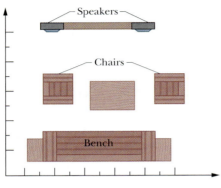

FIGURE 18-34 ■ Problem 58.

- Assume that the relevant objects lie on integer or half-integer grid points of the coordinate system.

- Take the speed of sound to be 343 m/s.

- Ignore the reflection of sound from the house, trees, and so on.

- The speakers are in phase.

(a) What will happen if you are sitting in the middle of the bench?
(b) If you are sitting in the lawn chair on the left, what will be the lowest frequency you will lose to destructive interference?
(c) Can you restore the frequency lost in part (a) by switching the leads to one of the speakers, thereby reversing the phase of that source?
(d) With the leads reversed, what will happen to the sound for a person sitting at the center of the bench?

59. Truthful Salesman? A salesperson claimed that a stereo system had a maximum audio power of 120 W. Testing the system with several speakers set up so as to simulate a point source, the consumer noted that she could get as close as 1.2 m with the volume full on before the sound hurt her ears. Was the salesperson truthful? Explain your answer with a calculation.

60. Experimenter An experimenter wishes to measure the speed of sound in an aluminum rod 10 cm long by measuring the time it takes for a sound pulse to travel the length of the rod. If results good to four significant figures are desired, how precisely must the length of the rod be known and how closely must the experimenter be able to resolve time intervals?

19 | The First Law of Thermodynamics

© Masato Ono, Tamagawa University.

The giant hornet *Vespa mandarinia japonica* preys on Japanese bees. However, if one of the hornets attempts to invade a bee hive, several hundred of the bees quickly form a compact ball around the hornet to stop it. After about 20 minutes the hornet is dead, although the bees do not sting, bite, crush, or suffocate it.

Why, then, does the hornet die?

The answer is in this chapter.

19-1 Thermodynamics

In the next three chapters we focus on a new subject—thermodynamics. The development of thermodynamic principles is one of humankind's most profound intellectual achievements. Why? The steam engine that powered the industrial revolution operates according to thermodynamic principles, as do many modern power plants. Thermodynamics has enriched our fundamental understanding of phenomena ranging from the metabolism of a lizard to the evolution of the universe.

Recall from Chapter 10 that the total mechanical energy of a system is the sum of the *macroscopic* kinetic and potential energies associated with the motion and configuration of the objects within the system. By macroscopic we mean that the particles in a system were large enough so that we could observe how fast they were moving (associated with kinetic energy) and also see their configuration (associated with potential energy). We found that when conservative forces act on the particles in an isolated system, the system's mechanical energy is conserved even if its potential energy is converted to kinetic energy or vice versa. Also recall that we can add mechanical energy to a system by doing work on it.

Furthermore, in Chapter 10 we observed that when a block slides along a surface and comes to a stop due to a nonconservative friction force, the temperatures of the block and surface rise. A decrease in the total mechanical energy of the block-surface system was accompanied by an increase in the temperature of the parts of the system. In response to this observation, we defined a new kind of energy, **thermal energy,** $E^{thermal}$, associated with the temperature of an object. Then, the overall system energy would still be conserved.

Our study of thermodynamics begins with learning how to quantify the hidden *internal* energy stored in ordinary matter on a microscopic (and hence invisible) scale. It also involves an examination of the role that temperature plays in determining whether a system's internal energy will increase or decrease when it comes into contact with another system. But, when hot steam is injected into a cylinder with a piston on top of it, the piston can be raised. So we believe that under other circumstances the internal energy in a system can be transformed back into mechanical energy.

In this chapter we consider temperature and various ways it can be measured. We then quantify the invisible transfer of thermal energy between objects of different temperatures. We also introduce the first law of thermodynamics, a statement of energy conservation, which relates internal energy change to both the thermal energy transferred to or from a system and the mechanical work done on or by it.

In Chapter 20 we introduce **kinetic theory** as an idealized model of how microscopic scale kinetic and potential energies associated with the motions and configurations of atoms and molecules can be added together to explain internal energy. The 19th century development of kinetic theory to explain internal energy is similar in character and importance to Einstein's 20th century discovery that matter also contains hidden energy by virtue of its mass. We then begin a study of heat engines by considering how the internal energy in hot expanding gases can be transformed into mechanical work.

In Chapter 21 we explain the principles that govern the operation of heat engines. We than introduce the concept of entropy that helps us understand why no one has ever invented a heat engine that can transform internal energy into mechanical work with anything close to 100% efficiency.

19-2 Thermometers and Temperature Scales

Temperature and its measurement are central to understanding the behavior of macroscopic systems that are heated and cooled. Although we have a natural ability to sense hot and cold, we can only use our sense of touch to tell whether an object is hot or cold over a relatively narrow range of temperatures. But, we will now need to *quantify* our intuitive sense of hotness. Recall that any characteristic of a material or object that is measurable can be referred to as a **quantity** or **measurable property.** In other words, a measurable property is one that can be quantified through physical comparison with a reference (Sections 1-1 and 1-2). Careful observation of our everyday world tells us that some objects such as a balloon full of air or a metal rod have characteristics that change as the object gets hotter or colder. For example, put a balloon full of air into the freezer and you can observe for yourself that it gets smaller. A metal rod may grow a little longer when heated. Volume, length, electrical resistance, and pressure are examples of measurable properties of a material object that can change with temperature. We can use any one of the properties of materials that change as an object gets hotter or colder to design crude (or not so crude) devices that quantify the hotness of an object. Such a device is called a **thermometer.**

Designing an accurate thermometer is not a trivial task. Nevertheless, thermometers are common devices and so many of us have a basic, "common sense" understanding of what a thermometer is and how to use one. We begin our study of thermodynamics with temperature. We will also reconsider this topic later in the chapter when we will refine and expand our understanding of the concept of temperature and its measurement.

Most, but not all, substances expand when heated. You can loosen a tight metal lid on a jar by holding it under a stream of hot water. Both the metal of the lid and the glass of the jar expand as the hot water transfers some of its hidden (or *internal*) thermal energy to both the jar and lid. This happens because with the added energy, the atoms in the lid can move a bit farther from each other than usual, pulling against the spring-like interatomic forces that hold every solid together. (See Chapter 6 if you need to remind yourself of our spring-atom model for solids.) However, the metal lid expands more than the glass, and so the lid is loosened. The familiar sealed liquid-in-glass thermometer (Fig. 19-1a) works in a similar way. The mercury or colored alcohol contained in the hollow glass bulb and tube expands more than the glass that surrounds it.

We call the transfer of energy from a "hotter" system to a "colder" system by invisible atomic and molecular collisions and chemical reactions **thermal energy transfer.** It is important to note that thermal energy transfers often involve no exchange of matter. For example, in the situation described above, no water penetrates the jar or its lid. How is this possible? Imagine a high kinetic energy cue ball colliding with a ball at rest in the middle of a billiard table. The cue ball loses all its energy and the other ball gains the amount that was lost—a massless transfer! So we can imagine that if the molecules in the hot water are vibrating more vigorously than the glass molecules in the jar, energy is transferred from the water molecules to the jar molecules.

Suppose you dip an unmarked liquid thermometer filled with red colored alcohol into a cup of cold water. Since the water feels cold to the touch you can make a scratch in the tube where the liquid stands and define that height as a cold temperature. Then you can transfer your unmarked thermometer to a cup of water that feels hot to the touch. You make another scratch where the liquid now stands and define that height as a hot temperature. What if you dip the thermometer into a cup of water that feels neither hot nor cold to your touch and the height of the alcohol is halfway between? Is it reasonable to assume that the new temperature is halfway between the two original temperatures? The assumption that this is true has guided the development of the historical Fahrenheit and Celsius temperature scales, and it is not too far from the truth. Thus, it is useful to start our study of thermodynamics with a very crude definition of temperature change as *a quantity that is proportional to changes of the height of the liquid inside a thermometer.* In order to quantify temperature, we need to assign numbers to various heights of the liquid in our glass tube. That is, we must set up a *temperature scale.*

(*a*)

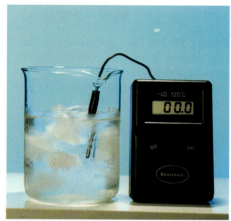

(*b*)

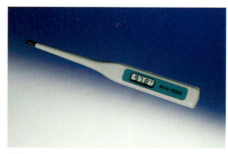

(*c*)

FIGURE 19-1 ■ Two types of thermometers based on changes in measurable properties of materials with hotness. (*a*) A liquid thermometer in which the liquid in a tube expands more than the glass that contains it when placed in hotter surroundings. (*b*) and (*c*) Electronic thermometers in which the electrical resistance of a sensing element embedded at the end of a thin rod changes when placed in hotter surroundings.

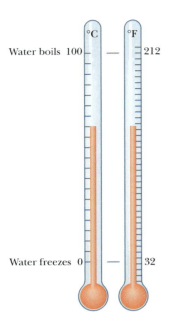

°C   °F

Water boils 100 ─── ─── 212

Water freezes 0 ─── ─── 32

FIGURE 19-2 ■ The Celsius and Fahrenheit temperature scales compared for the freezing and boiling points of water at standard atmospheric pressure at sea level. Each tic mark represents 5°.

The Fahrenheit and Celsius scales are the two temperature scales in common use today. You are probably familiar with the Fahrenheit scale from U.S. weather forecasts and you may have worked with the Celsius scale in other science courses. Each scale is set up using a reproducible low temperature "fixed point" and another at a higher temperature. A temperature is assigned to each of the two fixed points on the thermometer column. Then the distance between the two fixed points is divided into equally spaced **degrees.**

The Fahrenheit Scale

The scale set up by Gabriel Fahrenheit in 1714 starts with a zero point (0°F) defined to be the lowest temperature attainable by a mixture of ice and salt. His upper fixed point was set at human body temperature and defined as 96°F. So Fahrenheit put 96 divisions along the glass tube between his two fixed points. The Fahrenheit scale has proven to be quite awkward. For example, the freezing point of water at sea level turns out to be about 32°F and the boiling point of water at sea level turns out to be about 212°F as shown in Fig. 19-2.

The Celsius Scale

In 1742 a Swedish investigator named Celsius devised a more sensible scale that he called the centigrade scale. Celsius defined the freezing point of water at sea level as 0°C and the boiling point as 100°C. The modern Celsius scale was developed based on a degree that is almost the same "size" as the centigrade scale. However, it has been adjusted so one of its fixed points is the triple point of water. The triple point of water is the temperature and pressure at which solid ice, liquid water, and water vapor coexist. The triple point temperature has been defined as 0.01°C. So, a modern Celsius thermometer and the historical centigrade thermometer will give essentially the same reading. We discuss the triple point in more detail later in this chapter. Another refinement has been to define a standard value for the atmospheric pressure at sea level to help make the boiling point temperature more stable.

The Kelvin Scale

A 19th-century British physicist, Lord Kelvin, discovered that there is a natural limit to how cold any object can get. Kelvin defined an important temperature scale used in thermodynamics that is based on this natural zero point for temperature. We discuss the Kelvin scale used by scientists and engineers in more detail in Section 19-9.

Temperature Conversions

The Fahrenheit scale employs a smaller degree than the Celsius scale and a different zero of temperature. You can easily verify both these differences by examining an ordinary room thermometer on which both scales are marked as shown in Fig. 19-2. The equation for converting between these two scales can be derived quite easily by remembering a few corresponding fixed-point temperatures for each scale (see Table 19-1 or Fig. 19-2). The equation is

$$T_F = \tfrac{9}{5}T_C + 32°, \qquad (19\text{-}1)$$

where T_F is Fahrenheit temperature and T_C is Celsius temperature.

TABLE 19-1
Some Corresponding Temperatures

Temperature	°C	°F
Boiling point of water[a]	100	212
Normal body temperature	37.0	98.6
Accepted comfort level	20	68
Freezing point of water[a]	0	32
Zero of Fahrenheit scale	≈ -18	0
Scales coincide	-40	-40

[a]Strictly, the boiling point of water on the Celsius scale is 99.975°C, and the freezing point is 0.00°C. Thus there is slightly less than 100°C between those two points.

In general, people use the letters C and F to distinguish measurements and degrees on the two scales. Thus,

$$0°C = 32°F \qquad (19\text{-}2)$$

means that 0° on the Celsius scale measures the same temperature as 32° on the Fahrenheit scale. We might also say

$$\Delta T = 5 \, C° = 9 \, F°$$

which means that a temperature difference of 5 Celsius degrees (note the degree symbol appears *after* C) is equivalent to a temperature difference of 9 Fahrenheit degrees.

Problems with the Initial Definition of Temperature

There are several problems with basing our definition of temperature on the height of liquid in a scaled liquid thermometer: (1) the historically chosen fixed points are not highly reproducible since we will find that freezing and boiling points depend on pressure, (2) the height range over which liquids can vary in a glass tube is quite limited (liquids freeze when very cold or vaporize when very hot), (3) the assumption that liquids expand in proportion to temperature may not always be accurate and, the glass bulb thermometer must be small enough compared to the system that it doesn't affect it. We will return to our discussion of how to define temperature and design a better thermometer in Section 19-9. Meanwhile, we use our initial definition of temperature as the height of a liquid in a hollow glass thermometer as the starting point for our study of thermal interactions.

READING EXERCISE 19-1: List several additional measurable properties of an object. List several properties of an object that you believe are not measurable. ■

19-3 Thermal Interactions

Now that we have a means to measure temperature we can explore what happens to temperatures when two systems come into contact with each other. Consider the following two observations:

1. When a metal bucket of hot water is placed in a room, the reading of a thermometer placed in the water always decreases until, eventually, it matches the reading of a (assumed identical) thermometer in the room.

2. When a bucket of cold water is placed in a room, the reading of a thermometer in the bucket always increases until, eventually, it matches the reading of a thermometer in the room.

Our common sense tells us these observations indicate that there has been an interaction between the bucket of water and its surroundings. These interactions are examples of **thermal interactions.** We will call the bucket of water our *system.* We first introduced the term "system" in Chapter 8 in regard to the concept of conservation of momentum. Just as we did there, we will define a **system** to be the object or objects that are the primary focus of our interest. That, of course, means that we can make choices about what objects we include in our system. Just as in our study of conservation of energy and momentum, we will often be considering the interaction between two or more systems that can exchange energy with each other without exchanging matter. At other times we will consider the interaction between a single system and its surroundings.

The **environment** is defined as a system's surroundings or everything outside of it. Many thermal interactions of interest are interactions between a system (for example, the bucket of water) and its environment (for example, the air in the room). For practical purposes, the environment can often be taken as a system's immediate surroundings (without extending our consideration to the entire universe). All we must do is to be sure that the surroundings are much bigger than the system. For example, the environment for a bucket of water could be the room full of air that surrounds it, rather than the entire planet.

Let's now consider two additional observations for objects that are insulated from the surrounding environment:

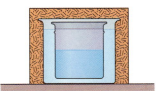

FIGURE 19-3 ■ Two vessels of water that have different temperatures at first are placed in thermal contact. They are surrounded by an insulating Styrofoam cup with a lid so that they don't interact significantly with their surroundings (room air). What happens to their temperatures?

3. When two similar objects at different temperatures are brought into contact with one another, the thermometer reading for the hotter object decreases and the thermometer reading of the colder object increases until the two readings are the same. (See Figs. 19-3 and 19-4).

4. When two similar objects at the same temperature are brought together, no changes in the thermometer readings occur.

All four of the observations discussed above include examples of a condition we will call *thermal equilibrium.* If two objects (for example, our system and the surrounding environment) produce the same thermometer readings (assuming identical thermometers) then the two objects are said to be in **thermal equilibrium.** That is, two objects are in thermal equilibrium if they have the same temperature. Observing the tendency of systems in contact to reach thermal equilibrium suggests that the hotter system is transferring thermal energy to the colder one until they both become warm.

The concept of thermal equilibrium is important in understanding temperature measurements. For example, as you likely know from common experience, it takes some time for a thermometer reading to stabilize. If you place a glass bulb thermometer under your tongue, it does not immediately measure your correct body temperature. This is (at least in part) because it takes some time for a thermometer to reach thermal equilibrium with another object like your body. When a thermometer comes to thermal equilibrium with an object that is not heating up or cooling down, the thermometer reading will reach a constant value. Only then do we have an accurate measurement of the object's temperature.

Thermal equilibrium is important in the measurement of temperature in another way. Consider three objects: object A, object B and object T. (Object T may be a thermometer, but it doesn't need to be.) Object A and object B are placed in separate, well-insulated environments as shown in Fig. 19-5. We measure the temperatures of the three objects, two at a time. Suppose we find that object A and object T are in thermal equilibrium. We then compare the temperatures of object B and object T and

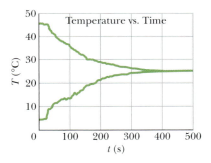

FIGURE 19-4 ■ A system consisting of $m_1 = 200$ g of cool water at about 5°C in a thin plastic cup is brought into thermal contact with another system consisting of $m_2 = 200$ g of warm water in a larger insulated container at about 45°C. A computer-based data acquisition system is used to monitor their temperatures. Even though no matter is exchanged between the systems (as shown in Fig. 19-3), the temperature of the cool water rises while that of the warm water falls until they reach thermal equilibrium at about 25°C, 400 seconds later.

find that they are in thermal equilibrium with each other as well (at the same thermometer reading as object A). Are object A and object B then necessarily in thermal equilibrium too? Experimentation provides the answer, which is referred to as the **zeroth law of thermodynamics:**

> If bodies A and B are each in thermal equilibrium with a third body T, then they are in thermal equilibrium with each other.

If we choose object T to be a thermometer, we see that we use the zeroth law constantly in science laboratories. It is the basis of our acceptance of the use of thermometers to compare temperatures. For example, if we want to know whether the liquids in two different students' beakers are at the same temperature, we can measure the two temperatures separately with a single thermometer and compare them. Often when measuring temperature, the thermometer starts at room temperature and its temperature must rise or fall when it is placed in contact with body A or body B. Again, we must be careful not to have the thermometer system be so large that either body loses or gains enough thermal energy to change its temperature noticeably.

19-4 Heating, Cooling, and Temperature

Is just measuring the initial and final temperature of an object that is changing temperature a good way to learn about how objects heat or cool? In order to answer this question, let us consider several examples of how water cools:

1. A small cup of very hot water is put in a room that is being maintained at a comfortable air temperature of 20°C. The water will cool down until it reaches thermal equilibrium with the surrounding air.

2. We place a large bucket of hot water in the same room, it will also cool down, but it will take longer than the cup of water does to cool.

3. We first place a small cup of hot water in a sealed thermos bottle and then place the bottle in the room, the water will still reach thermal equilibrium with the room but it will take much longer to cool down to room temperature than it did before.

4. We put a cup containing water at 30°C in a very cold freezer. We find the water freezes and becomes ice. During the freezing process we keep the water-ice mixture well-stirred and measure its temperature as a function of time. As shown in Fig. 19-6, once the mixture of ice and water are cooled to 0°C, the mixture undergoes no change in temperature until all the ice is frozen.

The examples above are evidence that our ability to analyze thermal interactions is significantly limited if we rely solely on initial and final temperature measurements. Hence, we are motivated to invent a new, broader concept that we can use in discussing thermal interactions—even in cases where temperature changes are not the significant feature. The name of the process we will introduce is "heating," which we can use to describe the interaction between a hotter body and colder body, *even if no temperature change occurs*. For example, if you surround ice water with air or water which has a higher temperature, the ice will melt. Or, heating cold water takes longer if it is insulated from its environment than if it is not. Observations such as those above form the basis of our understanding of the *process* called heating.

In order to gain some additional insight into the process of heating, consider two more observations.

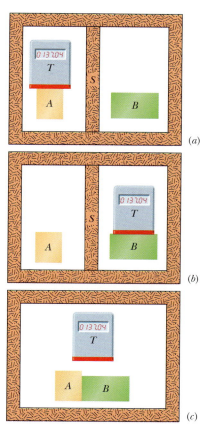

FIGURE 19-5 ■ (*a*) Body T (perhaps a thermometer) and body A are in thermal equilibrium. (Body S is a thermally insulating screen.) (*b*) Body T and body B are also in thermal equilibrium (*c*) If (*a*) and (*b*) are true, the zeroth law of thermodynamics states that body A and body B are also in thermal equilibrium.

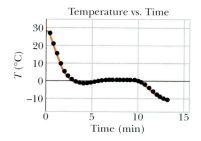

FIGURE 19-6 ■ A temperature vs. time graph recorded by a data acquisition system shows what happens to a cup of water after it is placed in a cold freezer. When the water temperature decreases to 0°C ice begins to form. While ice is forming the temperature does not change. Once all the water is changed to ice the temperature starts decreasing again.

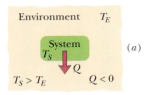

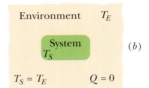

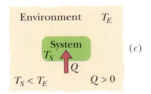

FIGURE 19-7 ▪ If the temperature of a system exceeds that of its environment as in (a), an amount of thermal energy Q is transferred by the system to the environment until thermal equilibrium as shown in part (b) is established. (c) If the temperature of the system is below that of the environment, thermal energy is absorbed by the system until thermal equilibrium is established.

FIGURE 19-8 ▪ In the United States we have not adopted the international system of units when referring to food energy, so the **food calorie** (denoted Cal rather than cal) is used in diet books and in all government-regulated food labels is actually a kilocalorie. Most other countries use *joules* (the accepted SI unit for energy) on food labels. The can in the photo above was purchased in Australia.

1. A container of water is placed on a burner to raise its temperature. It takes fuel (an energy source) for this process to occur.

2. If we start with two containers of water, one large and one small, at the same temperature and place them over identical burners, it takes a longer time (and so more fuel) to elevate the temperature of the larger amount of water to the same final value as the smaller.

Both of these observations are direct indications that there is energy involved in heating, just as we assumed in Chapter 10 when discussing energy conservation. Furthermore, the idea that the heating process is a *transfer of energy* from a hotter object to a colder object is consistent with all the other observations we have discussed in this chapter. Hence, we are led then to this important statement:

> **Heating** is the transfer of energy from a system with a higher temperature to one with lower temperature that is in contact with it that occurs simply because a temperature difference exists between the systems.

The net amount of transferred thermal energy as the result of microscopic energy exchanges between systems (or parts of a system) is denoted by the letter Q. The thermal energy transferred between systems is often called "heat." Because the term *heat* is sometimes used casually to mean the total thermal energy in a system, people often confuse heat with internal energy when describing thermal interactions. For this reason we try to avoid passive terms like *heat* or *stored heat* when describing thermal interactions. Since the phrase "heating a system" suggests an active process we will use terms like "heating," "cooling," or "thermal energy transfer" to refer to the additional energy that is added to (or subtracted from) a system through microscopic energy exchanges.

> The amount of transferred thermal energy Q is taken to be *positive* when it is transferred *to a system* (then we say that thermal energy is absorbed). The transferred thermal energy Q is *negative* when it is transferred *from the system* (we then say that thermal energy is released or lost by the system).

This transfer of energy is shown in Fig. 19-7. In the situation of Fig. 19-7a, in which the temperature of the system T_S is greater than the temperature of the environment T_E ($T_S > T_E$), energy is transferred from the system to the environment, so Q is negative. In Fig. 19-7b, in which $T_S = T_E$, no thermal energy transfer takes place, Q is zero, and energy is neither released nor absorbed. In Fig. 19-7c, in which $T_S < T_E$, the transfer is to the system from the environment, so Q is positive.

It took scientists a while to realize that "heating" was associated with the transfer of energy from one system to another. Hence, thermal energy ended up with its own unit. Since heating was initially considered strictly in terms of temperature change, the **calorie** (cal) was defined as the amount of energy that would raise the temperature of 1 g of water from 14.5°C to 15.5°C. In the British system, the corresponding unit of thermal energy was the **British thermal unit** (Btu), defined as the amount of energy that would raise the temperature of 1 lb of water from 63°F to 64°F.

In 1948, the scientific community decided that since heating (like work) is an energy transfer process, the SI unit for thermal energy transferred (or "heat added") should be the one we use for all other energy—namely, the **joule**. Joules are used instead of calories in most countries, as seen on the can of soda from Australia in Fig. 19-8. The calorie is now defined to be 4.1860 J (exactly), with no reference to the heating of water. (The "calorie" used in nutrition, sometimes called the Calorie with a capital C or Cal for short, is really a kilocalorie or 1000 calories.) The relations among the various thermal energy units are

$$1 \text{ cal } = 3.969 \times 10^{-3} \text{ Btu } = 4.1860 \text{ J} = 0.001 \text{ Cal (or food calorie).} \quad (19\text{-}3)$$

Mechanisms for Transfer of Thermal Energy

We have discussed heating, which is the transfer of energy between a system and its environment that takes place simply because a temperature difference exists between them, but we have not yet described how that transfer takes place. We will briefly describe the three heating mechanisms here and return to discuss them more fully at the end of this chapter.

If you leave the end of a metal poker in a fire for enough time, its handle will get hot. There are large vibrations of the atoms and electrons of the metal at the fire end of the poker because of the high temperature of their environment. These increased vibrational amplitudes, and thus the associated energy, are passed along the poker, from atom to atom, during collisions between adjacent atoms. In this way, a region of increasing temperature extends itself along the poker to the handle. It is important to note though that there has been no flow of matter—only energy that is transmitted along the poker. This type of heating process is called (thermal) **conduction.**

Conduction is, by definition, a transfer mechanism that requires direct contact between two objects at different temperatures. Consider our poker to be a series of systems that can have different temperatures. If there is not direct contact between the colder object and hotter object (one atom and the next in our poker example), there cannot be a transfer of thermal energy by conduction. As we mentioned above, *conduction does not involve any mass transfer*. However, in some real situations it is sometimes impossible to avoid the transfer of material from one object to the other during heating. In such cases, the process is no longer one of "pure" conduction.

When you open a low and a high window in a heated house, you can feel cold outside air rushing into the room through the low window and warm room air rushing outside through the high window. This is because the cold air is more dense than the warm air so it displaces the warm air at the bottom of the room. The warm air rises as a result of buoyant forces and flows out the top window. The room gets colder because of the exchange of cold and warm air. In this situation, thermal energy is being transported by the flow of matter (air currents in this case). The transfer by the exchange of hotter and cooler fluids is known as **convection.**

Examples of heating by convection are everywhere. This is how air circulation helps spread the warmth through a room when a radiator or heater gets hot. It is why *all* the water in a tea kettle gets hot (as opposed to only the water in contact with the hot kettle surface) when the kettle is placed on a hot stove. Convection is part of many other natural processes. Atmospheric convection plays a fundamental role in determining global climate patterns and daily weather variations. Glider pilots and birds alike seek rising thermals (convection currents of warm air) that keep them aloft. Huge energy transfers take place within the oceans by the same process. Finally, energy is transported to the surface of the sun from the nuclear furnace at its core by enormous cells of convection, in which hot gas rises to the surface along the cell core and cooler gas around the core descends below the surface. In all examples of heating by convection, gravitational forces play a vital role. Without gravitational forces, hotter materials would not rise above cooler materials. Hence, there is no heating by convection on a space station.

When you sit in the sun you can feel your skin getting warmer. If you put a shield between you and the sun, the sensation of warmth immediately disappears. How is this energy transfer taking place? We know that the light from the sun has to pass through over a hundred million kilometers of almost empty space. So, this thermal energy transfer can't be attributed to either convection or conduction. Solar energy transfer is attributed to a third transfer process—the absorption of **electromagnetic radiation.** Although visible light is one kind of electromagnetic radiation, the sun and a hot fire that can also warm you emit both visible light and invisible infrared radiation that has a longer characteristic wavelength than light. (See Chapter 34 for more details on electromagnetic waves.) No medium is required for energy transfer via electromagnetic radiation. Thermal energy transferred by infrared electromagnetic waves is often called **thermal radiation** or **radiant energy.**

READING EXERCISE 19-2: Explain the function of insulation in homes. ■

19-5 Thermal Energy Transfer to Solids and Liquids

If we start with two containers of water, one large and one small, at the same temperature and place them to heat over identical burners, it takes a longer time (and so more fuel and thus more thermal energy) to elevate the temperature of the larger amount of water to the same final value as the smaller. It takes almost twice as much energy to heat a cup of water as it does to heat a cup of motor oil to the same temperature. This is shown in Fig. 19-9a and b. A common electric immersion heater is placed in a container of motor oil and plugged in. It puts out a bit less than 200 W of power at a constant rate. Since energy is power × time the total amount of thermal energy transferred to the oil is directly proportional to the time the heater has been on. The graph in Fig. 19-9b shows that after 70 seconds, the change in water temperature, $T_f - T_i$, is about 15°C while the oil temperature has risen about 30°C. We can say the "water holds its heat" twice as well as motor oil. It would be more correct to say that water absorbs the thermal energy transferred to it without showing as much temperature change.

If we look at heating and cooling curves for a large number of different objects and different substances (see Fig. 19-9b for an example), we find that the relationship between thermal energy transfer and temperature is linear (as long as the material doesn't melt, freeze, or vaporize). However, the scaling factor (proportionality constant) changes from material to material. The **heat capacity** C of an object is the name that we give to the proportionality constant between the thermal energy transferred Q and the resulting temperature change ΔT of the object; that is,

$$Q = C \Delta T = C(T_f - T_i), \tag{19-4}$$

in which T_i and T_f are the initial and final temperatures of the object. Heat capacity C has the unit of energy per degree Celsius. The heat capacity C of, say, a marble slab used in a bun warmer might be 179 cal/C°.

The word "capacity" in this context is really misleading in that it suggests analogy with the capacity of a bucket to hold water. Since heat is not a substance, *that analogy is misleading,* and you should not think of the object as "containing" thermal energy or being limited in its ability to absorb thermal energy. Thermal energy transfer can proceed without limit as long as the necessary temperature difference between the object and its surroundings is maintained. The object may, of course, melt or vaporize during the process.

Specific Heat

Two objects made of the same material—say, marble—will have heat capacities proportional to their masses. It is therefore convenient to define a "heat capacity per unit mass" or **specific heat** c that refers not to an object but to a unit mass of the material of which the object is made. Equation 19-4 then becomes

$$Q = cm \Delta T = cm(T_f - T_i), \tag{19-5}$$

where $C = mc$.

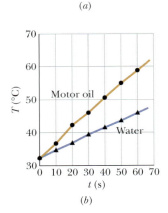

(a)

(b)

FIGURE 19-9 ■ (a) A computer-based data collection system with a digital temperature sensor can be used to monitor the temperature rise in a liquid while thermal energy is transferred to it at a constant rate by an immersion heater. The liquid in this photo is motor oil. (b) If thermal energy is transferred to two different liquids that have the same mass, the temperature does not usually rise at the same rate. So we define a different specific heat for each liquid. For example, a small immersion heater shown in Fig. 19-9a "heats up" 175 g of motor oil faster than 175 g of water. Data were taken once each second for 70 seconds with a computer-based data collection system.

Through experiment we would find that although the heat capacity of a particular marble slab might be 179 cal/C°, the specific heat of the marble (in that slab or in any other marble object) is 0.21 cal/g · C°.

From the way the calorie and the British thermal unit were initially defined, measurements show that the specific heat of water is

$$c = 1 \text{ cal/g} \cdot \text{C}° = 1 \text{ Btu/lb} \cdot \text{F}° = 4190 \text{ J/kg} \cdot \text{C}°. \tag{19-6}$$

Table 19-2 shows the specific heats of some substances at room temperature. Note that the value for water is relatively high. The specific heat of any substance actually depends somewhat on temperature, but the values in Table 19-2 apply reasonably well in a range of temperatures near room temperature.

TABLE 19-2
Specific Heats of Some Substances at Room Temperature and Constant Pressure

Substance	Specific Heat $\dfrac{\text{cal}}{\text{g} \cdot \text{C}°}$	$\dfrac{\text{J}}{\text{kg} \cdot \text{C}°}$	Molar Specific Heat $\dfrac{\text{J}}{\text{mol} \cdot \text{C}°}$
Elemental Solids			
Lead	0.0305	128	26.5
Tungsten	0.0321	134	24.8
Silver	0.0564	236	25.5
Copper	0.0923	386	24.5
Aluminum	0.215	900	24.4
Other Solids			
Brass	0.092	380	
Granite	0.19	790	
Glass	0.20	840	
Ice (-10°C)	0.530	2220	
Liquids			
Mercury	0.033	140	
Ethyl alcohol	0.58	2430	
Seawater	0.93	3900	
Water	1.00	4190	

Molar Specific Heat

In many instances the most convenient unit for specifying the amount of a substance is the mole (mol), where

$$1 \text{ mol} = 6.02 \times 10^{23} \text{ basic units}$$

of *any* substance. Thus 1 mol of aluminum means 6.02×10^{23} atoms (the atom being the elementary unit), and 1 mol of aluminum oxide means 6.02×10^{23} molecules of the oxide (because the molecule is the basic unit of a compound).

When quantities are expressed in moles, specific heats must also involve moles (rather than a mass unit); they are then called **molar specific heats.** Table 19-2 shows the values for some elemental solids (each consisting of a single element) at room temperature.

Specific Heats at Constant Pressure and Constant Volume

In determining and then using the specific heat of any substance, we need to know the conditions under which thermal energy is transferred. For solids and liquids, we usually assume that the sample is under constant pressure (usually atmospheric) during the transfer. It is also conceivable that the sample is held at constant volume while the thermal energy is absorbed. This means that thermal expansion of the sample is prevented by applying external pressure. For solids and liquids, this is very hard to arrange experimentally but the effect can be calculated, and it turns out that the specific heats under constant pressure and constant volume for any solid or liquid differ usually by no more than a few percent. Gases, as you will see, have quite different values for their specific heats under constant-pressure conditions and under constant-volume conditions.

Heats of Transformation

Recall our example above of the ice (solid water). "Solid" is a description of the water that we will call its **phase.** There are, in general, three phases of matter: solid, liquid, and vapor. As a *solid,* the molecules of a sample are locked into a fairly rigid structure by their mutual attraction. As a *liquid,* the molecules have more energy and move about more. They may form brief clusters, but the sample does not have a rigid structure and can flow or settle into a container. As a *gas* or *vapor,* the molecules have even more energy, are free of one another, and can fill up the full volume of a container. To fully describe a material for thermodynamic purposes we must specify not only the phase (solid, liquid, or vapor) but also the temperature, pressure, and volume. The phase of a material along with the temperature, pressure, and volume of the material specify the **state** of the material.

As we all know, we find that ice melts when exposed to a warm room and becomes liquid water. The melting is called a **change of phase.** When thermal energy is absorbed or lost by a solid, liquid, or vapor, the temperature of the sample does not necessarily change. Instead, the sample may change from one phase to another. Through experiments we have found that while a material is undergoing a change in phase additional transfers of thermal energy do not change the temperature of the material. We saw one example of this in Fig. 19-6.

To *melt* a solid means to change it from the solid phase to the liquid phase. The process requires energy because the molecules of the solid must be freed from their rigid structure. Melting an ice cube to form liquid water is a common example. To *freeze* a liquid to form a solid is the reverse of melting and requires that energy be removed from the liquid, so that the molecules can settle into a rigid structure. To *vaporize* a liquid means to change it from the liquid phase to the vapor or gas phase. This process, like melting, requires energy because the molecules must be freed from their clusters. Boiling liquid water transforms it to water vapor (or steam—a gas of individual water molecules) is a common example. *Condensing* a gas to form a liquid is the reverse of vaporizing; it requires that energy be removed from the gas, so that the molecules can cluster instead of flying away from one another.

The amount of energy per unit mass that must be transferred as thermal energy when a sample completely undergoes a phase change is called the heat of transformation L. Thus, when a sample of mass m completely undergoes a phase change, the total energy transferred is

$$Q = Lm. \tag{19-7}$$

When the phase change is from liquid to gas (then the sample must absorb thermal energy) or from gas to liquid (then the sample must release thermal energy), the thermal energy required for this transformation is called the **heat of vaporization L_V.*** For water at its normal boiling or condensation temperature,

$$L_V = 539 \text{ cal/g} = 40.7 \text{ kJ/mol} = 2256 \text{ kJ/kg} \qquad \text{(water} \rightarrow \text{steam).} \tag{19-8}$$

*Chemists often denote entropy as ΔH and call it enthalpy of vaporization.

When the phase change is from solid to liquid (then the sample must absorb thermal energy) or from liquid to solid (then the sample must release thermal energy), the heat of transformation is called the **heat of fusion** L_F. For water at its normal freezing or melting temperature,

$$L_F = 79.5 \text{ cal/g} = 6.01 \text{ kJ/mol} = 333 \text{ kJ/kg} \quad (\text{ice} \rightarrow \text{water}). \quad (19\text{-}9)$$

Table 19-3 shows the heat of transformation for some substances.

TABLE 19-3
Some Heats of Transformation

	Melting		Boiling	
Substance	Melting Point (°C)	Heat of Fusion L_F (kJ/kg)	Boiling Point (°C)	Heat of Vaporization L_V (kJ/kg)
Hydrogen	−259.2	58.0	−252.9	455
Oxygen	−218.4	13.9	−183.0	213
Mercury	−39.2	11.4	356.9	296
Water	−0.1	333	99.9	2256
Lead	327.9	23.2	1743.9	858
Silver	961.9	105	2049.9	2336
Copper	1082.9	207	2594.9	4730

Heating and Internal Energy

Consider melting a 0°C block of ice so that it melts and then becomes hot water. This system undergoes *both* a phase change and a temperature increase. What happens to the system's hidden internal energy that we mentioned in the introduction? Heating via thermal energy transfer increases its internal energy in two ways:

1. We change the system's microscopic configuration by allowing its atoms and molecules to move further away from each other (without a temperature change).

2. We increase its microscopic kinetic energy (or thermal energy) by increasing the motions of its atoms and molecules (which can be measured as a rise in temperature).

READING EXERCISE 19-3: If an amount of thermal energy Q is transferred to object A, it will cause each gram of A to rise in temperature by 3 C°. If the same amount of energy is transferred to object B, then the temperature of each gram of B will rise by 4 C°. If object A and B have the same mass, which one has the greater specific heat? ■

READING EXERCISE 19-4: Notice in Table 19-3 that water has a very large heat of vaporization. What is the significance of this to a firefighter who finds that the water sprayed on a very hot fire is converted to steam? Suppose that 1 g of steam comes in contact with 1 g of a firefighter's flesh and condenses. Assuming that flesh has the same heat capacity as water, what would be the temperature rise in 1 g of the firefighter's flesh? ■

TOUCHSTONE EXAMPLE 19-1: Melting Ice

(a) How much thermal energy must be absorbed by ice of mass $m = 720$ g at −10°C to take it to a liquid state at 15°C?

SOLUTION ■ The first **Key Idea** is that the heating process is accomplished in three steps.

Step 1. The **Key Idea** here is that the ice cannot melt at a temperature below the freezing point—so initially, any thermal energy transferred to the ice can only increase the temperature of the ice. The energy Q_1 needed to increase that temperature from the initial value $T_i = −10$°C to a final value $T_f = 0$°C (so that the ice can then melt) is given by Eq. 19-5 ($Q = cm \, \Delta T$). Using the specific heat of ice c_{ice} in Table 19-2 gives us

$$Q_1 = c_{ice}m(T_f - T_i)$$
$$= (2220 \text{ J/kg} \cdot {}^\circ\text{C})(0.720 \text{ kg})[0{}^\circ\text{C}-(-10{}^\circ\text{C})]$$
$$= 15 \, 984 \text{ J} = 15.98 \text{ kJ}.$$

Step 2. The next **Key Idea** is that the temperature cannot increase from 0°C until all the ice melts—so any energy transferred to the ice due to heating now can only change ice to liquid water. The thermal energy Q_2 needed to melt all the ice is given by Eq. 19-7 ($Q = Lm$). Here L is the heat of fusion L_F, with the value given in Eq. 19-9 and Table 19-3. We find

$$Q_2 = L_F m = (333 \text{ kJ/kg})(0.720 \text{ kg}) = 239.8 \text{ kJ}.$$

Step 3. Now we have liquid water at 0°C. The next **Key Idea** is that the energy transferred to the liquid water during heating now can only increase the temperature of the liquid water. The Q_3 needed to increase the temperature of the water from the initial value $T_i = 0°C$ to the final value $T_f = 15°C$ is given by Eq. 19-5 (with the specific heat of liquid water c_{liq}):

$$Q_3 = c_{liq}m(T_f - T_i)$$
$$= (4190 \text{ J/kg} \cdot {}^\circ\text{C})(0.720 \text{ kg})(15{}^\circ\text{C} - 0{}^\circ\text{C})$$
$$= 45 \, 252 \text{ J} \approx 45.25 \text{ kJ}.$$

The total required thermal energy transfer Q^{tot} is the sum of the amounts required in the three steps:

$$Q^{tot} = Q_1 + Q_2 + Q_3$$
$$= 15.98 \text{ kJ} + 239.8 \text{ kJ} + 45.25 \text{ kJ}$$
$$\approx 300 \text{ kJ}. \qquad \text{(Answer)}$$

Note that the energy transfer required to melt the ice is much greater than the energy transfer required to raise the temperature of either the ice or the liquid water.

(b) If we supply the ice with a total energy of only 210 kJ (as heat), what then are the final state and temperature of the water?

SOLUTION ■ From step 1, we know that 15.98 kJ is needed to raise the temperature of the ice to the melting point. The remaining energy required Q^{rem} is then 210 kJ − 15.98 kJ, or about 194 kJ. From step 2, we can see that this amount of energy is insufficient to melt all the ice. Then this **Key Idea** becomes important: Because the melting of the ice is incomplete, we must end up with a mixture of ice and liquid; the temperature of the mixture must be the freezing point, 0°C.

We can find the mass m of ice that is melted by the available energy Q^{rem} by using Eq. 19-7 with L_F:

$$m = \frac{Q^{rem}}{L_F} = \frac{194 \text{ kJ}}{333 \text{ kJ/kg}} = 0.583 \text{ kg} \approx 580 \text{ g}.$$

Thus, the mass of the ice that remains is 720 g − 580 g, or 140 g, and we have

$$580 \text{ g water and } 140 \text{ g ice at } 0°C. \qquad \text{(Answer)}$$

TOUCHSTONE EXAMPLE 19-2: Copper Slug

A copper slug whose mass m_c is 75 g is heated in a laboratory oven to a temperature T of 312°C. The slug is then dropped into a glass beaker containing a mass $m_w = 220$ g of water. The heat capacity C_b of the beaker is 45 cal/K. The initial temperature T_i of the water and the beaker is 12°C. Assuming that the slug, beaker, and water are an isolated system and the water does not vaporize, find the final temperature T_f of the system at thermal equilibrium.

SOLUTION ■ One **Key Idea** here is that, with the system isolated, only transfers of thermal energy can occur. There are three such transfers, all as thermal energy. The slug loses energy, the water gains energy, and the beaker gains energy. Another **Key Idea** is that, because these transfers do not involve a phase change, the energy transfers can only change the temperatures. To relate the transfers to the temperature changes, we can use Eqs. 19-4 and 19-5 to write

$$\text{for the water:} \quad Q_w = c_w m_w(T_f - T_i); \qquad (19\text{-}10)$$

$$\text{for the beaker:} \quad Q_b = C_b(T_f - T_i); \qquad (19\text{-}11)$$

$$\text{for the copper:} \quad Q_c = c_c m_c(T_f - T). \qquad (19\text{-}12)$$

A third **Key Idea** is that, with the system thermally isolated, the total energy of the system cannot change. This means that the sum of these three thermal energy transfers is zero:

$$Q_w + Q_b + Q_c = 0. \qquad (19\text{-}13)$$

Substituting Eqs. 19-10 through 19-12 into Eq. 19-13 yields

$$c_w m_w(T_f - T_i) + C_b(T_f - T_i) + c_c m_c(T_f - T) = 0. \qquad (19\text{-}14)$$

Temperatures are contained in Eq. 19-14 only as differences. Thus, because the differences on the Celsius and Kelvin scales are identical, we can use either of those scales in this equation. Solving it for T_f, we obtain

$$T_f = \frac{c_c m_c T + C_b T_i + c_w m_w T_i}{c_w m_w + C_b + c_c m_c}.$$

Using Celsius temperatures and taking values for c_c and c_w from Table 19-2, we find the numerator to be

$$(0.0923 \text{ cal/g} \cdot °C)(75 \text{ g})(312°C) + (45 \text{ cal/°C})(12°C)$$

$$+ (1.00 \text{ cal/g} \cdot °C)(220 \text{ g})(12°C) = 5339.8 \text{ cal},$$

and the denominator to be

$$(1.00 \text{ cal/g} \cdot °C)(220 \text{ g}) + 45 \text{ cal/°C} + (0.0923 \text{ cal/g} \cdot °C)(75 \text{ g})$$

$$= 271.9 \text{ cal/°C}.$$

We then have

$$T_f = \frac{5339.8 \text{ cal}}{271.9 \text{ cal/°C}} = 19.6°C \approx 20°C.$$

From the given data you can show that

$$Q_w \approx 1670 \text{ cal}, \qquad Q_b \approx 342 \text{ cal}, \qquad Q_c \approx -2020 \text{ cal}.$$

Apart from rounding errors, the algebraic sum of these three thermal energy transfers is indeed zero, as Eq. 19-13 requires.

19-6 Thermal Energy and Work

Here we look in some detail at how internal energy can be transferred into or out of both as thermal energy and as macroscopic physical work (involving the displacement of the system in the presence of net forces). Let us take as our system a gas confined to a cylinder with a movable piston, as in Fig. 19-10. This is the kind of device that is used to drive a steam engine, the engine of an automobile, and many other tools—all of which convert thermal energy into work.

In our piston–cylinder system, the upward force on the piston due to the pressure of the confined gas is equal to the weight of lead shot loaded onto the top of the piston. The walls of the cylinder are made of insulating material that does not allow any transfer of thermal energy. The bottom of the cylinder, however, rests on a reservoir of thermal energy—a *thermal reservoir* (perhaps a hot plate) whose temperature T can be held constant.

The system (the gas) starts from an *initial state i*, described by a pressure P_i, a volume V_i, and a temperature T_i. You want to change the system to a final state f, described by a pressure P_f, a volume V_f, and a temperature T_f. The procedure by which you change the system from its initial state to its final state is called a *thermodynamic process*. During such a process, an amount of energy Q may be transferred into or out of the system from the thermal reservoir. Also, work can be done by the system to raise the loaded piston (positive work) or lower it (negative work). *We assume that all such changes occur slowly, so that all parts of the system are always in (approximate) thermal equilibrium.*

Suppose that you remove a few of the lead shot from the piston of Fig. 19-10, allowing the gas to push the piston and remaining shot upward through a differential displacement $d\vec{s}$ with an upward force $\vec{F}$. Since the displacement is tiny, we can assume that $\vec{F}$ is constant during the displacement. Then $\vec{F}$ has a magnitude that is equal to PA, where P is the pressure of the gas and A is the face area of the piston. The differential work dW done by the gas during the displacement is

$$dW = \vec{F} \cdot d\vec{s} = (PA)(|d\vec{s}|) = P(A|d\vec{s}|)$$
$$= P\,dV, \tag{19-15}$$

in which dV is the differential change in the volume of the gas due to the movement of the piston. When you have removed enough shot to allow the gas to change its volume from V_i to V_f, the total work done by the gas is

$$W = \int dW = \int_{V_i}^{V_f} P\,dV. \tag{19-16}$$

During the change in volume, the pressure and temperature of the gas may also change. To evaluate the integral in this expression directly, we need to know how pressure varies with volume for the actual process by which the system changes from state i to state f.

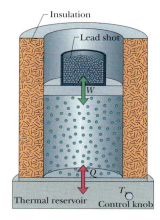

FIGURE 19-10 ■ Cross section of a system that confines a gas to a cylinder with a movable piston. (Its insulating lid is not shown.) Thermal energy Q can be transferred to or from the gas by regulating the temperature T of the adjustable thermal reservoir. Work W is done on the gas when the piston rises or falls.

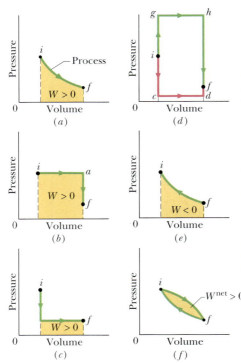

FIGURE 19-11 ■ (*a*) The shaded area represents the work *W* done by a system as it goes from an initial state *i* to a final state *f*. Work *W* is positive because the system's volume increases. (*b*) *W* is still positive, but now greater. (*c*) *W* is still positive, but now smaller. (*d*) *W* can be even smaller (path *icdf*) or larger (path *ighf*). (*e*) Here the system goes from state *i* to state *f*, as the gas is compressed to less volume by an external force. The work *W* done by the system is now negative. (*f*) The net work *W*^net done by the system during a complete cycle is represented by the shaded area.

There are actually many ways to take the gas from state *i* to state *f*. One way is shown in Fig. 19-11*a*, which is a plot of the pressure of the gas versus its volume and is called a *P-V* diagram. In Fig. 19-11*a*, the curve indicates that the pressure decreases as the volume increases. The integral in Eq. 19-16 (and thus the work *W* done by the gas) is represented by the shaded area under the curve between points *i* and *f*. Regardless of exactly what we do to take the gas along the curve, that work by the gas or system is positive, due to the fact that the gas increases its volume by forcing the piston upward.

Another way to get from state *i* to state *f* is shown in Fig. 19-11*b*: the change takes place in two steps—the first from state *i* to state *a* and the second from state *a* to state *f*.

Step *ia* of this process is carried out at constant pressure, which means that you leave undisturbed the lead shot that ride on top of the piston in Fig. 19-10. You cause the volume to increase (from V_i to V_f) by slowly turning up the temperature control knob, raising the temperature of the gas to some higher value T_a. (Increasing the temperature increases the force from the gas on the piston, moving it upward.) During this step, positive work is done by the expanding gas (to lift the loaded piston) and thermal energy is absorbed by the system from the thermal reservoir (in response to the arbitrarily small temperature differences that you create as you turn up the temperature). The thermal energy transferred (*Q*) is positive because it is added to the system.

Step *af* of the process of Fig. 19-11*b* is carried out at constant volume, so you must wedge the piston, preventing it from moving. Then as you use a control knob to decrease the reservoir temperature, you find that the pressure drops from P_a to its final value P_f. During this step, thermal energy is lost by the system to the thermal reservoir.

For the overall process *iaf*, the work *W*, which is positive and is carried out only during step *ia*, is represented by the shaded area under the curve. Energy is transferred as thermal energy during both steps *ia* and *af*, with a net thermal energy transfer *Q*.

Figure 19-11*c* shows a process in which the previous two steps are carried out in reverse order. The work *W* in this case is smaller than for Fig. 19-11*b*, as is the net thermal energy absorbed. Figure 19-11*d* suggests that you can make the work done by the gas as small as you want (by following a path like *icdf*) or as large as you want (by following a path like *ighf*).

To sum up: A system can be taken from a given initial state to a given final state by an infinite number of processes. Thermal energy transfers may or may not be involved, and in general, the work *W* and the thermal energy transfer *Q* will have different values for different processes. We say that thermal energy and work are *path-dependent* quantities.

Figure 19-11*e* shows an example in which negative work is done by a system as some external force compresses the system, reducing its volume. The absolute value of the work done is still equal to the area beneath the *P-V* curve, but because the gas is compressed, the work done by the gas is negative.

Figure 19-11*f* shows a thermodynamic cycle in which the system is taken from some initial state *i* to some other state *f* and then back to *i*. The net work done by the system during the cycle is the sum of the positive work done during the expansion and the negative work done during the compression as shown by the shaded area inside the path. In Fig. 19-11*f*, the net work is positive because the area under the expansion curve (*i* to *f*) is greater than the area under the compression curve (*f* to *i*).

READING EXERCISE 19-5: The *P-V* diagram here shows six curved paths (connected by vertical paths) that can be followed by a gas. Which two of them should be part of a closed cycle if the net work done by the gas is to be at its maximum positive value?

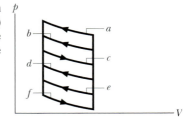

19-7 The First Law of Thermodynamics

You have just seen that when a system changes from a given initial state to a given final state, both the work W done by the system and the thermal energy transferred to the system Q depend on the nature of the process.* Experimentally, however, we find a surprising thing. The quantity $Q - W$ *is the same for all processes*. It depends only on the initial and final states and does not depend at all on how the system gets from one to the other. All other combinations of Q and W, including Q alone, W alone, $Q + W$, and $Q - 2W$, are path dependent; only the quantity $Q - W$ is not.

So, we come to believe that the quantity $Q - W$ must represent a change in some intrinsic property of the system. Since we know that both heating and work are energy transfer processes, we assume that the quantity $Q - W$ results in a change in energy. But a change in the energy of what? As we pointed out in the introduction to this chapter, matter has hidden within it energy associated with the motions and configurations of the microscopic atoms and molecules of which it is composed. We call this energy associated with the microscopic kinetic and potential energies of the system the **internal energy** E^{int}. So, we can then write

$$\Delta E^{\text{int}} = E_f^{\text{int}} - E_i^{\text{int}} = Q - W \qquad \text{(first law for } W = \text{work by system)} \qquad (19\text{-}17)$$

where Q represents the thermal or heat energy transferred *to the system* and W represents the work done on the surroundings *by the system*. This expression represents a very important relationship and is called the **first law of thermodynamics.** If the thermodynamic system undergoes only a differential change, we can write the first law as[†]

$$dE^{\text{int}} = dQ - dW \qquad \text{(first law).} \qquad (19\text{-}18)$$

> The internal energy E^{int} of a system increases if thermal energy Q is transferred to the system and decreases if thermal energy ($Q < 0$) is transferred from the system.
>
> The internal energy E^{int} of a system decreases if it does an amount of work $W > 0$ on its surroundings and increases when it has an amount of work $W < 0$ done on it by the surroundings. (Work done on a system is negative. Work done by a system is positive.)

The first law of thermodynamics should remind you of issues we first raised in Chapter 10. However, in Chapter 10, we discussed energy conservation as it applies to isolated systems—that is, to systems in which no energy enters or leaves the system. The first law of thermodynamics is an extension of that principle to systems that are *not* isolated. In such cases, energy may be transferred into or out of the system as some combination of macroscopic physical work W done by a system and transfer of the microscopic thermal energy Q into a system. In our statement of the first law of thermodynamics above, we assume that there are no changes in the *macroscopic* kinetic energy or the potential energy of the system as a whole; that is, $\Delta K = \Delta U = 0$.

Before this chapter, the term *work* and the symbol W always meant the work done on a system. However, starting with Eq. 19-15 above and continuing through the next two chapters about thermodynamics, we focus on the work done *by* a system, such as the gas in Fig. 19-10. This confusing reversal of focus is left over from the 19th

---

*Recall that a system's state is specified by its pressure, volume, temperature, and phase.

[†]Here dQ and dW, unlike dE^{int}, are not true differentials; that is, there are no such functions as $Q(P, V)$ and $W(P, V)$ that depend only on the system state. Both dQ and dW are inexact differentials and often represented by the symbols $đQ$ and $đW$. Here we can treat them simply as infinitesimally small energy transfers.

century when scientists and engineers were most interested in how much physical work could be done *by* an engine on its surroundings.

The work done *on* a system is always the negative of the work done by the system, so if we rewrite $\Delta E^{\text{int}} = E_f^{\text{int}} - E_i^{\text{int}} = Q - W$ in terms of the work W^{on} done *on* the system, we have $\Delta E^{\text{int}} = Q + W^{\text{on}}$. This tells us the following: the internal energy of a system tends to increase if thermal energy is absorbed by the system or if positive work is done *on* the system. Conversely, the internal energy tends to decrease if thermal energy is lost by the system or if negative work is done *on* the system. This can be a bit confusing, so it is a good idea to spend a little time now to make sure that this distinction is clear.

READING EXERCISE 19-6: The figure here shows four paths on a *P-V* diagram along which a gas can be taken from state *i* to state *f*. Rank the paths according to (a) the change ΔE^{int}, (b) the work W done by the gas, and (c) the magnitude of the thermal energy Q transferred to the gas, greatest first. ∎

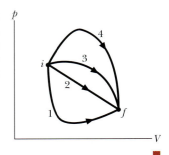

19-8 Some Special Cases of the First Law of Thermodynamics

Here we look at four different thermodynamic processes, in each of which a certain restriction is imposed on the system. We then see what consequences follow when we apply the first law of thermodynamics to the process. The results are summarized in Table 19-4.

TABLE 19-4
The First Laws of Thermodynamics: Four Special Cases

The First Law: $\Delta E^{\text{int}} = Q - W$ (Eq. 19-17)[a]		
Process	**Restriction**	**Consequence**
Adiabatic	$Q = 0$	$\Delta E^{\text{int}} = -W$
Constant volume	$W = 0$	$\Delta E^{\text{int}} = Q$
Closed cycle	$\Delta E^{\text{int}} = 0$	$Q = W$
Free expansion	$Q = W = 0$	$\Delta E^{\text{int}} = 0$

[a]$Q > 0$ represents thermal or heat energy added to the system; $W > 0$ represents work done by system.

1. ***Adiabatic processes.*** An adiabatic process is one in which *no transfer of thermal energy* occurs between the system and its environment. Adiabatic processes can occur if a system that is so well insulated that no thermal energy transfers can occur. A process can also be made adiabatic by having it occur so quickly that there is no time for the energy transfer. In either case, putting $Q = 0$ in the first law ($\Delta E^{\text{int}} = E_f^{\text{int}} - E_i^{\text{int}} = Q - W$) yields

$$\Delta E^{\text{int}} = -W \qquad \text{(adiabatic process).} \qquad (19\text{-}19)$$

This tells us that if the work done *by* the system on its surroundings is positive, the internal energy of the system decreases by an amount equal to the work it did. Conversely, if work is done *on* the system (that is, if W is negative), the internal energy of the system increases by that amount.

Figure 19-12 shows an idealized adiabatic process. Thermal energy cannot be transferred to or from a system because it is insulated from its surroundings. Thus, the only way energy can be transferred between the system and its environment is if work is either done by the system or on the system. If we remove shot from the piston and allow the gas to expand, work is done by the system (the gas) and thus is positive. Thus the internal energy of the gas must decrease. If, instead, we add shot and compress the gas, the work done by the system is negative and the internal energy of the gas increases.

2. ***Constant-volume processes.*** If the volume of a system (such as a gas in a container) is held constant, that system can do no work. Putting $W = 0$ in the first law ($\Delta E^{\text{int}} = \Delta E_f^{\text{int}} - E_i^{\text{int}} = Q - W$) yields

$$\Delta E^{\text{int}} = Q \qquad \text{(constant-volume process)}. \qquad (19\text{-}20)$$

Thus, all the thermal energy transferred to a system (so that Q is positive), contributes to an increase in the system's internal energy. Conversely, if thermal energy is transferred from the system to its surroundings so that ΔQ is negative, the internal energy of the system must decrease.

3. ***Cyclical processes.*** There are processes in which, after certain interchanges of thermal energy and work, the system is restored to its initial state. Engines undergo this type of process. Since the system is restored to its initial state, no intrinsic property of the system—including its internal energy—can possibly change. Putting $\Delta E^{\text{int}} = 0$ in the first law ($\Delta E^{\text{int}} = E_f^{\text{int}} - E_i^{\text{int}} = Q - W$) yields

$$Q(\text{in}) = W(\text{by}) \qquad \text{(cyclical process)}. \qquad (19\text{-}21)$$

Thus, the net work done by the system on its surroundings during the process must exactly equal the net amount of thermal energy transferred to the system. So the store of internal energy hidden in the system remains unchanged. Cyclical processes form a closed loop on a P-V plot, as shown in Fig. 19-11f.

4. ***Free expansions.*** These are adiabatic processes (that is, ones in which no transfer of thermal energy occurs between the system and its environment) and no work is done on or by the system. Thus, $Q = W = 0$ and the first law requires that

$$\Delta E^{\text{int}} = 0 \qquad \text{(free expansion)}. \qquad (19\text{-}22)$$

Figure 19-13 shows how such an expansion can be carried out. A gas, which is in thermal equilibrium within itself, is initially confined by a closed stopcock to one half of an insulated double chamber; the other half is evacuated. The stopcock is opened, and the gas expands freely to fill both halves of the chamber. No thermal energy is transferred to or from the gas because of the insulation. No work is done by the gas because it rushes into a vacuum; the motion of the gas atoms or molecules is not opposed by any pressure.

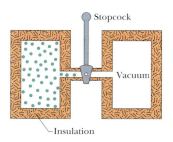

FIGURE 19-13 ▪ The initial stage of a free-expansion process. After the stopcock is opened, the gas fills both chambers and eventually reaches an equilibrium state.

A free expansion differs from all other processes we have considered because it cannot be done slowly and in a controlled way. As a result, at any given instant during the sudden expansion, the gas is not in thermal equilibrium and its

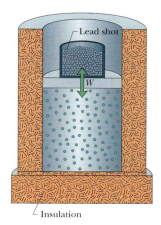

FIGURE 19-12 ▪ An adiabatic expansion can be carried out by removing lead shot from the top of the piston. Adding lead shot reverses the process at any stage. (The insulating lid is not shown.)

pressure is not the same everywhere. Therefore, although we can plot the initial and final states on a *P-V* diagram, we cannot plot the expansion itself.

READING EXERCISE 19-7: For one complete cycle as shown in the *P-V* diagram here, (a) is the internal energy change ΔE^{int} of the gas positive, negative, or zero? and (b) what about the net thermal energy, Q, transferred to the gas?

TOUCHSTONE EXAMPLE 19-3: Boiling Water

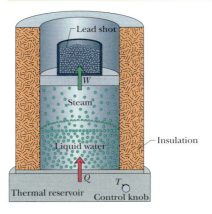

FIGURE 19-14 ■ Water boiling at constant pressure. Thermal energy is transferred from the thermal reservoir until the liquid water has changed completely into steam. Work is done by the expanding gas as it lifts the loaded piston. The insulating lid is not shown.

Let 1.00 kg of liquid water at 100°C be converted to steam at 100°C by boiling at standard atmospheric pressure (which is 1.00 atm or 1.01×10^5 Pa) in the arrangement of Fig. 19-14. The volume of that water changes from an initial value of 1.00×10^{-3} m³ as a liquid to 1.671 m³ as steam.

(a) How much work is done by the system during this process?

SOLUTION ■ The **Key Idea** here is that the system must do positive work because the volume increases. In the general case we would calculate the work W done by integrating the pressure with respect to the volume (Eq. 19-16). However, here the pressure is constant at 1.01×10^5 Pa, so we can take P outside the integral. We then have

$$W = \int_{V_i}^{V_f} P\,dV = P\int_{V_i}^{V_f} dV = P(V_f - V_i)$$

$$= (1.01 \times 10^5 \text{ Pa})(1.671 \text{ m}^3 - 1.00 \times 10^{-3} \text{ m}^3) \quad \text{(Answer)}$$

$$= 1.69 \times 10^5 \text{ J} = 169 \text{ kJ}.$$

(b) How much thermal energy is transferred as heat during the process?

SOLUTION ■ The **Key Idea** here is that the thermal energy causes only a phase change and not a change in temperature, so it is given fully by Eq. 19-7 ($Q = Lm$). Because the change is from a liquid to a gaseous phase, L is the heat of vaporization L_V, with the value given in Eq. 19-8 and Table 19-3. We find

$$Q = L_V m = (2256 \text{ kJ/kg})(1.00 \text{ kg})$$
$$= 2256 \text{ kJ} \approx 2260 \text{ kJ}. \quad \text{(Answer)}$$

(c) What is the change in the system's internal energy during the process?

SOLUTION ■ The **Key Idea** here is that the change in the system's internal energy is related to the thermal energy transferred into the system and the work done on the surroundings which transfers out of the system) by the first law of thermodynamics (Eq. 19-17). Thus, we can write

$$\Delta E^{int} = Q - W = 2256 \text{ kJ} - 169 \text{ kJ}$$
$$\approx 2090 \text{ kJ} = 2.09 \text{ MJ}. \quad \text{(Answer)}$$

This quantity is positive, indicating that the internal energy of the system has increased during the boiling process. This energy goes into separating the H_2O molecules, which strongly attract each other in the liquid state. We see that, when water is boiled, about 7.5% (= 169 kJ/2260 kJ) of the thermal energy added goes into the work of pushing back the atmosphere. The rest of the thermal energy added goes into the system's internal energy.

19-9 More on Temperature Measurement

In Section 1-4 we discussed the fact that there are only seven fundamental quantities in physics that serve as base units for the entire international system (or SI) of units. Temperature in kelvins is one of them. How is the kelvin unit defined? Why has a gas thermometer operating under constant volume become the standard thermometer? What are its fixed points? Why is it superior to the liquid thermometers we discussed in Section 19-2?

As we discussed in Section 19-2, there are several difficulties with defining temperature in terms of the height of a liquid column in a liquid thermometer. First, the historical fixed points (such as freezing and boiling points at sea level, body temperature, and lowest ice/salt mixture temperature) are not reproducible to a high accuracy. Second, liquid thermometers are only usable in a narrow range of temperatures between the freezing and boiling points of the liquid. Third, no two liquids expand and contract in exactly the same way as their temperatures change.

These difficulties prompted a search for highly reproducible fixed points and a way to measure temperature that is independent of the behavior of any one particular substance. Gases are more promising substances for temperature measurements, since they are already "boiling" and have no upper limit except at the melting point of their container. Also, gases do not tend to liquefy until they reach very low temperatures. For example, air liquefies at about −200°C. Generally a gas volume (at constant pressure) or a gas pressure (at constant volume) can be measured between two fixed points. The scales can be determined in the same way as they are for liquids. The various gas scales have been found to agree among themselves better than liquid scales do. Other thermometers are based on changes of the electrical properties or materials or changes in the light given off by glowing substances and so on. Although we have dozens of thermometric scales based on the behaviors of various substances, none of them can be proven exactly true. Here we present some methods for identifying better fixed points and designing more accurate thermometers.

Defining Standard Fixed Points

Using a low-density gas instead of liquid to measure low temperatures gives very interesting results. As we suggested in Section 19-2, when thermal energy is extracted from a gas held at constant pressure, its temperature and volume drop. A simple apparatus showing these results for a limited range of temperatures is shown in Fig. 19-15.*

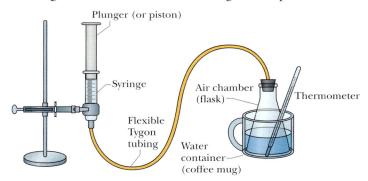

FIGURE 19-15 ■ Diagram of apparatus that can be used to measure volume changes as a function of the temperature of air or some other gas trapped at constant pressure. The pressure is a combination of atmospheric pressure plus the pressure exerted by the weight of the plunger of cross-sectional area A, so that $P = P^{\text{atm}} + mg/A$.

If you plot data for the volume of any low-density gas as a function of temperature, the graph is linear. By extrapolating the graph to zero volume, you can predict that the volume will go to zero at a temperature of approximately −273°C. In reality any gas will liquefy before its temperature gets that low. An alternative approach is to hold the volume of a gas constant and observe that its pressure will also approach zero at a temperature of about −273°C. A simple apparatus for doing this experiment is shown in Fig. 19-16.

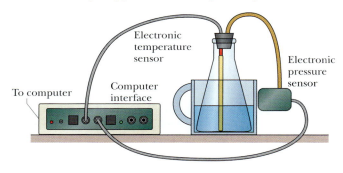

FIGURE 19-16 ■ Diagram of apparatus that uses a computer data acquisition system to measure pressure changes as a function of the temperature of air or some other gas trapped at constant volume.

*The temperature sensor should be placed inside the flask in Fig. 19-15.

Actually the extrapolation to zero pressure gives a slightly different result depending on how much gas is placed in the flask that holds it to a constant volume. However, as we place smaller and smaller masses of gas in the flask and retake the data we find that the temperature at zero pressure converges to a lower limit of −273.16°C. Student data for this experiment using different apparatus are shown in Fig. 19-17.

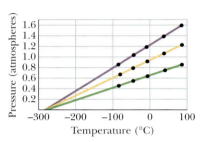

FIGURE 19-17 ■ Data from a student experiment measuring the pressure of a fixed volume of gas at various temperatures. The experimental data are represented by the points in the graphs and the lines are fits to these data. The three data sets are for three different masses of gas (air) in a container. Regardless of the amount of gas, the pressure is a linear function of temperature that extrapolates to zero at approximately −280°C. (More precise measurements show that the zero point does depend slightly on the amount of gas, but has a well-defined limit of −273.16°C as the density of the gas goes to zero.)

Since vanishingly small samples of all gases appear to approach the same minimum temperature regardless of their chemical composition, this temperature seems to be a fundamental property of nature. We call this minimum temperature **absolute zero.** We define absolute zero as the temperature of a body when it has the minimum possible internal energy. *Because of the universality of this minimum temperature of −273.16°C it has become our standard low temperature fixed point.*

But what should we use if we want a second standard fixed point? We could, for example, select the freezing point or the boiling point of water. However, water's change of phase, for example from liquid to vapor (boiling), depends not just on temperature, but also on pressure. This introduces some technical difficulties and reduces our confidence in using boiling or freezing as fixed, reproducible thermal phenomena. As we suggested in Section 19-2 a state known as the triple point of water is a good candidate for a fixed point.

We have found through extensive experimentation that three phases—liquid water, solid ice, and water vapor (gaseous water)—can coexist, in thermal equilibrium, at one and only one set of values of pressure and temperature. Thus, this temperature is called the **triple point of water.** Since both the temperature and pressure are well defined at this point, *the triple point of water is often used as a standard high temperature fixed point in designing a thermometer.*

Figure 19-18 shows a triple-point cell, in which the triple point of water can be set up in a laboratory. For reasons that we will explain when introducing the thermodynamic temperature scale, the triple point of water has been assigned a value of 0.01°C.

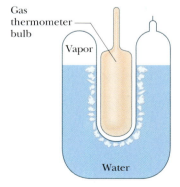

FIGURE 19-18 ■ A triple-point cell, in which solid ice, liquid water, and water vapor coexist in thermal equilibrium. By international agreement, the temperature of this mixture has been defined to be 273.16 K. The bulb of a constant-volume gas thermometer is shown inserted into the well of the cell.

The Thermodynamic Temperature Scale

The accepted SI unit of temperature is the kelvin. The kelvin is named after Lord Kelvin who first proposed that there is a natural limit to how cold any object can get. If we assume that temperature is an indicator of the amount of internal energy in a system, then the natural limit occurs when $E^{int} = 0$. Kelvin proposed that when $E^{int} = 0$ then $T = 0$. This zero is known as **absolute zero.** It is defined by using absolute zero and the triple point of water as its two fixed points. In order to tie in with the popular Celsius scale, the thermodynamic scale was set to have the same "size" degree as the Celsius scale. This yields a rather odd definition. Basically the minimum possible temperature is defined as zero kelvin or 0 K, and the triple point of water is defined as exactly +273.16 K.

Since the triple point of water is 0.01°C, the conversion between a Celsius scale temperature and the thermodynamic scale temperature is very simple since one merely needs to add 273.15 to the Celsius temperature, T_C, to get the thermodynamic temperature T. In other words,

$$T = T_C + 273.15. \qquad (19\text{-}23)$$

Figure 19-19 shows a wide range of temperatures in kelvin, either measured or conjectured. When expressing temperatures on the Fahrenheit or Celsius scales, we commonly say degrees Fahrenheit (°F) or degrees Celsius (°C), but thermodynamic temperatures are simply called "kelvin" (abbreviated K). Thus, we do not say "degrees kelvin" or write °K. Figure 19-20 compares the thermodynamic, Celsius, and Fahrenheit scales.

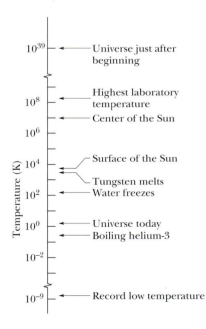

FIGURE 19-19 ▪ Some temperatures on the thermodynamic scale. Temperature $T = 0$ corresponds to $10^{-\infty}$ and cannot be plotted on this thermodynamic scale.

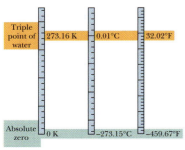

FIGURE 19-20 ▪ A comparison of the thermodynamic, Celsius, and Fahrenheit temperature scales.

When the universe began, some 10 to 20 billion years ago, its temperature was about 10^{39} K. As the universe expanded it cooled, and it has now reached an average temperature of about 3 K. We on Earth are a little warmer than that because we happen to live near a star. Without our sun, we too would be at 3 K (and we could not exist).

The Constant-Volume Gas Thermometer

Now that we have better fixed points we can use them as part of a standard thermometer that uses gas rather than a liquid as its medium. The standard thermometer, against which all other thermometers are calibrated, is based on the effect of temperature changes on the pressure of a gas occupying a fixed volume. Figure 19-21 shows such a **constant-volume gas thermometer;** it consists of a gas-filled bulb connected by a tube to a mercury manometer. By raising and lowering reservoir R, the mercury level on the left can always be brought to the zero of the scale to keep the gas volume constant (variations in the gas volume can affect temperature measurements).

The temperature of any body in thermal contact with the bulb (like the liquid in Fig. 19-21) is then defined to be

$$T = CP, \tag{19-24}$$

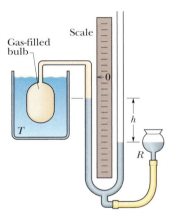

Figure 19-21 ■ A constant-volume gas thermometer, the gas-filled bulb on the left is immersed in a liquid whose temperature T is being measured. The mercury-filled bulb on the right is raised or lowered as P changes to keep the left-hand column of mercury at the zero point on the scale so the gas volume stays constant.

in which P is the pressure within the gas and C is a constant. From Eq. 15-10, the pressure P is

$$P = P^{\text{atm}} - \rho_{\text{Hg}}gh, \tag{19-25}$$

in which P^{atm} is the atmospheric pressure, ρ_{Hg} is the density of the mercury in the open-tube manometer (like that described in Section 15-5), and h is the measured difference between the mercury levels in the two arms of the tube.*

If we next put the bulb in a triple-point cell (Fig. 19-18), the temperature being measured is

$$T_3 = CP_3 \tag{19-26}$$

---

*For pressure units, we shall use units introduced in Section 15-3. The SI unit for pressure is the newton per square meter, which is called the pascal (Pa). The pascal is related to other common pressure units by

$$1.00 \text{ atm} = 1.01 \times 10^5 \text{ Pa} = 760 \text{ torr} \approx 14.7 \text{ lb/in.}^2$$

in which P_3 is the gas pressure now. Eliminating C between Eqs. 19-24 and 19-26 gives us the temperature as

$$T = T_3\left(\frac{P}{P_3}\right) = (273.16 \text{ K})\left(\frac{P}{P_3}\right) \quad \text{(provisional).} \quad (19\text{-}27)$$

We still have a problem with this thermometer. If we use it to measure a given temperature, we find that different gases in the bulb give slightly different results. However, as we use smaller and smaller masses of gas to fill the bulb, the readings converge nicely to a single temperature, no matter what gas we use.

Thus the recipe for measuring a temperature with a gas thermometer is

$$T = (273.16 \text{ K})\left(\lim_{\text{gas}\to 0}\frac{P}{P_3}\right). \quad (19\text{-}28)$$

The recipe instructs us to measure an unknown temperature T as follows: Fill the thermometer bulb with an arbitrary amount of *any* gas (for example, nitrogen) and measure P_3 (using a triple-point cell) and P, the gas pressure at the temperature being measured. (Keep the gas volume the same.) Calculate the ratio P/P_3. Then repeat both measurements with a smaller amount of gas in the bulb, and again calculate this ratio. Continue this way, using smaller and smaller amounts of gas, until you can extrapolate to the ratio P/P_3 that you would find if there were approximately no gas in the bulb. Calculate the temperature T by substituting that extrapolated ratio into Eq. 19-28. (The temperature is called the *ideal gas temperature*.)

19-10 Thermal Expansion

We already know that most materials expand when heated and contract when cooled. Indeed, we used this principle in our first definition of temperature. The design of thermometers and thermostats is often based on the differences in expansion between the components of a *bimetal strip* (Fig. 19-22). In aircraft manufacture, rivets and other fasteners are often cooled in dry ice before insertion and then allowed to expand to a tight fit. However, such **thermal expansion** is not always desirable, as Fig. 19-23 suggests. To prevent buckling, expansion slots must be placed in bridges to accommodate roadway expansion on hot days. Dental materials used for fillings must be matched in their thermal expansion properties to those of tooth enamel (otherwise consuming hot coffee or cold ice cream would be quite painful). Regardless of whether we might wish to exploit or avoid thermal expansion, this property has many important implications and we need to consider how it works in detail.

Linear Expansion

If the temperature of a metal rod of length L is raised by a small amount ΔT, its length is found to increase by an amount ΔL according to

$$\frac{\Delta L}{L} = \alpha\,\Delta T, \quad (19\text{-}29)$$

in which α is a constant called the coefficient of linear expansion. For small changes in temperature, the fractional length change ($\Delta L/L$) is proportional to the change in

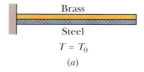

Brass

Steel

$T = T_0$

(a)

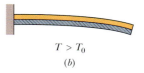

$T > T_0$

(b)

FIGURE 19-22 ■ (a) A bimetal strip, consisting of a strip of brass and a strip of steel welded together, at temperature T_0. (b) The strip bends as shown at temperatures above this reference temperature. Below the reference temperature the strip bends the other way. Many thermostats operate on this principle, making and breaking an electrical contact as the temperature rises and falls.

FIGURE 19-23 ■ Railroad tracks in Asbury Park, New Jersey, distorted because of thermal expansion on a very hot July day.

temperature. The coefficient α should be thought of as the ratio of how much the length changes per unit length given a certain change in temperature. It has the unit "per degree" or "per kelvin" and depends on the material. Although α varies somewhat with temperature, for most practical purposes it can be taken as constant for a particular material. Table 19-5 shows some coefficients of linear expansion. Note that the unit C° there could be replaced with the unit K.

TABLE 19-5
Some Coefficients of Linear Expansion^a

Substance	α (10^{-6}/C°)	Substance	α (10^{-6}/C°)
Ice (at 0°C)	51	Steel	11
Lead	29	Glass (ordinary)	9
Aluminum	23	Glass (Pyrex)	3.2
Brass	19	Diamond	1.2
Copper	17	Invar^b	0.7
Concrete	12	Fused quartz	0.5

^{a}Room temperature values except for the listing for ice.

^{b}This alloy was designed to have a low coefficient of expansion. The word is a shortened form of "invariable."

The thermal expansion of a solid is like a (three-dimensional) photographic enlargement. Figure 19-24b shows the (exaggerated) expansion of a steel ruler after its temperature is increased from that of Fig. 19-24a. Equation 19-29 applies to every linear dimension of the ruler, including its edge, thickness, diagonals, and the diameters of the circle etched on it and the circular hole cut in it. If the disk cut from that hole originally fits snugly in the hole, it will continue to fit snugly if it undergoes the same temperature increase as the ruler.

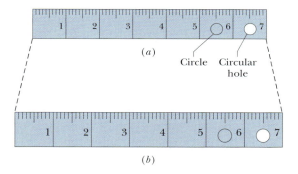

(a)

Circle   Circular hole

(b)

FIGURE 19-24 ■ The same steel ruler at two different temperatures. When the ruler expands, the scale, the numbers, the thickness, and the diameters of the circle and circular hole are all increased by the same factor. (The expansion has been exaggerated for clarity.)

Volume Expansion

If all dimensions of a solid expand with temperature, the volume of that solid must also expand. For liquids, volume expansion is the only meaningful expansion parameter. If the temperature of a solid or liquid whose volume is V is increased by a small amount ΔT, the increase in volume is found to be ΔV, where

$$\frac{\Delta V}{V} = \beta \, \Delta T \qquad (19\text{-}30)$$

and β is the **coefficient of volume expansion** of the solid or liquid.

The coefficients of volume expansion β and linear expansion α for a solid are related. To see how, consider a rectangular solid of height h, width w, and length l. The volume of the rectangular solid is hwl. If the solid is heated, each dimension expands linearly so that the height becomes $h + \Delta h$, the width becomes $w + \Delta w$, and the length becomes $l + \Delta l$. Hence, the new volume is $(h + \Delta h)(w + \Delta w)(l + \Delta l)$. If we multiply this out, ignoring all terms with two or more deltas (Δ) because those terms will be very small, we get $hwl + hw\,\Delta l + h\,\Delta wl + \Delta h\,wl$. Since each dimension expands linearly, this means that our new volume is equal to $hwl + hwl(3\alpha\,\Delta T)$. Hence,

$$\beta = 3\alpha. \tag{19-31}$$

The most common liquid, water, does not behave like other liquids. Above about 4°C, water expands as the temperature rises, as we would expect. Between 0 and about 4°C, however, water *contracts* with increasing temperature. Thus, at about 4°C, the density of water passes through a maximum. At all other temperatures, the density of water is less than this maximum value.

This behavior of water is the reason why lakes freeze from the top down rather than from the bottom up. As water on the surface is cooled from, say, 10°C toward the freezing point, it becomes denser ("heavier") than lower water and sinks to the bottom. Below 4°C, however, further cooling makes the water then on the surface *less* dense ("lighter") than the lower water, so it stays on the surface until it freezes. Thus the surface freezes while the lower water is still liquid. If lakes froze from the bottom up, the ice so formed would tend not to melt completely during the summer, because it would be insulated by the water above. After a few years, many bodies of open water in the temperate zones of Earth would be frozen solid all year round—and aquatic life as we know it could not exist.

READING EXERCISE 19-8: The figure below shows four rectangular metal plates, with sides of L, $2L$, or $3L$. They are all made of the same material, and their temperature is to be increased by the same amount. Rank the plates according to the expected increase in (a) their vertical heights and (b) their areas, greatest first.

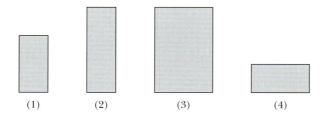

(1)   (2)   (3)   (4)   ■

READING EXERCISE 19-9: Suppose that one of the plates shown above has a round hole cut out of its center. The temperature is increased. Does the hole in the center get larger, smaller, or remain unchanged in size? Explain your reasoning.   ■

READING EXERCISE 19-10: Consider a cylindrical metal rod that stands on its base in a vertical orientation. The temperature increases uniformly. What happens to the pressure at the base of the rod? Does it increase, decrease, or remain unchanged? Explain your reasoning.   ■

TOUCHSTONE EXAMPLE 19-4: Hot Day in Vegas

On a hot day in Las Vegas, an oil trucker loaded 37 000 L of diesel fuel. He encountered cold weather on the way to Payson, Utah, where the temperature was 23.0 K lower than in Las Vegas, and where he delivered his entire load. How many liters did he deliver? The coefficient of volume expansion for diesel fuel is $9.50 \times 10^{-4}/°C$, and the coefficient of linear expansion for his steel truck tank is $11 \times 10^{-6}/°C$.

SOLUTION ■ The **Key Idea** here is that the volume of the diesel fuel depends directly on the temperature. Thus, because the temperature decreased, the volume of the fuel did also. From Eq. 19-25, the volume change is

$$\Delta V = V\beta \, \Delta T$$
$$= (37\ 000\ \text{L})(9.50 \times 10^{-4}/°C)(-23.0\ °C) = -808\ \text{L}.$$

Thus, the amount delivered was

$$V_{\text{del}} = V + \Delta V = 37\ 000\ \text{L} - 808\ \text{L}$$
$$= 36\ 192\ \text{L}. \qquad \text{(Answer)}$$

Note that the thermal expansion of the steel tank has nothing to do with the problem. Question: Who paid for the "missing" diesel fuel?

19-11 More on Thermal Energy Transfer Mechanisms

In Section 19-4, during our initial discussion of the transfer of thermal energy between a system and its environment, we qualitatively discussed the three transfer mechanisms (conduction, convection, and radiation). In order to enhance our understanding of transfer mechanisms, we expand upon our discussion of conduction and radiation here.

Conduction

We all have a natural ability to sense hot and cold. But unfortunately, our "temperature sense" is in fact not always reliable. On a cold winter day, for example, why does an iron railing seem much colder to the touch than a wooden fence post when both are at the same temperature? Why are frying pans made out of metal while pot holders are made out of cloth and other fibers? The answer is that some materials are much more effective than others at transferring thermal energy via conduction.

Consider a slab of face area A and thickness L, whose faces are maintained at temperatures T_H and T_C by a hot reservoir and a cold reservoir, as in Fig. 19-25. Let Q be the energy that is transferred as thermal energy through the slab, from its hot face to its cold face, in a time interval Δt. Experiment shows that the thermal energy *conduction rate* P^{cond} (the power or thermal energy transferred per unit time—not a pressure) is

$$P^{\text{cond}} = \frac{Q}{\Delta t} = kA\frac{T_H - T_C}{L}, \qquad (19\text{-}32)$$

in which k, called the *thermal conductivity,* is a constant that depends on the material of which the slab is made. That is, the thermal conductivity k is a *property* of the material. A material that readily transfers energy by conduction is a *good thermal conductor* and has a high value of k. Table 19-6 gives the thermal conductivities of some common metals, gases, and building materials.

Thermal Resistance to Conduction (R-Value)

If you are interested in insulating your house or in keeping cola cans cold on a picnic, you are more concerned with poor conductors of thermal energy than with good ones. For this reason, the concept of *thermal resistance R* has been introduced into engineering practice. The *R*-value of a slab of thickness L is defined as

$$R = \frac{L}{k}. \qquad (19\text{-}33)$$

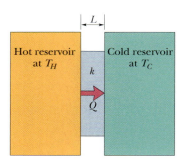

$T_H > T_C$

FIGURE 19-25 ■ Thermal conduction. Thermal energy Q is transferred from a reservoir at temperature T_H to a cooler reservoir at temperature T_C through a conducting slab of thickness L and thermal conductivity k.

The lower the thermal conductivity of the material of which a slab is made, the higher is the R-value of the slab, so something that has a high R-value is a *poor thermal conductor* and thus a *good thermal insulator*.

Note that R is a property attributed to a slab of a specified thickness, not to a material. That is, R is *not* a property of a material alone. The commonly used unit for R (which, in the United States at least, is almost never stated) is the square foot-fahrenheit degree-hour per British thermal unit (ft$^2 \cdot$ F$^\circ \cdot$ h/Btu).

Conduction Through a Composite Slab

Figure 19-26 shows a composite slab, consisting of two materials having different thicknesses L_1 and L_2 and different thermal conductivities k_1 and k_2. The temperatures of the outer surfaces of the slab are T_H and T_C. Each face of the slab has area A. Let us derive an expression for the conduction rate through the slab under the assumption that the transfer is a *steady-state* process; that is, the temperatures everywhere in the slab and the rate of energy transfer do not change with time.

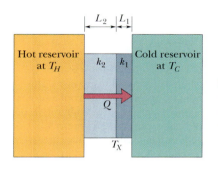

FIGURE 19-26 ■ Thermal energy Q is transferred from a hot reservoir to a cold reservoir at a steady rate through a composite slab made up of two different materials with different thicknesses and different thermal conductivities. The steady-state temperature at the interface of the two materials is denoted T_X.

TABLE 19-6	
Some Thermal Conductivities[a]	
Substance	**k(W/m $\cdot$ K)**
Metals	
Stainless steel	14
Lead	35
Aluminum	235
Copper	401
Silver	428
Gases	
Air (dry)	0.026
Helium	0.15
Hydrogen	0.18
Building Materials	
Polyurethane foam	0.024
Rock wool	0.043
Fiberglass	0.048
White pine	0.11
Window glass	1.0

[a]Conductivities change somewhat with temperature. The given values are at room temperature.

In the steady state, the conduction rates through the two materials must be equal. This is the same as saying that the energy transferred through one material in a certain time must be equal to that transferred through the other material in the same time. If this were not true, temperatures in the slab would be changing and we would not have a steady-state situation. Letting T_X be the temperature of the interface between the two materials, we can now use Eq. 19-33 to express the rate of thermal energy transfer as

$$P^{\text{cond}} = \frac{k_2 A(T_H - T_X)}{L_2} = \frac{k_1 A(T_X - T_C)}{L_1}. \tag{19-34}$$

Solving for T_X yields, after a little algebra,

$$T_X = \frac{k_1 L_2 T_C + k_2 L_1 T_H}{k_1 L_2 + k_2 L_1}. \tag{19-35}$$

Substituting this expression for T_X into either equality of Eq. 19-34 yields

$$P^{\text{cond}} = \frac{A(T_H - T_C)}{L_1/k_1 + L_2/k_2}. \tag{19-36}$$

We can extend Eq. 19-36 to apply to any number n of materials making up a slab:

$$P^{\text{cond}} = \frac{A(T_H - T_C)}{\sum\limits_{i=1}^{n}(L_i/k_i)}. \tag{19-37}$$

The summation sign in the denominator tells us to add the values of L/k for all the materials.

Radiation

The rate P^{rad} at which an object emits energy via electromagnetic radiation depends on the object's surface area A and the temperature T of that area in kelvins and is given by

$$P^{rad} = \sigma \varepsilon A T^4. \tag{19-38}$$

Here $\sigma = 5.6703 \times 10^{-8} \, \text{W/m}^2 \cdot \text{K}^4$ is called the *Stefan-Boltzmann constant* after Josef Stefan (who discovered Eq. 19-33 experimentally in 1879) and Ludwig Boltzmann (who derived it theoretically soon after). The symbol ε represents the *emissivity* of the object's surface, which has a value between 0 and 1, depending on the composition of the surface. A surface with the maximum emissivity of 1.0 is said to be a *blackbody radiator*, but such a surface is an ideal limit and does not occur in nature. Note again that the temperature in Eq. 19-38 must be in kelvins so that a temperature of absolute zero corresponds to no radiation. Note also that every object whose temperature is above 0 K—including you—emits thermal radiation. (See Fig. 19-27.)

FIGURE 19-27 ■ Thermogram of a house showing the distribution of heat over its surface. The color coding ranges from white to yellow for the warmest areas (greatest heat loss from windows, etc) through red to purple and green for the coolest areas (greatest insulation). This thermogram shows that the roof and windows (yellow) are poorly insulated, while the walls (red, purple, and green) are losing the least heat. Thermograms are often used to check houses for heat loss, so that they can be made more energy efficient through improved insulation.

The rate P^{abs} at which an object absorbs energy via thermal radiation from its environment, which we take to be at uniform temperature T_{env} (in kelvins), is

$$P^{abs} = \sigma \varepsilon A T_{env}^4. \tag{19-39}$$

The emissivity ε in Eq. 19-39 is the same as that in Eq. 19-38. An idealized blackbody radiator, with $\varepsilon = 1$, will absorb all the radiated energy it intercepts (rather than sending a portion back away from itself through reflection or scattering).

Because an object will radiate energy to the environment while it absorbs energy from the environment, the object's net rate P^{net} of energy exchange due to thermal radiation is

$$P^{net} = P^{abs} - P^{rad} = \sigma \varepsilon A (T_{env}^4 - T^4). \tag{19-40}$$

P^{net} is positive if net energy is being absorbed via radiation, and negative if it is being lost via radiation.

READING EXERCISE 19-11: The figure shows the face and interface temperatures of a composite slab consisting of four materials, of identical thicknesses, through which the thermal energy transfer is steady. Rank the materials according to their thermal conductivities, greatest first.

25°C ⎤ 15°C ⎤ 10°C ⎤ −5.0°C ⎤ −10°C ⎤

a *b* *c* *d*

TOUCHSTONE EXAMPLE 19-5: Thermal Conduction

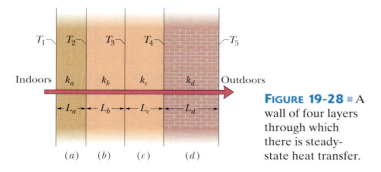

T_1 T_2 T_3 T_4 T_5

Indoors k_a k_b k_c k_d Outdoors

L_a L_b L_c L_d

(a) (b) (c) (d)

FIGURE 19-28 ■ A wall of four layers through which there is steady-state heat transfer.

Figure 19-28 shows the cross section of a wall made of white pine of thickness L_a and brick of thickness L_d ($= 2.0L_a$), sandwiching two layers of unknown material with identical thicknesses and thermal conductivities. The thermal conductivity of the pine is k_a and that of the brick is k_d ($= 5.0k_a$). The face area A of the wall is unknown. Thermal conduction through the wall has reached the steady state; the only known interface temperatures are $T_1 = 25°C$, $T_2 = 20°C$, and $T_5 = -10°C$. What is interface temperature T_4?

SOLUTION ■ One **Key Idea** here is that temperature T_4 helps determine the rate P_d^{cond} at which energy is conducted through the brick, as given by Eq. 19-32. However, we lack enough data to solve Eq. 19-32 for T_4. A second **Key Idea** is that because the conduction is steady, the conduction rate P_d^{cond} through the brick must equal the conduction rate P_a^{cond} through the pine. From Eq. 19-32 and Fig. 19-28, we can write

$$P_a^{cond} = k_a A \frac{T_1 - T_2}{L_a} \quad \text{and} \quad P_d^{cond} = k_d A \frac{T_4 - T_5}{L_d}.$$

Setting $P_a^{cond} = P_d^{cond}$ and solving for T_4 yield

$$T_4 = \frac{k_a L_d}{k_d L_a}(T_1 - T_2) + T_5.$$

Letting $L_d = 2.0L_a$ and $k_d = 5.0k_a$, and inserting the known temperatures, we find

$$T_4 = \frac{k_a(2.0L_a)}{(5.0k_a)L_a}(25°C - 20°C) + (-10°C)$$

$$= -8.0°C. \qquad \text{(Answer)}$$

TOUCHSTONE EXAMPLE 19-6: Japanese Bees

FIGURE 19-29 ■ The bees were unharmed by their increased body temperature, which the hornet could not withstand.

When hundreds of Japanese bees form a compact ball around a giant hornet that attempts to invade their hive, they can quickly raise their body temperature from the normal 35°C to 47°C or 48°C. That higher temperature is lethal to the hornet but not to the bees (Fig. 19-29). Assume the following: 500 bees form a ball of radius $R = 2.0$ cm for a time $\Delta t = 20$ min, the primary loss of energy by the ball is by thermal radiation, the ball's surface has emissivity $\varepsilon = 0.80$, and the ball has a uniform temperature. On average, how much additional energy must each bee produce during the 20 min to maintain 47°C?

SOLUTION ■ The **Key Idea** here is that, because the surface temperature of the bee ball increases after the ball forms, the rate at which energy is radiated by the ball also increases. Thus, the bees lose an additional amount of energy to thermal radiation. We can relate the surface temperature to the rate of radiation (energy per unit time) with Eq. 19-38 ($P^{rad} = \sigma\varepsilon A T^4$), in which A is the ball's surface area and T is the ball's surface temperature in kelvins. This rate is an energy per unit time; that is,

$$P^{rad} = \frac{\Delta E}{\Delta t}.$$

Thus, the amount of energy ΔE radiated in time t is $\Delta E = P^{rad}\Delta t$.

At the normal temperature $T_1 = 35°C$, the radiation rate is P_1^{rad} and the amount of energy radiated in time Δt is $\Delta E_1 = P_1^{rad}\Delta t$. At the increased temperature $T_2 = 47°C$, the (greater) radiation rate is P_2^{rad} and the (greater) amount of energy radiated in time Δt is $\Delta E_2 = P_2^{rad}\Delta t$. Thus, in maintaining the ball at T_2 for time Δt, the bees must (together) provide an additional amount of energy of $E = \Delta E_2 - \Delta E_1$.

We can now write

$$E = \Delta E_2 - \Delta E_1 = P_2^{rad}\Delta t - P_1^{rad}\Delta t$$

$$= (\sigma\varepsilon A T_2^4)\Delta t - (\sigma\varepsilon A T_1^4)\Delta t = \sigma\varepsilon A \Delta t(T_2^4 - T_1^4). \quad (19\text{-}41)$$

The temperatures here *must* be in kelvins; thus, we write them as

$$T_2 = 47°C + 273°C = 320 \text{ K}$$

and

$$T_1 = 35°C + 273°C = 308 \text{ K}.$$

The surface area A of the ball is

$$A = 4\pi R^2 = (4\pi)(0.020 \text{ m})^2 = 5.027 \times 10^{-3} \text{ m}^2,$$

and the time Δt is 20 min = 1200 s. Substituting these and other known values into Eq. 19-41, we find

$$E = (5.6703 \times 10^{-8} \text{ W/m}^2 \cdot \text{K}^4)(0.80)(5.027 \times 10^{-3} \text{ m}^2)$$

$$\times (1200 \text{ s})[(320 \text{ K})^4 - (308 \text{ K})^4] = 406.8 \text{ J}.$$

Thus, with 500 bees in the ball, each bee must produce an additional energy of

$$\frac{E}{500} = \frac{406.8 \text{ J}}{500} = 0.81 \text{ J}. \qquad \text{(Answer)}$$

Problems

SEC. 19-2 ■ THERMOMETERS AND TEMPERATURE SCALES

1. Fahrenheit and Celsius At what temperature is the Fahrenheit scale reading equal to (a) twice that of the Celsius and (b) half that of the Celsius?

2. Oymyakon (a) In 1964, the temperature in the Siberian village of Oymyakon reached −71°C. What temperature is this on the Fahrenheit scale? (b) The highest officially recorded temperature in the continental United States was 134°F in Death Valley, California. What is this temperature on the Celsius scale?

SEC. 19-4 ■ HEATING, COOLING, AND TEMPERATURE

3. Hot and Cold It is an everyday observation that hot and cold objects cool down or warm up to the temperature of their surroundings. If the temperature difference ΔT between an object and its surroundings ($\Delta T = T_{obj} - T_{sur}$) is not too great, the rate of cooling or warming of the object is proportional, approximately, to this temperature difference; that is,

$$\frac{d(\Delta T)}{dt} = -A(\Delta T),$$

where A is a constant. (The minus sign appears because ΔT decreases with time if ΔT is positive and increases if ΔT is negative.) This is known as *Newton's law of cooling*. (a) On what factors does A depend? What are its dimensions? (b) If at some instant $t_1 = 0$ the temperature difference is ΔT_1, show that it is

$$\Delta T = \Delta T_1 e^{-At_2}$$

at a later time t_2.

4. House Heater The heater of a house breaks down one day when the outside temperature is 7.0°C. As a result, the inside temperature drops from 22°C to 18°C in 1.0 h. The owner fixes the heater and adds insulation to the house. Now she finds that, on a similar day, the house takes twice as long to drop from 22°C to 18°C when the heater is not operating. What is the ratio of the new value of constant A in Newton's law of cooling (see Problem 3) to the previous value?

SEC. 19-5 ■ THERMAL ENERGY TRANSFER TO SOLIDS AND LIQUIDS

5. A Certain Substance A certain substance has a mass per mole of 50 g/mol. When 314 J of thermal energy is transferred to a 30.0 g sample, the sample's temperature rises from 25.0°C to 45.0°C. What are (a) the specific heat and (b) the molar specific heat of this substance? (c) How many moles are present?

6. Diet Doctor A certain diet doctor encourages people to diet by drinking ice water. His theory is that the body must burn off enough fat to raise the temperature of the water from 0.00°C to the body temperature of 37.0°C. How many liters of ice water would have to be consumed to burn off 454 g (about 1 lb) of fat, assuming that this much fat burning requires 3500 Cal be transferred to the ice water? Why is it not advisable to follow this diet? (One liter = 10^3 cm³. The density of water is 1.00 g/cm³.)

7. Minimum Energy Calculate the minimum amount of energy, in joules required to completely melt 130 g of silver initially at 15.0°C.

8. How Much Unfrozen How much water remains unfrozen after 50.2 kJ of thermal energy is transferred from 260 g of liquid water initially at its freezing point?

9. Energetic Athlete An energetic athlete can use up all the energy from a diet of 4000 food calories/day where a food calorie = 1000 cal. If he were to use up this energy at a steady rate, how would his rate of energy use compare with the power of a 100 W bulb? (The power of 100 W is the rate at which the bulb converts electrical energy to thermal energy and the energy of visible light.)

10. Four Lightbulbs A room is lighted by four 100 W incandescent lightbulbs. (The power of 100 W is the rate at which a bulb converts electrical energy to thermal energy and the energy of visible light.) Assuming that 90% of the energy is converted to thermal energy, how much thermal energy is transferred to the room in 1.00 h?

11. Drilling a Hole A power of 0.400 hp is required for 2.00 min to drill a hole in a 1.60 lb copper block. (a) If the full power is the rate at which thermal energy is generated, how much is generated in Btu? (b) What is the rise in temperature of the copper if the copper absorbs 75.0% of this energy? (Use the energy conversion 1 ft·lb = 1.285×10^{-3} Btu.)

12. How Much Butter How many grams of butter, which has a usable energy content of 6.0 Cal/g (= 6000 cal/g), would be equivalent to the change in gravitational potential energy of a 73.0 kg man who ascends from sea level to the top of Mt. Everest, at elevation 8.84 km? Assume that the average value of g is 9.80 m/s².

13. Immersion Heater A small electric immersion heater is used to heat 100 g of water for a cup of instant coffee. The heater is labeled "200 watts," so it converts electrical energy to thermal energy that is transferred to the water at this rate. Calculate the time required to bring the water from 23°C to 100°C ignoring any thermal energy that transfers out of the cup.

14. Tub of Water One way to keep the contents of a garage from becoming too cold on a night when a severe subfreezing temperature is forecast is to put a tub of water in the garage. If the mass of the water is 125 kg and its initial temperature is 20°C, (a) how much thermal energy must the water transfer to its surroundings in order to freeze completely and (b) what is the lowest possible temperature of the water and its surroundings until that happens?

15. A Chef A chef, on finding his stove out of order, decides to boil the water for his wife's coffee by shaking it in a thermos flask. Suppose that he uses tap water at 15°C and that the water falls 30 cm each shake, the chef making 30 shakes each minute. Neglecting any transfer of thermal energy out of the flask, how long must he shake the flask for the water to reach 100°C?

16. Copper Bowl A 150 g copper bowl contains 220 g of water, both at 20.0°C. A very hot 300 g copper cylinder is dropped into the water, causing the water to boil, with 5.00 g being converted to steam. The final temperature of the system is 100°C. Neglect energy transfers with the environment. (a) How much energy (in calories) is transferred to the water? (b) How much to the bowl? (d) What is the original temperature of the cylinder?

17. Ethyl Alcohol Ethyl alcohol has a boiling point of 78°C, a freezing point of −114°C, a heat of vaporization of 879 kJ/kg, a heat of fusion of 109 kJ/kg, and a specific heat of 2.43 kJ/kg · C°. How much thermal energy must be transferred out of 0.510 kg of ethyl alcohol that is initially a gas at 78°C so that it becomes a solid at −114°C?

18. Metric–Nonmetric *Nonmetric version:* How long does a 2.0×10^5 Btu/h water heater take to raise the temperature of 40 gal of water from 70°F to 100°F? *Metric version:* How long does a 59 kW water heater take to raise the temperature of 150 L of water from 21°C to 38°C?

19. Buick A 1500 kg Buick moving at 90 km/h brakes to a stop, at a uniform rate and without skidding, over a distance of 80 m. At what average rate is mechanical energy transformed into thermal energy in the brake system?

20. Solar Water Heater In a solar water heater, radiant energy from the Sun is transferred to water that circulates through tubes in a rooftop collector. The solar radiation enters the collector through a transparent cover and warms the water in the tubes; this water is pumped into a holding tank. Assume that the efficiency of the overall system is 20% (that is, 80% of the incident solar energy is lost from the system). What collector area is necessary to raise the temperature of 200 L of water in the tank from 20°C to 40°C in 1.0 h when the intensity of incident sunlight is 700 W/m²?

21. Steam What mass of steam at 100°C must be mixed with 150 g of ice at its melting point, in a thermally insulated container, to produce liquid water at 50°C?

22. Iced Tea A person makes a quantity of iced tea by mixing 500 g of hot tea (essentially water) with an equal mass of ice at its melting point. If the initial hot tea is at a temperature of (a) 90°C and (b) 70°C, what are the temperature and mass of the remaining ice when the tea and ice reach a common temperature? Neglect energy transfers with the environment.

23. Ice Cubes (a) Two 50 g ice cubes are dropped into 200 g of water in a thermally insulated container. If the water is initially at 25°C, and the ice comes directly from a freezer at −15°C, what is the final temperature of the drink when the drink reaches thermal equilibrium? (b) What is the final temperature if only one ice cube is used?

24. Thermos of Coffee An insulated Thermos contains 130 cm³ of hot coffee, at a temperature of 80.0°C. You put in a 12.0 g ice cube at its melting point to cool the coffee. By how many degrees has your coffee cooled once the ice has melted? Treat the coffee as though it were pure water and neglect energy transfers with the environment.

Sec. 19-8 ■ Some Special Cases of the First Law of Thermodynamics

25. Gas Expands A sample of gas expands from 1.0 m³ to 4.0 m³ while its pressure decreases from 40 Pa to 10 Pa. How much work is done by the gas if its pressure changes with volume via each of the three paths shown in the P-V diagram in Fig. 19-30?

26. Work Done Consider that 200 J of work is done on a system and 70.0 cal of thermal energy is trans-

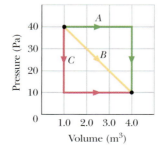

FIGURE 19-30 ■ Problem 25.

ferred out of the system. In the sense of the first law of thermodynamics, what are the values (including algebraic signs) of (a) W, (b) Q, and (c) ΔE^{int}?

27. Closed Chamber Gas within a closed chamber undergoes the cycle shown in the P-V diagram of Fig. 19-31. Calculate the net thermal energy added to the system during one complete cycle.

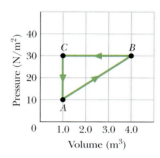

FIGURE 19-31 ■ Problem 27.

28. Thermodynamic A thermodynamic system is taken from an initial state A to another state B and back again to A, via state C, as shown by path $ABCA$ in the P-V diagram of Fig. 19-31a. (a) Complete the table in Fig. 19-32b by filling in either + or − for the sign of each thermodynamic quantity associated with each step of the cycle. (b) Calculate the numerical value of the work done by the system for the complete cycle $ABCA$.

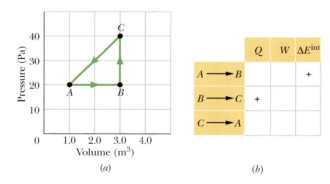

FIGURE 19-32 ■ Problem 28.

29. From i to f When a system is taken from state i to state f along path iaf in Fig. 19-33. $Q = 50$ cal and $W = 20$ cal. Along path ibf, $Q = 36$ cal. (a) What is W along path ibf? (b) If $W = -13$ cal for the return path fi, what is Q for this path? (c) Take $E_i^{int} = 10$ cal. What is E_f^{int}? (d) If $E_b^{int} = 22$ cal, what are the values of Q for path ib and path bf?

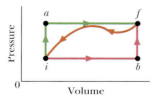

FIGURE 19-33 ■ Problem 29.

30. Gas Within Gas within a chamber passes through the cycle shown in Fig. 19-34. Determine the thermal energy transferred by the system during process CA if the thermal energy added Q_{AB}, during process *AB* is 20.0 J, no thermal energy is transferred during process *BC*, and the net work done during the cycle is 15.0 J.

FIGURE 19-34 ■ Problem 30.

SEC. 19-9 ■ MORE ON TEMPERATURE MEASUREMENT

31. Gas Thermometer A particular gas thermometer is constructed of two gas-containing bulbs, each of which is put into a water bath, as shown in Fig. 19-35. The pressure difference between the two bulbs is measured by a mercury manometer as shown. Appropriate reservoirs, not shown in the diagram, maintain constant gas volume in the two bulbs. There is no difference in pressure when both baths are at the triple point of water. The pressure difference is 120 torr when one bath is at the triple point and the other is at the boiling point of water. It is 90.0 torr when one bath is at the triple point and the other is at an unknown temperature to be measured. What is the unknown temperature?

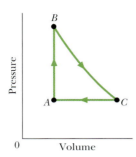

FIGURE 19-35 ■ Problem 31.

32. A Gas at Boiling Suppose the temperature of a gas at the boiling point of water is 373.15 K. What then is the limiting value of the ratio of the pressure of the gas at that boiling point to its pressure at the triple point of water? (Assume the volume of the gas is the same at both temperatures.)

33. Pairs of Scales At what temperature do the following pairs of scales read the same, if ever: (a) Fahrenheit and Celsius (verify the listing in Table 19-1), (b) Fahrenheit and Kelvin, and (c) Celsius and Kelvin?

SEC. 19-10 ■ THERMAL EXPANSION

34. Aluminum Flagpole An aluminum flagpole is 33 m high. By how much does its length increase as the temperature increases by 15 C°?

35. Pyrex Glass The Pyrex glass mirror in the telescope at the Mt. Palomar Observatory has a diameter of 200 in. The temperature ranges from −10°C to 50°C on Mt. Palomar. In micrometers, what is the maximum change in the diameter of the mirror, assuming that the glass can freely expand and contract?

36. Aluminum Alloy An aluminum-alloy rod has a length of 10.000 cm at 20.000°C and a length of 10.015 cm at the boiling point

of water. (a) What is the length of the rod at the freezing point of water? (b) What is the temperature if the length of the rod is 10.009 cm?

37. Circular Hole A circular hole in an aluminum plate is 2.725 cm in diameter at 0.000°C. What is its diameter when the temperature of the plate is raised to 100.0°C?

38. Lead Ball What is the volume of a lead ball at 30°C if the ball's volume at 60°C is 50 cm³?

39. Change in Volume Find the change in volume of an aluminum sphere with an initial radius of 10 cm when the sphere is heated from 0.0°C to 100°C.

40. Area Rectangular The area A of a rectangular plate is ab. Its coefficient of linear expansion is α. After a temperature rise ΔT, side a is longer by Δa and side b is longer by Δb (Fig. 19-36). Show that if the small quantity $(\Delta a\,\Delta b)/ab$ is neglected, then $\Delta A = 2\alpha A\,\Delta T$.

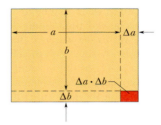

FIGURE 19-36 ■ Problem 40.

41. Aluminum Cup An aluminum cup of 100 cm³ capacity is completely filled with glycerin at 22°C. How much glycerin, if any, will spill out of the cup if the temperature of both the cup and glycerin is increased to 28°C? (The coefficient of volume expansion of glycerin is 5.1×10^{-4}/C°.)

42. Rod At 20°C, a rod is exactly 20.05 cm long on a steel ruler. Both the rod and the ruler are placed in an oven at 270°C, where the rod now measures 20.11 cm on the same ruler. What is the coefficient of thermal expansion for the material of which the rod is made?

43. Steel Rod A steel rod is 3.000 cm in diameter at 25°C. A brass ring has an interior diameter of 2.992 cm at 25°C. At what common temperature will the ring just slide onto the rod?

44. Metal Cylinder When the temperature of a metal cylinder is raised from 0.0°C to 100°C, its length increases by 0.23%. (a) Find the percent change in density. (b) What is the metal? Use Table 19-5.

45. Barometer Show that when the temperature of a liquid in a barometer changes by ΔT and the pressure is constant, the liquid's height h changes by $\Delta h = \beta h\,\Delta T$, where β is the coefficient of volume expansion. Neglect the expansion of the glass tube.

46. Copper Coin When the temperature of a copper coin is raised by 100 C°, its diameter increases by 0.18%. To two significant figures, give the percent increase in (a) the area of a face, (b) the thickness, (c) the volume, and (d) the mass of the coin. (e) Calculate the coefficient of linear expansion of the coin.

47. Pendulum Clock A pendulum clock with a pendulum made of brass is designed to keep accurate time at 20°C. If the clock operates at 0.0°C, what is the magnitude of its error, in seconds per hour, and does the clock run fast or slow?

48. Radioactive Source In a certain experiment, a small radioactive source must move at selected, extremely slow speeds. This motion is accomplished by fastening the source to one end of an aluminum rod and heating the central section of the rod in a controlled way. If the effective heated section of the rod in Fig. 19-37 is 2.00 cm, at what constant rate must the temperature of the rod be changed if the source is to move at a constant speed of 100 nm/s?

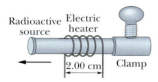

FIGURE 19-37 ■ Problem 48.

49. Temperature Rise As a result of a temperature rise of 32°C, a bar with a crack at its center buckles upward (Fig. 19-38). If the fixed distance L_0 is 3.77 m and the coefficient of linear expansion of the bar is $25 \times 10^{-6}/C°$, find the rise x of the center.

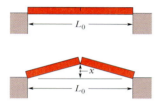

FIGURE 19-38 ■ Problem 49.

50. Copper Ring A 20.0 g copper ring has a diameter of 2.54000 cm at its temperature of 0.000°C. An aluminum sphere has a diameter of 2.54508 cm at its temperature of 100.0°C. The sphere is placed on top of the ring (Fig. 19-39), and the two are allowed to come to thermal equilibrium, with no thermal energy transferred to the surroundings. The sphere just passes through the ring at the equilibrium temperature. What is the mass of the sphere?

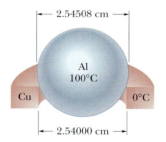

FIGURE 19-39 ■ Problem 50.

SEC. 19-11 ■ MORE ON THERMAL ENERGY TRANSFER MECHANISMS

51. Single-Family Dwelling The ceiling of a single-family dwelling in a cold climate should have an R-value of 30. To give such insulation, how thick would a layer of (a) polyurethane foam and (b) silver have to be?

52. North America The average rate at which energy is conducted outward through the ground surface in North America is 54.0 mW/m², and the average thermal conductivity of the near-surface rocks is 2.50 W/m · K. Assuming a surface temperature of 10.0°C, find the temperature at a depth of 35.0 km (near the base of the crust). Ignore the thermal energy transferred from the radioactive elements.

53. Slab Consider the slab shown in Fig. 19-25. Suppose that $L = 25.0$ cm, $A = 90.0$ cm², and the material is copper. If $T_H = 125$°C, $T_C = 10.0$°C, and a steady state is reached, find the conduction rate through the slab.

54. Body Heat (a) Calculate the rate at which body heat is conducted through the clothing of a skier in a steady-state process, given the following data: the body surface area is 1.8 m² and the clothing is 1.0 cm thick; the skin surface temperature is 33°C and the outer surface of the clothing is at 1.0°C; the thermal conductivity of the clothing is 0.040 W/m · K. (b) How would the answer to (a) change if, after a fall, the skier's clothes became soaked with water of thermal conductivity 0.60 W/m · K?

55. Copper Rod A cylindrical copper rod of length 1.2 m and cross-sectional area 4.8 cm² is insulated to prevent thermal energy from being transferred through its surface. The ends are maintained at a temperature difference of 100°C by having one end in a water–ice mixture and the other in boiling water and steam. (a) Find the rate at which thermal energy is conducted along the rod. (b) Find the rate at which ice melts at the cold end.

56. Without a Spacesuit If you were to walk briefly in space without a spacesuit while far from the Sun (as an astronaut does in the movie *2001*), you would feel the cold of space—while you radiated thermal energy, you would absorb almost none from your environment. (a) At what rate would you lose thermal energy? (b) How much thermal energy would you lose in 30 s? Assume that your emissivity is 0.90, and estimate other data needed in the calculations.

57. Rectangular Rods Two identical rectangular rods of metal are welded end to end as shown in Fig. 19-40*a*, and 10 J of thermal energy is conducted (in a steady-state process) through the rods in 2.0 min. How long would it take for 10 J to be conducted through the rods if they were welded together as shown in Fig. 19-40*b*?

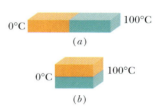

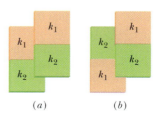

FIGURE 19-40 ■ Problem 57.

58. Four Squares Four square pieces of insulation of two different materials, all with the same thickness and area A, are available to cover an opening of area $2A$. This can be done in either of the two ways shown in Fig. 19-41. Which arrangement, (*a*) or (*b*), gives the lower thermal energy flow if $k_2 \neq k_1$?

FIGURE 19-41 ■ Problem 58.

59. Glass Window (a) What is the rate of thermal energy transfer in watts per square meter through a glass window 3.0 mm thick if the outside temperature is −20°F and the inside temperature is +72°F? (b) A storm window having the same thickness of glass is installed parallel to the first window, with an air gap of 7.5 cm between the two windows. What now is the rate of energy loss if conduction is the only important energy-transfer mechanism?

60. A Sphere A sphere of radius 0.500 m, temperature 27.0°C, and emissivity 0.850 is located in an environment of temperature 77.0°C. At what rate does the sphere (a) emit and (b) absorb thermal radiation? (c) What is the sphere's net rate of energy exchange?

61. Tank of Water A tank of water has been outdoors in cold weather, and a slab of ice 5.0 cm thick has formed on its surface (Fig. 19-42). The air above the ice is at −10°C.

FIGURE 19-42 ■ Problem 61.

Calculate the rate of formation of ice (in centimeters per hour) on the ice slab. Take the thermal conductivity and density of ice to be 0.0040 cal/s·cm·C° and 0.92 g/cm³. Assume that energy is not transferred through the walls or bottom of the tank.

62. A Wall Figure 19-43 shows (in cross section) a wall that consists of four layers. The thermal conductivities are $k_1 = 0.060$ W/m·K, $k_3 = 0.040$ W/m·K, and $k_4 = 0.12$ W/m·K (k_2 is not known). The layer thicknesses are $L_1 = 1.5$ cm, $L_3 = 2.8$ cm, and $L_4 = 3.5$ cm (L_2 is not known). Energy transfer through the wall is steady. What is the temperature of the interface indicated?

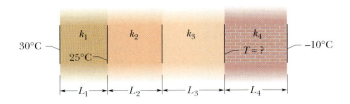

FIGURE 19-43 ■ Problem 62.

Additional Problems

63. 300 F Club You can join the semi-secret "300 F" club at the Amundsen–Scott South Pole Station only when the outside temperature is below −70°C. On such a day, you first bask in a hot sauna and then run outside wearing only your shoes. (This is, of course, extremely dangerous, but the rite is effectively a protest against the constant danger of the winter cold at the south pole.)

Assume that when you step out of the sauna, your skin temperature is 102°F and the walls, ceiling, and floor of the sauna room have a temperature of 30°C. Estimate your surface area, and take your skin emissivity to be 0.80. (a) What is the approximate net rate P^{net} at which you lose energy via thermal radiation transfer to the room? Next, assume that when you are outside half your surface area transfers thermal radiation to the sky at a temperature of −25°C and the other half transfers thermal radiation to the snow and ground at a temperature of −80°C. What is the approximate net rate at which you lose energy via thermal radiation exchanges with (b) the sky and (c) the snow and ground?

64. Shallow Pond Ice has formed on a shallow pond and a steady state has been reached, with the air above the ice at −5.0°C and the bottom of the pond at 4.0°C. If the total depth of *ice + water* is 1.4 m, how thick is the ice? (Assume that the thermal conductivities of ice and water are 0.40 and 0.12 cal/m·C°·s, respectively.)

65. Emperor Penguins Emperor penguins, those large penguins that resemble stuffy English butlers, breed and hatch their young even during severe Antarctic winters. Once an egg is laid, the father balances the egg on his feet to prevent the egg from freezing. He must do this for the full incubation period of 105 to 115 days, during which he cannot eat because his food is in the water. He can survive this long without food only if he can reduce his loss of internal food energy significantly. If he is alone, he loses that energy too quickly to stay warm, and eventually abandons the egg in order to eat. To

protect themselves and each other from the cold so as to reduce the loss of internal energy, penguin fathers huddle closely together, in groups of perhaps several thousand. In addition to providing other benefits, the huddling reduces the rate at which the penguins thermally radiate energy to their surroundings.

Assume that a penguin father is a circular cylinder with top surface area a, height h, surface temperature T, and emissivity ε. (a) Find an expression for the rate P_i at which an individual father would radiate energy to the environment from his top surface and his side surface were he alone with his egg.

If N identical fathers were well apart from one another, the total rate of energy loss via radiation would be NP_i. Suppose, instead, that they huddle closely to form a *huddled cylinder* with top surface area Na and height h. (b) Find an expression for the rate P_h at which energy is radiated by the top surface and the side surface of the huddled cylinder.

(c) Assuming $a = 0.34$ m² and $h = 1.1$ m and using the expressions you obtained for P_i and P_h, graph the ratio P_h/NP_i versus N_h. Of course, the penguins know nothing about algebra or graphing, but their instinctive huddling reduces this ratio so that more of their eggs survive to the hatching stage. From the graphs (as you will see, you probably need more than one version), approximate how many penguins must huddle so that P_h/NP_i is reduced to (d) 0.5, (e) 0.4, (f) 0.3, (g) 0.2, and (h) 0.15. (i) For the assumed data, what is the lower limiting value for P_h/NP_i?

66. The Penny and the Jelly Donut You see a penny lying on the ground. A penny won't buy much these days, so you think: "If I bend down to pick it up I will do work. To do that work I will have to burn some energy. It probably costs me more to buy the fuel (food) to provide that energy than I would gain by picking up the penny. It's not cost effective." You pass it by. Is the argument correct? Estimate the energy cost for picking up a penny. You may find the following information useful: A jelly donut contains about 250 Calories (1 Calorie = 1 Kcal).

67. Considering Changes For each of the situations described below, the object considered is undergoing some changes. Among the possible changes you should consider are: (Q) The object is absorbing or giving off thermal energy. (T) The object's temperature is changing. (E^{int}) The object's internal energy is changing. (W) The object is doing mechanical work or having work done on it. For each of the situations described below, identify which of the four changes are taking place and write as many of the letters Q, T, E^{int}, W, (or none) as are appropriate. (a) A cylinder with a piston on top contains a compressed gas and is sitting on a thermal reservoir (a large iron block). After everything has come to thermal equilibrium, the piston is moved upward somewhat (very slowly). The object to be considered is the gas in the cylinder. (b) Consider the same cylinder as in part (a), but it is wrapped in styrofoam, a very good thermal insulator, instead of sitting on a thermal reservoir. The piston is pressed downward (again, very slowly), compressing the gas. The object to be considered is the gas in the cylinder. (c) An ice cube that is sitting in the open air and is melting.

68. KE and Temperature Converting kinetic energy into thermal energy produces small rises in temperature. This was in part responsible for the difficulty in discovering the law of conservation of energy. It also implies that hot objects contain a lot of energy. (This latter comment is largely responsible for the industrial revolution in the 19th century.) To get some feel for these numbers, assume all mechanical energy is converted to thermal energy and carry out three estimates:

(a) A steel ball is dropped from a height of 3 m onto a concrete floor. It bounces a large number of times but eventually comes to rest. Estimate the ball's rise in temperature.

(b) Suppose the steel ball you used in part (a) is at room temperature. If you converted all its thermal energy to translational kinetic energy, how fast would it be moving? (Give your answer in units of miles per hour. Also, ignore the fact that you would have to create momentum.)

(c) Suppose a nickel–iron meteor falls to Earth from deep space. Estimate how much its temperature would rise on impact.

20 | The Kinetic Theory of Gases

When a container of cold champagne, soda pop, or any other carbonated drink is opened, a slight fog forms around the opening and some of the liquid sprays outward. (In the photograph, the fog is the white cloud that surrounds the stopper, and the spray has formed streaks within the cloud.)

What causes the fog?

The answer is in this chapter.

20-1 Molecules and Thermal Gas Behavior

In our studies of mechanics and thermodynamics we have found a number of strange and interesting results. In mechanics, we saw that moving objects tend to run down and come to a stop. We attributed this to the inevitable presence of friction and drag forces. Without these nonconservative forces mechanical energy would be conserved and perpetual motion would be possible. In thermodynamics, we discovered that ordinary objects, by virtue of their temperature, contain huge quantities of internal energy. This is where the "lost" energy resulting from friction forces is hidden. In this chapter, we will learn about some ways that matter can store internal energy. What you are about to learn may be counterintuitive. Instead of finding that the "natural state" of a system is to lose energy, you will find considerable evidence that the "natural state" of a system is quite the opposite. It is one in which its fundamental parts (atoms and molecules) are traveling every which way—in a state of perpetual motion.

Classical thermodynamics—the subject of the previous chapter—has nothing to say about atoms or molecules. Its laws are concerned only with such macroscopic variables as pressure, volume, and temperature. In this chapter we begin an exploration of the atomic and molecular basis of thermodynamics. As is usual in the development of new theories in physics, we start with a simple model. The fact that gases are fluid and compressible is evidence that their molecules are quite small relative to the average spacing between them. If so, we expect that gas molecules are relatively free and independent of one another. For this reason, we believe that the thermal behavior of gases will be easier to understand than that of liquids and solids. Thus, we begin an exploration of the atomic and molecular basis of thermodynamics by developing the kinetic theory of gases—a simplified model of gas behavior based on the laws of classical mechanics.

We start with a discussion of how the ideal gas law characterizes the macroscopic behavior of simple gases. This macroscopic law relates the amount of gas and its pressure, temperature, and volume to each other. Next we consider how kinetic theory, which provides us with a molecular (or microscopic) model of gas behavior, can be used to explain observed macroscopic relationships between gas pressure, volume, and temperature. We then move on to using kinetic theory as an underlying model of the characteristics of an ideal gas. The basic ideas of kinetic theory are that: (1) an ideal gas at a given temperature consists of a collection of tiny particles (atoms or molecules) that are in perpetual motion—colliding with each other and the walls of their container; and (2) the hidden internal energy of an ideal gas is directly proportional to the kinetic energy of its particles.

20-2 The Macroscopic Behavior of Gases

Any gas can be described by its macroscopic variables volume V, pressure P, and temperature T. Simple experiments were performed on low density gases in the 17th and 18th centuries to relate these variables. Robert Boyle (b. 1627) determined that at a constant temperature the product of pressure and volume remains constant. (See Fig. 20-1.)

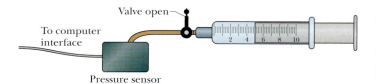

FIGURE 20-1 ■ A contemporary setup for determining the relationship between gas pressure and volume using an inexpensive medical syringe and an electronic pressure sensor attached to a computer data acquisition system. The volume is changed when the plunger is pushed or pulled. When temperature is held constant, P and V turn out to be inversely proportional to each other so that PV is constant.

French scientists Jacques Charles (b. 1746) and Joseph Gay-Lussac (b. 1778) found that as the Kelvin temperature of a fixed volume of gas is raised its pressure increases proportionally. (See Fig. 19-16.) Similarly, Charles, who was a hot-air balloonist, discovered that for a constant pressure (such as atmospheric pressure) the volume of a gas is proportional to its temperature. (See Fig. 19-15.) By combining the results of all three of these experiments we must conclude that there is a proportionality between PV and T:

$$PV \propto T.$$

The Molecular Form of the Ideal Gas Law

If we can find a constant of proportionality between PV and T for a relatively low density gas, then we will have formulated a gas law. An examination of the student-generated P vs. T data shown in Fig. 19-17 indicates that the constant of proportionality between the product PV and the variable T (determined by the slopes) of the graphs decreases as the mass of gas confined to the same volume decreases. Similar experiments have shown that the slope of a P vs. T graph will change if the same volume and mass of a different kind of gas is used. This suggests that the constant of proportionality we are looking for must be a function of *both* the mass and type of gas. It was puzzling to early investigators that the slopes of their P vs. T and V vs. T graphs were not just proportional to the mass of the different gases used in the experiments.

The key to finding a constant of proportionality that embodies both gas type and mass was a hypothesis developed in the early 19th century by the Italian scientist Amadeo Avogadro (1776–1856). In 1811, Avogadro proposed that equal volumes of any kind of gas at the same pressure would have the same number of molecules and occupy the same volume. Eventually it was discovered that the constant of proportionality needed for the fledgling gas law was one that is directly proportional to the number of molecules of a gas rather than its mass, so that

$$PV = Nk_BT \qquad \text{(molecular ideal gas law),} \qquad (20\text{-}1)$$

where N is the number of molecules of confined gas and k_B is a proportionality constant needed to shift from kelvins to joules, the SI units for the product PV. The experimentally determined value of k_B is known historically as the Boltzmann constant. Its measured value is

$$k_B = 1.38 \times 10^{-23} \text{ J/K} \qquad \text{(Boltzmann constant).} \qquad (20\text{-}2)$$

It turns out that common gases such as O_2, N_2, and Ar behave like ideal gases at relatively low pressure (< 10 atm) when their temperatures are well above their boiling points. For example, air near room temperature and 1 atm of pressure behaves like an ideal gas.

Avogadro's Number and the Mole

The problem with the molecular form of the ideal gas law we just presented is that it is hard to count molecules. It is much easier to measure the mass of a sample of gas or its volume at a standard pressure. In this subsection we will define two new quantities—*mole* and *molar mass*. Although these quantities are related to the number of molecules in a gas, they can be measured macroscopically, so it is useful to reformulate the ideal gas law in terms of moles.

Let's start our reformulation of the ideal gas law with definitions of mole and molar mass. In Section 1-7, we presented the SI definition of the *atomic mass unit* in terms of the mass of a carbon-12 atom. In particular, carbon-12 is assigned an atomic mass of exactly 12 u. Here the atomic mass unit u represents grams per mole (g/mol). In a related fashion, the SI definition of the *mole* (or *mol* for short) relates the number of particles in a substance to its macroscopic mass.

> A **mole** is defined as the amount of any substance that contains the same number of atoms or molecules as there are in *exactly* 12 g of carbon-12.

The results of many different types of experiments, including x-ray diffraction studies in crystals, have revealed that there are a very large number of atoms in 12 g of carbon-12. The number of atoms is known as Avogadro's number and is denoted as N_A.

$$N_A = 6.022137 \times 10^{23} \, \text{mol}^{-1} \qquad \text{(Avogadro's number)}. \qquad (20\text{-}3)$$

Here the symbol mol^{-1} represents the inverse mole or "per mole." Usually we round off the value to three significant figures so that $N_A = 6.02 \times 10^{23} \, \text{mol}^{-1}$.

The number of moles n contained in a sample of any substance is equal to the ratio of the number of atoms or molecules N in the sample to the number of atoms or molecules N_A in 1 mole of the same substance:

$$n = \frac{N}{N_A}. \qquad (20\text{-}4)$$

(*Caution:* The three symbols in this equation can easily be confused with one another, so you should sort them with their meanings now, before you end in "N-confusion.")

We can easily calculate the mass of one mole of atoms or molecules in any sample, defined as the **molar mass** (denoted as M), by looking in a table of atomic or molecular masses.

Note that if we refer to Appendix F to find the molar mass, M, in grams of a sample of matter, we can determine the number of moles in the sample by determining its mass M_{sam} and using the equation

$$n = \frac{M_{sam}}{M}. \qquad (20\text{-}5)$$

For atoms, the molar mass is just the atomic mass so that molar mass also has the unit g/mol, which is often denoted as u.

It is puzzling to note that the atomic mass of carbon that is listed in Appendix F is given as 12.01115 u rather than 12.00000 u. This is because a natural sample of carbon does not consist of only carbon-12. Instead it contains a relatively small percentage of carbon-13, which has an extra neutron in its nucleus. Nevertheless, by definition, a mole of pure carbon-12 and a mole of a naturally occurring mixture of carbon-12 and carbon-13 both contain Avogadro's number of atoms.

The Molar Form of the Ideal Gas Law

We can rewrite the molecular ideal gas law expressed in Eq. 20-1 in an alternative form by using Eq. 20-4, so that

$$PV = Nk_BT = nN_Ak_BT.$$

Since both Avogadro's number and the Boltzmann constant are constants, we can replace their product with a new constant R, which is called the **universal gas constant** because it has the same value for all ideal gases—namely,

$$R = N_A k_B = 6.02 \times 10^{23} \, \text{mol}^{-1}(1.38 \times 10^{-23} \, \text{J/K}) = 8.31 \, \text{J/mol} \cdot \text{K}. \quad (20\text{-}6)$$

This allows us to write

$$nR = Nk_B. \quad (20\text{-}7)$$

Substituting this into Eq. 20-1 gives a second expression for the **ideal gas law:**

$$PV = nRT \quad \text{(molar ideal gas law)}, \quad (20\text{-}8)$$

in which P is the absolute (not gauge) pressure, V is the volume, n is the number of moles of gas present, and T is the temperature in Kelvin. Provided the gas density is low, the ideal gas law as represented in either Eq. 20-1 or Eq. 20-8 holds for any single gas or for any mixture of different gases. (For a mixture, n is the total number of moles in the mixture.)

Note the difference between the two expressions for the ideal gas law—Eq. 20-8 involves the number of moles n and Eq. 20-1 involves the number of atoms N. That is, the Boltzmann constant k_B tells us about individual atomic particles, whereas the gas constant R tells us about moles of particles. Recall that moles are defined via macroscopic measurements that are easily done in the lab—such as 1 mol of carbon has a mass of 12 g. As a result, R is easily measured in the lab. On the other hand, since k_B is about individual atoms, to get to it from a lab measurement we have to count the number of molecules in a mole. This is a decidedly nontrivial task.

You may well ask, "What is an *ideal gas* and what is so 'ideal' about it?" The answer lies in the simplicity of the law (Eqs. 20-1 and 20-8) that describes the macroscopic properties of a gas. Using this law—as you will see—we can deduce many properties of the ideal gas in a simple way. There is no such thing in nature as a truly ideal gas. But *all* gases approach the ideal state at low enough densities—that is, under conditions in which their molecules are far enough apart that they do not interact with one another as much as they do with the walls of their containers. Thus, the two equivalent ideal gas equations allow us to gain useful insights into the behavior of most real gases at low densities.

TOUCHSTONE EXAMPLE 20-1: Final Pressure

A cylinder contains 12 L of oxygen at 20°C and 15 atm. The temperature is raised to 35°C, and the volume is reduced to 8.5 L. What is the final pressure of the gas in atmospheres? Assume that the gas is ideal.

SOLUTION ■ The **Key Idea** here is that, because the gas is ideal, its pressure, volume, temperature, and number of moles are related by the ideal gas law, both in the initial state i and in the final state f (after the changes). Thus, from Eq. 20-8 we can write $P_i V_i = nRT_i$ and $P_f V_f = nRT_f$. Dividing the second equation by the first equation and solving for P_f yields

$$P_f = \frac{P_i T_f V_i}{T_i V_f}. \quad (20\text{-}9)$$

Note here that if we converted the given initial and final volumes from liters to SI units of cubic meters, the multiplying conversion factors would cancel out of Eq. 20-9. The same would be true for conversion factors that convert the pressures from atmospheres to the more accepted SI unit of pascals. However, to convert the given temperatures to kelvins requires the addition of an amount that would not cancel and thus must be included. Hence, we must write

$$T_i = (273 + 20) \, \text{K} = 293 \, \text{K}$$

and

$$T_f = (273 + 35) \, \text{K} = 308 \, \text{K}.$$

Inserting the given data into Eq. 20-9 then yields

$$P_f = \frac{(15 \, \text{atm})(308 \, \text{K})(12 \, \text{L})}{(293 \, \text{K})(8.5 \, \text{L})} = 22 \, \text{atm}. \quad \text{(Answer)}$$

20-3 Work Done by Ideal Gases

Heat engines are devices that can absorb thermal energy and do useful work on their surroundings. As you will see in the next chapter, air, which is typically used as a working medium in heat engines, behaves like an ideal gas in some circumstances. For this reason engineers are interested in knowing how to calculate the work done by ideal gases. Before we turn our attention to how the action of molecules that make up an ideal gas can be used to explain the ideal gas law, we first consider how to calculate the work done by ideal gases under various conditions. We restrict ourselves to expansions that occur slowly enough that the gas is very close to thermal equilibrium throughout its volume.

Work Done by an Ideal Gas at Constant Temperature

Suppose we put an ideal gas in a piston–cylinder arrangement like those in Chapter 19. Suppose also that we allow the gas to expand from an initial volume V_i to a final volume V_f while we keep the temperature T of the gas constant. Such a process, at *constant temperature,* is called an **isothermal expansion** (and the reverse is called an **isothermal compression**).

On a P-V diagram, an *isotherm* is a curve that connects points that have the same temperature. Thus, it is a graph of pressure versus volume for a gas whose temperature T is held constant. For n moles of an ideal gas, it is a graph of the equation

$$P = nRT\,\frac{1}{V} = \text{ (a constant) } \frac{1}{V}. \tag{20-10}$$

Figure 20-2 shows three isotherms, each corresponding to a different (constant) value of T. (Note that the values of T for the isotherms increase upward to the right.) Superimposed on the middle isotherm is the path followed by a gas during an isothermal expansion from state i to state f at a constant temperature of 310 K.

To find the work done by an ideal gas during an isothermal expansion, we start with Eq. 19-16,

$$W = \int_{V_i}^{V_f} P\,dV. \tag{20-11}$$

This is a general expression for the work done during any change in volume of any gas. For an ideal gas, we can use Eq. 20-8 to substitute for P, obtaining

$$W = \int_{V_i}^{V_f} \frac{nRT}{V}\,dV. \tag{20-12}$$

Because we are considering an isothermal expansion, T is constant and we can move it in front of the integral sign to write

$$W = nRT \int_{V_i}^{V_f} \frac{dV}{V} = nRT[\ln V]_{V_i}^{V_f}. \tag{20-13}$$

By evaluating the expression in brackets at the limits and then using the relationship $\ln a - \ln b = \ln (a/b)$, we find that

$$W = nRT \ln \frac{V_f}{V_i} = Nk_B \ln \frac{V_f}{V_i} \qquad \text{(ideal gas, isothermal process)}. \tag{20-14}$$

Recall that the symbol ln specifies a *natural* logarithm, which has base e.

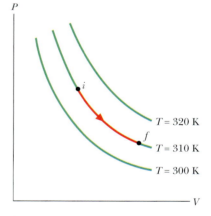

FIGURE 20-2 ■ Three isotherms on a P-V diagram. The path shown along the middle isotherm represents an isothermal expansion of a gas from an initial state i to a final state f. The path from f to i along the isotherm would represent the reverse process, an isothermal compression.

As we often do in science and engineering we have derived a mathematical relationship. Before using this relationship it's a good idea to check our equation to see whether it makes sense. Unless a gas is undergoing a free expansion into a vacuum, we know that an expanding gas does work on its surroundings. If the gas contracts we expect that the surroundings have done work on the gas instead. Is this what Eq. 20-14 tells us? For an expansion, V_f is greater than V_i, so the ratio V_f/V_i in Eq. 20-14 is greater than unity. The natural logarithm of a quantity greater than unity is positive, and so the work W done by an ideal gas during an isothermal expansion is positive, as we expect. For a compression, V_f is less than V_i, so the ratio of volumes in Eq. 20-14 is less than unity. The natural logarithm in that equation—hence the work W—is negative, again as we expect.

Work Done at Constant Volume and at Constant Pressure

Equation 20-14 does not give the work W done by an ideal gas during *every* thermodynamic process. Instead, it gives the work only for a process in which the temperature is held constant. If the temperature varies, then the symbol T in Eq. 20-12 cannot be moved in front of the integral symbol as in Eq. 20-13, and thus we do not end up with Eq. 20-14.

However, we can go back to Eq. 20-11 to find the work W done by an ideal gas (or any other gas) during two more processes—a constant-volume process and a constant-pressure process. If the volume of the gas is constant, then Eq. 20-11 yields

$$W = 0 \quad \text{(constant-volume process).} \tag{20-15}$$

If, instead, the volume changes while the pressure P of the gas is held constant, then Eq. 20-11 becomes

$$W = P(V_f - V_i) = P\Delta V \quad \text{(constant-pressure process).} \tag{20-16}$$

READING EXERCISE 20-1: An ideal gas has an initial pressure of 3 pressure units and an initial volume of 4 volume units. The table gives the final pressure and volume of the gas (in those same units) in five processes. Which processes start and end on the same isotherm?

	a	b	c	d	e
P	12	6	5	4	1
V	1	2	7	3	12

■

TOUCHSTONE EXAMPLE 20-2: Work Done by Expansion

One mole of oxygen (assume it to be an ideal gas) expands at a constant temperature T of 310 K from an initial volume V_i of 12 L to a final volume V_f of 19 L. How much work is done by the gas during the expansion?

SOLUTION ■ The **Key Idea** is this: Generally we find the work by integrating the gas pressure with respect to the gas volume, using Eq. 20-11. However, because the gas here is ideal and the expansion is isothermal, that integration leads to Eq. 20-14. Therefore, we can write

$$W = nRT \ln\frac{V_f}{V_i}$$

$$= (1 \text{ mol})(8.31 \text{ J/mol·K})(310 \text{ K}) \ln\left(\frac{19 \text{ L}}{12 \text{ L}}\right)$$

$$= 1180 \text{ J.} \qquad \text{(Answer)}$$

The expansion is graphed in the *P-V* diagram of Fig. 20-3. The work done by the gas during the expansion is represented by the area beneath the curve between *i* and *f*.

You can show that if the expansion is now reversed, with the gas undergoing an isothermal compression from 19 L to 12 L, the work done by the gas will be −1180 J. Thus, an external force would have to do 1180 J of work on the gas to compress it.

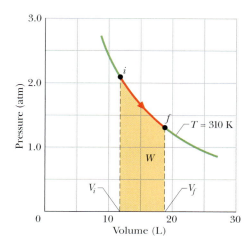

FIGURE 20-3 ■ The shaded area represents the work done by 1 mol of oxygen in expanding from V_i to V_f at a constant temperature *T* of 310 K.

20-4 Pressure, Temperature, and Molecular Kinetic Energy

In terms of our everyday experiences, molecules and atoms are invisible. Only in the past 40 years or so have scientists been able to "see" molecules using electron microscopes and field ion microscopes. But long before atoms and molecules could be "seen," 19th-century scientists such as James Clerk Maxwell and Ludwig Boltzmann in Europe and Josiah Willard Gibbs in the United States constructed models that made the description and prediction of the *macroscopic* (visible to the naked eye) behavior of thermodynamic systems possible. Their models were based on the yet unseen *microscopic* atoms and molecules.

Is it possible to describe the behavior of an ideal gas that obeys the first law of thermodynamics microscopically as a collection of moving molecules? To answer this question, let's observe the pressure exerted by a hypothetical molecule undergoing perfectly elastic collisions with the walls of a cubical box. By using the laws of mechanics we can derive a mathematical expression for the pressure exerted by just one of the molecules as a function of the volume of the box. Next we can extend our "ideal gas" so it is a low-density collection of molecules all having the same mass. By low density we mean that the volume occupied by the molecules is negligible compared to the volume of their container. This means that the molecules are far enough apart on the average that attractive interactions between molecules are also negligible. For this reason an ideal gas has internal energy related to its configuration. If we then define temperature as being related to the average kinetic energy of the molecules in an ideal gas, we can show that kinetic theory is a powerful construct for explaining both the ideal gas law and the first law of thermodynamics.

We start developing our idealized kinetic theory model by considering *N* molecules of an ideal gas that are confined in a cubical box of volume *V*, as in Fig. 20-4. The walls of the box are held at temperature *T*. How is the pressure *P* exerted by the gas on the walls related to the speeds of the molecules? Remember from our discussions of fluids in Chapter 15 that pressure is a scalar defined as the ratio of the magnitude of force (exerted normal to a surface) and the area of the surface. In the example at hand, a gas confined to a box, the pressure results from the motion of molecules in all directions resulting in elastic collisions between gas molecules and the walls of the box. We ignore (for the time being) collisions of the molecules with one another and consider only elastic collisions with the walls.

Figure 20-4 shows a typical gas molecule, of mass *m* and velocity $\vec{v}$, which is about to collide with the shaded wall. Because we assume that any collision of a molecule

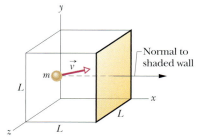

FIGURE 20-4 ■ We assume a cubical box of edge *L* contains *N* ideal gas molecules (not shown) that move around perpetually without losing energy. One of the molecules of mass *m* and velocity $\vec{v}$ is shown heading for a collision with the shaded wall of area L^2. The normal to the shaded wall points in the positive *x* direction.

with a wall is elastic, when this molecule collides with the shaded wall, the only component of its velocity that is changed by the collision is its x-component. That x-component has the same magnitude after collision but its sign is reversed. This means that the only change in the particle's momentum is along the x axis, so

$$(\Delta p_x)_{\text{molecule}} = p_{fx} - p_{ix} = (-m|v_x|) - (+m|v_x|) = -2m|v_x|.$$

But the law of conservation of momentum tells us that the momentum change $(\Delta p_x)_{\text{wall}}$ that the wall experiences after a molecule collides with it is $+2m|v_x|$. Remember that in this book $\vec{p}, p_x, p_y,$ and p_z denote momentum vectors or vector components and capital P represents pressure. *Be careful not to confuse them.*

The molecule of Fig. 20-4 will hit the shaded wall repeatedly. The time Δt between collisions is the time the molecule takes to travel to the opposite wall and back again (a distance of $2L$) at speed $|v_x|$. Thus, Δt is equal to $2L/|v_x|$. (Note that this result holds even if the molecule bounces off any of the other walls along the way, because those walls are parallel to x and so cannot change $|v_x|$.) Therefore, the average rate at which momentum is delivered to the shaded wall by this single molecule is

$$\frac{(\Delta p_x)_{\text{wall}}}{\Delta t} = \frac{+2m|v_x|}{2L/|v_x|} = \frac{mv_x^2}{L}.$$

From Newton's Second Law ($\vec{F} = d\vec{p}/dt$), the rate at which momentum is delivered to the wall is the force acting on that wall. To find the total force, we must add up the contributions of all of the N molecules that strike the wall during a short time interval Δt. We will allow for the possibility that all the molecules have different velocities. Then we can divide the magnitude of the total force acting normal to the shaded wall $|F_x|$ by the area of the wall (L^2) to determine the pressure P on that wall. Thus,

$$P = \frac{|F_x|}{L^2} = \frac{mv_{x\,1}^2/L + mv_{x\,2}^2/L + \cdots + mv_{x\,N}^2/L}{L^2} \tag{20-17}$$

$$= \left(\frac{m}{L^3}\right)(v_{x\,1}^2 + v_{x\,2}^2 + \cdots + v_{x\,N}^2).$$

Since by definition $\langle v_x^2 \rangle = (v_{x\,1}^2 + v_{x\,2}^2 + \cdots + v_{x\,N}^2)/N$ we can replace the sum of squares of the velocities in the second parentheses of Eq. 20-17 by $N\langle v_x^2 \rangle$, where $\langle v_x^2 \rangle$ is the average value of the square of the x-components of all the speeds. Equation 20-17 for the pressure on the container wall then reduces to

$$P = \frac{Nm}{L^3}\langle v_x^2 \rangle = \frac{Nm}{V}\langle v_x^2 \rangle, \tag{20-18}$$

since the volume V of the cubical box is just L^3.

It is reasonable to assume that molecules are moving at random in three dimensions rather than just in the x direction that we considered initially, so that $\langle v_x^2 \rangle = \langle v_y^2 \rangle = \langle v_z^2 \rangle$ and

$$\langle v^2 \rangle = \langle v_x^2 + v_y^2 + v_z^2 \rangle = \langle v_x^2 \rangle + \langle v_y^2 \rangle + \langle v_z^2 \rangle = 3\langle v_x^2 \rangle,$$

or

$$\langle v_x^2 \rangle = \langle v^2 \rangle/3.$$

Thus, we can rewrite the expression above as

$$P = \frac{Nm\langle v^2 \rangle}{3V}. \tag{20-19}$$

The square root of $\langle v^2 \rangle$ is a kind of average speed, called the **root-mean-square speed** of the molecules and symbolized by v^{rms}. Its name describes it rather well: You *square* each speed, you find the *mean* (that is, the average) of all these squared speeds, and then you take the square *root* of that mean. With $\sqrt{\langle v^2 \rangle} = v^{rms}$, we can then write Eq. 20-19 as

$$P = \frac{Nm(v^{rms})^2}{3V}. \tag{20-20}$$

Equation 20-20 is very much in the spirit of kinetic theory. It tells us how the pressure of the gas (a purely macroscopic quantity) depends on the speed of the molecules (a purely microscopic quantity). We can turn Eq. 20-20 around and use it to calculate v^{rms} as

$$v^{rms} = \sqrt{\frac{3PV}{Nm}}.$$

Combining this with the molecular form of the ideal gas law in Eq. 20-1 ($PV = Nk_BT$) gives us

$$v^{rms} = \sqrt{\frac{3k_BT}{m}} \qquad \text{(ideal gas)}, \tag{20-21}$$

where m is the mass of a single molecule in kilograms.

Table 20-1 shows some rms speeds calculated from Eq. 20-21. The speeds are surprisingly high. For hydrogen molecules at room temperature (300 K), the rms speed is 1920 m/s or 4300 mi/h—faster than a speeding bullet! Remember too that the rms speed is only a kind of average speed; some molecules move much faster than this, and some much slower.

TABLE 20-1
Some Molecular Speeds at Room Temperature ($T = 300$ K)[a]

Gas	Molar Mass $M = mN_A$ $(10^{-3}$ kg/mol)	v^{rms} (m/s)
Hydrogen (H_2)	2.02	1920
Helium (He)	4.0	1370
Water vapor (H_2O)	18.0	645
Nitrogen (N_2)	28.0	517
Oxygen (O_2)	32.0	483
Carbon dioxide (CO_2)	44.0	412
Sulfur dioxide (SO_2)	64.1	342

[a]For convenience, we often set room temperature at 300 K even though (at 27°C or 81°F) that represents a fairly warm room.

The speed of sound in a gas is closely related to the rms speed of the molecules of that gas. In a sound wave, the disturbance is passed on from molecule to molecule by means of collisions. The wave cannot move any faster than the "average" speed of the molecules. In fact, the speed of sound must be somewhat less than this "average" molecular speed because not all molecules are moving in exactly the same direction as the wave. As examples, at room temperature, the rms speeds of hydrogen and nitrogen molecules are 1920 m/s and 517 m/s, respectively. The speeds of sound in these two gases at this temperature are 1350 m/s and 350 m/s, respectively.

Translational Kinetic Energy

Let's again consider a single molecule of an ideal gas as it moves around in the box of Fig. 20-4, but we now assume that its speed changes when it collides with other molecules. Its translational kinetic energy at any instant is $\frac{1}{2}mv^2$. Its *average* translational kinetic energy over the time that we watch it is

$$\langle K \rangle = \tfrac{1}{2}\langle mv^2 \rangle = \tfrac{1}{2}m\langle v^2 \rangle = \tfrac{1}{2}m(v^{\text{rms}})^2, \tag{20-22}$$

in which we make the assumption that the average speed of the molecule during our observation is the same as the average speed of all the molecules at any given instant. (Provided the total energy of the gas is not changing and we observe our molecule for long enough, this assumption is appropriate.) Substituting for v^{rms} from Eq. 20-21 leads to

$$\langle K \rangle = (\tfrac{1}{2}m)\frac{3k_B T}{m}$$

so that
$$\langle K \rangle = \tfrac{3}{2}k_B T \qquad \text{(one ideal gas molecule)}. \tag{20-23}$$

This equation tells us something unexpected:

> At a given temperature T, all ideal gas molecules—no matter what their mass—have the same average translational kinetic energy—namely, $(\tfrac{3}{2})k_B T$. When we measure the temperature of a gas, we are also measuring the average translational kinetic energy of its molecules.

READING EXERCISE 20-2: What happens to the average translational kinetic energy of each molecule in a gas when its temperature in kelvin: (a) doubles and (b) is reduced to zero? ∎

READING EXERCISE 20-3: A gas mixture consists of molecules of types 1, 2, and 3, with molecular masses $m_1 > m_2 > m_3$. Rank the three types according to (a) average kinetic energy and (b) rms speed, greatest first. ∎

20-5 Mean Free Path

In considering the motion of molecules, a question often arises: If molecules move so fast (hundreds of meters per second), why does it take as long as a minute or so before you can smell perfume when someone opens a bottle across a room (only a few meters away)? To answer this question, we continue to examine the motion of molecules in an ideal gas. Figure 20-5 shows the path of a typical molecule as it moves through the gas, changing both speed and direction abruptly as it collides elastically with other molecules. Between collisions, our typical molecule moves in a straight line at constant speed. Although the figure shows all the other molecules as stationary, they too are moving similarly.

One useful parameter to describe this random motion is the **mean free path** λ of the molecules. As its name implies, λ is the average distance traversed by a molecule between collisions. We expect λ to vary inversely with N/V, the number of molecules per unit volume (or "number density" of molecules). The larger N/V is, the more collisions there should be and the smaller the mean free path. We also expect λ to vary inversely with the size of the molecules, say, with their diameter d. (If the molecules were points, as we have assumed them to be, they would never collide and the mean

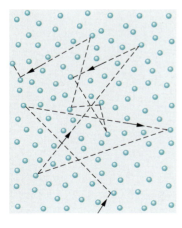

FIGURE 20-5 ∎ A molecule traveling through a gas, colliding with other gas molecules in its path. Although the other molecules are shown as stationary, we believe they are also moving in a similar fashion.

free path would be infinite.) Thus, the larger the molecules are, the smaller the mean free path. We can even predict that λ should vary (inversely) as the *square* of the molecular diameter because the cross section of a molecule—not its diameter—determines its effective target area.

The expression for the mean free path does, in fact, turn out to be

$$\lambda = \frac{1}{\sqrt{2}\,\pi d^2\, N/V} \qquad \text{(ideal gas mean free path).} \qquad (20\text{-}24)$$

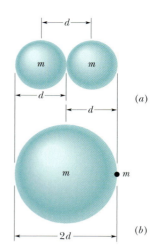

To justify Eq. 20-24, we focus attention on a single molecule and assume—as Fig. 20-5 suggests—that our molecule is traveling with a constant speed v and that all the other molecules are at rest. Later, we shall relax this assumption.

We assume further that the molecules are spheres of diameter d. A collision will then take place if the centers of the molecules come within a distance d of each other, as in Fig. 20-6a. Another, more helpful way to look at the situation is to consider our single molecule to have a *radius* of d and all the other molecules to be *points,* as in Fig. 20-6b. This does not change our criterion for a collision.

As our single molecule zigzags through the gas, it sweeps out a short cylinder of cross-sectional area πd^2 between successive collisions. If we watch this molecule for a time interval Δt, it moves a distance $v\Delta t$, where v is its assumed speed. Thus, if we align all the short cylinders swept out in Δt, we form a composite cylinder (Fig. 20-7) of length $v\Delta t$ and volume $(\pi d^2)(v\Delta t)$. The number of collisions that occur in time Δt is then equal to the number of (point) molecules that lie within this cylinder.

Since N/V is the number of molecules per unit volume, the number of molecules in the cylinder is N/V times the volume of the cylinder, or $(N/V)(\pi d^2 v\Delta t)$. This is also the number of collisions in time Δt. The mean free path is the length of the path (and of the cylinder) divided by this number:

FIGURE 20-6 ■ (*a*) A collision occurs when the centers of two molecules come within a distance d of each other, d being the molecular diameter. (*b*) An equivalent but more convenient representation is to think of the moving molecule of interest as having a *radius* d and all other molecules as being points. The condition for a collision is unchanged.

$$\lambda = \frac{\text{length of path}}{\text{number of collisions}} \approx \frac{v\Delta t}{\pi d^2 v\Delta t\, N/V}$$

$$= \frac{1}{\pi d^2\, N/V}. \qquad (20\text{-}25)$$

This equation is only approximate because it is based on the assumption that all the molecules except one are at rest. In fact, *all* the molecules are moving; when this is taken properly into account, Eq. 20-24 results. Note that it differs from the (approximate) Eq. 20-25 only by a factor of $1/\sqrt{2}$.

We can even get a glimpse of what is "approximate" about Eq. 20-25. The v in the numerator and that in the denominator are—strictly—not the same. The v in the numerator is $\langle v \rangle$, the mean speed of the molecule *relative to the container.* The v in the denominator is $\langle v^{\text{rel}} \rangle$, the mean speed of our single molecule *relative to the other molecules,* which are moving. It is this latter average speed that determines the number of collisions. A detailed calculation, taking into account the actual speed distribution of the molecules, gives $\langle v^{\text{rel}} \rangle = \sqrt{2}\,\langle v \rangle$ and thus the factor $\sqrt{2}$.

The mean free path of air molecules at sea level is about 0.1 μm. At an altitude of 100 km, the density of air has dropped to such an extent that the mean free path rises to about 16 cm. At 300 km, the mean free path is about 20 km. A problem faced by those who would study the physics and chemistry of the upper atmosphere in the laboratory is the unavailability of containers large enough to hold gas samples that simulate upper atmospheric conditions. Yet studies of the concentrations of freon, carbon dioxide, and ozone in the upper atmosphere are of vital public concern.

Recall the question that began this section: If molecules move so fast, why does it take as long as a minute or so before you can smell perfume when someone opens a bottle across a room? We now know part of the answer. In still air, each perfume molecule moves away from the bottle only very slowly because its repeated collisions with other molecules prevent it from moving directly across the room to you.

FIGURE 20-7 ■ In time Δt the moving molecule effectively sweeps out a cylinder of length $v\,\Delta t$ and radius d.

TOUCHSTONE EXAMPLE 20-3: Mean Free Path

(a) What is the mean free path λ for oxygen molecules at temperature $T = 300$ K and pressure $P = 1.00$ atm? Assume that the molecular diameter is $d = 290$ pm and the gas is ideal.

SOLUTION ■ The **Key Idea** here is that each oxygen molecule moves among other *moving* oxygen molecules in a zigzag path due to the resulting collisions. Thus, we use Eq. 20-24 for the mean free path, for which we need the number of molecules per unit volume, N/V. Because we assume the gas is ideal, we can use the ideal gas law of Eq. 20-1 ($PV = Nk_BT$) to write $N/V = P/k_BT$. Substituting this into Eq. 20-24, we find

$$\lambda = \frac{1}{\sqrt{2}\pi d^2\, N/V} = \frac{k_BT}{\sqrt{2}\pi d^2\, P}$$

$$= \frac{(1.38 \times 10^{-23}\ \text{J/K})(300\ \text{K})}{\sqrt{2}\pi(2.9 \times 10^{-10}\ \text{m})^2\, (1.01 \times 10^5\ \text{Pa})} \quad \text{(Answer)}$$

$$= 1.1 \times 10^{-7}\ \text{m}.$$

This is about 380 molecular diameters.

(b) Assume the average speed of the oxygen molecules is $\langle v \rangle = 450$ m/s. What is the average time interval Δt between successive

collisions for any given molecule? At what rate does the molecule collide; that is, what is the frequency f of its collisions?

SOLUTION ■ To find the time interval Δt between collisions, we use this **Key Idea**: Between collisions, the molecule travels, on average, the mean free path λ at average speed $\langle v \rangle$. Thus, the average time between collisions is

$$\langle \Delta t \rangle = \frac{(\text{distance})}{(\text{average speed})} = \frac{\lambda}{\langle v \rangle} = \frac{1.1 \times 10^{-7}\ \text{m}}{450\ \text{m/s}} \quad \text{(Answer)}$$

$$= 2.44 \times 10^{-10}\ \text{s} \approx 0.24\ \text{ns}.$$

This tells us that, on average, any given oxygen molecule has less than a nanosecond between collisions.

To find the frequency f of the collisions, we use this **Key Idea**: The average rate or frequency at which the collisions occur is the inverse of the average time $\langle \Delta t \rangle$ between collisions. Thus,

$$f = \frac{1}{2.44 \times 10^{-10}\ \text{s}} = 4.1 \times 10^9\ \text{s}^{-1}. \quad \text{(Answer)}$$

This tells us that, on average, any given oxygen molecule makes about 4 billion collisions per second.

20-6 The Distribution of Molecular Speeds

The root-mean-square speed v^{rms} gives us a general idea of molecular speeds in a gas at a given temperature. We often want to know more. For example, what fraction of the molecules have speeds greater than the rms value? Greater than twice the rms value? To answer such questions, we need to know how the possible values of speed are distributed among the molecules. Figure 20-8a shows this distribution for oxygen molecules at room temperature ($T = 300$ K); Fig. 20-8b compares it with the distribution at $T = 80$ K.

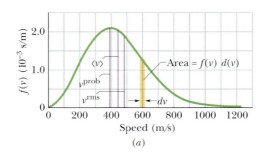
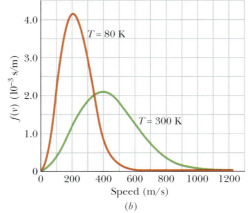

FIGURE 20-8 ■ (a) The Maxwell speed distribution for oxygen molecules at $T = 300$ K. The three characteristic speeds are marked. (b) The curves for 300 K and 80 K. Note that the molecules move more slowly at the lower temperature. Because these are probability distributions, the area under each curve has a numerical value of unity.

In 1852, Scottish physicist James Clerk Maxwell first solved the problem of finding the speed distribution of gas molecules. His result, known as **Maxwell's speed distribution law,** is

$$f(v) = 4\pi\left(\frac{m}{2\pi k_B T}\right)^{3/2} v^2 e^{-mv^2/2k_B T}. \tag{20-26}$$

Here v is the molecular speed, T is the gas temperature, m is the mass of a single gas molecule, and k_B is Boltzmann's constant. It is this equation that is plotted in Fig. 20-8a, b. The quantity $f(v)$ in Eq. 20-26 and Fig. 20-8 is a *probability distribution function:* For any speed v, the product $f(v)dv$ (a dimensionless quantity) is the fraction of molecules whose speeds lie in the interval of width dv centered on speed v.

As Fig. 20-8a shows, this fraction is equal to the area of a strip with height $f(v)$ and width dv. The total area under the distribution curve corresponds to the fraction of the molecules whose speeds lie between zero and infinity. All molecules fall into this category, so the value of this total area is unity; that is,

$$\int_0^\infty f(v)\,dv = 1. \tag{20-27}$$

The fraction of molecules with speeds in an interval of, say, v_1 to v_2 is then

$$\text{fraction} = \int_{v_1}^{v_2} f(v)\,dv. \tag{20-28}$$

Average, RMS, and Most Probable Speeds

In principle, we can find the **average speed** $\langle v \rangle$ of the molecules in a gas with the following procedure: We *weight* each value of v in the distribution; that is, we multiply it by the fraction $f(v)\,dv$ of molecules with speeds in a differential interval dv centered on v. Then we add up all these values of $vf(v)\,dv$. The result is $\langle \vec{v} \rangle$. In practice, we do all this by evaluating

$$\langle v \rangle = \int_0^\infty v\,f(v)\,dv. \tag{20-29}$$

Substituting for $f(v)$ from Eq. 20-26 and using definite integral 20 from the list of integrals in Appendix E, we find

$$\langle v \rangle = \sqrt{\frac{8k_B T}{\pi m}} \qquad \text{(average speed).} \tag{20-30}$$

Similarly, we can find the average of the square of the speeds $\langle v^2 \rangle$ with

$$\langle v^2 \rangle = \int_0^\infty v^2\,f(v)\,dv. \tag{20-31}$$

Substituting for $f(v)$ from Eq. 20-27 and using generic integral 16 from the list of integrals in Appendix E, we find

$$\langle v^2 \rangle = \frac{3k_B T}{m}. \tag{20-32}$$

The square root of $\langle v^2 \rangle$ is the **root-mean-square speed** v^{rms}. Thus,

$$v^{\text{rms}} = \sqrt{\frac{3k_B T}{m}} \qquad \text{(rms speed),} \tag{20-33}$$

which agrees with Eq. 20-21.

The **most probable speed** v^{prob} is the speed at which $f(v)$ is maximum (see Fig. 20-8a). To calculate v^{prob}, we set $df/dv = 0$ (the slope of the curve in Fig. 20-8a is zero at the maximum of the curve) and then solve for v. Doing so, we find

$$v^{\text{prob}} = \sqrt{\frac{2k_B T}{m}} \qquad \text{(most probable speed).} \qquad (20\text{-}34)$$

What is the relationship between the most probable speed, the average speed, and the rms speed of a molecule? The relationship is fixed.

> The most probable speed v^{prob} is always less than the average speed $\langle v \rangle$ which in turn is less than the rms speed v^{rms}. More specifically, $v^{\text{prob}} = 0.82\ v^{\text{rms}}$ and $\langle v \rangle = 0.92\ v^{\text{rms}}$.

This is consistent with the idea that a molecule is more likely to have speed v^{prob} than any other speed, but some molecules will have speeds that are many times v^{prob}. These molecules lie in the *high-speed tail* of a distribution curve like that in Fig. 20-8a. We should be thankful for these few, higher speed molecules because they make possible both rain and sunshine (without which we could not exist). We next see why.

Rain: The speed distribution of water molecules in, say, a pond at summertime temperatures can be represented by a curve similar to that of Fig. 20-8a. Most of the molecules do not have nearly enough kinetic energy to escape from the water through its surface. However, small numbers of very fast molecules with speeds far out in the tail of the curve can do so. It is these water molecules that evaporate, making clouds and rain a possibility.

As the fast water molecules leave the surface, carrying energy with them, the temperature of the remaining water is maintained by thermal energy transfer from the surroundings. Other fast molecules—produced in particularly favorable collisions—quickly take the place of those that have left, and the speed distribution is maintained.

Sunshine: Let the distribution curve of Fig. 20-8a now refer to protons in the core of the Sun. The Sun's energy is supplied by a nuclear fusion process that starts with the merging of two protons. However, protons repel each other because of their electrical charges, and protons of average speed do not have enough kinetic energy to overcome the repulsion and get close enough to merge. Very fast protons with speeds in the tail of the distribution curve can do so, however, and thus the Sun can shine.

20-7 The Molar Specific Heats of an Ideal Gas

Up to now, we have taken the specific heat of a substance as a quantity to be measured. But now, with the kinetic theory of gases, we know something about the structure of matter and where its energy is stored. With this additional information, we can actually calculate and make predictions about what we expect the specific heats of different kinds of gases to be. If we compare our predictions based on kinetic theory to experimental measurements, we get some good agreement and also some surprises. The surprises are among the first hints that the laws of matter at the atomic level are not just Newton's laws scaled down. In other words, we begin to notice that atoms aren't just little billiard balls but something different from any macroscopic object with which we have experience.

To explore this idea, we derive here (from molecular considerations) an expression for the internal energy E^{int} of an ideal gas. In other words, we find an expression for the energy associated with the random motions of the atoms or molecules in the gas. We shall then use that expression to derive the molar specific heats of an ideal gas.

Internal Energy E^{int}

Let us first assume that our ideal gas is a *monatomic gas* (which has individual atoms rather than molecules), such as helium, neon, or argon. Let us also assume that the internal energy E^{int} of our ideal gas is simply the sum of the translational kinetic energies of its atoms.

The average translational kinetic energy of a single atom depends only on the gas temperature and is given by Eq. 20-23 as $\langle K \rangle = \frac{3}{2}k_B T$. A sample of n moles of such a gas contains nN_A atoms. The internal energy E^{int} of the sample is then

$$E^{int} = (nN_A)\langle K \rangle = (nN_A)(\tfrac{3}{2}k_B T). \tag{20-35}$$

Using Eq. 20-6 ($k_B = R/N_A$), we can rewrite this as

$$E^{int} = \tfrac{3}{2}Nk_B T = \tfrac{3}{2}nRT \quad \text{(monatomic ideal gas).} \tag{20-36}$$

Thus,

> The internal energy E^{int} of an ideal gas is a function of the gas temperature *only;* it does not depend on any other variable.

With Eq. 20-36 in hand, we are now able to derive an expression for the molar specific heat of an ideal gas. Actually, we shall derive two expressions. One is for the case in which the volume of the gas remains constant as thermal energy is transferred to or from it. The other is for the case in which the pressure of the gas remains constant as thermal energy is transferred to or from it. The symbols for these two molar specific heats are C_V and C_P, respectively. (By convention, the capital letter C is used in both cases, even though C_V and C_P represent types of specific heat and not heat capacities.)

Molar Specific Heat at Constant Volume

Figure 20-9a shows n moles of an ideal gas at pressure P and temperature T, confined to a cylinder of fixed volume V. This *initial state i* of the gas is marked on the P-V diagram of Fig. 20-9b. Suppose that you add a small amount of thermal energy Q to the gas by slowly turning up the temperature of the thermal reservoir. The gas temperature rises a small amount to $T + \Delta T$, and its pressure rises to $P + \Delta P$, bringing the gas to *final state f*.

In such experiments, we would find that the thermal energy transferred Q is related to the temperature change ΔT by

$$Q = nC_V \Delta T \quad \text{(constant volume),} \tag{20-37}$$

where C_V is a constant called the **molar specific at constant volume.** Substituting this expression for Q into the first law of thermodynamics as given by Eq. 19-17 ($\Delta E^{int} = Q - W$) yields

$$\Delta E^{int} = nC_V \Delta T - W. \tag{20-38}$$

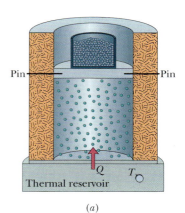

(a)

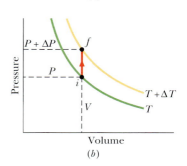

Volume

(b)

FIGURE 20-9 ■ (a) The temperature of an ideal gas is raised from T to $T + \Delta T$ in a constant-volume process. Thermal energy is added, but no work is done. (b) The process on a P-V diagram. (The system's insulated lid is not shown.)

With the volume held constant, the gas cannot expand and thus cannot do any work. Therefore, $W = 0$, and Eq. 20-38 gives us

$$C_V = \frac{\Delta E^{\text{int}}}{n\,\Delta T}. \tag{20-39}$$

From Eq. 20-36 we know that $E^{\text{int}} = \frac{3}{2}nRT$, so the change in internal energy must be

$$\Delta E^{\text{int}} = \frac{3}{2}nR\,\Delta T. \tag{20-40}$$

Substituting this result into Eq. 20-39 yields

$$C_V = \frac{3}{2}R = 12.5 \text{ J/mol} \cdot \text{K} \qquad \text{(monatomic gas).} \tag{20-41}$$

As Table 20-2 shows, this prediction that $C_V = \frac{3}{2}R$ based on ideal gas kinetic theory agrees very well with experiment for the real monatomic gases (the case that we have assumed). The experimental values of C_V for *diatomic gases* and *polyatomic gases* (which have molecules with more than two atoms) are greater than the predicted value of $\frac{3}{2}R$. Reasons for this will be discussed in Section 20-8.

TABLE 20-2
Molar Specific Heats

Molecule		Example	C_V (J/mol · K)
	Ideal		$\frac{3}{2}R = 12.5$
Monatomic	Real	He	12.5
(1 atom)	Real	Ar	12.6
	Ideal		$\frac{5}{2}R = 20.8^*$
Diatomic	Real	N_2	20.7
(2 atoms)	Real	O_2	20.8
	Ideal		$3R = 24.9^*$
Polyatomic	Real	NH_4	29.0
(> 2 atoms)	Real	CO_2	29.7

*The presentation of the $\frac{5}{2}R$ and $3R$ will be explained in the next section.

We can now generalize Eq. 20-36 for the internal energy of any ideal gas by substituting C_V for $\frac{3}{2}R$; we get

$$E^{\text{int}} = nC_VT \qquad \text{(any ideal gas).} \tag{20-42}$$

This equation applies not only to an ideal monatomic gas but also to diatomic and polyatomic ideal gases, provided the experimentally determined value of C_V is used. Just as with Eq. 20-37, we see that the internal energy of a gas depends on the temperature of the gas but not on its pressure or density.

When an ideal gas that is confined to a container undergoes a temperature change ΔT, then from either Eq. 20-39 or Eq. 20-42 we can write the resulting change in its internal energy as

$$\Delta E^{\text{int}} = nC_V\,\Delta T \qquad \text{(any ideal gas, any process).} \tag{20-43}$$

This equation tells us:

> A change in the internal energy E^{int} of a confined ideal gas depends on the change in the gas temperature only; it does *not* depend on what type of process produces the change in the temperature.

As examples, consider the three paths between the two isotherms in the *P-V* diagram of Fig. 20-10. Path 1 represents a constant-volume process. Path 2 represents a constant-pressure process (that we are about to examine). Path 3 represents a process in which no thermal energy is exchanged with the system's environment (we discuss this in Section 20-11). Although the values of Q and work W associated with these three paths differ, as do P_f and V_f, the values of ΔE^{int} associated with the three paths are identical and are all given by Eq. 20-43, because they all involve the same temperature change ΔT. Therefore, no matter what path is actually taken between T and $T + \Delta T$, we can *always* use path 1 and Eq. 20-43 to compute ΔE^{int} easily.

Molar Specific Heat at Constant Pressure

We now assume that the temperature of the ideal gas is increased by the same small amount ΔT as previously, but that the necessary thermal energy (Q) is added with the gas under constant pressure. An experiment for doing this is shown in Fig. 20-11*a*; the *P-V* diagram for the process is plotted in Fig. 20-11*b*. From such experiments we find that the transferred thermal energy Q is related to the temperature change ΔT by

$$Q = nC_P \, \Delta T \qquad \text{(constant pressure),} \qquad (20\text{-}44)$$

where C_P is a constant called the **molar specific heat at constant pressure.** This C_P is *greater* than the molar specific heat at constant volume C_V, because energy must now be supplied not only to raise the temperature of the gas but also for the gas to do work—that is, to lift the weighted piston of Fig. 20-11*a*.

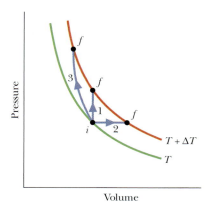

Volume

FIGURE 20-10 ■ Three paths representing three different processes that take an ideal gas from an initial state *i* at temperature *T* to some final state *f* at temperature $T + \Delta T$. The change ΔE^{int} in the internal energy of the gas is the same for these three processes and for any others that result in the same change of temperature.

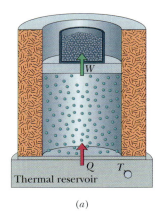

(*a*)

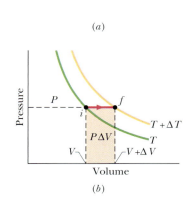

(*b*)

FIGURE 20-11 ■ (*a*) The temperature of an ideal gas is raised from *T* to $T + \Delta T$ in a constant-pressure process. Thermal energy is added and work is done in lifting the loaded piston. (*b*) The process on a *P-V* diagram. The work $P\,\Delta V$ is given by the shaded area.

To relate molar specific heats C_P and C_V, we start with the first law of thermodynamics (Eq. 19-17):

$$\Delta E^{int} = Q - W. \qquad (20\text{-}45)$$

We next replace each term in Eq. 20-45. For ΔE^{int}, we substitute from Eq. 20-43. For Q, we substitute from Eq. 20-44. To replace W, we first note that since the pressure remains constant, Eq. 20-16 tells us that $W = P\Delta V$. Then we note that, using the ideal gas equation ($PV = nRT$), we can write

$$W = P\,\Delta V = nR\,\Delta T. \tag{20-46}$$

Making these substitutions in Eq. 20-45, we find

$$nC_V\Delta T = nC_P\Delta T - nR\Delta T$$

and then dividing through by $n\,\Delta T$,

$$C_V = C_P - R,$$

so
$$C_P = C_V + R \quad \text{(any ideal gas)}. \tag{20-47}$$

This relationship between C_P and C_V predicted by kinetic theory agrees well with experiment, not only for monatomic gases but for gases in general, as long as their density is low enough so that we may treat them as ideal. As we discuss in Section 19-5, there is very little difference between C_P and C_V for liquids and solids because of their relative incompressibility.

READING EXERCISE 20-4: The figure here shows five paths traversed by a gas on a P-V diagram. Rank the paths according to the change in internal energy of the gas, greatest first.

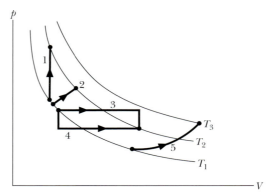

TOUCHSTONE EXAMPLE 20-4: Helium Bubble

A bubble of 5.00 mol of helium is submerged at a certain depth in liquid water when the water (and thus the helium) undergoes a temperature increase ΔT of 20.0 C° at constant pressure. As a result, the bubble expands. The helium is monatomic and ideal.

(a) How much thermal energy is added to the helium during the increase and expansion?

SOLUTION ■ One **Key Idea** here is that the thermal energy transferred Q is related to the temperature change ΔT by the molar

specific heat of the gas. Because the pressure P is held constant during the addition of energy, we use the molar specific heat at constant pressure C_P and Eq. 20-44,

$$Q = nC_P\,\Delta T, \tag{20-48}$$

to find Q. To evaluate C_P we go to Eq. 20-47, which tells us that for any ideal gas, $C_P = C_V + R$. Then from Eq. 20-41, we know that for any *monatomic* gas (like helium), $C_V = \tfrac{3}{2}R$. Thus, Eq. 20-48 gives us

$$Q = n(C_V + R)\,\Delta T = n(\tfrac{3}{2}R + R)\,\Delta T = n(\tfrac{5}{2}R)\,\Delta T$$

$$= (5.00 \text{ mol})(2.5)(8.31 \text{ J/mol} \cdot \text{K})(20.0 \text{ C}°) \qquad \text{(Answer)}$$

$$= 2077.5 \text{ J} \approx 2080 \text{ J}.$$

(b) What is the change ΔE^{int} in the internal energy of the helium during the temperature increase?

SOLUTION ■ Because the bubble expands, this is not a constant-volume process. However, the helium is nonetheless confined (to the bubble). Thus, a **Key Idea** here is that the change ΔE^{int} is the same as *would occur* in a constant-volume process with the same temperature change ΔT. We can easily find the constant-volume change ΔE^{int} with Eq. 20-43:

$$\Delta E^{\text{int}} = nC_V\,\Delta T = n(\tfrac{3}{2}R)\,\Delta T$$

$$= (5.00 \text{ mol})(1.5)(8.31 \text{ J/mol} \cdot \text{K})(20.0 \text{ C}°) \qquad \text{(Answer)}$$

$$= 1246.5 \text{ J} \approx 1250 \text{ J}.$$

(c) How much work W is done by the helium as it expands against the pressure of the surrounding water during the temperature increase?

SOLUTION ■ One **Key Idea** here is that the work done by *any* gas expanding against the pressure from its environment is

given by Eq. 20-11, which tells us to integrate $P\,dV$. When the pressure is constant (as here), we can simplify that to $W = P\,\Delta V$. When the gas is *ideal* (as here), we can use the ideal gas law (Eq. 20-8) to write $P\,\Delta V = nR\,\Delta T$. We end up with

$$W = nR\,\Delta T$$

$$= (5.00 \text{ mol})(8.31 \text{ J/mol} \cdot \text{K})(20.0 \text{ C}°) \qquad \text{(Answer)}$$

$$= 831 \text{ J}.$$

Because we happen to know Q and ΔE^{int}, we can work this problem another way. The **Key Idea** now is that we can account for the energy changes of the gas with the first law of thermodynamics, writing

$$W = Q - \Delta E^{\text{int}} = 2077.5 \text{ J} - 1246.5 \text{ J}$$

$$\text{(Answer)}$$

$$= 831 \text{ J}.$$

Note that during the temperature increase, only a portion (1250 J) of the thermal energy (2080 J) that is transferred to the helium goes to increasing the internal energy of the helium and thus the temperature of the helium. The rest (831 J) is transferred out of the helium as work that the helium does during the expansion. If the water were frozen, it would not allow that expansion. Then the same temperature increase of 20.0 C° would require only 1250 J of energy, because no work would be done by the helium.

20-8 Degrees of Freedom and Molar Specific Heats

As Table 20-2 shows, the prediction that $C_V = \tfrac{3}{2}R$ agrees with experiment for monatomic gases. But it fails for diatomic and polyatomic gases. Let us try to explain the discrepancy by considering the possibility that molecules with more than one atom can store internal energy in forms other than *translational* kinetic energy.

Figure 20-12 shows common models of helium (a *monatomic* molecule, containing a single atom), oxygen (a *diatomic* molecule, containing two atoms), and methane (a *polyatomic* molecule). From such models, we would assume that all three types of molecules can have translational motions (say, moving left–right and up–down) and rotational motions (spinning about an axis like a top). However, due to their highly symmetric nature, rotational motions in a monatomic molecule need special consideration. We will return to this point shortly. In addition, we would assume that the diatomic and polyatomic molecules can have oscillatory motions, with the atoms oscillating slightly toward and away from one another, as if attached to opposite ends of a spring.

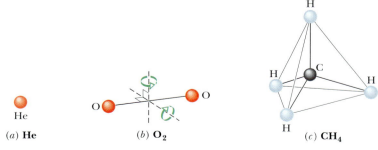

(a) **He**   (b) **O$_2$**   (c) **CH$_4$**

FIGURE 20-12 ■ Models of molecules as used in kinetic theory: (*a*) helium, a typical monatomic molecule; (*b*) oxygen, a typical diatomic molecule; and (*c*) methane, a typical polyatomic molecule. The spheres represent atoms, and the lines between them represent bonds. Two rotation axes are shown for the oxygen molecule.

To keep account of the various ways in which energy can be stored in a gas, James Clerk Maxwell introduced the theorem of the **equipartition of energy:**

> Every kind of molecule has a certain number f of **degrees of freedom,** which are independent ways in which the molecule can store energy. Each such degree of freedom has associated with it—on average—an energy of $\frac{1}{2}k_B T$ per molecule (or $\frac{1}{2}RT$ per mole).

Let us apply the theorem to the translational and rotational motions of the molecules in Fig. 20-12. (We discuss oscillatory motion in the next section.) For the translational motion, superimpose an xyz coordinate system on any gas. The molecules will, in general, have velocity components along all three axes. Thus, gas molecules of all types have three degrees of translational freedom (three ways to move in translation) and, on average, an associated energy of $3(\frac{1}{2}k_B T)$ per molecule.

For the rotational motion, imagine the origin of our xyz coordinate system at the center of each molecule in Fig. 20-12. In a gas, each molecule should be able to rotate with an angular velocity component along each of the three axes, so each gas should have three degrees of rotational freedom and, on average, an additional energy of $3(\frac{1}{2}k_B T)$ per molecule. *However,* experiment shows this is true only for the polyatomic molecules.

A possible solution to this dilemma is that rotations about an axis of symmetry don't count as a degree of freedom. For example, as seen in Fig. 20-12, a single-atom molecule is symmetric about all three (mutually perpendicular) axes through the molecule. Hence, according to our proposed solution, these rotations are not additional degrees of freedom. A diatomic molecule is symmetric about only one axis (the axis through the center of both atoms). Accordingly, a diatomic molecule would have two rather than three degrees of freedom associated with rotation of the molecule.

It appears that modifying our theory in this manner brings us more in alignment with the experimental results. However, one should ask what reasoning (other than experimental evidence) supports this modification of the theory. One thing is clear. If a molecule were rotating about an axis of symmetry, it would be impossible to tell. Unlike a baseball (which has stitches or other marks) molecules have no characteristics that allow us to sense the rotation. Although classical physics gives us no real foundation for ignoring the motion simply because it is indistinguishable from no motion at all, this is what quantum theory would suggest.

So, according to our new model, a monatomic molecule has zero degrees of freedom associated with rotation because any rotation would be about an axis of symmetry. A diatomic molecule has two degrees of freedom associated with rotations about the two axes perpendicular to the line connecting the atoms (the axes are shown in Fig. 20-12b) but no degree of freedom for rotation about that line itself. Therefore, a diatomic molecule can have a rotational energy of only $2(\frac{1}{2}k_B T)$ per molecule. A polyatomic molecule has a full three degrees of freedom associated with rotational motion.

To extend our analysis of molar specific heats (C_P and C_V, in Section 20-7) to ideal diatomic and polyatomic gases, it is necessary to retrace the derivations of that analysis in detail. First, we replace Eq. 20-36 ($E^{\text{int}} = \frac{3}{2}nRT$) with $E^{\text{int}} = (f/2)nRT$, where f is the number of degrees of freedom listed in Table 20-3. Doing so leads to the prediction

$$C_V = \left(\frac{f}{2}\right)R = 4.16\,f \quad \text{J/mol·K}, \tag{20-49}$$

which agrees—as it must—with Eq. 20-41 for monatomic gases ($f = 3$). As Table 20-3 shows, this prediction also agrees with experiment for diatomic gases ($f = 5$), but it is too low for polyatomic gases. *Note*: The symbol f used here to denote degrees of freedom should not be confused with $f(v)$ used to describe the velocity distribution function for molecules.

TABLE 20-3
Degrees of Freedom for Various Molecules

| Molecule | Example | Degrees of Freedom | | | Predicted Molar Specific Heats | |
		Translational	Rotational	Total (f)	C_V (Eq. 20-47)	$C_P = C_V + R$
Monatomic	He	3	0	3	$\frac{3}{2}R$	$\frac{5}{2}R$
Diatomic	O_2	3	2	5	$\frac{5}{2}R$	$\frac{7}{2}R$
Polyatomic	CH_4	3	3	6	$3R$	$4R$

TOUCHSTONE EXAMPLE 20-5: Internal Energy Change

A cabin of volume V is filled with air (which we consider to be an ideal diatomic gas) at an initial low temperature T_1. After you light a wood stove, the air temperature increases to T_2. What is the resulting change ΔE^{int} in the internal energy of the air in the cabin?

SOLUTION ■ As the air temperature increases, the air pressure P cannot change but must always be equal to the air pressure outside the room. The reason is that, because the room is not airtight, the air is not confined. As the temperature increases, air molecules leave through various openings and thus the number of moles n of air in the room decreases. Thus, one **Key Idea** here is that we *cannot* use Eq. 20-43 ($\Delta E^{int} = nC_V \Delta T$) to find ΔE^{int}, because it requires constant n.

A second **Key Idea** is that we *can* relate the internal energy E^{int} at any instant to n and the temperature T with Eq. 20-42 ($E^{int} = nC_V T$). From that equation we can then write

$$\Delta E^{int} = \Delta(nC_V T) = C_V \Delta(nT).$$

Next, using Eq. 20-8 ($PV = nRT$), we can replace nT with PV/R, obtaining

$$\Delta E^{int} = C_V \Delta\left(\frac{PV}{R}\right). \tag{20-50}$$

Now, because P, V, and R are all constants, Eq. 20-50 yields

$$\Delta E^{int} = 0, \qquad \text{(Answer)}$$

even though the temperature changes.

Why does the cabin feel more comfortable at the higher temperature? There are at least two factors involved: (1) You exchange electromagnetic radiation (thermal radiation) with surfaces inside the room, and (2) you exchange energy with air molecules that collide with you. When the room temperature is increased, (1) the amount of thermal radiation emitted by the surfaces and absorbed by you is increased, and (2) the amount of energy you gain through the collisions of air molecules with you is increased.

20-9 A Hint of Quantum Theory

We can improve the agreement of kinetic theory with experiment by including the oscillations of the atoms in a gas of diatomic or polyatomic molecules. For example, the two atoms in the O_2 molecule of Fig. 20-12b can oscillate toward and away from each other, with the interconnecting bond acting like a spring. However, experiment shows that such oscillations occur only at relatively high temperatures of the gas—the motion is "turned on" only when the gas molecules have relatively large energies. Rotational motion is also subject to such "turning on," but at a lower temperature.

Figure 20-13 is of help in seeing this turning on of rotational motion and oscillatory motion. The ratio C_V/R for diatomic hydrogen gas (H_2) is plotted there against temperature, with the temperature scale logarithmic to cover several orders of magnitude. Below about 80 K, we find that $C_V/R = 1.5$. This result implies that only the three translational degrees of freedom of hydrogen are involved in the specific heat.

FIGURE 20-13 ■ A plot of C_V/R versus temperature for (diatomic) hydrogen gas. Because rotational and oscillatory motions begin at certain energies, only translation is possible at very low temperatures. As the temperature increases, rotational motion can begin. At still higher temperatures, oscillatory motion can begin.

As the temperature increases, the value of C_V/R gradually increases to 2.5, implying that two additional degrees of freedom have become involved. Quantum theory shows that these two degrees of freedom are associated with the rotational motion of the hydrogen molecules and that this motion requires a certain minimum amount of energy. At very low temperatures (below 80 K), the molecules do not have enough energy to rotate. As the temperature increases from 80 K, first a few molecules and then more and more obtain enough energy to rotate, and C_V/R increases, until all of them are rotating and $C_V/R = 2.5$.

Similarly, quantum theory shows that oscillatory motion of the molecules requires a certain (higher) minimum amount of energy. This minimum amount is not met until the molecules reach a temperature of about 1000 K, as shown in Fig. 20-13. As the temperature increases beyond 1000 K, the number of molecules with enough energy to oscillate increases, and C_V/R increases, until all of them are oscillating and $C_V/R = 3.5$. (In Fig. 20-13, the plotted curve stops at 3200 K because at that temperature, the atoms of a hydrogen molecule oscillate so much that they overwhelm their bond, and the molecule then *dissociates* into two separate atoms.)

The observed fact that rotational degrees of freedom are not excited until sufficiently high temperatures are reached implies that rotational kinetic energy is not a continuous function of angular velocity. Instead, a discrete, quantized energy level must be attained before rotation is excited. This discreteness of energy levels is a hallmark of quantum mechanical behavior. It is interesting to note that some of the issues discussed in this chapter are the first examples (with many more to come) that macroscopic properties of matter, which are easily measured in the laboratory, depend critically on (and provide strong evidence for) the quantum theory we will develop later.

The compatibility between microscopic theory and macroscopic observations when coupled with quantum theory and other phenomena in physics and chemistry provided additional support for the theory that matter is composed of atoms and molecules.

20-10 The Adiabatic Expansion of an Ideal Gas

We saw in Section 18-2 that sound waves are propagated through air and other gases as a series of compressions and expansions; these variations in the transmission medium take place so rapidly that there is no time for thermal energy to be transferred from one part of the medium to another. As we saw in Section 19-8, a process for which $Q = 0$ is an *adiabatic process*. We can ensure that $Q = 0$ either by carrying out the process very quickly (as in sound waves) or by doing it (at any rate) in a well-insulated container. Let us see what the kinetic theory has to say about adiabatic processes.

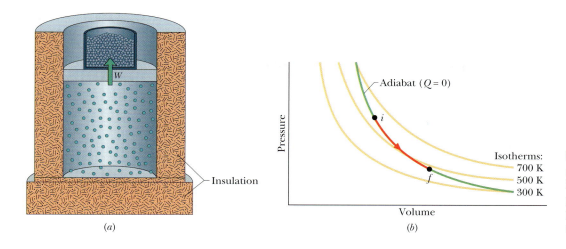

Figure 20-14*a* shows our usual insulated cylinder. Its insulating lid is not shown. It now contains an ideal gas and rests on an insulating stand. By removing mass from the piston, we can allow the gas to expand adiabatically (in a slow process rather than a free expansion). As the volume increases, both the pressure and the temperature drop. We shall prove next that the relation between the pressure and the volume during such an adiabatic process is

$$PV^\gamma = \text{a constant} \qquad \text{(ideal gas adiabatic process)}, \qquad (20\text{-}51)$$

in which $\gamma = C_P/C_V$, the ratio of the molar specific heats for the gas. On a *P-V* diagram such as that in Fig. 20-14*b*, the process occurs along a line (called an *adiabat*) that has the equation $P = (\text{a constant})/V^\gamma$. Since the gas goes from an initial state *i* to a final state *f*, we can rewrite Eq. 20-51 as

$$P_i V_i^\gamma = P_f V_f^\gamma \qquad \text{(ideal gas adiabatic process)}. \qquad (20\text{-}52)$$

We can also write an equation for an adiabatic process in terms of *T* and *V*. To do so, we use the ideal gas equation ($PV = nRT$) to eliminate *P* from Eq. 20-51, finding

$$\left(\frac{nRT}{V}\right) V^\gamma = \text{a constant}.$$

Because *n* and *R* are constants, we can rewrite this in the alternative form

$$TV^{\gamma-1} = \text{a constant} \qquad \text{(ideal gas adiabatic process)}. \qquad (20\text{-}53)$$

in which the constant is different from that in Eq. 20-51. When the gas goes from an initial state *i* to a final state *f*, we can rewrite Eq. 20-53 as

$$T_i V_i^{\gamma-1} = T_f V_f^{\gamma-1} \qquad \text{(adiabatic process)}. \qquad (20\text{-}54)$$

We can now answer the question that opens this chapter. At the top of an unopened carbonated drink, there is a gas of carbon dioxide and water vapor. Because the pressure of the gas is greater than atmospheric pressure, the gas expands out into the atmosphere when the container is opened. Thus, the gas increases its volume, but that means it must do work to push against the atmosphere. Because the expansion is so rapid, it is adiabatic and the only source of energy for the work is the internal energy of the gas. Because the internal energy decreases, the temperature of the gas must also decrease, which can cause the water vapor in the gas to condense into tiny drops, forming the fog. (Note that Eq. 20-54 also tells us that the temperature must decrease during an adiabatic expansion: Since V_f is greater than V_i, then T_f must be less than T_i.)

Proof of Eq. 20-51

Suppose that you remove some shot from the piston of Fig. 20-14a, allowing the ideal gas to push the piston and the remaining shot upward and thus to increase the volume by a differential amount dV. Since the volume change is tiny, we may assume that the pressure P of the gas on the piston is constant during the change. This assumption allows us to say that the work dW done by the gas during the volume increase is equal to $P\,dV$. From Eq. 19-18, the first law of thermodynamics can then be written as

$$dE^{\text{int}} = Q - P\,dV. \tag{20-55}$$

Since the gas is thermally insulated (and thus the expansion is adiabatic), we substitute 0 for Q. Then we use Eq. 20-43 to substitute $nC_V\,dT$ for dE^{int}. With these substitutions, and after some rearranging, we have

$$n\,dT = -\left(\frac{P}{C_V}\right)dV. \tag{20-56}$$

Now using the ideal gas law ($PV = nRT$) and derivative rule 3 in Appendix E we have

$$P\,dV + V\,dP = nR\,dT. \tag{20-57}$$

Replacing R with its equal, $C_P - C_V$, in Eq. 20-57 yields

$$n\,dT = \frac{P\,dV + V\,dP}{C_P - C_V}. \tag{20-58}$$

Equating Eqs. 20-56 and 20-58 and rearranging them give

$$\frac{dP}{P} + \left(\frac{C_P}{C_V}\right)\frac{dV}{V} = 0.$$

Replacing the ratio of the molar specific heats with γ and integrating (see integral 5 in Appendix E) yield

$$\ln P + \gamma \ln V = \text{a constant.}$$

Rewriting the left side as $\ln PV^\gamma$ and then taking the antilog of both sides, we find

$$PV^\gamma = \text{a constant,} \tag{20-59}$$

which is what we set out to prove.

Free Expansions

Recall from Section 19-8 that a free expansion of a gas is an adiabatic process that involves no work done on or by the gas, and no change in the internal energy of the gas.

A free expansion is thus quite different from the type of adiabatic process described by Eqs. 20-51 through 20-59, in which work is done and the internal energy changes. Those equations then do *not* apply to a free expansion, even though such an expansion is adiabatic.

Also recall that in a free expansion, a gas is in equilibrium only at its initial and final points; thus, we can plot only those points, but not the expansion itself, on a P-V diagram. In addition, because $\Delta E^{int} = 0$, the temperature of the final state must be that of the initial state. Thus, the initial and final points on a P-V diagram must be on the same isotherm, and instead of Eq. 20-54 we have

$$T_i = T_f \qquad \text{(free expansion).} \qquad (20\text{-}60)$$

If we next assume that the gas is ideal (so that $PV = nRT$), because there is no change in temperature, there can be no change in the product PV. Thus, instead of Eq. 20-51 a free expansion involves the relation

$$P_i V_i = P_f V_f \qquad \text{(free expansion).} \qquad (20\text{-}61)$$

TOUCHSTONE EXAMPLE 20-6: Final Temperature

In Touchstone Example 20-2, 1 mol of oxygen (assumed to be an ideal gas) expands isothermally (at 310 K) from an initial volume of 12 L to a final volume of 19 L.

(a) What would be the final temperature if the gas had expanded adiabatically to this same final volume? Oxygen (O_2) is diatomic and here has rotation but not oscillation.

SOLUTION ■ The **Key Ideas** here are as follows:

1. When a gas expands against the pressure of its environment, it must do work.

2. When the process is adiabatic (no thermal energy is transferred as heat), then the energy required for the work can come only from the internal energy of the gas.

3. Because the internal energy decreases, the temperature T must also decrease.

We can relate the initial and final temperatures and volumes with Eq. 20-54:

$$T_i V_i^{\gamma-1} = T_f V_f^{\gamma-1}. \qquad (20\text{-}62)$$

Because the molecules are diatomic and have rotation but not oscillation, we can take the molar specific heats from Table 20-3. Thus,

$$\gamma = \frac{C_P}{C_V} = \frac{\frac{7}{2}R}{\frac{5}{2}R} = 1.40.$$

Solving Eq. 20-62 for T_f and inserting known data then yield

$$T_f = \frac{T_i V_i^{\gamma-1}}{V_f^{\gamma-1}} = \frac{(310 \text{ K})(12 \text{ L})^{1.40-1}}{(19 \text{ L})^{1.40-1}}$$

$$= (310 \text{ K})(\tfrac{12}{19})^{0.40} = 258 \text{ K}. \qquad \text{(Answer)}$$

(b) What would be the final temperature and pressure if, instead, the gas had expanded freely to the new volume, from an initial pressure of 2.0 Pa?

SOLUTION ■ Here the **Key Idea** is that the temperature does not change in a free expansion:

$$T_f = T_i = 310 \text{ K}. \qquad \text{(Answer)}$$

We find the new pressure using Eq. 20-61, which gives us

$$P_f = P_i \frac{V_i}{V_f} = (2.0 \text{ Pa}) \frac{12 \text{ L}}{19 \text{ L}} = 1.3 \text{ Pa.} \qquad \text{(Answer)}$$

Problems

SEC. 20-2 ■ THE MACROSCOPIC BEHAVIOR OF GASES

1. Arsenic Find the mass in kilograms of 7.50×10^{24} atoms of arsenic, which has a molar mass of 74.9 g/mol.

2. Gold Gold has a molar mass of 197 g/mol. (a) How many moles of gold are in a 2.50 g sample of pure gold? (b) How many atoms are in the sample?

3. Water If the water molecules in 1.00 g of water were distributed uniformly over the surface of Earth, how many such molecules would there be on 1.00 cm² of the surface?

4. It Is Written A distinguished scientist has written: "There are enough molecules in the ink that makes one letter of this sentence to provide not only one for every inhabitant of Earth, but one for every creature if each star of our galaxy had a planet as populous as Earth." Check this statement. Assume the ink sample (molar mass = 18 g/mol) to have a mass of 1 µg, the population of Earth to be 5 × 10⁹, and the number of stars in our galaxy to be 10¹¹.

5. Compute Compute (a) the number of moles and (b) the number of molecules in 1.00 cm³ of an ideal gas at a pressure of 100 Pa and a temperature of 220 K.

6. Best Vacuum The best laboratory vacuum has a pressure of about 1.00×10^{-18} atm, or 1.01×10^{-13} Pa. How many gas molecules are there per cubic centimeter in such a vacuum at 293 K?

7. Oxygen Gas Oxygen gas having a volume of 1000 cm³ at 40.0°C and 1.01×10^5 Pa expands until its volume is 1500 cm³ and its pressure is 1.06×10^5 Pa. Find (a) the number of moles of oxygen present and (b) the final temperature of the sample.

8. Tire An automobile tire has a volume of 1.64×10^{-2} m³ and contains air at a gauge pressure (pressure above atmospheric pressure) of 165 kPa when the temperature is 0.00°C. What is the gauge pressure of the air in the tires when its temperature rises to 27.0°C and its volume increases to 1.67×10^{-2} m³? Assume atmospheric pressure is 1.00×10^5 Pa.

9. A Quantity of Ideal Gas A quantity of ideal gas at 10.0°C and 100 kPa occupies a volume of 2.50 m³. (a) How many moles of the gas are present? (b) If the pressure is now raised to 300 kPa and the temperature is raised to 30.0°C, how much volume does the gas occupy? Assume no leaks.

SEC. 20-3 ■ WORK DONE BY IDEAL GASES

10. Work Done by External Agent Calculate the work done by an external agent during an isothermal compression of 1.00 mol of oxygen from a volume of 22.4 L at 0°C and 1.00 atm pressure to 16.8 L.

11. P, V, T Pressure P, volume V, and temperature T for a certain non-ideal material are related by

$$P = \frac{AT - BT^2}{V},$$

where A and B are constants. Find an expression for the work done by the material if the temperature changes from T_1 to T_2 while the pressure remains constant.

12. A Container Encloses A container encloses two ideal gases. Two moles of the first gas are present, with molar mass M_1. The second gas has molar mass $M_2 = 3M_1$, and 0.5 mol of this gas is present. What fraction of the total pressure on the container wall is attributable to the second gas? (The kinetic theory explanation of pressure leads to the experimentally discovered law of partial pressures for a mixture of gases that do not react chemically: *The total pressure exerted by the mixture is equal to the sum of the pressures*

that the several gases would exert separately if each were to occupy the vessel alone.)

13. Air Initially Occupies Air that initially occupies 0.14 m³ at a gauge pressure of 103.0 kPa is expanded isothermally to a pressure of 101.3 kPa and then cooled at constant pressure until it reaches its initial volume. Compute the work done by the air. (Gauge pressure is the difference between the actual pressure and atmospheric pressure.)

14. A Sample A sample of an ideal gas is taken through the cyclic process *abca* shown in Fig. 20-15; at point a, $T = 200$ K. (a) How many moles of gas are in the sample? What are (b) the temperature of the gas at point b, (c) the temperature of the gas at point c, and (d) the net thermal energy transferred to the gas during the cycle?

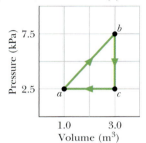

FIGURE 20-15 ■
Problem 14.

15. Air Bubble An air bubble of 20 cm³ volume is at the bottom of a lake 40 m deep where the temperature is 4.0°C. The bubble rises to the surface, which is at a temperature of 20°C. Take the temperature of the bubble's air to be the same as that of the surrounding water. Just as the bubble reaches the surface, what is its volume?

16. Pipe of Length L A pipe of length $L = 25.0$ m that is open at one end contains air at atmospheric pressure. It is thrust vertically into a freshwater lake until the water rises halfway up in the pipe, as shown in Fig. 20-16. What is the depth h of the lower end of the pipe? Assume that the temperature is the same everywhere and does not change.

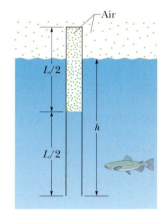

FIGURE 20-16 ■
Problem 16.

17. Container A Container A in Fig. 20-17 holds an ideal gas at a pressure of 5.0×10^5 Pa and a temperature of 300 K. It is connected by a thin tube (and a closed valve) to container B, with four times the volume of A. Container B holds the same ideal gas at a pressure of 1.0×10^5 Pa and a temperature of 400 K. The valve is opened to allow the pressures to equalize, but the temperature of each container is kept constant at its initial value. What then is the pressure in the two containers?

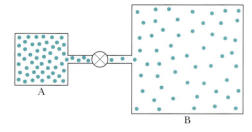

FIGURE 20-17 ■ Problem 17.

SEC. 20-4 ■ PRESSURE, TEMPERATURE, AND MOLECULAR KINETIC ENERGY

18. Helium Atoms Calculate the rms speed of helium atoms at 1000 K. See Appendix F for the molar mass of helium atoms.

19. Lowest Possible The lowest possible temperature in outer space is 2.7 K. What is the root-mean-square speed of hydrogen molecules at this temperature? (The molar mass of hydrogen molecules (H_2) is given in Table 20-1)

20. Speed of Argon Find the rms speed of argon atoms at 313 K. See Appendix F for the molar mass of argon atoms.

21. Sun's Atmosphere The temperature and pressure in the Sun's atmosphere are 2.00×10^6 K and 0.0300 Pa. Calculate the rms speed of free electrons (mass = 9.11×10^{-31} kg) there, assuming they are an ideal gas.

22. Nitrogen Molecule (a) Compute the root-mean-square speed of a nitrogen molecule at 20.0°C. The molar mass of nitrogen molecules (N_2) is given in Table 20-1. At what temperatures will the root-mean-square speed be (b) half that value and (c) twice that value?

23. Hydrogen Molecules A beam of hydrogen molecules (H_2) is directed toward a wall, at an angle of 55° with the normal to the wall. Each molecule in the beam has a speed of 1.0 km/s and a mass of 3.3×10^{-24} g. The beam strikes the wall over an area of 2.0 cm², at the rate of 10^{23} molecules per second. What is the beam's pressure on the wall?

24. Density of Gas At 273 K and 1.00×10^{-2} atm, the density of a gas is 1.24×10^{-5} g/cm³. (a) Find v^{rms} for the gas molecules. (b) Find the molar mass of the gas and identify the gas. (*Hint*: The gas is listed in Table 20-1.)

25. Translational Kinetic Energy What is the average translational kinetic energy of nitrogen molecules at 1600 K?

26. Average Value Determine the average value of the translational kinetic energy of the molecules of an ideal gas at (a) 0.00°C and (b) 100°C. What is the translational kinetic energy per mole of an ideal gas at (c) 0.00°C and (d) 100°C?

27. Evaporating Water Water standing in the open at 32.0°C evaporates because of the escape of some of the surface molecules. The heat of vaporization (539 cal/g) is approximately equal to εn, where ε is the average energy of the escaping molecules and n is the number of molecules per gram. (a) Find ε. (b) What is the ratio of ε to the average kinetic energy of H_2O molecules, assuming the latter is related to temperature in the same way as it is for gases?

28. Alternative Form Show that the ideal gas equation, Eq. 20-8, can be written in the alternative form $P = \rho RT/M$, where ρ is the mass density of the gas and M is the molar mass.

29. Avogadro's Law *Avogadro's law* states that under the same conditions of temperature and pressure, equal volumes of gas contain equal numbers of molecules. Is this law equivalent to the ideal gas law? Explain.

SEC. 20-5 ■ MEAN FREE PATH

30. Nitrogen Molecules The mean free path of nitrogen molecules at 0.0°C and 1.0 atm is 0.80×10^{-5} cm. At this temperature and pressure there are 2.7×10^{19} molecules/cm³. What is the molecular diameter?

31. Earth's Surface At 2500 km above Earth's surface, the number density of the atmosphere is about 1 molecule/cm³. (a) What mean free path is predicted by Eq. 20-24 and (b) what is its significance under these conditions? Assume a molecular diameter of 2.0×10^{-8} cm.

32. What Frequency At what frequency would the wavelength of sound in air be equal to the mean free path of oxygen molecules at 1.0 atm pressure and 0.00°C? Take the diameter of an oxygen molecule to be 3.0×10^{-8} cm.

33. Mean Free Path of Jelly Beans Assuming that jelly beans in a bag could behave like ideal gas particles, what is the mean free path for 15 spherical jelly beans in a bag that is vigorously shaken? The volume of the bag is 1.0 L, and the diameter of a jelly bean is 1.0 cm. (Consider bean-bean collisions, not bean-bag collisions.)

34. Mean Free Path for Argon At 20°C and 750 torr pressure, the mean free paths for argon gas (Ar) and nitrogen gas (N_2) are $\lambda_{Ar} = 9.9 \times 10^{-6}$ cm and $\lambda_{N_2} = 27.5 \times 10^{-6}$ cm. (a) Find the ratio of the effective diameter of argon to that of nitrogen. What is the mean free path of argon at (b) 20°C and 150 torr, and (c) −40°C and 750 torr?

35. Particle Accelerator In a certain particle accelerator, protons travel around a circular path of diameter 23.0 m in an evacuated chamber, whose residual gas is at 295 K and 1.00×10^{-6} torr pressure. (a) Calculate the number of gas molecules per cubic centimeter at this pressure. (b) What is the mean free path of the gas molecules if the molecular diameter is 2.00×10^{-8} cm?

SEC. 20-6 ■ THE DISTRIBUTION OF MOLECULAR SPEEDS

36. Twenty-Two Particles Twenty-two particles have speeds as follows (N_i represents the number of particles that have speed v_i):

N_i	2	4	6	8	2
v_i (cm/s)	1.0	2.0	3.0	4.0	5.0

(a) Compute their average speed $\langle v \rangle$. (b) Compute their root-mean-square speed v^{rms}. (c) Of the five speeds shown, which is the most probable speed v^{prob}?

37. Ten Molecules The speeds of 10 molecules are 2.0, 3.0, 4.0 . . . , 11 km/s. (a) What is their average speed? (b) What is their root-mean-square speed?

38. Ten Particles (a) Ten particles are moving with the following speeds: four at 200 m/s, two at 500 m/s, and four at 600 m/s. Calculate their average and root-mean-square speeds. Is $v^{rms} > \langle v \rangle$? (b) Make up your own speed distribution for the 10 particles and show that $v^{rms} \geq \langle v \rangle$ for your distribution. (c) Under what condition (if any) does $v^{rms} = \langle v \rangle$?

39. Compute Temperature (a) Compute the temperatures at which the rms speed for (a) molecular hydrogen and (b) molecular oxygen is equal to the speed of escape from Earth. (c) Do the same for the speed of escape from the Moon, assuming the local gravitational constant on its surface to be 0.16*g*. (d) The temperature high in Earth's upper atmosphere (in the thermosphere) is about 1000 K. Would you expect to find much hydrogen there? Much oxygen? Explain.

40. Most Probable Speed It is found that the most probable speed of molecules in a gas when it has (uniform) temperature T_2 is the same as the rms speed of the molecules in this gas when it has (uniform) temperature T_1. Calculate T_2/T_1.

41. Hydrogen Molecule A molecule of hydrogen (diameter 1.0×10^{-8} cm), traveling with the rms speed, escapes from a furnace ($T = 4000$ K) into a chamber containing atoms of *cold* argon (diameter 3.0×10^{-8} cm) at a number density of 4.0×10^{19} atoms/cm³. (a) What is the speed of the hydrogen molecule? (b) If the H₂ molecule collides with an argon atom, what is the closest their centers can be, considering each as spherical? (c) What is the initial number of collisions per second experienced by the hydrogen molecule? (*Hint:* Assume that the cold argon atoms are stationary. Then the mean free path of the hydrogen molecule is given by Eq. 20-25, and not Eq. 20-24.)

42. Two Containers—Same Temperature Two containers are at the same temperature. The first contains gas with pressure P_1, molecular mass m_1, and root-mean-square speed v_1^{rms}. The second contains gas with pressure $2P_1$, molecular mass m_2, and average speed $\langle v_2 \rangle = 2v_1^{rms}$. Find the mass ratio m_1/m_2.

43. Hypothetical Speeds Figure 20-18 shows a hypothetical speed distribution for a sample of N gas particles (note that $f(v) = 0$ for $v > 2v_0$). (a) Express a in terms of N and v_0. (b) How many of the particles have speeds between $1.5v_0$ and $2.0v_0$? (c) Express the average speed of the particles in terms of v_0. (d) Find v^{rms}.

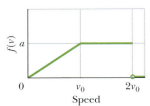

FIGURE 20-18 ■ Problem 43.

SEC. 20-7 ■ THE MOLAR SPECIFIC HEATS OF AN IDEAL GAS

44. Internal Energy What is the internal energy of 1.0 mol of an ideal monatomic gas at 273 K?

45. Isothermal Expansion One mole of an ideal gas undergoes an isothermal expansion. Find the thermal energy Q added to the gas in terms of the initial and final volumes and the temperature. (*Hint:* Use the first law of thermodynamics.)

46. Added as Heat When 20.9 J of thermal energy was added to a particular ideal gas, the volume of the gas changed from 50.0 cm³ to 100 cm³ while the pressure remained constant at 1.00 atm. (a) By how much did the internal energy of the gas change? If the quantity of gas present is 2.00×10^{-3} mol, find the molar specific heat of the gas at (b) constant pressure and (c) constant volume.

47. Three Nonreacting Gases A container holds a mixture of three nonreacting gases: n_1 moles of the first gas with molar specific heat at constant volume C_1, and so on. Find the molar specific heat at constant volume of the mixture, in terms of the molar specific heats and quantities of the separate gases.

48. Ideal Diatomic Gas One mole of an ideal diatomic gas goes from a to c along the diagonal path in Fig. 20-19. During the transition, (a) what is the change in internal energy of the gas, and (b) how much thermal energy is added to the gas? (c) How much thermal energy is required if the gas goes from a to c along the indirect path abc?

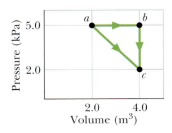

FIGURE 20-19 ■ Problem 48.

49. Gas Molecule The mass of a gas molecule can be computed from its specific heat at constant volume C_V. Take $C_V = 0.075$ cal/g · C° for argon and calculate (a) the mass of an argon atom and (b) the molar mass of argon.

SEC. 20-8 ■ DEGREES OF FREEDOM AND MOLAR SPECIFIC HEATS

50. Heating a Diatomic Gas We give 70 J of thermal energy to a diatomic gas, which then expands at constant pressure. The gas molecules rotate but do not oscillate. By how much does the internal energy of the gas increase?

51. One Mole of Oxygen One mole of oxygen (O₂) is heated at constant pressure starting at 0°C. How much energy Q must be added to the gas to double its volume? (The molecules rotate but do not oscillate.)

52. Oxygen Heating Suppose 12.0 g of oxygen (O₂) is heated at constant atmospheric pressure from 25.0°C to 125°C. (a) How many moles of oxygen are present? (See Table 20-1 for the molar mass.) (b) How much thermal energy is transferred to the oxygen? (The molecules rotate but do not oscillate.) (c) What fraction of the thermal energy absorbed by the oxygen is used to raise the internal energy of the oxygen?

53. Molecular Rotation Suppose 4.00 mol of an ideal diatomic gas, with molecular rotation but not oscillation, experienced a temperature increase of 60.0 K under constant-pressure conditions. (a) How much thermal energy was transferred to the gas? (b) How much did the internal energy of the gas increase? (c) How much work was done by the gas? (d) How much did the translational kinetic energy of the gas increase?

SEC. 20-10 ■ THE ADIABATIC EXPANSION OF AN IDEAL GAS

54. Liter of Gas (a) One liter of a gas with $\gamma = 1.3$ is at 273 K and 1.0 atm pressure. It is suddenly compressed adiabatically to half its original volume. Find its final pressure and temperature. (b) The gas is now cooled back to 273 K at constant pressure. What is its final volume?

55. A Certain Gas A certain gas occupies a volume of 4.3 L at a pressure of 1.2 atm and a temperature of 310 K. It is compressed adiabatically to a volume of 0.76 L. Determine (a) the final pressure and (b) the final temperature, assuming the gas to be an ideal gas for which $\gamma = 1.4$.

56. Adiabatic Process We know that for an adiabatic process $PV^\gamma = a$ constant. Evaluate the constant for an adiabatic process involving exactly 2.0 mol of an ideal gas passing through the state having exactly $P = 1.0$ atm and $T = 300$ K. Assume a diatomic gas whose molecules have rotation but not oscillation.

57. Let *n* Moles Let n moles of an ideal gas expand adiabatically from an initial temperature T_1 to a final temperature T_2. Prove that the work done by the gas is $nC_V(T_1 - T_2)$, where C_V is the molar specific heat at constant volume. (*Hint*: Use the first law of thermodynamics.)

58. Bulk Modulus For adiabatic processes in an ideal gas, show that (a) the bulk modulus is given by

$$B = -V\frac{dP}{dV} = \gamma P,$$

and therefore (b) the speed of sound in the gas is

$$v_{wave} = \sqrt{\frac{\gamma P}{\rho}} = \sqrt{\frac{\gamma RT}{M}}.$$

See Eqs. 18-1 and 18-2. Here M is the molar mass and the total mass of the gas is $m = nM$.

59. Molar Specific Heats Air at $0.000°C$ and 1.00 atm pressure has a density of 1.29×10^{-3} g/cm^3, and the speed of sound in air is 331 m/s at that temperature. Use those data to compute the ratio γ of the molar specific heats of air. (*Hint*: See Problem 58.)

60. Free Expansion (a) An ideal gas initially at pressure P_0 undergoes a free expansion until its volume is 3.00 times its initial volume. What then is its pressure? (b) The gas is next slowly and adiabatically compressed back to its original volume. The pressure after compression is $(3.00)^{1/3}P_0$. Is the gas monatomic, diatomic, or polyatomic? (c) How does the average kinetic energy per molecule in this final state compare with that in the initial state?

61. The Cycle One mole of an ideal monatomic gas traverses the cycle of Fig. 20-20. Process $1 \rightarrow 2$ occurs at constant volume, process $2 \rightarrow 3$ is adiabatic, and process $3 \rightarrow 1$ occurs at constant pressure. (a) Compute the thermal energy Q absorbed by the gas, the change in its internal energy ΔE^{int}, and the work done by the gas W, for each of the three processes and for the cycle as a whole. (b) The initial pressure at point 1 is 1.00 atm. Find the pressure and the volume at points 2 and 3. Use 1.00 atm $= 1.013 \times 10^5$ Pa and $R = 8.314$ J/mol · K.

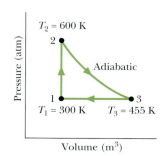

FIGURE 20-20 ▪ Problem 61.

Additional Problems

62. Extensive vs. Intensive An intensive variable is one that can be defined locally within a system. Its magnitude does not depend on whether we select the whole system or a part of the system. An extensive variable is one that is defined for the system as a whole; its magnitude does depend on how much of the system we choose to select. Which of the following variables are intensive and which are extensive? Explain your reasoning in each case. (a) density, (b) pressure, (c) volume, (d) temperature, (e) mass, (f) internal energy, (g) number of moles, and (h) molecular weight.

63. Changes in Gas Molecules An ideal gas is contained in an airtight box. Complete each of the following five statements below to show the quantitative change that will occur. For example, if you want to say that the volume, initially equal to V, quadruples, complete the statement with "$4V$".

(a) If the absolute temperature of the gas is halved, the average speed of a gas molecule, $\langle v \rangle$, becomes_____.

(b) If the average speed of a gas molecule doubles, the pressure, P, on a the wall of the box becomes_____.

(c) If the absolute temperature of the gas is halved, the pressure, P, on a wall of the box becomes_____.

(d) If the absolute temperature of the gas is increased by 25%, the total internal energy of the gas, E^{int}, becomes_____.

(e) If the number of gas molecules inside the box is doubled, but the temperature is kept the same, the pressure, P, on a wall of the box becomes_____.

64. Scales in a Gas The actual diameter of an atom is about 1 angstrom (10^{-10} m). In order to develop some intuition for the molecular scale of a gas, assume that you are considering a liter of air

(mostly N_2 and O_2) at room temperature and a pressure of 10^5 Pa. (a) Calculate the number of molecules in the sample of gas. (b) Estimate the average spacing between the molecules. (c) Estimate the average speed of a molecule using the Maxwell-Boltzmann distribution. (d) Suppose that the gas were rescaled upward so that each atom was the size of a tennis ball (but we don't change the time scale). What would be the average spacing between molecules and the average speed of the molecules in miles/hour?

65. Is the M-B Distribution Wrong? A good student makes the following observation. "If I try to accelerate a small sphere through air, it will be resisted by air drag. If I drop an object, it will eventually reach a terminal velocity where the air is resisting as much as gravity is trying to accelerate (Eq. 6-25). The smaller the ball, the slower is the terminal velocity. But you tell me the Maxwell distribution says that the molecules of air move very rapidly. I estimate that this is much faster than the molecule's terminal velocity, so it can't move that fast. The Maxwell distribution must be wrong." (a) The Newton drag law for a sphere moving through air is calculated by a molecular model to be $D = \rho\pi R^2 v^2$, where D is the magnitude of the drag force, R is the radius of the sphere, ρ is the density of the air, and v is the velocity of the object through the air. Calculate the terminal velocity for a sphere the size and mass of an air molecule falling through a fluid the density of air ($\rho = 1$ kg/m^3). (b) From your estimate in part (a), is this speed greater or less than the average speed the molecule should have given the M-B distribution? What is wrong with the student's argument?

66. Avogadro's Hypothesis Why does Avogadro's hypothesis (that a given volume of gas at a given temperature and pressure has the same number of molecules no matter what kind of gas it is) not hold for liquids and solids?

67. Boiling Molecules When a molecule of a liquid approaches the surface, it experiences a force barrier that tries to keep it in the liquid. Thus it has to do work to escape and loses some of its kinetic energy when it leaves.

(a) Assume that a water molecule can evaporate from the liquid if it hits the surface from the inside with a kinetic energy greater than the thermal energy corresponding to the temperature of boiling water, 100°C. Use this to estimate the numerical value of the work W required to remove a water molecule from the liquid.

(b) Even though the average speed of a molecule in water below the boiling point corresponds to a kinetic energy less than W, some molecules leave anyway and the water evaporates. Explain why this happens.

21 | Entropy and the Second Law of Thermodynamics

An anonymous graffito on a wall of the Pecan Street Cafe in Austin, Texas, reads: "Time is God's way of keeping things from happening all at once." Time also has direction—some things happen in a certain sequence and could never happen on their own in a reverse sequence. As an example, an accidentally dropped egg splatters in a cup. The reverse process, a splattered egg re-forming into a whole egg and jumping up to an outstretched hand, will never happen on its own—but why not? Why can't that process be reversed, like a videotape run backward?

What in the world gives direction to time?

The answer is in this chapter.

21-1 Some One-Way Processes

Suppose you come indoors on a very cold day and wrap your cold hands around a warm mug of cocoa. Then your hands get warmer and the mug gets cooler. However, it never happens the other way around. That is, your cold hands never get still colder while the warm mug gets still warmer.

If we assume that the rate of thermal energy transfer from the warm mug and your hands to the room is slow, the system consisting of your hands and the mug is approximately a *closed system,* one that is more or less isolated from (does not interact with) its environment. Here are some other one-way processes that we observe to occur in closed systems: (1) A crate sliding over a horizontal surface eventually stops—but you never see an initially stationary crate on a horizontal surface start to move all by itself. (2) If you drop a glob of putty, it falls to the floor—but an initially motionless glob of putty on the floor never leaps spontaneously into the air. (3) If you puncture a helium-filled balloon in a closed room, the helium gas spreads throughout the room—but the individual helium atoms will never migrate back out of the room and refill the balloon. We call such one-way processes **irreversible,** meaning that they cannot be reversed by means of only small changes in their environment.

Many chemical transformations are also irreversible. For example, when methane gas is burned, each methane molecule mixes with an oxygen molecule. Water vapor and carbon dioxide are given off as shown in the following chemical equation:

$$CH_4 + O_2 \longrightarrow CO_2 + H_2O.$$

This combustion process is irreversible. We don't find water and carbon dioxide spontaneously reacting to produce methane and oxygen gas.

The one-way character of such thermodynamic processes is so pervasive that we take it for granted. If these processes were to occur spontaneously (on their own) in the "wrong" direction, we would be astonished beyond belief. *Yet none of these "wrong-way" events would violate the law of conservation of energy.* In the "cold hands-warm mug" example, energy would be conserved even for a "wrong-way" thermal energy transfer between hands and mug. Conservation of energy would be obeyed even if a stationary crate or a stationary glob of putty suddenly were to transfer internal energy to macroscopic kinetic energy and begin to move. Energy would still be conserved if the helium atoms released from a balloon were, on their own, to clump together again.

Changes in energy within a closed system do not determine the direction of irreversible processes. So, we conclude that the direction must be set by another property that we have not yet considered. We shall discuss this new property quite a bit in this chapter. It is called the *entropy S* of the system. Knowing the *change in entropy ΔS* of a system turns out to be a useful quantity in analyzing thermodynamic processes. It is defined in the next section, but here we can state the central property of entropy change (often called the *entropy postulate*):

> If an irreversible process occurs in a *closed* system, the entropy S of the system always increases; it never decreases.

Entropy, unlike energy, does *not* obey a conservation law. The energy of a closed system is conserved. It always remains constant. For irreversible processes, the *entropy* of a closed system always increases. Because of this property, the change in entropy is sometimes called "the arrow of time." For example, we associate the irreversible breaking of the egg in our opening photograph with the forward direction of time and also with an increase in entropy. The backward direction of time (a videotape run backward) would correspond to the broken egg re-forming into a whole egg and rising into the air. This backward process would result in an entropy decrease and so it never happens.

There are two equivalent ways to define the change in entropy of a system. The first is macroscopic. It is in terms of the system's temperature and any thermal energy transfers between the system and its surroundings. The second is macroscopic or statistical. It is defined by counting the ways in which the atoms or molecules that make up the system can be arranged. We use the first approach in the next section, and the second in Section 21-7.

READING EXERCISE 21-1: Consider the irreversible process of dropping a glob of putty on the floor. Describe what energy transformations are taking place that allow us to believe that energy is conserved. ∎

READING EXERCISE 21-2: List additional everyday phenomena that illustrate irreversibility without violating energy conservation. ∎

21-2 Change in Entropy

In this section, we will try to develop a definition for *change in entropy* by looking again at the macroscopic process that we described in Sections 19-8 and 20-10—namely, the free expansion of an ideal gas. Figure 21-1a shows the gas in its initial equilibrium state *i,* confined by a closed stopcock to the left half of a thermally insulated container. If we open the stopcock, the gas rushes to fill the entire container, eventually reaching the final equilibrium state *f* shown in Fig. 21-1b. Unless the number of gas molecules is small (which is very hard to accomplish), this is an *irreversible* process. Again, what we mean by irreversible is that it is extremely improbable that all the gas particles would return, by themselves, to the left half of the container.

The *P-V* plot of the process in Fig. 21-2 shows the pressure and volume of the gas in its initial state *i* and final state *f.* As we discuss in Section 19-1, the pressure and volume of the gas depend only on the state that the gas is in, and not on the process by which it arrived in that state. Therefore, pressure and volume are examples of *state properties.* **State properties** are properties that depend only on the state of the gas and not on how it reached that state. Other state properties are temperature and internal energy. We now assume that the gas has still another state property—its entropy. Furthermore, we define the **change in entropy** $S_f - S_i$ of a system during a process that takes the system from an initial state *i* to a final state *f* as

$$\Delta S = S_f - S_i \equiv \int_i^f \frac{dQ}{T} \qquad \text{(change in entropy defined).} \qquad (21\text{-}1)$$

Here *Q* is the *thermal energy transferred* to or from the system during a heating or cooling process, and *T* is the temperature of the system in kelvin. Thus, an entropy change depends not only on the thermal energy transferred *Q*, but also on the temperature at which the transfer takes place. Because *T* is always positive, the sign of ΔS is the same as that of *Q* (positive if thermal energy is transferred to the system and negative if thermal energy is transferred from the system). We see from this relation (Eq. 21-1) that the SI unit for entropy and entropy change is the joule per kelvin.

There is a problem, however, in applying Eq. 21-1 to the free expansion of Fig. 21-1. As the gas rushes to fill the entire container, the pressure, temperature, and volume of the gas fluctuate unpredictably. In other words, they do not have a sequence of well-defined equilibrium values during the intermediate stages of the change from initial equilibrium state *i* to final equilibrium state *f.* Thus, we cannot trace a pressure–volume path for the free expansion on the *P-V* plot of Fig. 21-2. More importantly, that means that we cannot find a relation between thermal energy transfer *Q* and temperature *T* that allows us to integrate as Eq. 21-1 requires.

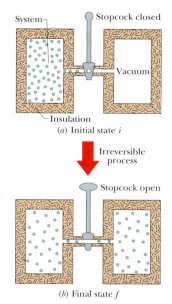

FIGURE 21-1 ∎ The free expansion of an ideal gas consisting of a large number of molecules. (*a*) The gas is confined to the left half of an insulated container by a closed stopcock. (*b*) When the stopcock is opened, the gas rushes to fill the entire container. This process is irreversible; that is, it is never observed to occur in reverse, with the gas spontaneously collecting itself in the left half of the container.

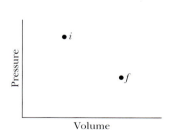

FIGURE 21-2 ∎ A *P-V* diagram showing the initial state *i* and the final state *f* of the free expansion of Fig. 21-1. The intermediate states of the gas cannot be shown because they are not equilibrium states.

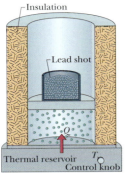

(a) Initial state i

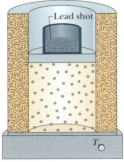

(b) Final state f

FIGURE 21-3 ▪ The isothermal expansion of an ideal gas, done in a reversible way. The gas has the same initial state i and same final state f as in the irreversible process of Figs. 21-1 and 21-2.

However, if our assumption is correct and entropy is truly a state property, the difference in entropy between states i and f must depend *only on those states* and not at all on the way the system went from one state to the other. That means that we can replace the irreversible free expansion of Fig. 21-1 with a *reversible* process that connects states i and f. With a reversible process we can trace a pressure–volume path on a P-V plot, and we can find a relation between thermal energy transfer Q and temperature T that allows us to use Eq. 21-1 to obtain the entropy change.

We saw in Section 20-10 that the temperature of an ideal gas does not change during a free expansion. So, $T_i = T_f = T$. Thus, points i and f in Fig. 21-2 must be on the same isotherm. A convenient replacement process is then a reversible isothermal expansion from state i to state f, which actually proceeds *along* that isotherm. Furthermore, because T is constant throughout a reversible isothermal expansion, the integral of Eq. 21-1 is greatly simplified.

Figure 21-3 shows how to produce such a reversible isothermal expansion. We confine the gas to an insulated cylinder that rests on a thermal reservoir maintained at the temperature T. We begin by placing just enough lead shot on the movable piston so that the pressure and volume of the gas are those of the initial state i of Fig. 21-1a. We then remove shot slowly (piece by piece) until the pressure and volume of the gas are those of the final state f of Fig. 21-1b. The temperature of the gas does not change because the gas remains in thermal contact with the reservoir throughout the process.

The reversible isothermal expansion of Fig. 21-3 is physically quite different from the irreversible free expansion of Fig. 21-1. However, *both processes have the same initial state and the same final state*. Thus, if entropy is a state property, these two processes *must result in the same change in entropy*.

Because we removed the lead shot slowly, the intermediate states of the gas are equilibrium states, so we can plot them on a P-V diagram (Fig. 21-4). To apply Eq. 21-1 to the isothermal expansion, we take the constant temperature T outside the integral, obtaining

$$\Delta S = S_f - S_i = \frac{1}{T}\int_i^f dQ.$$

Because $\int dQ = Q$, where Q is the thermal energy transferred during the process, we have

$$\Delta S = S_f - S_i = \frac{Q}{T} \qquad \text{(change in entropy, isothermal process).} \qquad (21\text{-}2)$$

To keep the temperature T of the gas constant during the isothermal expansion of Fig. 21-3, the thermal energy transferred *from* the reservoir to the gas must have been Q. Thus, Q is positive and the entropy of the gas *increases* during the isothermal process and during the free expansion of Fig. 21-1.

To summarize:

> Assuming entropy is a state property, we can find the entropy change for an irreversible process occurring in a *closed* system by replacing that process with any reversible process that connects the same initial and final states. We can then calculate the entropy change for this reversible process with Eq. 21-1. The change in entropy for an irreversible process connecting the same two states would be the same.

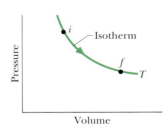

FIGURE 21-4 ▪ A P-V diagram for the reversible isothermal expansion of Fig. 21-3. The intermediate states, which are now equilibrium states, are shown.

We can even use this approach if the temperature of the system is not quite constant. That is, if the temperature change ΔT of a system is small relative to the temperature (in kelvin) before and after the process, the entropy change can be approximated as

$$\Delta S = S_f - S_i \approx \frac{Q}{\langle T \rangle} \qquad (21\text{-}3)$$

where $\langle T \rangle$ is the average kelvin temperature of the system during the process.

Entropy as a State Property

In the previous section, we assumed that entropy, like pressure, internal energy, and temperature, is a property of the state of a system and is independent of how that state is reached. The fact that entropy is indeed a state property (or *state function* as state properties are sometimes called) can really only be deduced by careful experiment. However, we will prove entropy is a state property for the special and important case of an ideal gas undergoing a reversible process. This proof will serve two purposes. First, it will verify (for at least this one case) that entropy is a state property (or state function) as we assumed in the section above. Second, it will allow us to develop an expression for the entropy change in an ideal gas as it goes from some initial state i to some final state f via a reversible process.

To make the process reversible, we must make changes slowly in a series of small steps, with the ideal gas in an equilibrium state at the end of each step. For each small step, the thermal energy transfer to or from the gas is dQ, the work done by or on the gas is dW, and the change in internal energy is dE^{int}. These are related by the first law of thermodynamics in differential form (Eq. 19-18):

$$dE^{\text{int}} = dQ - dW.$$

Because the steps are reversible, with the gas in equilibrium states, we can replace dW with $P\,dV$ (Eq. 19-15). Since we are dealing with an ideal gas, we can also replace dE^{int} with $nC_V\,dT$ (Eq. 20-43). Solving for the thermal energy transferred to or from the system in a single small step of the process dQ then leads to

$$dQ = P\,dV + nC_V\,dT.$$

We replace the pressure P in this equation with nRT/V (using the ideal gas law). Then we divide each term in the resulting equation by the temperature T, obtaining

$$\frac{dQ}{T} = nR\frac{dV}{V} + nC_V\frac{dT}{T}.$$

Now let us integrate each term of this equation between an arbitrary initial state i and an arbitrary final state f to get

$$\int_i^f \frac{dQ}{T} = \int_i^f nR\frac{dV}{V} + \int_i^f nC_V\frac{dT}{T}.$$

The quantity on the left is the entropy change $\Delta S (= S_f - S_i)$ as we defined it in Eq. 21-1. Substituting this and integrating the quantities on the right yields an expression for the entropy change in an ideal gas undergoing a reversible process:

$$\Delta S = S_f - S_i = nR\ln\frac{V_f}{V_i} + nC_V\ln\frac{T_f}{T_i}. \tag{21-4}$$

Note that we did not have to specify a particular reversible process when we integrated. Therefore, the integration must hold for all reversible processes that take the gas from state i to state f. Thus, we see that the change in entropy ΔS between the initial and final states of an ideal gas does depend only on properties of the initial state (V_i and T_i) and properties of the final states (V_f and T_f); ΔS does not depend on how the gas changes between the two equilibrium states. Therefore, in at least this one case, we know that entropy must be a state property. In the work that follows in this chapter, we will accept without further proof that entropy is in fact a state property for any system undergoing any process.

TOUCHSTONE EXAMPLE 21-1: Nitrogen

One mole of nitrogen gas is confined to the left side of the container of Fig. 21-1a. You open the stopcock and the volume of the gas doubles. What is the entropy change of the gas for this irreversible process? Treat the gas as ideal.

SOLUTION ∎ We need two **Key Ideas** here. One is that we can determine the entropy change for the irreversible process by calculating it for a reversible process that provides the same change in volume. The other is that the temperature of the gas does not change in the free expansion. Thus, the reversible process should be an isothermal expansion—namely, the one of Figs. 21-3 and 21-4.

Since the internal energy of an ideal gas depends only on temperature, $\Delta E^{\text{int}} = 0$ here and so $Q = W$ from the first law. Combining this result with Eq. 20-14 gives

$$Q = W = nRT \ln\left(\frac{V_f}{V_i}\right),$$

in which n is the number of moles of gas present. From Eq. 21-2 the entropy change for this isothermal reversible process is

$$\Delta S^{\text{rev}} = \frac{Q}{T} = \frac{nRT \ln(V_f/V_i)}{T} = nR \ln \frac{V_f}{V_i}.$$

Substituting $n = 1.00$ mol and $V_f/V_i = 2$, we find

$$\Delta S^{\text{rev}} = nR \ln \frac{V_f}{V_i} = (1.00 \text{ mol})(8.31 \text{ J/mol} \cdot \text{K})(\ln 2)$$

$$= +5.76 \text{ J/K}. \qquad \text{(Answer)}$$

Thus, the entropy change for the free expansion (and for all other processes that connect the initial and final states shown in Fig. 21-2) is

$$\Delta S^{\text{irrev}} = \Delta S^{\text{rev}} = +5.76 \text{ J/K}.$$

ΔS is positive, so the entropy increases, in accordance with the entropy postulate of Section 21-1.

TOUCHSTONE EXAMPLE 21-2: Copper Blocks

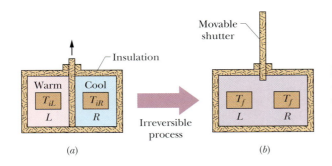

(a)   (b)

FIGURE 21-5 ∎ (a) In the initial state, two copper blocks L and R, identical except for their temperatures, are in an insulating box and are separated by an insulating shutter. (b) When the shutter is removed, the blocks exchange thermal energy and come to a final state, both with the same temperature T_f. The process is irreversible.

Figure 21-5a shows two identical copper blocks of mass $m = 1.5$ kg: block L at temperature $T_{iL} = 60°C$ and block R at temperature $T_{iR} = 20°C$. The blocks are in a thermally insulated box and are separated by an insulating shutter. When we lift the shutter, the blocks eventually come to the equilibrium temperature $T_f = 40°C$ (Fig. 21-5b). What is the net entropy change of the two-block system during this irreversible process? The specific heat of copper is 386 J/kg·K.

SOLUTION ∎ The **Key Idea** here is that to calculate the entropy change, we must find a reversible process that takes the system from the initial state of Fig. 21-5a to the final state of Fig. 21-5b. We can calculate the net entropy change ΔS^{rev} of the reversible process using Eq. 21-1, and then the entropy change for the irreversible process is equal to ΔS^{rev}. For such a reversible process we need a thermal reservoir whose temperature can be changed slowly (say, by turning a knob). We then take the blocks through the following two steps, illustrated in Fig. 21-6.

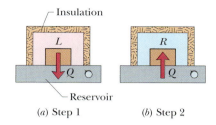

(a) Step 1   (b) Step 2

FIGURE 21-6 ■ The blocks of Fig. 21-5 can proceed from their initial state to their final state in a reversible way if we use a reservoir with a controllable temperature (a) to transfer thermal energy reversibly from block L and (b) to transfer thermal energy reversibly to block R.

Step 1. With the reservoir's temperature set at 60°C, put block L on the reservoir. (Since block and reservoir are at the same temperature, they are already in thermal equilibrium.) Then slowly lower the temperature of the reservoir and the block to 40°C. As the block's temperature changes by each increment dT during this process, thermal energy dQ is transferred *from* the block to the reservoir. Using Eq. 19-5, we can write this transferred energy as $dQ = mc\ dT$, where c is the specific heat of copper. According to Eq. 21-1, the entropy change ΔS_L of block L during the full temperature change from initial temperature $T_{iL}\ (= 60°C = 333$ K) to final temperature $T_f\ (= 40°C = 313$ K) is

$$\Delta S_L = \int_i^f \frac{dQ}{T} = \int_{T_{iL}}^{T_f} \frac{mc\ dT}{T} = mc \int_{T_{iL}}^{T_f} \frac{dT}{T}$$

$$= mc\ \ln\frac{T_f}{T_{iL}}.$$

Inserting the given data yields

$$\Delta S_L = (1.5\text{ kg})(386\text{ J/[kg}\cdot\text{K]})\ln\frac{313\text{ K}}{333\text{ K}}$$

$$= -35.86\text{ J/K}.$$

Step 2: With the reservoir's temperature now set at 20°C, put block R on the reservoir. Then slowly raise the temperature of the reservoir and the block to 40°C. With the same reasoning used to find ΔS_L, you can show that the entropy change ΔS_R of block R during this process is

$$\Delta S_R = (1.5\text{ kg})(386\text{ J/[kg}\cdot\text{K]})\ \ln\frac{313\text{ K}}{293\text{ K}}$$

$$= +38.23\text{ J/K}.$$

The net entropy change ΔS^{rev} of the two-block system undergoing this two-step reversible process is then

$$\Delta S^{\text{rev}} = \Delta S_L + \Delta S_R$$

$$= -35.85\text{ J/K} + 38.23\text{ J/K} = 2.4\text{ J/K}.$$

Thus, the net entropy change ΔS^{irrev} for the two-block system undergoing the actual irreversible process is $\Delta S^{\text{irrev}} = \Delta S^{\text{rev}} = 2.4$ J/K. This result is positive, in accordance with the entropy postulate of Section 21-1.

21-3 The Second Law of Thermodynamics

Here is a puzzle. We saw in Touchstone Example 21-1 that if we cause the reversible, isothermal process of Fig. 21-3 to proceed from (a) to (b) in that figure, the change in entropy of the gas—which we take as our system—is positive. However, because the process is reversible, we can just as easily make it proceed from (b) to (a), simply by slowly adding lead shot to the piston of Fig. 21-3b until the original volume of the gas is restored. In this reverse isothermal process, the gas must keep its temperature from increasing and so must transfer thermal energy to its surroundings to make up for the work done via the lead shot. Since this thermal energy transfer is *from* our system (the gas), Q is negative. So, from $\Delta S = S_f - S_i = Q/T$ (Eq. 21-2) we find that ΔS must be negative and hence the entropy of the gas must decrease.

Doesn't this decrease in the entropy of the gas violate the entropy postulate of Section 21-1, which states that entropy always increases? No, because that postulate holds only for *irreversible* processes occurring in *closed* systems. The procedure suggested here does not meet these requirements. The process is *not* irreversible and (because there is a heat transfer from the gas to the reservoir) the system—which is the gas alone—is not closed.

However, if we include the reservoir, along with the gas, as part of the system, then we do have a closed system. Let's check the change in entropy of the enlarged system *gas + reservoir* for the process that takes it from (*b*) to (*a*) in Fig. 21-3. During this reversible process, thermal energy is transferred from the gas to the reservoir—that is, from one part of the enlarged system to another. Let $|Q|$ represent the amount of energy transferred. With $\Delta S = S_f - S_i = Q/T$ (Eq. 21-2), we can then calculate separately the entropy changes for the gas (which loses $|Q|$ of thermal energy) and the reservoir (which gains $|Q|$ in thermal energy). We get

$$\Delta S_{gas} = -\frac{|Q|}{T}$$

and

$$\Delta S_{res} = +\frac{|Q|}{T}.$$

The entropy change of the closed system is the sum of these two quantities, *which* (since the process is isothermal) *is zero*.

With this result, we can modify the entropy postulate of Section 21-1 to include both reversible and irreversible processes:

> If a process occurs in a *closed* system, the entropy of the system increases for irreversible processes and remains constant for reversible processes. It never decreases.

Although entropy may decrease in part of a closed system, there will always be an equal or larger entropy increase in another part of the system, so that the entropy of the system as a whole never decreases. This fact is one form of the **second law of thermodynamics** and can be written as

$$\Delta S \geq 0 \qquad \text{(second law of thermodynamics)}, \qquad (21\text{-}5)$$

where the greater-than sign applies to irreversible processes, and the equals sign to reversible processes. But remember, this relation applies only to closed systems.

In the real world almost all processes are irreversible to some extent because of friction, turbulence, and other factors. So, the entropy of real closed systems undergoing real processes always increases. Processes in which the system's entropy remains constant are idealizations.

21-4 Entropy in the Real World: Engines

Engines, which are fundamentally thermodynamic devices, are everywhere around us and are a big part of what makes modern life possible. However, not all engines are the same. For example, the engine in your car is different from the engine in a typical power plant. Nevertheless, these engines are similar in that they function through the use of a *working substance* that can expand and contract as it exchanges energy with its surroundings. In a power plant, the working substance is often water, in both its vapor and liquid forms. In an automobile engine the working substance is a gasoline-air mixture. If an engine is to do work on a sustained basis, the working substance must operate in a *cycle*. That is, the working substance must pass through a repeating series of thermodynamic processes, called **strokes,** returning again and again to each state in its **cycle.** The fundamental difference between the engine in your car and that in a power plant is that these two engines use different working substances and different types of thermodynamic cycles. That is, the working substances in these two engines undergo different thermodynamic processes.

Most engines that we meet in everyday life are some version of what we will call a *heat engine*. **Heat engines** are devices that take the thermal energy transfers (or heat transfers) Q that result from temperature differences and convert them to useful work W. Internal combustion engines (like in your car) are complicated heat engines that convert chemical energy (from gasoline or diesel fuel) and thermal energy to work. We will not discuss internal combustion engines here. Instead, we will focus on a simpler class of engines in which the working substance simply cycles between two constant temperatures (isotherms) and the engine converts a portion of the resulting thermal energy transfers directly to mechanical work.

The Carnot Engine

We have seen that we can learn much about real gases by analyzing an ideal gas, which obeys the simple law $PV = nRT$. This is a useful plan because, although an ideal gas does not exist, any real gas approaches ideal behavior as closely as you wish if its density is low enough. In much the same spirit we choose to study real engines by analyzing the behavior of an **ideal engine.**

> In an ideal engine, all processes are reversible and no wasteful energy transfers occur due to friction, turbulence, or other processes.

We shall focus here on a particular ideal engine called an ideal **Carnot engine** named after the French scientist and engineer N. L. Sadi Carnot (pronounced "car-no"), who first proposed the engine's concept in 1824. A Carnot engine is an example of a heat engine. It operates between two constant temperatures and uses the resulting thermal energy transfers directly to do useful work. The ideal Carnot engine is especially important because it turns out to be the best engine of this type.

Figure 21-7 shows the operation of a Carnot engine schematically. During each cycle of the engine, energy $|Q_H|$ is transferred to the working substance from a thermal reservoir at constant temperature T_H and energy $|Q_L|$ is transferred from the working substance to a second thermal reservoir at a constant and lower temperature T_L. The Carnot engine converts the difference between the amount of energy transferred into the system $|Q_H|$ and the amount of energy transferred out of the system $|Q_L|$ into useful work.

The cycle followed by the working substance in a Carnot engine is called a **Carnot cycle.** Figure 21-8a shows a P-V plot of the Carnot cycle. As indicated by the arrows, the cycle is traversed in the clockwise direction. Imagine the working substance to be a gas, confined to an insulating cylinder with a weighted, movable piston. Figure 21-8b shows how the Carnot cycle might be accomplished. The cylinder may be placed at will on either of the two thermal reservoirs, or on an insulating slab. If we place the cylinder in contact with the high-temperature reservoir at temperature T_H, $|Q_H|$ represents the thermal energy transfer *to* the working substance *from* this reservoir as the gas undergoes an isothermal *expansion* from volume V_a to volume V_b. Similarly, when the working substance is in contact with the low-temperature reservoir at temperature T_L, the gaseous substance undergoes an isothermal *compression* from volume V_c to volume V_d. At the same time energy $|Q_L|$ is transferred *from* the working substance *to* this reservoir. Note that in our engine thermal energy transfers to or from the working substance can take place *only* during the isothermal processes *ab* and *cd* of Fig. 21-8b. Thermal energy transfers do not occur in processes *bc* and *da* in that figure, which connect the two isotherms at temperatures T_H and T_L. Therefore, those two processes must be (reversible) adiabatic processes. To ensure this, during processes *bc* and *da*, the cylinder is placed on an insulating slab as the volume of the working substance is changed.

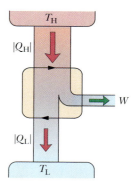

FIGURE 21-7 ■ The elements of a Carnot engine. The two black arrowheads on the central loop suggest the working substance operating in a cycle, as if on a P-V plot. Thermal energy $|Q_H|$ is transferred from the high-temperature reservoir at temperature T_H to the working substance. Thermal energy $|Q_L|$ is transferred from the working substance to the low-temperature reservoir at temperature T_L. Work W is done by the engine (actually by the working substance) on something in the environment.

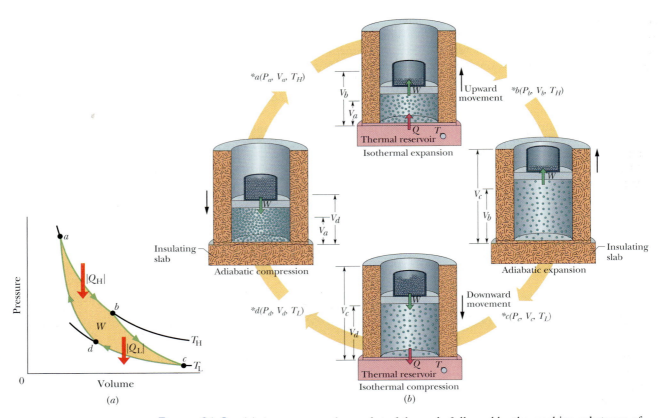

FIGURE 21-8 ■ (a) A pressure–volume plot of the cycle followed by the working substance of the Carnot engine in Fig. 21-7. The cycle consists of two isothermal processes (*ab* and *cd*) and two adiabatic processes (*bc* and *da*). The shaded area enclosed by the cycle is equal to the work *W* per cycle done by the Carnot engine. (b) An example of how this set of cycles could be accomplished. The upward motions of a piston during processes *ab* and *bc* are accomplished by slowly removing weight from the piston. The downward motions of the piston during processes *cd* and *da* are accomplished by slowing adding weight to the piston.

Work Done: During the consecutive processes *ab* and *bc* of Fig. 21-8, the working substance is expanding and thus doing positive work as it raises the weighted piston. This work is represented in Fig. 21-8*a* by the area under curve *abc*. During the consecutive processes *cd* and *da*, the working substance is being compressed, which means that it is doing negative work on its environment (the environment is doing positive work on it). This work is represented by the area under curve *cda*. The *net work per cycle*, which is represented by *W* in Figs. 21-7 and 21-8*a*, is the difference between these two areas. It is a positive quantity equal to the area enclosed by cycle *abcda* in Fig. 21-8*a*. This work *W* is performed on some outside object. The engine might, for example, be used to lift a weight.

To calculate the net work done by a Carnot engine during a cycle, let us apply the first law of thermodynamics ($\Delta E^{\text{int}} = Q - W$), to the working substance of a Carnot engine. That substance must return again and again to any arbitrarily selected state in that cycle. Thus, if X represents any state property of the working substance, such as pressure, temperature, volume, internal energy, or entropy, we must have $\Delta X = 0$ for every cycle. It follows that $\Delta E^{\text{int}} = 0$ for a complete cycle of the working substance. Recall that Q in Eq. 19-18 is the *net* thermal energy transfer per cycle and W is the net work. We can then write the first law of thermodynamics ($\Delta E^{\text{int}} = Q - W$) for the Carnot cycle as

$$W = |Q_{\text{H}}| - |Q_{\text{L}}|. \qquad (21\text{-}6)$$

Entropy Changes: Equation 21-1 ($\Delta S = \int dQ/T$) tells us that any thermal energy transfer between a system and its surroundings must involve a change in entropy. To illustrate the entropy changes for a Carnot engine, we can plot the Carnot cycle on a temperature–entropy (*T-S*) diagram as in Fig. 21-9. The lettered points *a*, *b*, *c*, and *d* in Fig. 21-9 correspond to the lettered points in the *P-V* diagram in Fig. 21-8. The two horizontal lines in Fig. 21-9 correspond to the two isothermal processes of the Carnot cycle (because the temperature is constant). Process *ab* is the isothermal expansion stroke of the cycle. As the high temperature reservoir transfers thermal energy $|Q_H|$ reversibly to the working substance at temperature T_H, its entropy increases. Similarly, during the isothermal compression *cd*, the working substance transfers thermal energy $|Q_L|$ reversibly to the low temperature reservoir at temperature T_L. In this process the entropy of the working substance decreases.

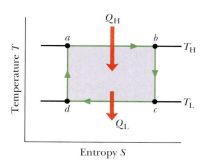

The two vertical lines in Fig. 21-9 correspond to the two adiabatic processes of the Carnot cycle. Because no thermal energy transfers occur during the adiabatic processes, the entropy of the working substance does not change during either of these processes. So, in a Carnot engine, there are *two* (and only two) reversible thermal energy transfers, and thus two changes in entropy—one at temperature T_H and one at T_L. The net entropy change per cycle is then

Figure 21-9 ■ The Carnot cycle of Fig. 21-8 plotted on a temperature–entropy diagram. During processes *ab* and *cd* the temperature remains constant. During processes *bc* and *da* the entropy remains constant.

$$\Delta S = \Delta S_H + \Delta S_L = \frac{|Q_H|}{T_H} - \frac{|Q_L|}{T_L}. \tag{21-7}$$

Here ΔS_H is positive because energy $|Q_H|$ is *transferred to* the working substance from the surroundings (an increase in entropy) and ΔS_L is negative because energy $|Q_L|$ is *transferred from* the working substance to the surroundings (a decrease in entropy). Because entropy is a state property, we must have $\Delta S = 0$ for a complete cycle. Putting $\Delta S = 0$ in above (Eq. 21-7) requires that

$$\frac{|Q_H|}{T_H} = \frac{|Q_L|}{T_L}. \tag{21-8}$$

Note that, because $T_H > T_L$, we must have $|Q_H| > |Q_L|$. That is, more energy is transferred from the high-temperature reservoir to the engine than the engine transfers to the low-temperature reservoir.

We shall now use our findings on the work done (Eq. 21-6) and entropy change (Eq. 21-8) in an ideal Carnot cycle to derive an expression for the efficiency of an ideal Carnot engine.

Efficiency of an Ideal Carnot Engine

The purpose of any heat engine is to transform as much of the thermal energy, Q_H, transferred to the engine's working medium into useful mechanical work as possible. We measure its success in doing so by its **thermal efficiency** ε, defined as the work the engine does per cycle ("energy we get") divided by the thermal energy transferred to it per cycle ("energy we pay for"):

$$\varepsilon = \frac{\text{energy we get}}{\text{energy we pay for}} = \frac{|W|}{|Q_H|} \qquad \text{(efficiency, any engine).} \tag{21-9}$$

For a Carnot engine we can substitute $W = |Q_H| - |Q_L|$ from Eq. 21-6 to write

$$\varepsilon_C = \frac{|Q_H| - |Q_L|}{|Q_H|} = 1 - \frac{|Q_L|}{|Q_H|}. \tag{21-10}$$

Using $|Q_H|/T_H = |Q_L|/T_L$ (Eq. 21-8) we can write this as

$$\varepsilon_C = 1 - \frac{T_L}{T_H} \qquad \text{(efficiency, ideal Carnot engine)}, \qquad (21\text{-}11)$$

where the temperatures T_L and T_H are in kelvin. Because $T_L < T_H$, the Carnot engine necessarily has a thermal efficiency less than unity—that is, less than 100%. This is indicated in Fig. 21-7, which shows that only part of the energy transferred to the engine from the high-temperature reservoir causes the engine's working substance to expand and do physical work on the surroundings. The rest of the energy absorbed by the engine provides for the heat transfer to the low-temperature reservoir. We will show in Section 21-6 that no real engine can have a thermal efficiency greater than that calculated for the ideal Carnot engine (Eq. 21-11).

Other Types of Cycles and Real Engines

Efficiency is typically our main concern when designing an engine. For an engine that operates on an ideal Carnot cycle, the efficiency is

$$\varepsilon_C = 1 - \frac{T_L}{T_H}.$$

But remember, an ideal Carnot cycle means that the cycle is composed of the following four processes: a perfectly isothermal (constant temperature) expansion of the working substance, a perfectly adiabatic (zero thermal energy transfer) expansion of the working substance, a perfectly isothermal compression of the working substance, and a perfectly adiabatic compression of the working substance. Perfectly isothermal means the temperature of the working substance cannot change at all during these strokes. Perfectly adiabatic means that there can be no thermal energy transfer at all. These tasks are not easy to accomplish. If you do accomplish them, then you have an *ideal* Carnot cycle and the efficiency of the engine is given by the equation above. Most engines built on the Carnot cycle have efficiencies that are measurably lower than this.

It is important to note that even an ideal Carnot engine cannot have an efficiency of one. That is, it does not do a perfect job of converting thermal energy transferred to it into work. Inspection of the Carnot efficiency expression $\varepsilon_C = 1 - T_L/T_H$ (Eq. 21-11) shows that we can achieve 100% engine efficiency (that is, $\varepsilon = 1$) only if $T_L = 0$ K or $T_H \to \infty$. These requirements are impossible to meet. So, decades of practical engineering experience have led to the following alternative version of the second law of thermodynamics:

> It is impossible to design an engine that converts thermal energy transferred to it from a thermal reservoir to useful work with 100% efficiency.

As we mentioned earlier, Carnot engines are not the only type of heat engine in which the working substance cycles between two constant temperatures and converts some of the associated heat transferred to the engine's working medium to useful work. For example, Fig. 21-10 shows the operating cycle of an ideal **Stirling engine.** Comparison with the Carnot cycle of Fig. 21-8 shows that each cycle includes isothermal energy transfers at temperatures T_H and T_L. However, the two isotherms of the Stirling engine cycle of Fig. 21-10 are connected, not by adiabatic processes (no thermal energy transfer) as for the Carnot engine, but by constant-volume processes. To reversibly increase the temperature of a gas at constant volume from T_L to T_H (as in process da of Fig. 21-10) requires a thermal energy transfer to the working substance from a thermal reservoir whose temperature can be varied smoothly between those limits.

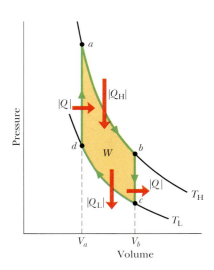

FIGURE 21-10 ■ A P-V plot for the working substance of an ideal Stirling engine, assumed for convenience to be an ideal gas. Processes ab and cd are isothermal while bc and da are constant volume.

Note that reversible thermal energy transfers (and corresponding entropy changes) occur in all four of the processes that form the cycle of a Stirling engine, not just two processes as in a Carnot engine. Thus, the derivation that led to the efficiency expression for the Carnot engine (Eq. 21-11) does not apply to an ideal Stirling engine. More important, the efficiency of an ideal Stirling engine, or any other heat engine based on operation between two isotherms, is lower than that of a Carnot engine operating between the same two temperatures. This makes the ideal Carnot engine an ideal version of the ideal type of this class of engine! Of course, common Stirling engines have even lower efficiencies than the ideal Stirling engine discussed here.

Many engines important in our lives operate based on cycles between two isotherms and convert thermal energy transfers to work. For example, consider the nuclear power plant shown in Fig. 21-11. It, like most power plants, is an engine when taken in its entirety. A reactor core (or perhaps a coal-powered furnace) provides the high-temperature reservoir. Thermal energy transfer to the working substance (usually water) is converted to work done on a turbine (which often results in electricity production). The remaining energy $|Q_L|$ is transferred to a low-temperature reservoir, which is usually a nearby river, or the atmosphere (if cooling towers are used). If the power plant shown in Fig. 21-11 operated as an ideal Carnot engine, its efficiency would be about 40%. Its actual efficiency is about 30%.

How does the efficiency of the internal combustion engine compare to that of the ideal Carnot engine? Well, this is a bit like comparing apples and oranges since the internal combustion engine does not operate between two isotherms like the Carnot, Stirling, or power plant engines do. However, we can estimate that if your car could be powered by a Carnot engine, it would have an efficiency of about 55% according to $\varepsilon_C = 1 - T_L/T_H$ (Eq. 21-11). Its actual efficiency (with an internal combustion engine) is probably about 25%.

FIGURE 21-11 ■ The North Anna nuclear power plant near Charlottesville, Virginia, which generates electrical energy at the rate of 900 MW. At the same time, by design, it discards energy into the nearby river at the rate of 2100 MW. This plant—and all others like it—throws away more energy than it delivers in useful form. It is a real counterpart to the ideal engine of Fig. 21-7.

READING EXERCISE 21-4: Three Carnot engines operate between reservoir temperatures of (a) 400 and 500 K, (b) 600 and 800 K, and (c) 400 and 600 K. Rank the engines according to their thermal efficiencies, greatest first. ■

TOUCHSTONE EXAMPLE 21-3: Carnot Engine

Imagine an ideal Carnot engine that operates between the temperatures $T_H = 850$ K and $T_L = 300$ K. The engine performs 1200 J of work each cycle, the duration of each cycle being 0.25 s.

(a) What is the efficiency of this engine?

SOLUTION ■ The **Key Idea** here is that the efficiency ε of an ideal Carnot engine depends only on the ratio T_L/T_H of the temperatures (in kelvins) of the thermal reservoirs to which it is connected. Thus, from Eq. 21-11, we have

$$\varepsilon = 1 - \frac{T_L}{T_H} = 1 - \frac{300 \text{ K}}{850 \text{ K}} = 0.647 \approx 65\% \quad \text{(Answer)}$$

(b) What is the average power of this engine?

SOLUTION ■ Here the **Key Idea** is that the average power P of an engine is the ratio of the work W it does per cycle to the time Δt that each cycle takes. For this Carnot engine, we find

$$P = \frac{W}{\Delta t} = \frac{1200 \text{ J}}{0.25 \text{ s}} = 4800 \text{ W} = 4.8 \text{ kW}. \quad \text{(Answer)}$$

(c) How much thermal energy Q_H is extracted from the high-temperature reservoir every cycle?

SOLUTION ■ Now the **Key Idea** is that, for any engine including a Carnot engine, the efficiency ε is the ratio of the work W that is done per cycle to the thermal energy Q_H that is extracted from the high-temperature reservoir per cycle. This relation, $\varepsilon = |W|/|Q_H|$ (Eq. 21-9), gives us

$$Q_H = \frac{W}{\varepsilon} = \frac{1200 \text{ J}}{0.647} = 1855 \text{ J}. \quad \text{(Answer)}$$

(d) How much thermal energy Q_L is delivered to the low-temperature reservoir every cycle?

SOLUTION ■ The **Key Idea** here is that for a Carnot engine, the work W done per cycle is equal to the difference in energy transfers $|Q_H| - |Q_L|$. (See Eq. 21-6.) Thus, we have

$$|Q_L| = |Q_H| - W = 1855 \text{ J} - 1200 \text{ J} = 655 \text{ J}. \quad \text{(Answer)}$$

(e) What entropy change is associated with the energy transfer to the working substance from the high-temperature reservoir? From the working substance to the low-temperature reservoir?

SOLUTION ■ The **Key Idea** here is that the entropy change ΔS during a transfer of thermal energy Q at constant temperature T is given by Eq. 21-2 ($\Delta S = Q/T$). Thus, for the transfer of energy Q_H from the high-temperature reservoir at T_H, we have

$$\Delta S_H = \frac{Q_H}{T_H} = \frac{1855 \text{ J}}{850 \text{ K}} = +2.18 \text{ J/K}.$$

For the transfer of energy Q_L to the low-temperature reservoir at T_L, we have

$$\Delta S_L = \frac{Q_L}{T_L} = \frac{-655 \text{ J}}{300 \text{ K}} = -2.18 \text{ J/K}.$$

Note that the algebraic signs of the two thermal energy transfers are different. Note also that, as Eq. 21-8 requires, the net entropy change of the working substance for one cycle (which is the algebraic sum of the two quantities calculated above) is zero.

TOUCHSTONE EXAMPLE 21-4: Better Than the Ideal?

An inventor claims to have constructed a heat engine that has an efficiency of 75% when operated between the boiling and freezing points of water. Is this possible?

SOLUTION ■ The **Key Idea** here is that the efficiency of a real engine (with its irreversible processes and wasteful energy transfers) must be less than the efficiency of an ideal Carnot engine operating between the same two temperatures. From Eq. 21-11, we find that the efficiency of an ideal Carnot engine operating between

the boiling and freezing points of water is

$$\varepsilon = 1 - \frac{T_L}{T_H} = 1 - \frac{(0 + 273) \text{ K}}{(100 + 273) \text{ K}}$$

$$= 0.268 \approx 27\% \quad \text{(Answer)}$$

Thus, the claimed efficiency of 75% for a real heat engine operating between the given temperatures is impossible.

21-5 Entropy in the Real World: Refrigerators

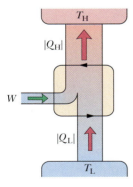

FIGURE 21-12 ■ The elements of a refrigerator. The two black arrowheads on the central loop suggest the working substance operating in a cycle, as if on a P-V plot. Thermal energy Q_L is transferred to the working substance from the low-temperature reservoir. Thermal energy Q_H is transferred to the high-temperature reservoir from the working substance. Work W is done on the refrigerator (on the working substance) by something in the environment.

A heat engine operated in a reverse cycle would require an input of work and transfer thermal energy from a low-temperature reservoir to a high-temperature reservoir as it continuously repeats a set series of thermodynamic processes. We call such a device a **refrigerator.** In a household refrigerator, for example, an electrical compressor does work in order to transfer thermal energy from the food storage compartment (a low-temperature reservoir) to the room (a high-temperature reservoir). Air conditioners and heat pumps are also refrigerators. The differences are only in the nature of the high- and low-temperature reservoirs. For an air conditioner, the low-temperature reservoir is the room that is to be cooled, and the high-temperature reservoir is the (presumably warmer) outdoors. A heat pump is an air conditioner that can also be operated in such a way as to transfer thermal energy to the air in a room from the (presumably cooler) outdoors.

Let us now consider an *ideal refrigerator:*

> In an ideal refrigerator, all processes are reversible and no wasteful energy transfers occur between the refrigerator and its surroundings due to friction, turbulence, or other processes.

Figure 21-12 shows the basic elements of an ideal refrigerator that operates based on a Carnot cycle. That is, it is the Carnot engine of Fig. 21-8 operating in reverse. All the energy transfers, either thermal energy or work, are reversed from those of a Carnot engine. Thus, we call such an ideal refrigerator an ideal **Carnot refrigerator.**

The designer of a refrigerator would like to do amount of work W (that we pay for) and cause as large a thermal energy transfer $|Q_L|$ as possible from the low-temperature reservoir (for example, the storage space in a kitchen refrigerator or the room to be cooled by the air conditioner). A measure of the efficiency of a refrigerator, then, is

$$K = \frac{\text{what we want}}{\text{what we pay for}} = \frac{|Q_L|}{|W|} \qquad \text{(coefficient of performance, any refrigerator),} \qquad (21\text{-}12)$$

where K is called the *coefficient of performance*. For a Carnot refrigerator, the first law of thermodynamics gives $|W| = |Q_H| - |Q_L|$, where $|Q_H|$ is the amount of the thermal energy transfer to the high-temperature reservoir. The coefficient of performance for our ideal Carnot refrigerator then becomes

$$K_C = \frac{|Q_L|}{|Q_H| - |Q_L|}. \qquad (21\text{-}13)$$

Because an ideal Carnot refrigerator is an ideal Carnot engine operating in reverse, we can again use $|Q_H|/T_H = |Q_L|/T_L$ (Eq. 21-8) and rewrite this expression as

$$K_C = \frac{T_L}{T_H - T_L} \qquad \text{(coefficient of performance, Carnot refrigerator).} \qquad (21\text{-}14)$$

For typical room air conditioners, $K \approx 2.5$. For household refrigerators, $K \approx 5$. Unfortunately, but logically, the efficiency (and so the value of K) of a given refrigerator is higher the closer the temperatures of the two reservoirs are to each other. For example, a given Carnot air conditioner is more efficient on a warm day than when it is very hot outside.

It would be nice to own a refrigerator that did not require an input of work—that is, one that would run without being plugged in. Figure 21-13 represents an "inventor's dream," of a *perfect refrigerator* that transfers thermal energy Q from a cold reservoir to a warm reservoir without the need for work. Because the unit returns to the same state at the end of each cycle, and entropy is a state property, we know that the change in entropy of the working substance for this imagined refrigerator would be zero for a complete cycle. The entropies of the two reservoirs, however, would change. The entropy change for the low temperature reservoir would be $-|Q|/T_L$, and that for the high temperature reservoir would be $+|Q|/T_H$. Thus, the net entropy change for the entire system is

$$\Delta S = -\frac{|Q|}{T_L} + \frac{|Q|}{T_H}.$$

Because $T_H > T_L$, the right side of this equation would be negative and thus the net change in entropy per cycle for the closed system *refrigerator + reservoirs* would also be negative. Because such a decrease in entropy violates the second law of thermodynamics $\Delta S \geq 0$ (Eq. 21-5), it must be that a perfect refrigerator cannot exist. That is, if you want your refrigerator to operate, you must plug it in!

This result leads us to another (equivalent) formulation of the second law of thermodynamics:

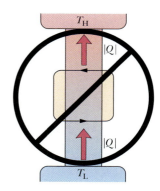

FIGURE 21-13 ■ The elements of a perfect (but *impossible*) refrigerator—that is, one that transfers energy from a low-temperature reservoir to a high-temperature reservoir without any input of work.

It is impossible to design a refrigerator that can cause a thermal energy transfer from a reservoir at a lower temperature to one at a higher temperature without the input of work (that is, with 100% efficiency).

In short, *there are no perfect refrigerators.*

21-6 Efficiency Limits of Real Engines

As we have just seen, a "perfect" Carnot refrigerator would violate the second law of thermodynamics which states that entropy must always either remain constant or increase. Therefore, we accept that a search for a 100% efficient Carnot refrigerator is futile. They do not exist. But what about Carnot engines? Can we have a "perfect" (that is, 100% efficient) engine?

Fundamentally, the inefficiency in an ideal Carnot engine is associated with the thermal energy transfer at the low temperature reservoir interface. Naive inventors continually try to improve Carnot engine efficiency by reducing the waste energy $|Q_L|$ transferred to the low-temperature reservoir and, hence, "thrown away" during each cycle. The inventor's dream is to produce the *perfect engine,* diagrammed in Fig. 21-14, in which $|Q_L|$ is reduced to zero and $|Q_H|$ is converted completely into work. For example, if we could do it, a perfect engine on an ocean liner could use thermal energy transferred to it from seawater to drive the propellers, with no fuel cost. An automobile, fitted with such a perfect engine, could use energy transferred from the surrounding air to turn its wheels, again with no fuel cost.

Alas, what seems too good to be true usually is. A perfect engine cannot exist. We already noted in Section 21-4 that since the efficiency of an ideal Carnot engine is given by $\varepsilon_C = 1 - T_L/T_H$ (Eq. 21-11), even an ideal Carnot engine cannot have 100% efficiency. This is clear since we cannot have $T_L = 0$ K or $T_H \to \infty$. So, if a real engine is to have 100% efficiency, it will have to have a higher efficiency than our ideal Carnot engine. Let's see if that is possible.

Let us assume for a moment that an inventor, working in her garage, has constructed an engine X, which she claims has an efficiency ε_X that is greater than ε_C, where ε_C is the efficiency of an ideal Carnot engine operating between two temperatures. Then,

$$\varepsilon_X > \varepsilon_C \quad \text{(a claim).} \tag{21-15}$$

Let us connect the inventor's engine X to a Carnot refrigerator, as in Fig. 21-15a. We adjust the strokes of the refrigerator so that the work it requires per cycle is just equal to that provided by the engine X. Thus, no (external) work is needed for the operation

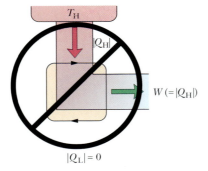

FIGURE 21-14 ■ The elements of a perfect (and *impossible*) engine — that is, one that converts thermal energy transfer Q_H from a high-temperature reservoir directly into work W with 100% efficiency.

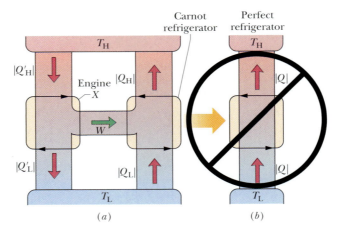

(a) (b)

FIGURE 21-15 ■ (*a*) Engine X drives a Carnot refrigerator. (*b*) If, as claimed, engine X is more efficient than a Carnot engine, then the combination shown in (*a*) is equivalent to the perfect refrigerator shown here. This violates the second law of thermodynamics, so we conclude that engine X cannot be more efficient than a Carnot engine.

of the combination *engine + refrigerator* of Fig. 21-15a, which we take as our system. Since the efficiency of an engine is ε = energy we get/energy we pay for = $|W|/|Q_H|$ (Eq. 21-9), and $\varepsilon_X > \varepsilon_C$, we must have

$$\frac{|W|}{|Q'_H|} > \frac{|W|}{|Q_H|},$$

where Q'_H is the heat transfer at the high-temperature reservoir in engine X. The right side of the inequality is the efficiency of the Carnot refrigerator shown in Fig. 21-15a when it operates (in reverse) as an engine. Since $|W|$ is the same in both cases, this inequality requires that

$$|Q_H| > |Q'_H|. \tag{21-16}$$

Because the work done by engine X is equal to the work done on the Carnot refrigerator, which is (from Eq. 21-6) $W = |Q_H| - |Q_L|$, we have

$$|Q_H| - |Q_L| = |Q'_H| - |Q'_L|.$$

We can rearrange terms and write this as

$$|Q_H| - |Q'_H| = |Q_L| - |Q'_L| = Q. \tag{21-17}$$

Here Q'_L is the thermal energy transfer at the low-temperature reservoir in engine X. Because we found that $|Q_H| > |Q'_H|$ (Eq. 21-16), the quantity Q must be positive. What does Q represent? Careful evaluation of this expression (Eq. 21-17) tells us that Q is the net thermal energy transfer at the high-temperature reservoir as a result of our combined engine plus refrigerator. That is, the net effect of engine X and the Carnot refrigerator, working in combination, is to transfer thermal energy Q from a low-temperature reservoir to a high-temperature reservoir. Notably, this is done with no work input to the combined engine–refrigerator system. Thus, the combination acts like the perfect refrigerator of Fig. 21-13, whose existence is a violation of the second law of thermodynamics.

Thus, we conclude that engine X cannot be more efficient than the ideal Carnot engine. In general, *no real engine can have an efficiency greater than that of a Carnot engine when both engines work between the same two temperatures.* At most, it can have an efficiency equal to that of an ideal Carnot engine. In that case, engine X is an ideal Carnot engine. Since ideal Carnot engines cannot be 100% efficient, this means that *"perfect" (100% efficient) engines are physically impossible.*

21-7 A Statistical View of Entropy

In Chapter 20 we saw that the macroscopic properties of gases can be explained in terms of their microscopic, or molecular, behavior. For one example, recall that we were able to account for the pressure exerted by a gas on the walls of its container in terms of the momentum transferred to those walls by rebounding gas molecules. Such explanations are part of a study called **statistical mechanics.**

Here we shall focus our attention on a single problem, involving the distribution of gas molecules between the two halves of an insulated box. This problem is reasonably simple to analyze, and it allows us to use statistical mechanics to calculate the entropy change for the free expansion of an ideal gas. You will see in Touchstone Example 21-6 that statistical mechanics leads to the same entropy change we obtain in Example 21-2 using thermodynamics.

Figure 21-16 shows a box that contains six identical (and thus indistinguishable) molecules of a gas. At any instant, a given molecule will be in either the left or the

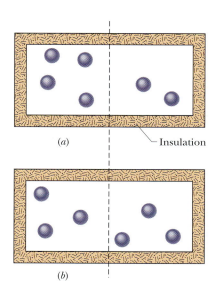

FIGURE 21-16 ■ An insulated box contains six gas molecules. Each molecule has the same probability of being in the left half of the box as in the right half. The arrangement in (a) corresponds to configuration III in Table 21-1, and that in (b) corresponds to configuration IV.

right half of the box; because the two halves have equal volumes, the molecule has the same likelihood, or probability, of being in either half.

Table 21-1 shows four of the seven possible *configurations* of the six molecules, each configuration labeled with Roman numerals. For example, in configuration I, all six molecules are in the left half of the box ($n_1 = 6$), and none are in the right half ($n_2 = 0$). The three configurations not shown are V with a (2, 4) split, VI with a (1, 5) split, and VII with a (0, 6) split. In configuration II, five molecules are in one half of the box, leaving one molecule in the other half. We see that, in general, a given configuration can be achieved in a number of different ways. We call these different arrangements of the molecules *microstates*. Let us see how to calculate the number of microstates that correspond to a given configuration.

TABLE 21-1
Six Molecules in a Box

Configuration[a]			Multiplicity W (number of microstates)	Calculation of W (Eq. 21-18)	Entropy 10^{-23} J/K (Eq. 21-19)
Label	n_1	n_2			
I	6	0	1	$6!/(6! \, 0!) = 1$	0
II	5	1	6	$6!/(5! \, 1!) = 6$	2.47
III	4	2	15	$6!/(4! \, 2!) = 15$	3.74
IV	3	3	20	$6!/(3! \, 3!) = 20$	4.13

Total number of microstates = 64

[a] The configurations not listed are $n_1 = 0$, $n_2 = 6$; $n_1 = 1$, $n_2 = 5$; and $n_1 = 2$, $n_2 = 4$. These have the same multiplicities as I, II and III respectively.

Suppose we have N molecules, distributed with n_1 molecules in one half of the box and n_2 in the other half. (Thus $n_1 + n_2 = N$.) Let us imagine that we distribute the molecules "by hand," one at a time. If $N = 6$, we can select the first molecule in six independent ways; that is, we can pick any one of the six molecules. We can pick the second molecule in five ways, by picking any one of the remaining five molecules, and so on. The total number of ways in which we can select all six molecules is the product of these independent ways, or $6 \times 5 \times 4 \times 3 \times 2 \times 1 = 720$. In mathematical shorthand we write this product as $6! = 720$, where $6!$ is pronounced "six factorial." Your hand-held calculator can probably calculate factorials. For later use you will need to know that $0! = 1$. (Check this on your calculator.)

However, because the molecules are indistinguishable, these 720 arrangements are not all different. In the case that $n_1 = 4$ and $n_2 = 2$ (which is configuration III in Table 21-1), for example, the order in which you put four molecules in one half of the box does not matter, because after you have put all four in, there is no way that you can tell the order in which you did so. The number of ways in which you can order the four molecules is $4!$ or 24. Similarly, the number of ways in which you can order two molecules for the other half of the box is simply $2!$ or 2. To get the number of different arrangements that lead to the 4, 2 split of configuration III, we must divide 720 by 24 and also by 2. We call the resulting quantity, which is the number of microstates that correspond to a given configuration, the *multiplicity* W of that configuration. Thus, for configuration III,

$$W_{\text{III}} = \frac{6!}{4! \, 2!} = \frac{720}{24 \times 2} = 15.$$

Thus, Table 21-1 tells us there are 15 independent microstates that correspond to configuration III. Note that, as the table also tells us, the total number of microstates for six molecules distributed over four configurations is 42.

Extrapolating from six molecules to the general case of N molecules, we have

$$W = \frac{N!}{n_1!\, n_2!} \qquad \text{(multiplicity of configuration)}. \qquad (21\text{-}18)$$

You should verify that Eq. 21-18 gives the multiplicities for all the configurations listed in Table 21-1.

The basic assumption of statistical mechanics is:

> All microstates are equally probable.

In other words, if we were to take a great many snapshots of the six molecules as they jostle around in the box of Fig. 21-16 and then count the number of times each microstate occurred, we would find that all 42 microstates will occur equally often. In other words, the system will spend, on average, the same amount of time in each of the 42 microstates listed in Table 21-1.

Because the microstates are equally probable, but different configurations have different numbers of microstates, the configurations are *not* equally probable. In Table 21-1 configuration IV, with 20 microstates, is the *most probable configuration*, with a probability of 20/64 = 0.313. This means that the system is in configuration IV 31.3% of the time. Configurations I and VII, in which all the molecules are in one half of the box, are the least probable, each with a probability of 1/64 = 0.016 or 1.6%. It is not surprising that the most probable configuration is the one in which the molecules are evenly divided between the two halves of the box, because that is what we expect at thermal equilibrium. However, it *is* surprising that there is any probability, however small, of finding all six molecules clustered in half of the box, with the other half empty. In Touchstone Example 21-5 we show that this state can occur because six molecules is an extremely small number.

For large values of N there are extremely large numbers of microstates, but nearly all the microstates belong to the configuration in which the molecules are divided equally between the two halves of the box, as Fig. 21-17 indicates. Even though the measured temperature and pressure of the gas remain constant, the gas is churning away endlessly as its molecules "visit" all probable microstates with equal probability. However, because so few microstates lie outside the very narrow central configuration peak of Fig. 21-17, we might as well assume that the gas molecules are always divided equally between the two halves of the box. As we shall see, this is the configuration with the greatest entropy.

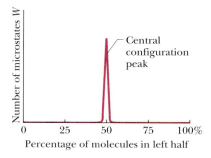

FIGURE 21-17 ■ For a large number of molecules in a box, a plot of the number of microstates that require various percentages of the molecules to be in the left half of the box. Nearly all the microstates correspond to an approximately equal sharing of the molecules between the two halves of the box; those microstates form the central configuration peak on the plot. For $N \approx 10^{22}$ molecules, the central configuration peak is much too narrow to be drawn on this plot.

Probability and Entropy

In 1877, Austrian physicist Ludwig Boltzmann (the Boltzmann of Boltzmann's constant k_B) derived a relationship between the entropy S of a configuration of a gas and the multiplicity W of that configuration. That relationship is

$$S = k_B \ln W \qquad \text{(Boltzmann's entropy equation)}. \qquad (21\text{-}19)$$

This famous formula is engraved on Boltzmann's tombstone.

It is natural that S and W should be related by a logarithmic function. The total entropy of two systems is the *sum* of their separate entropies. The probability of occurrence of two independent systems is the *product* of their separate probabilities. Because $\ln ab = \ln a + \ln b$, the logarithm seems the logical way to connect these quantities.

Table 21-1 displays the entropies of the configurations of the six-molecule system of Fig. 21-16, computed using Eq. 21-19. Configuration IV, which has the greatest multiplicity, also has the greatest entropy.

When you use Eq. 21-18 to calculate W, your calculator may signal "OVERFLOW" if you try to find the factorial of a number greater than a few hundred. Fortunately, there is a very good approximation, known as **Stirling's approximation,** not for $N!$ but for $\ln N!$, which as it happens is exactly what is needed in Eq. 21-19. Stirling's approximation is

$$\ln N! \approx N(\ln N) - N \qquad \text{(Stirling's approximation)}. \qquad (21\text{-}20)$$

The Stirling of this approximation is not the Stirling of the Stirling engine.

READING EXERCISE 21-6: A box contains one mole of a gas. Consider two configurations: (a) each half of the box contains one-half of the molecules, and (b) each third of the box contains one-third of the molecules. Which configuration has more microstates? ∎

TOUCHSTONE EXAMPLE 21-5: Indistinguishable

Suppose that there are 100 indistinguishable molecules in the box of Fig. 21-16. How many microstates are associated with the configuration $n_1 = 50$ and $n_2 = 50$? How many are associated with the configuration $n_1 = 100$ and $n_2 = 0$? Interpret the results in terms of the relative probabilities of the two configurations.

SOLUTION ∎ The **Key Idea** here is that the multiplicity W of a configuration of indistinguishable molecules in a closed box is the number of independent microstates with that configuration, as given by Eq. 21-18. For the (n_1, n_2) configuration $(50, 50)$, that equation yields

$$W = \frac{N!}{n_1! n_2!} = \frac{100!}{50! \, 50!}$$

$$= \frac{9.33 \times 10^{157}}{(3.04 \times 10^{64})(3.04 \times 10^{64})} \qquad \text{(Answer)}$$

$$= 1.01 \times 10^{29}.$$

Similarly, for the configuration of $(100, 0)$, we have

$$W = \frac{N!}{n_1! n_2!} = \frac{100!}{100! \, 0!} = \frac{1}{0!} = \frac{1}{1} = 1 \qquad \text{(Answer)}$$

Thus, a 50-50 distribution is more likely than a 100-0 distribution by the enormous factor of about 1×10^{29}. If you could count, at one per nanosecond, the number of microstates that correspond to the 50-50 distribution, it would take you about 3×10^{12} years, which is about 750 times longer than the age of the universe. Even 100 molecules is *still* a very small number. Imagine what these calculated probabilities would be like for a mole of molecules—say, about $N = 10^{24}$. You need never worry about suddenly finding all the air molecules clustering in one corner of your room!

TOUCHSTONE EXAMPLE 21-6: Entropy Increase

In Touchstone Example 21-1 we showed that when n moles of an ideal gas doubles its volume in a free expansion, the entropy increase from the initial state i to the final state f is $S_f - S_i = nR \ln 2$. Derive this result with statistical mechanics.

SOLUTION ■ One **Key Idea** here is that we can relate the entropy S of any given configuration of the molecules in the gas to the multiplicity W of microstates for that configuration, using Eq. 21-19 ($S = k_B \ln W$). We are interested in two configurations: the final configuration f (with the molecules occupying the full volume of their container in Fig. 21-1b) and the initial configuration i (with the molecules occupying the left half of the container).

A second **Key Idea** is that, because the molecules are in a closed container, we can calculate the multiplicity W of their microstates with Eq. 21-18. Here we have N molecules in the n moles of the gas. Initially, with the molecules all in the left half of the container, their (n_1, n_2) configuration is $(N, 0)$. Then, Eq. 21-18 gives their multiplicity as

$$W_i = \frac{N!}{N!\,0!} = 1.$$

Finally, with the molecules spread through the full volume, their (n_1, n_2) configuration is $(N/2, N/2)$. Then, Eq. 21-18 gives their multiplicity as

$$W_f = \frac{N!}{(N/2)!(N/2)!}.$$

From Eq. 21-19, the initial and final entropies are

$$S_i = k \ln W_i = k \ln 1 = 0$$

and

$$S_f = k_B \ln W_f = k \ln(N!) - 2k_B \ln[(N/2)!]. \quad (21\text{-}21)$$

In writing Eq. 21-21, we have used the relation

$$\ln \frac{a}{b^2} = \ln a - 2 \ln b.$$

Now, applying Eq. 21-20 to evaluate Eq. 21-21, we find that

$$\begin{aligned}
S_f &= k_B \ln (N!) - 2k_B \ln [(N/2)!] \\
&= k_B[N(\ln N) - N] - 2k_B[(N/2) \ln (N/2) - (N/2)] \\
&= k_B[N(\ln N) - N - N \ln (N/2) + N] \quad (21\text{-}22) \\
&= k_B[N(\ln N) - N(\ln N - \ln 2)] = Nk_B \ln 2.
\end{aligned}$$

From Eq. 20-7 we can substitute nR for Nk_B, where R is the universal gas constant. Equation 21-22 then becomes

$$S_f = nR \ln 2.$$

The change in entropy from the initial state to the final is thus

$$\begin{aligned}
S_f - S_i &= nR \ln 2 - 0 \\
&= nR \ln 2, \quad (\text{Answer})
\end{aligned}$$

which is what we set out to show. In Touchstone Example 21-1 we calculated this entropy increase for a free expansion with thermodynamics by finding an equivalent reversible process and calculating the entropy change for *that* process in terms of temperature and heat transfer. Here we have calculated the same increase with statistical mechanics using the fact that the system consists of molecules.

Problems

SEC. 21-2 ■ CHANGE IN ENTROPY

1. Expands Reversibly A 2.50 mol sample of an ideal gas expands reversibly and isothermally at 360 K until its volume is doubled. What is the increase in entropy of the gas?

2. Reversible Isothermal How much thermal energy must be transferred for a reversible isothermal expansion of an ideal gas at 132 °C if the entropy of the gas increases by 46.0 J/K?

3. Four Moles Four moles of an ideal gas undergo a reversible isothermal expansion from volume V_1 to volume $V_2 = 2V_1$ at temperature $T = 400$ K. Find (a) the work done by the gas and (b) the entropy change of the gas. (c) If the expansion is reversible and adiabatic instead of isothermal, what is the entropy change of the gas?

4. Reversible Isothermal Expansion An ideal gas undergoes a reversible isothermal expansion at 77.0°C, increasing its volume from 1.30 L to 3.40 L. The entropy change of the gas is 22.0 J/K. How many moles of gas are present?

5. Energy Absorbed Find (a) the thermal energy transfer and (b) the change in entropy of a 2.00 kg block of copper whose temperature is increased reversibly from 25°C to 100°C. The specific heat of copper is 386 J/kg·K.

6. Initial Temperature An ideal monatomic gas at initial temperature T_0 (in kelvins) expands from initial volume V_0 to volume $2V_0$ by each of the five processes indicated in the T-V diagram of Fig. 21-18. In which process is the expansion (a) isothermal, (b) isobaric (constant pressure), and (c) adiabatic? Explain your answers. (d) In which processes does the entropy of the gas decrease?

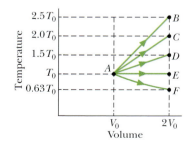

FIGURE 21-18 ■ Problem 6.

7. Entropy Change (a) What is the entropy change of a 12.0 g ice cube that melts completely in a bucket of water whose temperature is just above the freezing point of water? (b) What is the entropy change of a 5.00 g spoonful of water that evaporates completely on a hot plate whose temperature is slightly above the boiling point of water?

8. Ideal Gas Undergoes A 2.0 mol sample of an ideal monatomic gas undergoes the reversible process shown in Fig. 21-19. (a) How much thermal energy is transferred to the gas? (b) What is the change in the internal energy of the gas? (c) How much work is done by the gas?

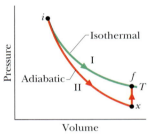

(Figure: Temperature (K) vs Entropy (J/K) graph, with points at 400 K / 5 and 200 K / 20)

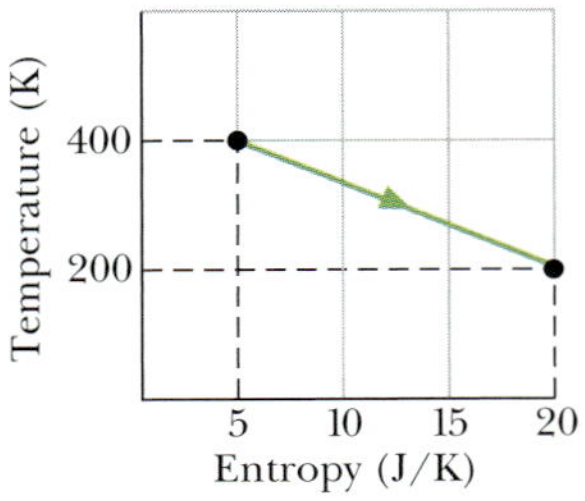

FIGURE 21-19 ■ Problem 8.

9. Aluminum and Water In an experiment, 200 g of aluminum (with a specific heat of 900 J/kg·K) at 100°C is mixed with 50.0 g of water at 20.0°C, with the mixture thermally isolated. (a) What is the equilibrium temperature? What are the entropy changes of (b) the aluminum, (c) the water and (d) the aluminum–water system?

10. Irreversible Process In the irreversible process of Fig. 21-5, let the initial temperatures of identical blocks L and R be 305.5 K and 294.5 K, respectively. Let 215 J be the thermal energy transfer between the blocks required to reach equilibrium. Then for the reversible processes of Fig. 21-6, what are the entropy changes of (a) block L, (b) its reservoir, (c) block R, (d) its reservoir, (e) the two-block system, and (f) the system of the two blocks and the two reservoirs?

11. Reversible Apparatus Use the reversible apparatus of Fig. 21-6 to show that, if the process of Fig. 21-5 happened in reverse, the entropy of the system would decrease, a violation of the second law of thermodynamics.

12. Rotating Not Oscillating An ideal diatomic gas, whose molecules are rotating but not oscillating, is taken through the cycle in Fig. 21-20. Determine for all three processes, in terms of P_1, V_1, T_1, and R: (a) P_2, P_3, and T_3 and (b) W, Q, ΔE^{int}, and ΔS per mole?

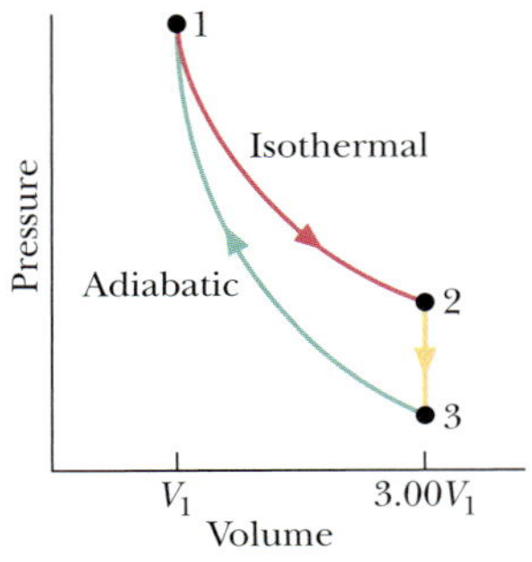

13. Copper in Box A 50.0 g block of copper whose temperature is 400 K is placed in an insulating box with a 100 g block of lead whose temperature is 200 K. (a) What is the equilibrium temperature of the two-block system? (b) What is the change in the internal energy of the two-block system between the initial state and the equilibrium state? (c) What is the change in the entropy of the two-block system? (See Table 19-2.)

FIGURE 21-20 ■
Problem 12.

(Figure: Pressure vs Volume, isothermal and adiabatic curves, points 1, 2, 3 between V_1 and $3.00V_1$)

14. Initial to Final One mole of a monatomic ideal gas is taken from an initial pressure P and volume V to a final pressure $2P$ and volume $2V$ by two different processes: (I) It expands isothermally until its volume is doubled, and then its pressure is increased at constant volume to the final pressure. (II) It is compressed isothermally until its pressure is doubled, and then its volume is increased at constant pressure to the final volume. (a) Show the path of each process on a P-V diagram. For each process calculate, in terms of P and V, (b) the thermal energy absorbed by the gas in each part of the process, (c) the work done by the gas in each part of the process, (d) the change in internal energy of the gas, ΔE^{int}, and (e) the change in entropy of the gas, ΔS.

15. Ice Cube in a Lake A 10 g ice cube at $-10°C$ is placed in a lake whose temperature is 15°C. Calculate the change in entropy of the cube–lake system as the ice cube comes to thermal equilibrium with the lake. The specific heat of ice is 2220 J/kg·K. (*Hint:* Will the ice cube affect the temperature of the lake?)

16. Ice Cube in Thermos An 8.0 g ice cube at $-10°C$ is put into a Thermos flask containing 100 cm³ of water at 20°C. By how much has the entropy of the cube–water system changed when a final equilibrium state is reached? The specific heat of ice is 2220 J/kg·K.

17. Water and Ice A mixture of 1773 g of water and 227 g of ice is in an initial equilibrium state at 0.00°C. The mixture is then, in a reversible process, brought to a second equilibrium state where the water–ice ratio, by mass, is 1:1 at 0.00°C. (a) Calculate the entropy change of the system during this process. (The heat of fusion for water is 333 kJ/kg.) (b) The system is then returned to the initial equilibrium state in an irreversible process (say, by using a Bunsen burner). Calculate the entropy change of the system during this process. (c) Are your answers consistent with the second law of thermodynamics?

18. Cylinder of Gas A cylinder contains n moles of a monatomic ideal gas. If the gas undergoes a reversible isothermal expansion from initial volume V_i to final volume V_f along path I in Fig. 21-21, its change in entropy is $\Delta S = nR \ \ln(V_f/V_i)$. (See Touchstone Example 21-1) Now consider path II in Fig. 21-21, which takes the gas from the same initial state i to state x by a reversible adiabatic expansion, and then from that state x to the same final state f by a reversible constant volume process. (a) Describe how you would carry out the two reversible processes for path II. (b) Show that the temperature of the gas in state x is

$$T_x = T_i(V_i/V_f)^{2/3}.$$

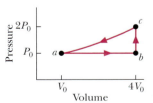

(Figure: Pressure vs Volume graph with Isothermal and Adiabatic curves, points i, f, x, T, paths I and II)

FIGURE 21-21 ■ Problem 18.

(c) What are the thermal energy transferred along path I (Q_1) and the thermal energy transferred along path II (Q_{II})? Are they equal? (d) What is the entropy change ΔS for path II? Is the entropy change for path I equal to it? (e) Evaluate T_x, Q_I, Q_{II}, and ΔS for $n = 1$, $T_i = 500$ K, and $V_f/V_i = 2$.

19. Through the Cycle One mole of an ideal monatomic gas is taken through the cycle in Fig. 21-22. (a) How much work is done by the gas in going from state a to state c along path abc? What are the changes in internal energy and entropy in going (b) from b to c and (c) through one complete cycle? Express all answers in terms of the pressure P_0, volume V_0, and temperature T_0 of state a.

(Figure: Pressure vs Volume graph, points a, b, c, with $2P_0$, P_0, V_0, $4V_0$)

FIGURE 21-22 ■ Problem 19.

20. Initial Pressure and Temperature One mole of an ideal monatomic gas, at an initial pressure of 5.00 kPa and initial temperature of 600 K, expands from initial volume $V_i = 1.00$ m³ to final volume $V_f = 2.00$ m³. During the expansion, the pressure P and volume V of the gas are related by $P = (5.00 \text{ kPa}) \exp[(V_i - V)/a]$, where $a = 1.00$ m³. What are (a) the final pressure and (b) the final temperature of the gas? (c) How much work is done by the gas during the expansion? (d) What is the change in entropy of the gas during the expansion? (*Hint*: Use two simple reversible processes to find the entropy change.)

SEC. 21-4 ■ ENTROPY IN THE REAL WORLD: ENGINES

Consider all Carnot engines discussed in these problems to be ideal.

21. Work and Efficiency in a Cycle A Carnot engine absorbs 52 kJ of thermal energy and exhausts 36 kJ as thermal energy in each cycle. Calculate (a) the engine's efficiency and (b) the work done per cycle in kilojoules.

22. Low-Temp Reservoir A Carnot engine whose low-temperature reservoir is at 17°C has an efficiency of 40%. By how much should the temperature of the high-temperature reservoir be increased to increase the efficiency to 50%?

23. Operates Between A Carnot engine operates between 235°C and 115°C, absorbing 6.30×10^4 J per cycle at the higher temperature. (a) What is the efficiency of the engine? (b) How much work per cycle is this engine capable of performing?

24. Reactor In a hypothetical nuclear fusion reactor, the fuel is deuterium gas at temperature of about 7×10^8 K. If this gas could be used to operate a Carnot engine with $T_L = 100$°C, what would be the engine's efficiency?

25. Carnot Engine A Carnot engine has an efficiency of 22.0%. It operates between constant-temperature reservoirs differing in temperature by 75.0°C. What are the temperatures of the two reservoirs?

26. Engine Has Power A Carnot engine has a power of 500 W. It operates between constant-temperature reservoirs at 100°C and 60.0°C. What are (a) the rate of thermal energy input and (b) the rate of exhaust heat output, in kilojoules per second?

27. Process *bc* One mole of a monatomic ideal gas is taken through the reversible cycle shown in Fig. 21-23. Process *bc* is an adiabatic expansion, with $P_b = 10.0$ atm and $V_b = 1.00 \times 10^{-3}$ m³. Find (a) the thermal energy transferred to the gas, (b) the thermal energy transferred from the gas, (c) the net work done by the gas, and (d) the efficiency of the cycle.

28. Enclosed Area Show that the area enclosed by the Carnot cycle on the temperature–entropy plot of Fig. 21-9 represents the net thermal energy transfer per cycle to the working substance.

29. Assume $P = 2P_0$ One mole of an ideal monatomic gas is taken through the cycle shown in Fig. 21-24. Assume that $P = 2P_0$, $V = 2V_0$, $P_0 = 1.01 \times 10^5$ Pa, and $V_0 = 0.0225$ m³. Calculate (a) the work done during the cycle, (b) the thermal energy added during stroke *abc*, and (c) the efficiency of the cycle. (d) What is the efficiency of a Carnot engine operating between the highest and lowest temperatures that occur in the cycle? How does this compare to the efficiency calculated in (c)?

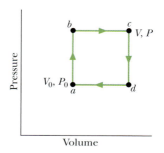

FIGURE 21-24 ■ Problem 29.

30. First Stage In the first stage of a two-stage Carnot engine, thermal energy Q_1 is absorbed at temperature T_1, work W_1 is done, and thermal energy Q_2 is expelled at a lower temperature T_2. The second stage absorbs that energy Q_2, does work W_2, and expels energy Q_3 at a still lower temperature T_3. Prove that the efficiency of the two-stage engine is $(T_1 - T_3)/T_1$.

31. Deep Shaft Suppose that a deep shaft were drilled in Earth's crust near one of the poles, where the surface temperature is -40°C, to a depth where the temperature is 800°C. (a) What is the theoretical limit to the efficiency of an engine operating between these temperatures? (b) If all the thermal energy released as heat into the low-temperature reservoir were used to melt ice that was initially at -40°C, at what rate could liquid water at 0°C be produced by a 100 MW power plant (treat it as an engine)? The specific heat of ice is 2220 J/kg · K; water's heat of fusion is 333 kJ/kg. (Note that the engine can operate only between 0°C and 800°C in this case. Energy exhausted at -40°C cannot be used to raise the temperature of anything above -40°C)

32. Working Substance One mole of an ideal gas is used as the working substance of an engine that operates on the cycle shown in Fig. 21-25. *BC* and *DA* are reversible adiabatic processes. (a) Is the gas monatomic, diatomic, or polyatomic? (b) What is the efficiency of the engine?

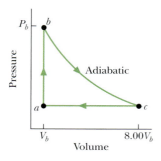

FIGURE 21-23 ■ Problem 27.

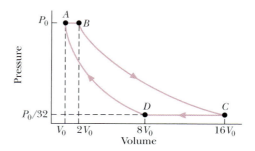

FIGURE 21-25 ■ Problem 32.

33. Gasoline Engine The operation of a gasoline internal combustion engine is represented by the cycle in Fig. 21-26. Assume the gasoline–air intake mixture is an ideal gas and use a compression ratio of 4:1 ($V_4 = 4V_1$). Assume that $P_2 = 3P_1$. (a) Determine the pressure and temperature at each of the vertex points of the P-V diagram in terms of P_1, T_1, and the ratio γ of the molar specific heats of the gas. (b) What is the efficiency of the cycle?

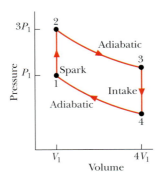

FIGURE 21-26 ■ Problem 33.

SEC. 21-5 ■ ENTROPY IN THE REAL WORLD: REFRIGERATORS

34. Carnot Refrigerator A Carnot refrigerator does 200 J of work to remove 600 J of thermal energy from its cold compartment. (a) What is the refrigerator's coefficient of performance? (b) How much thermal energy per cycle is exhausted to the kitchen?

35. Carnot Air Conditioner A Carnot air conditioner takes energy from the thermal energy of a room at 70°F and transfers it to the outdoors, which is at 96°F. For each joule of electric energy required to operate the air conditioner, how many joules are removed from the room?

36. Heat Pump Transfers The electric motor of a heat pump transfers thermal energy from the outdoors, which is at −5.0°C, to a room, which is at 17°C. If the heat pump were a Carnot heat pump (a Carnot engine working in reverse), how many joules would be transferred to the thermal energy of the room for each joule of electric energy consumed?

37. Heat Pump to Heat Building A heat pump is used to heat a building. The outside temperature is −5.0°C, and the temperature inside the building is to be maintained at 22°C. The pump's coefficient of performance is 3.8, and the heat pump delivers 7.54 MJ of thermal energy to the building each hour. If the heat pump is a Carnot engine working in reverse, at what rate must work be done to run the heat pump?

38. How Much Work How much work must be done by a Carnot refrigerator to transfer 1.0 J of thermal energy (a) from a reservoir at 7.0°C to one at 27°C, (b) from a reservoir at −73°C to one at 27°C, (c) from a reservoir at −173°C to one at 27°C, and (d) from a reservoir at −223°C to one at 27°C?

39. Air Conditioner An air conditioner operating between 93°F and 70°F is rated at 4000 Btu/h cooling capacity. Its coefficient of performance is 27% of that of a Carnot refrigerator operating between the same two temperatures. What horsepower is required of the air conditioner motor?

40. Motor in Refrigerator The motor in a refrigerator has a power of 200 W. If the freezing compartment is at 270 K and the outside air is at 300 K, and assuming the efficiency of a Carnot refrigerator, what is the maximum amount of thermal energy that can be extracted from the freezing compartment in 10.0 min?

41. Engine Driving Refrigerator A Carnot engine works between temperatures T_1 and T_2. It drives a Carnot refrigerator that works between temperatures T_3 and T_4 (Fig. 21-27). Find the ratio $|Q_3|/|Q_1|$ in terms of T_1, T_2, T_3, and T_4.

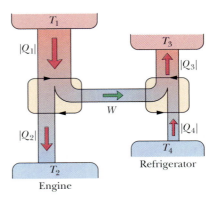

FIGURE 21-27 ■ Problem 41.

SEC. 21-7 ■ A STATISTICAL VIEW OF ENTROPY

42. Construct a Table Construct a table like Table 21-1 for eight molecules.

43. Show for N Molecules Show that for N molecules in a box, the number of possible microstates is 2^N when microstates are defined by whether a given molecule is in the left half of the box or the right half. Check this for the situation of Table 21-1.

44. A Box of N Gas Molecules A box contains N gas molecules, equally divided between its two halves. For $N = 50$: (a) What is the multiplicity of this central configuration? (b) What is the total number of microstates for the system? (*Hint:* See Problem 43.) (c) What percentage of the time does the system spend in its central configuration? (d) Repeat (a) through (c) for $N = 100$. (e) Repeat (a) through (c) for $N = 200$. (f) As N increases, you will find that the system spends *less* time (not more) in its central configuration. Explain why this is so.

45. Three Equal Parts A box contains N gas molecules. Consider the box to be divided into three equal parts. (a) By extension of Eq. 21-18, write an equation for the multiplicity of any given configuration. (b) Consider two configurations: configuration A with equal numbers of molecules in all three thirds of the box, and configuration B with equal numbers of molecules in both halves of the box. What is the ratio W_A/W_B of the multiplicity of configuration A to that of configuration B? (c) Evaluate W_A/W_B for $N = 100$. (Because 100 is not evenly divisible by 3, put 34 molecules into one of the three box parts and 33 in each of the other parts for configuration A.)

46. Four Particles in a Box Four particles are in the insulated box of Fig. 21-16. What are (a) the least multiplicity, (b) the greatest multiplicity, (c) the least entropy, and (d) the greatest entropy of the four-particle system?

Additional Problems

47. Velocity Spread A sample of nitrogen gas (N_2) undergoes a temperature increase at constant volume. As a result, the distribution of molecular speeds increases. That is, the probability distribution function $f(v)$ for the nitrogen molecules spreads to higher values of speed, as suggested in Fig. 20-8b. One way to report the spread in $f(v)$ is to measure the difference Δv between the most probable speed v^{prob} and the rms speed v^{rms}. When $f(v)$ spreads to higher speeds, Δv increases. (a) Write an equation relating the change ΔS in the entropy of the nitrogen gas to the initial difference Δv_i and the final difference Δv_f. Assume that the gas is an ideal diatomic gas with rotation but not oscillation of its molecules. Let the number of moles be 1.5 mol, the initial temperature be 250 K, and the final temperature be 500 K. What are (b) the initial difference Δv_i, (c) the final difference Δv_f, and (d) the entropy change ΔS for the gas?

48. Carnot Graph A Carnot engine is set up to produce a certain work W per cycle. In each cycle, thermal energy Q_H is transferred to the working substance of the engine from the higher-temperature thermal reservoir, which is at an adjustable temperature T_H. The lower-temperature thermal reservoir is maintained at temperature $T_L = 250$ K. Figure 21-28 gives Q_H for a range of T_H, if T_H is set at 550 K, what is Q_H?

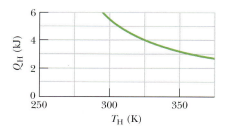

FIGURE 21-28 ■ Problem 48.

49. Gas Sample A gas sample undergoes a reversible isothermal expansion. Figure 21-29 gives the change ΔS in entropy of the gas versus the final volume V_f of the gas. How many moles are in the gas?

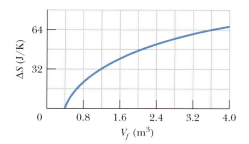

FIGURE 21-29 ■ Problem 49.

50. ΔS Graph A 364 g block is put in contact with a thermal reservoir. The block is initially at a lower temperature than the reservoir. Assume that the consequent heating of the block by the reservoir is reversible. Figure 21-30 gives the change in entropy of the block ΔS until thermal equilbrium is reached. What is the specific heat of the block?

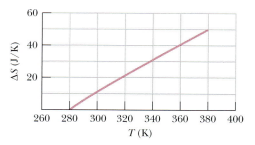

FIGURE 21-30 ■ Problem 50.

51. C/T Graph An object of (constant) heat capacity C is heated from an initial temperature T_i to a final temperature T_f by a constant-temperature reservoir at T_f. (a) Represent the process on a graph of C/T versus T, and show graphically that the total change in entropy ΔS of the object–reservoir system is positive. (b) Explain how the use of reservoirs at intermediate temperatures would allow the process to be carried out in a way that makes ΔS as small as desired.

52. Friction A moving block is slowed to a stop by friction. Both the block and the surface along which it slides are in an insulated enclosure. (a) Devise a reversible process to change the system from its initial state (block moving) to its final state (block stationary; temperature of the block and surface slightly increased). Show that this reversible process results in an entropy increase for the closed block–surface system. For the process, you can use an ideal engine and a constant-temperature reservoir. (b) Show, using the same process, but in reverse, that if the temperature of the system were to decrease spontaneously and the block were to start moving again, the entropy of the system would decrease (a violation of the second law of thermodynamics).

53. T-S Diagram A diatomic gas of 2 mol is taken reversibly around the cycle shown in the T-S diagram of Fig. 21-31. The molecules rotate but do not oscillate. What are the thermal energy transfers Q for (a) the path from point 1 to point 2, (b) the path from point 2 to point 3, and (c) the full cycle? (d) What is the work W for the isothermal process? The volume V_1 at point 1 is 0.200 m³. What are the volumes at (e) point 2 and (f) point 3?

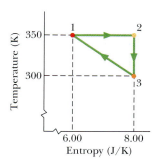

FIGURE 21-31 ■ Problem 53.

What are the changes ΔE^{int} in internal energy for (g) the path from point 1 to point 2, (h) the path from point 2 to point 3, and (i) the full cycle? (*Hint*: Part (h) can be done with one or two lines of calculation using Section 20-8 or with a page of calculation using Section 20-11.) (j) What is the work W for the adiabatic process?

54. Inventor's Engine An inventor has built an engine (engine X) and claims that its efficiency ε_X is greater than the efficiency ε of an ideal engine operating between the same two temperatures. Suppose that you couple engine X to an ideal refrigerator (Fig. 21-32a) and adjust the cycle of engine X so that the work per cycle that it provides equals the work per cycle required by the ideal refrigerator. Treat this combination as a single unit and show that if the inventor's claim were true (if $\varepsilon_X > \varepsilon$), the combined unit would act as a perfect refrigerator (Fig. 21-32b), transferring thermal energy from the low temperature reservoir to the high-temperature reservoir without the need of work.

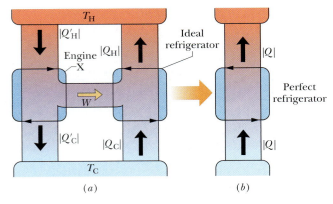

FIGURE 21-32 ■ Problem 54.

55. Ideal Diatomic Gas An ideal diatomic gas with 3.4 mol is taken through a cycle of three processes:

1. its temperature is increased from 200 K to 500 K at constant volume;

2. it is then isothermally expanded to its original pressure;

3. it is then contracted at constant pressure back to its original volume.

Throughout the cycle, the molecules rotate but do not oscillate. What is the efficiency of the cycle?

The International System of Units (SI)*

1 SI Base Units

1. The SI Base Units

Quantity	Name	Symbol	Definition
length	meter	m	"...the length of the path traveled by light in vacuum in 1/299 792 458 of a second." (1983)
mass	kilogram	kg	"...this prototype [a certain platinum–iridium cylinder] shall henceforth be considered to be the unit of mass." (1889)
time	second	s	"...the duration of 9 192 631 770 periods of the radiation corresponding to the transition between the two hyperfine levels of the ground state of the cesium-133 atom." (1967)
electric current	ampere	A	"...that constant current which, if maintained in two straight parallel conductors of infinite length, of negligible circular cross section, and placed 1 meter apart in vacuum, would produce between these conductors a force equal to 2×10^{-7} newton per meter of length." (1946)
thermodynamic temperature	kelvin	K	"...the fraction 1/273.16 of the thermodynamic temperature of the triple point of water." (1967)
amount of substance	mole	mol	"...the amount of substance of a system which contains as many elementary entities as there are atoms in 0.012 kilogram of carbon-12." (1971)
luminous intensity	candela	cd	"...the luminous intensity, in a given direction, of a source that emits monochromatic radiation of frequency 540×10^{12} hertz and that has a radiant intensity in that direction of 1/683 watt per steradian." (1979)

2 The SI Supplementary Units

2. The SI Supplementary Units

Quantity	Name of Unit	Symbol
plane angle	radian	rad
solid angle	steradian	sr

*Adapted from "The International System of Units (SI)," National Bureau of Standards Special Publication 330, 2001 edition. The definitions above were adopted by the General Conference of Weights and Measures, an international body, on the dates shown. In this book we do not use the candela.

3 Some SI Derivations

3. Some SI Derived Units

Quantity	Name of Unit	Symbol	In Terms of other SI Units
area	square meter	m²	
volume	cubic meter	m³	
frequency	hertz	Hz	s⁻¹
mass density (density)	kilogram per cubic meter	kg/m³	
speed, velocity	meter per second	m/s	
rotational velocity	radian per second	rad/s	
acceleration	meter per second per second	m/s²	
rotational acceleration	radian per second per second	rad/s²	
force	newton	N	kg·m/s²
pressure	pascal	Pa	N/m²
work, energy, quantity of heat	joule	J	N·m
power	watt	W	J/s
quantity of electric charge	coulomb	C	A·s
potential difference, electromotive force	volt	V	W/A
electric field strength	volt per meter (or newton per coulomb)	V/m	N/C
electric resistance	ohm	Ω	V/A
capacitance	farad	F	A·s/V
magnetic flux	weber	Wb	V·s
inductance	henry	H	V·s/A
magnetic flux density	tesla	T	Wb/m²
magnetic field strength	ampere per meter	A/m	
entropy	joule per kelvin	J/K	
specific heat	joule per kilogram kelvin	J/(kg·K)	
thermal conductivity	watt per meter kelvin	W/(m·K)	
radiant intensity	watt per steradian	W/sr	

4 Mathematical Notation

Poorly chosen mathematical notation can be a source of considerable confusion to those trying to learn and to do physics. For example, ambiguity in the meaning of a mathematical symbol can prevent a reader from understanding the meaning of a crucial relationship. It is also difficult to solve problems when the symbols used ot represent different quantities are not distinctive. In this text we have taken special care to use mathematical notation in ways that allow important distinctions to be easily visible both on the printed page and in handwritten work.

An excellent starting point for clear mathematical notation is the U.S. National Institute of Standard and Technology's Special Publication 811 (SP 811), *Guide for the Use of the International System of Units (SI)*, available at http://physics.nist.gov/cuu/Units/bibliography.html. In addition to following the National Institute guidelines, we have made a number of systematic choices to facilitate the translation of printed notation into handwritten mathematics. For example:

- Instead of making vectors bold, vector quantities (even in one dimension) are denoted by an arrow above the symbol. So printed equations look like handwritten equations. Example: $\vec{v}$ rather than **v** is used to denote an instantaneous velocity.

- In general, each vector component has an explicit subscript denoting that it represents the component along a chosen coordinate axis. The one exception is the position vector, $\vec{r}$, whose components are simply written as x, y, and z. For example, $\vec{r} = x\hat{i} + y\hat{j} + z\hat{k}$, whereas, $\vec{v} = v_x\hat{i} + v_y\hat{j} + v_z\hat{k}$.

- To emphasize the distinction between a vector's components and its magnitude, we write the magnitude of a vector, such as $\vec{F}$, as $|\vec{F}|$. However, when it is obvious that a magnitude is being described, we use the plain symbol (such as F with no coordinate subscript) to denote a vector's magnitude.

- We often choose to spell out the names of objects that are associated with mathematical variables—writing, for example, $\vec{v}_{ball}$ and not $\vec{v}_b$ for the velocity of a ball.

- Numerical subscripts most commonly denote sequential times, positions, velocities, and so on. For example, x_1 is the x-component of the position of some object at time t_1, whereas x_2 is the value of that parameter at some later time t_2. We have avoided using the subscript zero to denote initial values, as in x_0 to denote "the initial position along the x axis," to emphasize that *any* time can be chosen as the initial time for consideration of the subsequent time evolution of a system.

- To avoid confusing the numerical time sequence labels with object labels, we prefer to use capital letters as object labels. For example, we would label two particles A and B rather than 1 and 2. Thus, $\vec{p}_{A\,1}$ and $\vec{p}_{B\,1}$ would represent the translational momenta of two particles before a collision whereas $\vec{p}_{A\,2}$ and $\vec{p}_{B\,2}$ would be their momenta after a collision.

- To avoid excessively long strings of subscripts, we have made the unconventional choice to write all adjectival labels as *super*scripts. Thus, Newton's Second Law is written $\vec{F}^{net} = m\vec{a}$ whereas the sum of the forces acting on a certain object might be written as $\vec{F}^{net} = \vec{F}^{grav} + \vec{F}^{app}$. To avoid confusion with mathematical exponents, an adjectival label is never a single letter.

- Following a usage common in contemporary physics, the time average of a variable $\vec{v}$ is denoted as $\langle \vec{v} \rangle$ and not as $\vec{v}_{avg}$.

- Physical constants such as e, c, g, G, are all **positive** scalar quantities.

5 Significant Figures and the Precision of Numerical Results

Quoting the result of a calculation or a measurement to the correct number of significant figures is merely a way of telling your reader roughly how precise you believe the result to be. Quoting too many significant figures overstates the precision of your result and quoting too few implies less precision than the result may actually possess. So how many significant figures should you quote when reporting your result.

Determining Significant Figures

Before answering the question of how many significant figures to quote, we need to have a clear method for determining how many significant figures a reported number has. The standard method is quite simple:

> **METHOD FOR COUNTING SIGNIFICANT FIGURES:** Read the number from left to right, and count the first nonzero digit and all the digits (zero or not) to the right of it as significant.

Using this rule, 350 mm, 0.000350 km, and 0.350 m each has *three* significant figures. In fact, each of these numbers merely represents the same distance, expressed in different units. As you can see from this example, the number of *decimal places* that a number has is *not* the same as its number of *significant figures*. The first of these distances has zero decimal places, the second has six decimal places, and the third has three, yet all three of these numbers have three significant figures.

One consequence of this method is especially worth noting. Trailing zeros count as significant figures. For example, 2700 m/s has four significant figures. If you really meant it to have only three significant figures, you would have to write it either as 2.70 km/s (changing the unit) or 2.70×10^3 m/s (using scientific notation.)

A Simple Rule for Reporting Significant Figures in a Calculated Result

Now that you know how to count significant figures, how many should the result of a calculation have? A simple rule that will work in most calculations is:

> **SIGNIFICANT FIGURES IN A CALCULATED RESULT:** The common practice is to quote the result of a calculation to the number of significant figures of the *least* precise number used in the calculation.

Although this simple rule will often either understate or (less frequently) overstate the precision of a result, it still serves as a good rule-of-thumb for everyday numerical work. In introductory physics you will only rarely encounter data that are known to better than two, three, or four significant figures. This simple rule then tells you that you can't go very far wrong if you round off all your final results to three significant figures.

There are two situations in which the simple rule should *not* be applied to a calculation. One is when an exact number is involved in the calculation and another is when a calculation is done in parts so that intermediate results are used.

1. *Using Exact Data*   There are some obvious situations in which a number used in a calculation is exact. Numbers based on counting items are exact. For example, if you are told that there are 5 people on an elevator, there are exactly 5 people, not 4.7 or 5.1. Another situation arises when a number is exact by definition. For example, the conversion factor 2.54 cm/inch does *not* have three significant figures because the inch is *defined* to be exactly 2.5400000 . . . cm. *Data that are known exactly should not be included when deciding which of the original data has the fewest significant figures.*

2. *Significant Figures in Intermediate Results* Only the final result at the end of your calculation should be rounded using the simple rule. Intermediate results should never be rounded. Spreadsheet software takes care of this for you, as does your calculator if you store your intermediate results in its memory rather than writing them down and then rekeying them. If you must write down intermediate results, keep a few more significant figures than your final result will have.

Understanding and Refining the Simple Significant Figure Rule

Quoting the result of a calculation or measurement to the correct number of significant figures is a way of indicating its precision. You need to understand what limits the precision of data before you fully understand how to use the simple rule or its exceptions.

Absolute Precision There are two ways of talking about precision. First there is *absolute precision*, which tells you explicitly the smallest scale division of the measurement. It's always quoted in the same units as the measured quantity. For example, saying "I measured the length of the table to the nearest centimeter" states the absolute precision of the measurement. The absolute precision tells you how many *decimal places* the measurement has; it alone does not determine the number of significant figures. Example: if a table is 235 cm long, then 1 cm of absolute precision translates into three significant figures. On the other hand, if a table is for a doll's house and is only 8 cm long, then the same 1 cm of absolute precision has only one significant figure.

Relative Precision Because of this problem with absolute precision, scientists often prefer to describe the precision of data *relative* to the size of the quantity being measured. To use the previous examples, the *relative precision* of the length of the real table in the previous example is 1 cm out of 235 cm. This is usually stated as a ratio (1 part in 235) or as a percentage ($1/235 = 0.004255 \approx 0.4\%$). In the case of the toy table, the same 1 cm of absolute precision yields a relative precision of only 1 part in 8 or $1/8 = 0.125 = 12.5\%$.

Inconsistencies between Significant Figures and Relative Precision There is an inconsistency that goes with using a certain number of significant figures to express relative precision. Quoted to the same number of significant figures, the relative precision of results can be quite different. For example, 13 cm and 94 cm both have two significant figures. Yet the first is specified to only 1 part in 13 or $1/13 \approx 10\%$, whereas the second is known to 1 part in 94 or $1/94 \approx 1\%$. This bias toward greater relative precision for results with larger first significant figures is one weakness of using significant figures to track the precision of calculated results. You can partially address this problem, by including one more significant figure than the simple rule suggests, when the final result of a calculation has a 1 as its first significant figure.

Multiplying and Dividing When multiplying or dividing numbers, the *relative* precision of the result cannot exceed that of the least precise number used. Since the number of significant figures in the result tells us its relative precision, the simple rule is all that you need when you multiply or divide. For example, the area of a strip of paper of measured size is 280 cm by 2.5 cm would be correctly reported, according to the simple rule, as 7.0×10^2 cm^2. This result has only two significant figures since the less precise measurement, 2.5 cm, that went into the calculation had only two significant figures. Reporting this result as 700 cm^2 would not be correct since this result has three significant figures, exceeding the relative precision of the 2.5 cm measurement.

Addition and Subtraction When adding or subtracting, you line up the decimal points before you add or subtract. This means that it's the *absolute* precision of the least precise number that limits the precision of the sum or the difference. This can lead to some exceptions to the simple rule. For example, adding 957 cm and 878 cm yields 1835 cm. Here the result is reliable to an absolute precision of about 1 cm since both of the original distances had this reliability. But the result then has four significant figures whereas each of the original numbers had only three. If, on the other hand, you take the difference between these two distances you get 79 cm. The difference is still reliable to about 1 cm, but that absolute precision now translates into only two significant figures worth of relative precision. So, you should be careful when adding or subtracting, since addition can actually increase the relative precision of your result and, more important, subtraction can reduce it.

Evaluating Functions What about the evaluation of functions? For example, how many significant figures does the $\sin(88.2°)$ have? You can use your calculator to answer this question. First use your calculator to note that $\sin(88.2°) = 0.999506$. Now add 1 to the least significant decimal place of the argument of the function and evaluate it again. Here this gives $\sin(88.3°) = 0.999559$. Take the last significant figure in the result to be *the first one from the left that changed* when you repeated the calculation. In this example the first digit that changed was the 0; it became a 5 (the second 5) in the recalculation. So, using the empirical approach gives you five significant figures.

Some Fundamental Constants of Physics*

Constant	Symbol	Computational Value	Best (1998) Value	
			Value[a]	Uncertainty[b]
Speed of light in a vacuum	c	3.00×10^8 m/s	2.997 924 58	exact
Elementary charge	e	1.60×10^{-19} C	1.602 176 462	0.039
Gravitational constant	G	6.67×10^{-11} m^3/s$^2 \cdot$ kg	6.673	1500
Universal gas constant	R	8.31 J/mol $\cdot$ K	8.314 472	1.7
Avogadro constant	N_A	6.02×10^{23} mol^{-1}	6.022 141 99	0.079
Boltzmann constant	k_B	1.38×10^{-23} J/K	1.380 650 3	1.7
Stefan–Boltzmann constant	σ	5.67×10^{-8} W/m$^2 \cdot$ K^4	5.670 400	7.0
Molar volume of ideal gas at STP[d]	V_m	2.27×10^{-2} m^3/mol	2.271 098 1	1.7
Electric constant (permittivity)	ϵ_0	8.85×10^{-12} C^2/N $\cdot$ m^2	8.854 187 817 62	exact
Coulomb constant	$k = 1/4\pi\epsilon_0$	8.99×10^9 N $\cdot$ m^2/C^2	8.987 551 787	5×10^{-10}
Magnetic constant (permeability)	μ_0	1.26×10^{-6} N/A^2	1.256 637 061 43	exact
Planck constant	h	6.63×10^{-34} J $\cdot$ s	6.626 068 76	0.078
Electron mass[c]	m_e	9.11×10^{-31} kg	9.109 381 88	0.079
		5.49×10^{-4} u	5.485 799 110	0.0021
Proton mass[c]	m_p	1.67×10^{-27} kg	1.672 621 58	0.079
		1.0073 u	1.007 276 466 88	$1.3 \times .10^{-4}$
Ratio of proton mass to electron mass	m_p/m_e	1840	1836.152 667 5	0.0021
Electron charge-to-mass ratio	e/m_e	1.76×10^{11} C/kg	1.758 820 174	0.040
Neutron mass[c]	m_n	1.68×10^{-27} kg	1.674 927 16	0.079
		1.0087 u	1.008 664 915 78	5.4×10^{-4}
Hydrogen atom mass[c]	m_{1H}	1.0078 u	1.007 825 031 6	0.0005
Deuterium atom mass[c]	m_{2H}	2.0141 u	2.014 101 777 9	0.0005
Helium atom mass[c]	m_{4He}	4.0026 u	4.002 603 2	0.067
Muon mass	m_μ	1.88×10^{-28} kg	1.883 531 09	0.084
Electron magnetic moment	μ_e	9.28×10^{-24} J/T	9.284 763 62	0.040
Proton magnetic moment	μ_p	1.41×10^{-26} J/T	1.410 606 663	0.041
Bohr magneton	μ_B	9.27×10^{-24} J/T	9.274 008 99	0.040
Nuclear magneton	μ_N	5.05×10^{-27} J/T	5.050 783 17	0.040
Bohr radius	r_B	5.29×10^{-11} m	5.291 772 083	0.0037
Rydberg constant	R	1.10×10^7 m^{-1}	1.097 373 156 854 8	7.6×10^{-6}
Electron Compton wavelength	λ_C	2.43×10^{-12} m	2.426 310 215	0.0073

[a]Values given in this column should be given the same unit and power of 10 as the computational value.
[b]Parts per million.
[c]Masses given in u are in unified atomic mass units, where 1 u = 1.660 538 73 $\times 10^{-27}$ kg.
[d]STP means standard temperature and pressure: 0°C and 1.0 atm (0.1 MPa).

*The values in this table were selected from the 1998 CODATA recommended values (www.physics.nist.gov).

Some Astronomical Data

Some Distances from Earth

To the Moon*	3.82×10^8 m	To the center of our galaxy	2.2×10^{20} m
To the Sun*	1.50×10^{11} m	To the Andromeda Galaxy	2.1×10^{22} m
To the nearest star (Proxima Centauri)	4.04×10^{16} m	To the edge of the observable universe	$\sim 10^{26}$ m

* Mean distance.

The Sun, Earth, and the Moon

Property	Unit	Sun		Earth	Moon
Mass	kg	1.99×10^{30}		5.98×10^{24}	7.36×10^{22}
Mean radius	m	6.96×10^8		6.37×10^6	1.74×10^6
Mean density	kg/m^3	1410		5520	3340
Free-fall acceleration at the surface	m/s^2	274		9.81	1.67
Escape velocity	km/s	618		11.2	2.38
Period of rotation[a]	—	37 d at poles[b]	26 d at equator[b]	23 h 56 min	27.3 d
Radiation power[c]	W	3.90×10^{26}			

[a] Measured with respect to the distant stars, [b] The Sun, a ball of gas, does not rotate as a rigid body; [c] Just outside Earth's atmosphere solar energy is received, assuming normal incidence, at the rate of 1340 W/m^2.

Some Properties of the Planets

	Mercury	Venus	Earth	Mars	Jupiter	Saturn	Uranus	Neptune	Pluto
Mean distance from Sun, 10^6 km	57.9	108	150	228	778	1430	2870	4500	5900
Period of revolution, y	0.241	0.615	1.00	1.88	11.9	29.5	84.0	165	248
Period of rotation,[a] d	58.7	-243^b	0.997	1.03	0.409	0.426	-0.451^b	0.658	6.39
Orbital speed, km/s	47.9	35.0	29.8	24.1	13.1	9.64	6.81	5.43	4.74
Equatorial diameter, km	4880	12 100	12 800	6790	143 000	120 000	51 800	49 500	2300
Mass (Earth = 1)	0.0558	0.815	1.000	0.107	318	95.1	14.5	17.2	0.002
Surface value of g,[c] m/s^2	3.78	8.60	9.78	3.72	22.9	9.05	7.77	11.0	0.5
Escape velocity,[c] km/s	4.3	10.3	11.2	5.0	59.5	35.6	21.2	23.6	1.1

[a] Measured with respect to the distant stars.
[b] Venus and Uranus rotate opposite their orbital motion.
[c] Gravitational acceleration measured at the planet's equator.

APPENDIX D

Conversion Factors

Conversion factors may be read directly from these tables. For example, 1 degree = 2.778×10^{-3} revolutions, so $16.7° = 16.7 \times 2.778 \times 10^{-3}$ rev. The SI units are fully capitalized. Adapted in part from G. Shortley and D. Williams, *Elements of Physics*, 1971, Prentice-Hall, Englewood Cliffs, N.J.

Solid Angle

1 sphere
= 4π steradians
= 12.57 steradians

Plane Angle

	°	′	″	RADIAN	rev
1 degree =	1	60	3600	1.745×10^{-2}	2.778×10^{-3}
1 minute =	1.667×10^{-2}	1	60	2.909×10^{-4}	4.630×10^{-5}
1 second =	2.778×10^{-4}	1.667×10^{-2}	1	4.848×10^{-6}	7.716×10^{-7}
1 RADIAN =	57.30	3438	2.063×10^5	1	0.1592
1 revolution =	360	2.16×10^4	1.296×10^6	6.283	1

Length

	cm	METER	km	in.	ft	mi
1 centimeter =	1	10^{-2}	10^{-5}	0.3937	3.281×10^{-2}	6.214×10^{-6}
1 METER =	100	1	10^{-3}	39.37	3.281	6.214×10^{-4}
1 kilometer =	10^5	1000	1	3.937×10^4	3281	0.6214
1 inch =	2.540	2.540×10^{-2}	2.540×10^{-5}	1	8.333×10^{-2}	1.578×10^{-5}
1 foot =	30.48	0.3048	3.048×10^{-4}	12	1	1.894×10^{-4}
1 mile =	1.609×10^5	1609	1.609	6.336×10^4	5280	1

1 angström = 10^{-10} m   1 fermi = 10^{-15} m   1 light-year = 9.460×10^{12} km   1 fathom = 6 ft   1 yard = 3 ft   1 mil = 10^{-3} in.
1 nautical mile = 1852 m   1 parsec = 3.084×10^{13} km   1 Bohr radius = 5.292×10^{-11} m   1 rod = 16.5 ft   1 nm = 10^{-9} m
= 1.151 miles = 6076 ft

Area

	METER²	cm²	ft²	in.²
1 SQUARE METER =	1	10^4	10.76	1550
1 square centimeter =	10^{-4}	1	1.076×10^{-3}	0.1550
1 square foot =	9.290×10^{-2}	929.0	1	144
1 square inch =	6.452×10^{-4}	6.452	6.944×10^{-3}	1

key: 1 square mile = 2.788×10^7 ft² = 640 acres; 1 barn = 10^{-28} m²; 1 acre = 43 560 ft²; 1 hectare = 10^4 m² = 2.471 acres.

Volume

	METER³	cm³	L	ft³	in.³
1 CUBIC METER =	1	10^6	1000	35.31	6.102×10^4
1 cubic centimeter =	10^{-6}	1	1.000×10^{-3}	3.531×10^{-5}	6.102×10^{-2}
1 liter =	1.000×10^{-3}	1000	1	3.531×10^{-2}	61.02
1 cubic foot =	2.832×10^{-2}	2.832×10^4	28.32	1	1728
1 cubic inch =	1.639×10^{-5}	16.39	1.639×10^{-2}	5.787×10^{-4}	1

key: 1 U.S. fluid gallon = 4 U.S. fluid quarts = 8 U.S. pints = 128 U.S. fluid ounces = 231 in.³ 1 British imperial gallon = 277.4 in.³ = 1.201 U.S. fluid gallons.

Mass

Quantities in the colored areas are not mass units but are often used as such. When we write, for example, 1 kg "=" 2.205 lb, this means that a kilogram is a *mass* that *weighs* 2.205 pounds at a location where *g* has the standard value of 9.80665 m/s².

g	KILOGRAM	slug	u	oz	lb	ton
1 gram = 1	0.001	6.852×10^{-5}	6.022×10^{23}	3.527×10^{-2}	2.205×10^{-3}	1.102×10^{-6}
1 KILOGRAM = 1000	1	6.852×10^{-2}	6.022×10^{26}	35.27	2.205	1.102×10^{-3}
1 slug = 1.459×10^4	14.59	1	8.786×10^{27}	514.8	32.17	1.609×10^{-2}
1 atomic mass unit = 1.661×10^{-24}	1.661×10^{-27}	1.138×10^{-28}	1	5.857×10^{-26}	3.662×10^{-27}	1.830×10^{-30}
1 ounce = 28.35	2.835×10^{-2}	1.943×10^{-3}	1.718×10^{25}	1	6.250×10^{-2}	3.125×10^{-5}
1 pound = 453.6	0.4536	3.108×10^{-2}	2.732×10^{26}	16	1	0.0005
1 ton = 9.072×10^5	907.2	62.16	5.463×10^{29}	3.2×10^4	2000	1

1 metric ton = 1000 kg

Time

y	d	h	min	SECOND
1 year = 1	365.25	8.766×10^3	5.259×10^5	3.156×10^7
1 day = 2.738×10^{-3}	1	24	1440	8.640×10^4
1 hour = 1.141×10^{-4}	4.167×10^{-2}	1	60	3600
1 minute = 1.901×10^{-6}	6.944×10^{-4}	1.667×10^{-2}	1	60
1 SECOND = 3.169×10^{-8}	1.157×10^{-5}	2.778×10^{-4}	1.667×10^{-2}	1

Speed

ft/s	km/h	METER/SECOND	mi/h	cm/s
1 foot per second = 1	1.097	0.3048	0.6818	30.48
1 kilometer per hour = 0.9113	1	0.2778	0.6214	27.78
1 METER per SECOND = 3.281	3.6	1	2.237	100
1 mile per hour = 1.467	1.609	0.4470	1	44.70
1 centimeter per second = 3.281×10^{-2}	3.6×10^{-2}	0.01	2.237×10^{-2}	1

1 knot = 1 nautical mi/h = 1.688 ft/s   1 mi/min = 88.00 ft/s = 60.00 mi/h

Force

dyne	NEWTON	lb	pdl
1 dyne = 1	10^{-5}	2.248×10^{-6}	7.233×10^{-5}
1 NEWTON = 10^5	1	0.2248	7.233
1 pound = 4.448×10^5	4.448	1	32.17
1 poundal = 1.383×10^4	0.1383	3.108×10^{-2}	1

1 ton = 2000 lb

Pressure

atm	dyne/cm^2	inch of water	cm Hg	PASCAL	lb/in.^{2}	lb/ft^2
1 atmosphere = 1	1.013×10^6	406.8	76	1.013×10^5	14.70	2116
1 dyne per centimeter^2 = 9.869×10^{-7}	1	4.015×10^{-4}	7.501×10^{-5}	0.1	1.405×10^{-5}	2.089×10^{-3}
1 inch of water[a] at 4°C = 2.458×10^{-3}	2491	1	0.1868	249.1	3.613×10^{-2}	5.202
1 centimeter of mercury[a] at 0°C = 1.316×10^{-2}	1.333×10^4	5.353	1	1333	0.1934	27.85
1 PASCAL = 9.869×10^{-6}	10	4.015×10^{-3}	7.501×10^{-4}	1	1.450×10^{-4}	2.089×10^{-2}
1 pound per inch^2 = 6.805×10^{-2}	6.895×10^4	27.68	5.171	6.895×10^3	1	144
1 pound per foot^2 = 4.725×10^{-4}	478.8	0.1922	3.591×10^{-2}	47.88	6.944×10^{-3}	1

[a] Where the acceleration of gravity has the standard value of 9.80665 m/s^2.

1 bar = 10^6 dyne/cm^2 = 0.1 MPa   1 millibar = 10^3 dyne/cm^2 = 10^2 Pa   1 torr = 1 mm Hg

Energy, Work, Heat

Btu	erg	ft·lb	hp·h	JOULE	cal	kW·h	eV	MeV
1 British thermal unit = 1	1.055×10^{10}	777.9	3.929×10^{-4}	1055	252.0	2.930×10^{-4}	6.585×10^{21}	6.585×10^{15}
1 erg = 9.481×10^{-11}	1	7.376×10^{-8}	3.725×10^{-14}	10^{-7}	2.389×10^{-8}	2.778×10^{-14}	6.242×10^{11}	6.242×10^5
1 foot-pound = 1.285×10^{-3}	1.356×10^7	1	5.051×10^{-7}	1.356	0.3238	3.766×10^{-7}	8.464×10^{18}	8.464×10^{12}
1 horsepower-hour = 2545	2.685×10^{13}	1.980×10^6	1	2.685×10^6	6.413×10^5	0.7457	1.676×10^{25}	1.676×10^{19}
1 JOULE = 9.481×10^{-4}	10^7	0.7376	3.725×10^{-7}	1	0.2389	2.778×10^{-7}	6.242×10^{18}	6.242×10^{12}
1 calorie = 3.969×10^{-3}	4.186×10^7	3.088	1.560×10^{-6}	4.186	1	1.163×10^{-6}	2.613×10^{19}	2.613×10^{13}
1 kilowatt hour = 3413	3.600×10^{13}	2.655×10^6	1.341	3.600×10^6	8.600×10^5	1	2.247×10^{25}	2.247×10^{19}
1 electron-volt = 1.519×10^{-22}	1.602×10^{-12}	1.182×10^{-19}	5.967×10^{-26}	1.602×10^{-19}	3.827×10^{-20}	4.450×10^{-26}	1	10^{-6}
1 million electron-volts = 1.519×10^{-16}	1.602×10^{-6}	1.182×10^{-13}	5.967×10^{-20}	1.602×10^{-13}	3.827×10^{-14}	4.450×10^{-20}	10^{-6}	1

Power

Btu/h	ft·lb/s	hp	cal/s	kW	WATT
1 British thermal unit per hour = 1	0.2161	3.929×10^{-4}	6.998×10^{-2}	2.930×10^{-4}	0.2930
1 foot-pound per second = 4.628	1	1.818×10^{-3}	0.3239	1.356×10^{-3}	1.356
1 horsepower = 2545	550	1	178.1	0.7457	745.7
1 calorie per second = 14.29	3.088	5.615×10^{-3}	1	4.186×10^{-3}	4.186
1 kilowatt = 3413	737.6	1.341	238.9	1	1000
1 WATT = 3.413	0.7376	1.341×10^{-3}	0.2389	0.001	1

Magnetic Field

gauss	TESLA	milligauss
1 gauss = 1	10^{-4}	1000
1 TESLA = 10^4	1	10^7
1 milligauss = 0.001	10^{-7}	1

Magnetic Flux

maxwell	WEBER
1 maxwell = 1	10^{-8}
1 WEBER = 10^8	1

1 tesla = 1 weber/meter^2

Mathematical Formulas

Geometry

Circle of radius r: circumference $= 2\pi r$; area $= \pi r^2$.

Sphere of radius r: area $= 4\pi r^2$; volume $= \frac{4}{3}\pi r^3$.

Right circular cylinder of radius r and height h:
area $= 2\pi r^2 + 2\pi rh$; volume $= \pi r^2 h$.

Triangle of base a and altitude h: area $= \frac{1}{2}ah$.

Quadratic Formula

If $ax^2 + bx + c = 0$, then $x = \dfrac{-b \pm \sqrt{b^2 - 4ac}}{2a}$.

Trigonometric Functions of Angle θ

$\sin\theta = \dfrac{y}{r}$ $\cos\theta = \dfrac{x}{r}$

$\tan\theta = \dfrac{y}{x}$ $\cot\theta = \dfrac{x}{y}$

$\sec\theta = \dfrac{r}{x}$ $\csc\theta = \dfrac{r}{y}$

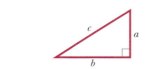

Pythagorean Theorem

In this right triangle,
$a^2 + b^2 = c^2$

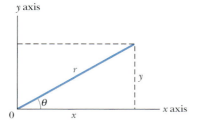

Triangles

Angles are A, B, C

Opposite sides are a, b, c

Angles $A + B + C = 180°$

$\dfrac{\sin A}{a} = \dfrac{\sin B}{b} = \dfrac{\sin C}{c}$

$c^2 = a^2 + b^2 - 2ab\cos C$

Exterior angle $D = A + C$

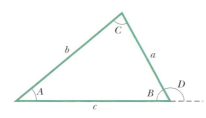

Mathematical Signs and Symbols

$=$ equals

$\approx$ equals approximately

$\sim$ is the order of magnitude of

$\neq$ is not equal to

$\equiv$ is identical to, is defined as

$>$ is greater than ($\gg$ is much greater than)

$<$ is less than ($\ll$ is much less than)

$\geq$ is greater than or equal to (or, is no less than)

$\leq$ is less than or equal to (or, is no more than)

$\pm$ plus or minus

$\propto$ is proportional to

Σ the sum of

$\langle x \rangle$ the average value of x

Trigonometric Identities

$\sin(90° - \theta) = \cos\theta$

$\cos(90° - \theta) = \sin\theta$

$\sin\theta/\cos\theta = \tan\theta$

$\sin^2\theta + \cos^2\theta = 1$

$\sec^2\theta - \tan^2\theta = 1$

$\csc^2\theta - \cot^2\theta = 1$

$\sin 2\theta = 2\sin\theta\cos\theta$

$\cos 2\theta = \cos^2\theta - \sin^2\theta = 2\cos^2\theta - 1 = 1 - 2\sin^2\theta$

$\sin(\alpha \pm \beta) = \sin\alpha\cos\beta \pm \cos\alpha\sin\beta$

$\cos(\alpha \pm \beta) = \cos\alpha\cos\beta \mp \sin\alpha\sin\beta$

$\tan(\alpha \pm \beta) = \dfrac{\tan\alpha \pm \tan\beta}{1 \mp \tan\alpha\tan\beta}$

$\sin\alpha \pm \sin\beta = 2\sin\frac{1}{2}(\alpha \pm \beta)\cos\frac{1}{2}(\alpha \mp \beta)$

$\cos\alpha + \cos\beta = 2\cos\frac{1}{2}(\alpha + \beta)\cos\frac{1}{2}(\alpha - \beta)$

$\cos\alpha - \cos\beta = -2\sin\frac{1}{2}(\alpha + \beta)\sin\frac{1}{2}(\alpha - \beta)$

Binomial Theorem

$$(1 + x)^n = 1 + \frac{nx}{1!} + \frac{n(n-1)x^2}{2!} + \cdots \qquad (x^2 < 1)$$

Exponential Expansion

$$e^x = 1 + x + \frac{x^2}{2!} + \frac{x^3}{3!} + \cdots$$

Logarithmic Expansion

$$\ln(1 + x) = x - \tfrac{1}{2}x^2 + \tfrac{1}{3}x^3 - \cdots \qquad (|x| < 1)$$

Trigonometric Expansions (θ in radians)

$$\sin\theta = \theta - \frac{\theta^3}{3!} + \frac{\theta^5}{5!} - \cdots$$

$$\cos\theta = 1 - \frac{\theta^2}{2!} + \frac{\theta^4}{4!} - \cdots$$

$$\tan\theta = \theta + \frac{\theta^3}{3} + \frac{2\theta^5}{15} + \cdots$$

Cramer's Rule

Two simultaneous equations in unknowns x and y,

$$a_1 x + b_1 y = c_1 \quad \text{and} \quad a_2 x + b_2 y = c_2,$$

have the solutions

$$x = \frac{\begin{vmatrix} c_1 & b_1 \\ c_2 & b_2 \end{vmatrix}}{\begin{vmatrix} a_1 & b_1 \\ a_2 & b_2 \end{vmatrix}} = \frac{c_1 b_2 - c_2 b_1}{a_1 b_2 - a_2 b_1}$$

and

$$y = \frac{\begin{vmatrix} a_1 & c_1 \\ a_2 & c_2 \end{vmatrix}}{\begin{vmatrix} a_1 & b_1 \\ a_2 & b_2 \end{vmatrix}} = \frac{a_1 c_2 - a_2 c_1}{a_1 b_2 - a_2 b_1}.$$

Products of Vectors

Let $\hat{i}, \hat{j},$ and $\hat{k}$ and be unit vectors in the $x, y,$ and z directions. Then

$$\hat{i} \cdot \hat{i} = \hat{j} \cdot \hat{j} = \hat{k} \cdot \hat{k} = 1, \quad \hat{i} \cdot \hat{j} = \hat{j} \cdot \hat{k} = \hat{k} \cdot \hat{i} = 0,$$

$$\hat{i} \times \hat{i} = \hat{j} \times \hat{j} = \hat{k} \times \hat{k} = 0,$$

$$\hat{i} \times \hat{j} = \hat{k}, \quad \hat{j} \times \hat{k} = \hat{i}, \quad \hat{k} \times \hat{i} = \hat{j}.$$

Any vector $\vec{a}$ with components $a_x, a_y,$ and a_z along the $x, y,$ and z axes can be written as

$$\vec{a} = a_x \hat{i} + a_y \hat{j} + a_z \hat{k}.$$

Let $\vec{a}, \vec{b},$ and $\vec{c}$ be arbitrary vectors with magnitudes $a, b,$ and c. Then

$$\vec{a} \times (\vec{b} + \vec{c}) = (\vec{a} \times \vec{b}) + (\vec{a} \times \vec{c})$$

$$(s\vec{a}) \times \vec{b} = \vec{a} \times (s\vec{b}) = s(\vec{a} \times \vec{b}) \quad (s = \text{a scalar}).$$

Let θ be the smaller of the two angles between $\vec{a}$ and $\vec{b}$. Then

$$\vec{a} \cdot \vec{b} = \vec{b} \cdot \vec{a} = a_x b_x + a_y b_y + a_z b_z = ab \cos \theta$$

$$\vec{a} \times \vec{b} = -\vec{b} \times \vec{a} = \begin{vmatrix} \hat{i} & \hat{j} & \hat{k} \\ a_x & a_y & a_z \\ b_x & b_y & b_z \end{vmatrix}$$

$$= \hat{i} \begin{vmatrix} a_y & a_z \\ b_y & b_z \end{vmatrix} - \hat{j} \begin{vmatrix} a_x & a_z \\ b_x & b_z \end{vmatrix} + \hat{k} \begin{vmatrix} a_x & a_y \\ b_x & b_y \end{vmatrix}$$

$$= (a_y b_z - b_y a_z)\hat{i} + (a_z b_x - b_z a_x)\hat{j} + (a_x b_y - b_x a_y)\hat{k}$$

$$|\vec{a} \times \vec{b}| = ab \sin \theta$$

$$\vec{a} \cdot (\vec{b} \times \vec{c}) = \vec{b} \cdot (\vec{c} \times \vec{a}) = \vec{c} \cdot (\vec{a} \times \vec{b})$$

$$\vec{a} \times (\vec{b} \times \vec{c}) = (\vec{a} \cdot \vec{c})\vec{b} - (\vec{a} \cdot \vec{b})\vec{c}$$

Derivatives and Integrals

In what follows, the letters u and v stand for any functions of x, and a and m are constants. To each of the indefinite integrals should be added an arbitrary constant of integration. The *Handbook of Chemistry and Physics* (CRC Press Inc.) gives a more extensive tabulation.

Derivatives

1. $\dfrac{dx}{dx} = 1$

2. $\dfrac{d}{dx}(au) = a\dfrac{du}{dx}$

3. $\dfrac{d}{dx}(u + v) = \dfrac{du}{dx} + \dfrac{dv}{dx}$

4. $\dfrac{d}{dx} x^m = mx^{m-1}$

5. $\dfrac{d}{dx} \ln x = \dfrac{1}{x}$

6. $\dfrac{d}{dx}(uv) = u\dfrac{dv}{dx} + v\dfrac{du}{dx}$

7. $\dfrac{d}{dx} e^x = e^x$

8. $\dfrac{d}{dx} \sin x = \cos x$

9. $\dfrac{d}{dx} \cos x = -\sin x$

10. $\dfrac{d}{dx} \tan x = \sec^2 x$

11. $\dfrac{d}{dx} \cot x = -\csc^2 x$

12. $\dfrac{d}{dx} \sec x = \tan x \sec x$

13. $\dfrac{d}{dx} \csc x = -\cot x \csc x$

14. $\dfrac{d}{dx} e^u = e^u \dfrac{du}{dx}$

15. $\dfrac{d}{dx} \sin u = \cos u \dfrac{du}{dx}$

16. $\dfrac{d}{dx} \cos u = -\sin u \dfrac{du}{dx}$

Integrals

1. $\int dx = x$

2. $\int au\, dx = a\int u\, dx$

3. $\int (u + v)\, dx = \int u\, dx + \int v\, dx$

4. $\int x^m dx = \dfrac{x^{m+1}}{m + 1}$ $(m \neq -1)$

5. $\int \dfrac{dx}{x} = \ln |x|$

6. $\int u\, \dfrac{dv}{dx}\, dx = uv - \int v\, \dfrac{du}{dx}\, dx$

7. $\int e^x dx = e^x$

8. $\int \sin x\, dx = -\cos x$

9. $\int \cos x\, dx = \sin x$

10. $\int \tan x\, dx = \ln |\sec x|$

11. $\int \sin^2 x\, dx = \frac{1}{2}x - \frac{1}{4}\sin 2x$

12. $\int e^{-ax} dx = -\dfrac{1}{a}\, e^{-ax}$

13. $\int xe^{-ax} dx = -\dfrac{1}{a^2}\, (ax + 1)e^{-ax}$

14. $\int x^2 e^{-ax} dx = -\dfrac{1}{a^3}\, (a^2 x^2 + 2ax + 2)e^{-ax}$

15. $\int_0^\infty x^n e^{-ax} dx = \dfrac{n!}{a^{n+1}}$

16. $\int_0^\infty x^{2n} e^{-ax^2} dx = \dfrac{1 \cdot 3 \cdot 5 \cdots (2n - 1)}{2^{n+1} a^n} \sqrt{\dfrac{\pi}{a}}$

17. $\int \dfrac{dx}{\sqrt{x^2 + a^2}} = \ln(x + \sqrt{x^2 + a^2})$

18. $\int \dfrac{x\, dx}{(x^2 + a^2)^{3/2}} = -\dfrac{1}{(x^2 + a^2)^{1/2}}$

19. $\int \dfrac{dx}{(x^2 + a^2)^{3/2}} = \dfrac{x}{a^2(x^2 + a^2)^{1/2}}$

20. $\int_0^\infty x^{2n+1} e^{-ax^2} dx = \dfrac{n!}{2a^{n+1}}$ $(a > 0)$

21. $\int \dfrac{x\, dx}{x + d} = x - d\ln(x + d)$

Properties of Common Elements

All physical properties are for a pressure of 1 atm unless otherwise specified.

Element	Symbol	Atomic Number Z	Molar Mass, g/mol	Density, g/cm³ at 20°C	Melting Point, °C	Boiling Point, °C	Specific Heat, J/(g·°C) at 25°C
Aluminum	Al	13	26.9815	2.699	660	2450	0.900
Antimony	Sb	51	121.75	6.691	630.5	1380	0.205
Argon	Ar	18	39.948	1.6626×10^{-3}	-189.4	-185.8	0.523
Arsenic	As	33	74.9216	5.78	817 (28 atm)	613	0.331
Barium	Ba	56	137.34	3.594	729	1640	0.205
Beryllium	Be	4	9.0122	1.848	1287	2770	1.83
Bismuth	Bi	83	208.980	9.747	271.37	1560	0.122
Boron	B	5	10.811	2.34	2030	—	1.11
Bromine	Br	35	79.909	3.12 (liquid)	-7.2	58	0.293
Cadmium	Cd	48	112.40	8.65	321.03	765	0.226
Calcium	Ca	20	40.08	1.55	838	1440	0.624
Carbon	C	6	12.01115	2.26	3727	4830	0.691
Cesium	Cs	55	132.905	1.873	28.40	690	0.243
Chlorine	Cl	17	35.453	3.214×10^{-3} (0°C)	-101	-34.7	0.486
Chromium	Cr	24	51.996	7.19	1857	2665	0.448
Cobalt	Co	27	58.9332	8.85	1495	2900	0.423
Copper	Cu	29	63.54	8.96	1083.40	2595	0.385
Fluorine	F	9	18.9984	1.696×10^{-3} (0°C)	-219.6	-188.2	0.753
Gadolinium	Gd	64	157.25	7.90	1312	2730	0.234
Gallium	Ga	31	69.72	5.907	29.75	2237	0.377
Germanium	Ge	32	72.59	5.323	937.25	2830	0.322
Gold	Au	79	196.967	19.32	1064.43	2970	0.131
Hafnium	Hf	72	178.49	13.31	2227	5400	0.144
Helium	He	2	4.0026	0.1664×10^{-3}	-269.7	-268.9	5.23
Hydrogen	H	1	1.00797	0.08375×10^{-3}	-259.19	-252.7	14.4
Indium	In	49	114.82	7.31	156.634	2000	0.233
Iodine	I	53	126.9044	4.93	113.7	183	0.218
Iridium	Ir	77	192.2	22.5	2447	(5300)	0.130
Iron	Fe	26	55.847	7.874	1536.5	3000	0.447
Krypton	Kr	36	83.80	3.488×10^{-3}	-157.37	-152	0.247
Lanthanum	La	57	138.91	6.189	920	3470	0.195
Lead	Pb	82	207.19	11.35	327.45	1725	0.129
Lithium	Li	3	6.939	0.534	180.55	1300	3.58
Magnesium	Mg	12	24.312	1.738	650	1107	1.03
Manganese	Mn	25	54.9380	7.44	1244	2150	0.481
Mercury	Hg	80	200.59	13.55	-38.87	357	0.138
Molybdenum	Mo	42	95.94	10.22	2617	5560	0.251
Neodymium	Nd	60	144.24	7.007	1016	3180	0.188

Element	Symbol	Atomic Number Z	Molar Mass, g/mol	Density, g/cm³ at 20°C	Melting Point, °C	Boiling Point, °C	Specific Heat, J/(g·°C) at 25°C
Neon	Ne	10	20.183	0.8387×10^{-3}	−248.597	−246.0	1.03
Nickel	Ni	28	58.71	8.902	1453	2730	0.444
Niobium	Nb	41	92.906	8.57	2468	4927	0.264
Nitrogen	N	7	14.0067	1.1649×10^{-3}	−210	−195.8	1.03
Osmium	Os	76	190.2	22.59	3027	5500	0.130
Oxygen	O	8	15.9994	1.3318×10^{-3}	−218.80	−183.0	0.913
Palladium	Pd	46	106.4	12.02	1552	3980	0.243
Phosphorus	P	15	30.9738	1.83	44.25	280	0.741
Platinum	Pt	78	195.09	21.45	1769	4530	0.134
Plutonium	Pu	94	(244)	19.8	640	3235	0.130
Polonium	Po	84	(210)	9.32	254	—	—
Potassium	K	19	39.102	0.862	63.20	760	0.758
Radium	Ra	88	(226)	5.0	700	—	—
Radon	Rn	86	(222)	9.96×10^{-3} (0°C)	(−71)	−61.8	0.092
Rhenium	Re	75	186.2	21.02	3180	5900	0.134
Rubidium	Rb	37	85.47	1.532	39.49	688	0.364
Scandium	Sc	21	44.956	2.99	1539	2730	0.569
Selenium	Se	34	78.96	4.79	221	685	0.318
Silicon	Si	14	28.086	2.33	1412	2680	0.712
Silver	Ag	47	107.870	10.49	960.8	2210	0.234
Sodium	Na	11	22.9898	0.9712	97.85	892	1.23
Strontium	Sr	38	87.62	2.54	768	1380	0.737
Sulfur	S	16	32.064	2.07	119.0	444.6	0.707
Tantalum	Ta	73	180.948	16.6	3014	5425	0.138
Tellurium	Te	52	127.60	6.24	449.5	990	0.201
Thallium	Tl	81	204.37	11.85	304	1457	0.130
Thorium	Th	90	(232)	11.72	1755	(3850)	0.117
Tin	Sn	50	118.69	7.2984	231.868	2270	0.226
Titanium	Ti	22	47.90	4.54	1670	3260	0.523
Tungsten	W	74	183.85	19.3	3380	5930	0.134
Uranium	U	92	(238)	18.95	1132	3818	0.117
Vanadium	V	23	50.942	6.11	1902	3400	0.490
Xenon	Xe	54	131.30	5.495×10^{-3}	−111.79	−108	0.159
Ytterbium	Yb	70	173.04	6.965	824	1530	0.155
Yttrium	Y	39	88.905	4.469	1526	3030	0.297
Zinc	Zn	30	65.37	7.133	419.58	906	0.389
Zirconium	Zr	40	91.22	6.506	1852	3580	0.276

The values in parentheses in the column of molar masses are the mass numbers of the longest-lived isotopes of those elements that are radioactive. Melting points and boiling points in parentheses are uncertain. The data for gases are valid only when these are in their usual molecular state, such as H_2, He, O_2, Ne, etc. The specific heats of the gases are the values at constant pressure. *Primary source*: Adapted fron J. Emsley, *The Elements*, 3rd ed., 1998, Clarendon Press, Oxford (www.webelements.com). Data on newest elements are current.

Periodic Table of the Elements

Legend:
- Metals
- Metalloids
- Nonmetals

Period	IA	IIA	IIIB	IVB	VB	VIB	VIIB	VIIIB			IB	IIB	IIIA	IVA	VA	VIA	VIIA	0
1	1 H																	2 He
2	3 Li	4 Be											5 B	6 C	7 N	8 O	9 F	10 Ne
3	11 Na	12 Mg											13 Al	14 Si	15 P	16 S	17 Cl	18 Ar
4	19 K	20 Ca	21 Sc	22 Ti	23 V	24 Cr	25 Mn	26 Fe	27 Co	28 Ni	29 Cu	30 Zn	31 Ga	32 Ge	33 As	34 Se	35 Br	36 Kr
5	37 Rb	38 Sr	39 Y	40 Zr	41 Nb	42 Mo	43 Tc	44 Ru	45 Rh	46 Pd	47 Ag	48 Cd	49 In	50 Sn	51 Sb	52 Te	53 I	54 Xe
6	55 Cs	56 Ba	57-71 *	72 Hf	73 Ta	74 W	75 Re	76 Os	77 Ir	78 Pt	79 Au	80 Hg	81 Tl	82 Pb	83 Bi	84 Po	85 At	86 Rn
7	87 Fr	88 Ra	89-103 †	104 Rf	105 Db	106 Sg	107 Bh	108 Hs	109 Mt	110 Ds	111 Uua	112 Uub	113	114 Uuq	115	116	117	118

Alkali metals IA — Transition metals — Noble gases 0 — THE HORIZONTAL PERIODS

Inner transition metals

Lanthanide series *

57 La	58 Ce	59 Pr	60 Nd	61 Pm	62 Sm	63 Eu	64 Gd	65 Tb	66 Dy	67 Ho	68 Er	69 Tm	70 Yb	71 Lu

Actinide series †

89 Ac	90 Th	91 Pa	92 U	93 Np	94 Pu	95 Am	96 Cm	97 Bk	98 Cf	99 Es	100 Fm	101 Md	102 No	103 Lr

The names of elements 104 through 109 (Rutherfordium, Dubnium, Seaborgium, Bohrium, Hassium, and Meitnerium, respectively) were adopted by the International Union of Pure and Applied Chemistry (IUPAC) in 1997. As of May 2003, elements 110, 111, 112, and 114 have been discovered. See www.webelements.com for the latest information and newest elements.

Answers to Reading Exercises and Odd-Numbered Problems

(Answers that involve a proof, graph, or otherwise lengthy solution are not included.)

Chapter 13

RE 13-1: Situations (c), (e), and (f) can yield static equilibrium, since in each case both the net force and the net torque can be zero. In (a), (b), and (d) the net force can be zero but the net torque cannot.

RE 13-2: The apple's center of gravity will end up directly below the rod, since only in that position is the net torque on the apple *stably* zero. The net torque on the apple is also zero when the apple's center of gravity is directly *above* the rod, but this is an *unstable* equilibrium point and the slightest rotation will cause the apple to rotate away from this position.

RE 13-3: You are better off if there is no friction between the ladder and the wall. With no friction between the ladder and the ground, the ground cannot exert any *horizontal* force to counter the horizontal force that the wall must exert on the ladder to keep it in place.

RE 13-4: In each of these three cases, the net horizontal force is zero independent of the magnitudes of the three unknown forces. This leaves only two independent equations for equilibrium— namely, net vertical force equals zero and net torque equals zero. But we have three unknowns to solve for. Since we can't do this, each of these three situations is indeterminate.

RE 13-5: Equation 13-29 tells us that, for elastic stretching, Young's modulus is just the stress (F/A) divided by the strain ($\Delta L/L$). Relative to rod 1, rod 2 has the same stress and twice the strain, and so its Young's modulus is half that of rod 1. By the same reasoning, rod 3 also has half the Young's modulus of rod 1, and rod 4 has a Young's modulus that is four times larger than that of rod 1. So, from higher to lower Young's modulus, rod 4 is the largest, rod 1 is next, and rods 2 and 3 tie for smallest.

RE 13-6: During bending, the particles on the inside of the bend are pushed closer together while those on the outside of the bend are pulled farther apart. During a shear deformation, adjacent planes of particles shift laterally with respect to one another. While the planes remain the same distance from one another, the "springs" (bonds) between adjacent planes are each stretched by the same amount.

Problems

1. (a) 2; (b) 7 **3.** (a) $(-27 \text{ N})\hat{i} + (2 \text{ N})\hat{j}$; (b) 176° counterclockwise from $+x$ direction **5.** 7920 N **7.** (a) $(mg/L)\sqrt{L^2 + r^2}$; (b) mgr/L **9.** (a) 1160 N, down; (b) 1740 N, up; (c) left; (d) right **11.** 74 g **13.** (a) 280 N; (b) 880 N, 71° above the horizontal **15.** (a) 8010 N; (b) 3.65 kN; (c) 5.66 kN **17.** 71.7 N **19.** (a) 5.0 N; (b) 30 N; (c) 1.3 m

21. $mg\sqrt{\dfrac{2rh - h^2}{r - h}}$ **23.** (a) 192 N; (b) 96.1 N; (c) 55.5 N **25.** (a) 6630 N; (b) 5740 N; (c) 5960 N **27.** 2.20 m **29.** 0.34 **31.** (a) 211 N; (b) 534 N; (c) 320 N **33.** (a) 445 N; (b)0.50; (c) 315 N **35.** (a) slides at 31°; (b) tips at 34° **37.** (a) 6.5×10^6 N/m²; (b) 1.1×10^{-5} m **39.** (a) 867 N; (b) 143 N; (c) 0.165 **41.** 44 N

Chapter 14

RE 14-1: The ratio (relative amount) of the magnitudes of these two forces depends only on the square of the ratio of the two center-to-center distances between the Earth and the other mass.

RE 14-2: Equation 14-2 tells us that g (the acceleration of a freely falling body) is just (Gm^{Earth}/r^2) where r is the distance to the center of the Earth to the point where g is measured. This is consistent with the model that assumes that the Moon stays in its orbit simply because it is in free fall. Although the observations are consistent with this model, this does not "prove" that the model is "true." It only establishes that this model is "good enough" to account for the data at hand.

RE 14-3: Since the location of the particle lies outside each of the spheres at the same distance from the center of the sphere in each case, each of the spheres will exert exactly the *same* magnitude force on the particle.

RE 14-4: $\vec{F}^{\text{grav}}$ due to the Earth *always* points directly toward the center of the Earth. However, the object's apparent weight associated with $\vec{N}$, (the "normal" force exerted on an object "at rest" on the surface of the rotating Earth), is *not* always directed exactly *away* from the center of the Earth! In fact, $\vec{N}$ points directly away from the center of the Earth only at the Earth's poles and at its equator. Why?

RE 14-5: In each case the direction of $\vec{F}^{\text{grav}}$ would be toward the center of the Earth. Case A: The magnitude of $\vec{F}^{\text{grav}}$ would decrease as $1/r^2$ where r is the distance to the center of the Earth. Case B: The magnitude of $\vec{F}^{\text{grav}}$ would be proportional to r, *and hence decrease*. Case C: Because of the considerably higher density of the Earth's core compared with its surface crust, the magnitude of $\vec{F}^{\text{grav}}$ would increase at first but then decrease to zero as we moved toward the center of the Earth.

RE 14-6: (a) The gravitational potential energy of the ball–sphere system increases. (b) The gravitational force between the ball and the sphere is attractive (inward), and the displacement is outward. Since the force and displacement are in opposite directions, the work done by the gravitational force is negative.

Problems

1. 19 m **3.** 29 pN **5.** 1/2 **7.** 2.60×10^5 km **9.** 0.017 N, toward the 300 kg sphere **11.** 3.2×10^{-7} N **13.** $\dfrac{GmM}{d^2}\left[1 - \dfrac{1}{8(1 - R/2d)^2}\right]$

15. 2.6×10^6 m **17.** (b) 1.9 h **21.** (a) $0.414R$; (b) $0.5R$ **23.** (a) $(3.0 \times 10^{-7}$ N/kg)m; (b) $(3.3 \times 10^{-7}$ N/kg)m; (c) $(6.7 \times 10^{-7}$ N/kg·m)mr **25.** (a) 9.83 m/s²; (b) 9.84 m/s²; (c) 9.79 m/s² **27.** (a) -1.3×10^{-4} J; (b) less; (c) positive; (d) negative **29.** (a) 0.74; (b) 3.7 m/s²; (c) 5.0 km/s **31.** (a) 5.0×10^{-11} J; (b) -5.0×10^{-11} J **35.** (a) 1700 m/s; (b) 250 km; (c) 1400 m/s **37.** (a) 82 km/s; (b) 1.8×10^4 km/s **39.** 2.5×10^4 km

Chapter 15

RE 15-1: Half the weight of the woman, (125 lb/2)(9.8 N/2.2 lb) ≈ 300 N, is supported by her two spike heels. Let's say that each heel makes contact with 1 cm² = 10^{-4} m² of the floor. Then the pressure of her heels on the floor is $P = F/A = (300$ N/10^{-4} m²) = 3×10^6 Pa. This estimate is close to that presented in the table. This pressure is high because of the small contact area over which this otherwise modest force is applied. An automobile has a much larger contact area.

RE 15-2: If air and water are made up of molecules that are about the same size and mass, then the average distance between the molecules in air at sea level must be about $1000^{1/3} = 10$ times larger than those of the water. This suggests that there is significantly more empty space around each air molecule, allowing them to be compressed closer together by quite a bit before they fill all of the available volume.

RE 15-3: The force the air exerts on the book is about $(10^5$ N/m²) $(2.54 \times 10^{-2}$ m/in)² (2.2 lb/9.8 N) (8 in)(10 in) ≅ 1200 lb. The close fit and the flexibility of the rubber mat prevents air from leaking into the space between the mat and the smooth tabletop, holding the mat down against the table with close to the full 1200 lb of force the air exerts on the top surface of the mat. The rougher surface of the book, as well as its rigidity, let air readily leak into the space between the book and the table when you start to pick up the book. This "equalizes" the pressure on each side of the book, reducing the net force that the air exerts on the book to a negligible amount.

RE 15-4: The density of air is only about one-thousandth that of water.

RE 15-5: The pressure at a depth Δy is the *same* in each container of oil. The shape of the container does not matter.

RE 15-6: Compressible fluids, like compressible springs, can "absorb" work and store it as elastic potential energy. Increasing the pressure of the compressible fluid in the hydraulic jack will thus store

some of the work done on the fluid as elastic potential energy and slightly reduce the amount of work that the fluid does on the output by that amount.

RE 15-7: The pressure at the bottom of this container is determined solely by the depth of the fluid above the bottom and the pressure that the air exerts on the surface of that fluid. In particular, the weight of the "extra" fluid that lies outside the central column is *not* carried by the horizontal bottom of the container and does *not* increase the pressure there.

RE 15-8: (a) Since the penguin floats in each of the three fluids, each fluid supplies a buoyant force exactly equal to the penguin's weight, so each fluid supplies the *same* buoyant force ($A = B = C$). (b) The penguin must displace the amount of fluid that matches her weight. Thus she *displaces* more of the least dense fluid B than of A, and even less of the most dense fluid C ($B > A > C$).

RE 15-9: You need to make sure that the weight of the canoe and its load is less than that of the water it displaces before the water starts coming in over the top edge of the hull. Although a chunk of concrete cannot displace its weight with water, a thin concrete canoe and its riders can.

RE 15-10: The net flow into (+) and out of (−) the entire system must be zero. So: $+x + (4 + 8 + 4 - 6 + 5 - 2)$ cm³/s = 0 cm³/s or $x = -13$ cm³/s, so fluid flows out of the unlabeled pipe at a rate of 13 cm³/s.

RE 15-11: (a) The area of face 1 is 4.0 cm² = 4.0×10^{-4} m². Face 2 has an area of 5.7 cm² = 5.7×10^{-4} m². Face 3 has an area of 8.0 cm² = 8.0×10^{-4} m². (b) The total surface area of all 6 faces is 69.7 cm² = 69.7×10^{-4} m².

RE 15-12: (a) The flux through any face is $v \, \Delta A \cos (\theta)$. Thus the flux through face 1 is (0.5 m/s) $(4.0 \times 10^{-4}$ m²) $(\cos (0)) = +2.0 \times 10^{-4}$ m³/s. The flux through face 2 is (0.5 m/s) $(5.7 \times 10^{-4}$ m²) $(\cos (45°)) = +2.0 \times 10^{-4}$ m³/s. The flux through face 3 is (0.5 m/s) $(8.0 \times 10^{-4}$ m²) $(\cos (180°)) = -4.0 \times 10^{-4}$ m³/s. (b) The flux through the front, back, and bottom faces is zero because $\theta = 90°$ for each of these faces and so the $\cos (\theta)$ term in the expression for the flux is zero. (c) Adding the contributions from all six faces yields zero net flux through this closed surface, as expected.

RE 15-13: (a) The volume flow rate is the *same* through each of the four sections. (b) The flow speed is largest in section 1, followed by section 2 and section 3, where it will be the same, and finally section 4 has the smallest flow speed. Recall that the flow speed is inversely proportional to the local cross-sectional area of the pipe. (c) The pressure will be greatest in section 4, less in section 3, still less in section 2, and least in section 1. The pressure difference between sections 2 and 3 is due to their difference in altitude. The pressure differences between sections at the same altitude are due to differences in the flow speed.

Problems

1. 1.1×10^5 Pa or 1.1 atm **3.** 2.9×10^4 N **5.** 0.074 **7.** (b) 26 kN **9.** 5.4×10^4 Pa **11.** (a) 5.3×10^6 N; (b) 2.8×10^5N; (c) 7.4×10^5 N; (d) no **13.** 7.2×10^5 N **15.** $\frac{1}{4}\rho gA(h_2 - h_1)^2$ **17.** 1.7 km **19.** (a) $\rho gWD^2/2$; (b) $\rho gWD^3/6$; (c) $D/3$ **21.** (a) 7.9 km; (b) 16 km **23.** 4.4 mm

25. (a) 2.04×10^{-2} m³; (b) 1570 N **27.** (a) 670 kg/m³; (b) 740 kg/m³ **29.** (a) 1.2 kg; (b) 1300 kg/m³ **31.** 57.3 cm **33.** 0.126 m³ **35.** (a) 45 m²; (b) car should be over center of slab if slab is to be level **37.** (a) 9.4 N; (b) 1.6 N **39.** 8.1 m/s **41.** 66 W **43.** (a) 2.5 m/s; (b) 2.6×10^5 Pa **45.** (a) 3.9 m/s; (b) 88 kPa **47.** (a) 1.6×10^{-3} m³/s; (b) 0.90 m **49.** 116 m/s **51.** (a) 6.4 m³; (b) 5.4 m/s; (c) 9.8×10^4 Pa **53.** (a) 74 N; (b) 150 m³ **55.** (b) 2.0×10^{-2} m³/s **57.** (b) 63.3 m/s

Chapter 16

RE 16-1: The amplitude and the angular frequency will stay the same. The initial phase will differ from ϕ_0 by 90° or $\pi/2$ since you can think of a cosine as a sine that has been shifted 90° to the left.

RE 16-2: (a) When $t = 2.00\,T$ the particle will have moved through two full oscillations and will be back where it started from—namely, at $x = -X$. (b) When $t = 3.50T$ the particle will have moved through three full oscillations and an additional half oscillation and so will be at $x = +X$. (c) When $t = 5.25T$ the particle will have moved through five full oscillations and an additional quarter oscillation and so will be at $x = 0$.

RE 16-3: Equation 16-12 tells us that the period of a mass on a given spring increases as the amount of oscillating mass increases. The fact that the mass of the spring itself oscillates along with the mass on its end suggests that some of the spring's mass should be included in the mass that appears in Eq. 16-12. Since the spring oscillates with a progressively smaller amplitude as we go from its moving end to its fixed end, only some fraction of the spring's mass needs to be included in this corrected total oscillating mass.

RE 16-4: Only (a) implies simple harmonic motion. Although (b) is a restoring type of force, it is quadratic, not linear in x. Force (c) is repulsive rather than attractive, driving the particle away from $x = 0$ rather than back toward it. Force (d) is both repulsive and nonlinear.

RE 16-5: The particle's velocity component is zero at $t = t_2$ and t_4. The particle is moving to the left at its greatest speed at $t = t_1$ and it is moving to the right at its greatest speed at $t = t_3$. Considering v_x as a mathematical function of t, we can indeed say that v_x is a minimum at t_1 and a maximum at t_3, but do remember that it is actually moving at its fastest speed when the velocity component is both a minimum and a maximum.

RE 16-6: The vertical component a_x of the acceleration is *increasing* in regions 1 and 2 and it is *decreasing* in regions 3 and 4. Note, however, that the *magnitude* of this acceleration is actually *decreasing* in regions 1 and 3 while the magnitude is *increasing* in regions 2 and 4. Pause and reflect on this!

RE 16-7: In each of these cases, the net force acting on the pendulum mass is proportional to the mass itself. Since Newton's Second Law tells us that acceleration is net force divided by mass, the mass cancels out here and so acceleration will be independent of the mass in these cases.

RE 16-8: "Same shape and size" for these three pendula means that the rotational inertia of each is simply proportional to its mass *with the same constant of proportionality in each case*. "Suspended at the same point" means the same distance from the point of suspension to the center of mass in each case. Since I/m and h are the same

for each of the three, Eq. 16-26 tells us that each will have the same period.

RE 16-9: Since $K = 3$ J and $U = 2$ J at *a* given point, then the total mechanical energy of this system is $E = K + U = 5$ J *at every point* in its motion. Conservation of mechanical energy rules! (a) In particular, when the block is at $x = 0$, the system's potential energy is zero and so its kinetic energy must be 5 J. (b) At $x = -X$, the system's kinetic energy is zero so then $U = 5$ J.

RE 16-10: From Eq. 16-39 the time it takes for the mechanical energy of a damped oscillator to fall to one-fourth (or to any given fraction, for that matter) of its initial value is proportional to m/b. The ratio of m/b for set 2 is $4/6 = 2/3$ that of set 1, and for set 3 it is $1/3$ of that for set 1. Thus set 1 takes the longest time to lose one-fourth of its mechanical energy, followed by set 2, then by set 3. (set 1 > set 2 > set 3)

Problems

1. (a) 0.50 s; (b) 2.0 Hz; (c) 18 cm **3.** (a) 0.500 s; (b) 2.00 Hz; (c) 12.6 rad/s; (d) 79.0 N/m; (e) 4.40 m/s; (f) 27.6 N **5.** $f > 500$ Hz **7.** (a) 6.28×10^5 rad/s; (b) 1.59 mm **9.** (a) 1.0 mm; (b) 0.75 m/s; (c) 570 m/s² **11.** (a) 1.29×10^5 N/m; (b) 2.68 Hz **13.** 7.2 m/s **15.** 2.08 h **17.** 3.1 cm **19.** (a) 5.58 Hz; (b) 0.325 kg; (c) 0.400 m **21.** (a) 2.2 Hz; (b) 56 cm/s; (c) 0.10 kg; (d) 20.0 cm below y_i 23. (a) $0.183A$; (b) same direction **29.** (a) $(n + 1)k/n$; (b) $(n + 1)k$; (c) $\sqrt{(n + 1)/n}f$; (d) $\sqrt{n + 1}\,f$ **31.** (a) 39.5 rad/s; (b) 34.2 rad/s; (c) 124 rad/s² **33.** 99 cm **35.** 5.6 cm

37. (a) $2\pi\sqrt{\dfrac{L^2 + 12d^2}{12gd}}$; (b) increases for d $< L/\sqrt{12}$, decreases for $d > L/\sqrt{12}$; (c) increases; (d) no change **39.** (a) 0.205 kg·m²; (b) 47.7 cm; (c) 1.50 s **41.** $2\pi\sqrt{m/3k}$ **43.** (a) 0.35 Hz; (b) 0.39 Hz; (c) 0 **45.** (b) smaller **47.** (a) $(r/R)\sqrt{k/m}$; (b) $\sqrt{k/m}$; (c) no oscillation **49.** 37 mJ **51.** (a) 2.25 Hz; (b) 125 J; (c) 250 J; (d) 86.6 cm **53.** (a) $\frac{3}{4}$; (b) $\frac{1}{4}$; (c) $x^{max}/\sqrt{2}$ **55.** (a) 16.7 cm; (b) 1.23% **57.** 0.39 **59.** (a) 14.3 s; (b) 5.27 **61.** (a) $F^{max}/b\omega$; (b) F^{max}/b

Chapter 17

RE 17-1: (a) None of these graphs correctly shows the displacement of the rope versus position along the rope at $t = 0$ s. Graph (b) is the closest to correct of the four but fails to show the considerable length of undisturbed rope that lies between $x = 0$ and the trailing (left) edge of the pulse at $t = 0$ s. (b) Graph (a) correctly shows the displacement of the rope versus time at $x = x_1$ as the pulse passes by.

RE 17-2: Realizing that each of these phase expressions is of the form $(kx - \omega t)$ and that the wavelength $\lambda = 2\,\pi/k$, we see that wave 1 has the smallest wavelength and thus the largest k so it must correspond to case (c). Wave 2 has the smallest k and so must go with case (a), and wave 3 has the middle value of k and so matches case (b).

RE 17-3: The velocity v_y^{string} describes the up and down (transverse) motion of a particular small segment of the string as the wave passes by that location. Typically the velocity varies rapidly between positive and negative values as the wave goes by. The magnitude of the maximum of this velocity increases with increasing amplitude of passing wave having the same wavelength and frequency. The other velocity, v_x^{wave} tells us how fast and in what direction any given *crest* of the wave itself moves along the rope. For a uniform rope its time does not vary at all and it does not depend on the amplitude of the wave. In the derivation in this section, v_x^{wave} is used to obtain the mass of the segment of string under study and v_y^{string} is used to obtain the momentum change of the segment.

RE 17-4: If you increase the frequency of the oscillations driving waves in a string, holding the tension constant, then (a) the speed of the waves remains the same and (b) the wavelength decreases. If you instead increase the tension keeping the driving frequency constant, then (c) the wave speed increases and (d) the wavelength also increases.

RE 17-5: (a) Equation (1) represents the interference of a pair of waves traveling in the positive x direction. (b) Equation (3) represents the interference of a pair of waves traveling in the negative x direction. (c) Equation (2) represents the interference of a pair of waves traveling in opposite directions.

RE 17-6: (a) The missing frequency is 75 Hz. (b) The seventh harmonic has a frequency of 7×75 Hz $= 525$ Hz.

Problems

1. (a) 3.49 m^{-1}; (b) 31.5 m/s **3.** (a) 0.68 s; (b) 1.47 Hz; (c) 2.06 m/s **7.** (a) $y(x, t) = 2.0 \sin 2\pi(0.10x - 400t)$, with x and y in cm and t in s; (b) 50 m/s; (c) 40 m/s **9.** (a) 11.7 cm; (b) π rad **11.** 129 m/s **13.** (a) 15 m/s; (b) 0.036 N **15.** $y(x, t) = 0.12 \sin(141x + 628t)$, with y in mm, x in m, and t in s **17.** (a) $2\pi y^{max}/\lambda$; (b) no **19.** (a) 5.0 cm; (b) 40 cm; (c) 12 m/s; (d) 0.033 s; (e) 9.4 m/s; (f) $5.0 \sin(16x + 190t + 0.93)$, with x in m, y in cm, and t in s **21.** 2.63 m from the end of the wire from which the later pulse originates **25.** $1.4y^{max}$ **27.** (a) 0.31 m; (b) 1.64 rad; (c) 2.2 mm **29.** (a) 140 m/s; (b) 60 cm; (c) 240 Hz **31.** (a) 82.0 m/s; (b) 16.8 m; (c) 4.88 Hz **33.** 7.91 Hz, 15.8 Hz, 23.7 Hz **35.** (a) 105 Hz; (b) 158 m/s **37.** (a) 0.25 cm; (b) 120 cm/s; (c) 3.0 cm; (d) zero **39.** (a) 50 Hz; (b) $y = 0.50 \sin[\pi(x \pm 100t)]$, with x in m, y in cm, and t in s **41.** (a) 1.3 m; (b) $y = 0.002 \sin(9.4x) \cos(3800t)$, with x and y in m and t in s **43.** (a) 2.0 Hz; (b) 200 cm; (c) 400 cm/s; (d) 50 cm, 150 cm, 250 cm, etc.; (e) 0, 100 cm, 200 cm, etc. **47.** (a) 323 Hz; (b) eight **49.** 5.0 cm

Chapter 18

RE 18-1: We express units in terms of very basic elements of length [L], mass [M] and time [T]. Since B is a force per unit area, its units are $B \sim \dfrac{[M][L]/[T^2]}{[L^2]} = [M]/[L][T^2]$. ρ is a mass per unit volume so $\rho \sim [M]/[L^3]$. $B/\rho \sim [L^2]/[T^2]$ and $\sqrt{B/\rho} \sim [L]/[T]$ or a "velocity" given by [m]/[s] in SI units.

RE 18-2: The measured wave speed for the round trip is $v^{wave} = (2)(2.4 \text{ m})/(.0133 - .0002)(\text{s}) = 366$ m/s, which is in reasonable agreement with the stated 343 m/s in room-temperature air.

RE 18-3: Since energy per unit time passing through a surface that faces a source of sound is just the product of the sound intensity there and the area of the surface, and since sound intensity falls off with distance from the source, (a) the intensity of the sound is the same at surfaces 1 and 2 and is smaller at surface 3, and (b) the areas of surfaces 1 and 2 are equal while that of surface 3 is larger.

RE 18-4: The second harmonic of the longer pipe B has the same frequency as the first harmonic of the shorter pipe A.

RE 18-5: (a) and (e) have greater detected frequency than emitted frequency. (b) and (f) have reduced detected frequencies. (c) and (d) are indeterminate.

Problems

1. divide the time by 3 **3.** (a) 79 m, 41 m; (b) 89 m **5.** 1900 km **7.** 40.7 m **9.** (a) 0.0762 mm; (b) 0.333 mm **11.** (a) 343 $(1 + 2m)$ Hz, with m being an integer from 0 to 28; (b) $686m$ Hz, with m being an integer from 1 to 29 **13.** (a) 143 Hz, 429 Hz, 715 Hz; (b) 286 Hz, 572 Hz, 858 Hz **15.** 17.5 cm **17.** 15.0 mW **19.** (a) 1000; (b) 32 **21.** (a) 59.7; (b) 2.81×10^{-4} **23.** (a) 5000; (b) 71; (c) 71 **25.** (a) 5200 Hz; (b) amplitude$_{SAD}$/amplitude$_{SBD}$ $= 2$ **27.** (a) 57.2 cm; (b) 42.9 cm **29.** (a) 405 m/s; (b) 596 N; (c) 44.0 cm; (d) 37.3 cm **31.** (a) 1129, 1506, and 1882 Hz **33.** 12.4 m **35.** (a) node; (c) 22 s **37.** 45.3 N **39.** 387 Hz **41.** 0.02 **43.** 17.5 kHz **45.** (a) 526 Hz; (b) 555 Hz **47.** (a) 1.02 kHz; (b) 1.04 kHz **49.** 155 Hz **51.** (a) 485.8 Hz; (b) 500.0 Hz; (c) 486.2 Hz; (d) 500.0 Hz **53.** (a) 598 Hz; (b) 608 Hz; (c) 589 Hz **55.** (a) 42°; (b) 11 s

Chapter 19

RE 19-1: Some properties that are measurable include mass, volume, hardness, elasticity, and breaking strength. Flavor and color, for example, are less easily quantified.

RE 19-2: For comfort we often want to maintain the temperature inside our homes significantly higher (in winter) or lower (in summer) than that of the environment outside. Thermal insulation inside the walls of our homes reduces the amount of heat energy that would otherwise flow out of the house in winter or into the house in summer, reducing the expenditure of energy needed to maintain a comfortable interior temperature.

RE 19-3: Equation 19-5 tells us that, for the same amount of heat energy added to the same mass, the temperature increase is inversely proportional to the specific heat of the material being heated. Thus object A has a greater specific heat than object B.

RE 19-4: The good news for the firefighter is that each kilogram of water sprayed on the fire can remove a relatively large amount of heat from the burning object. The bad news is that this heat can easily be transferred to the firefighter's body if the steam condenses on her skin. One gram of steam at 100 °C condensing on one gram of (water-like) flesh at 37 °C will yield two grams of water-like material at a temperature of (100 °C + 37 °C)/2 = 69 °C.

RE 19-5: For the net work done by the gas on its environment to be positive, the top curve must go from left (lower pressure) to right (higher pressure.) For maximum positive work each cycle that area on the P-V diagram enclosed by the cycle must be as large as possible. So curves c and e yield the maximum possible positive work here.

RE 19-6: (a) The change in the internal energy of the gas is the same in each case. (b) The work done by the gas is greatest for path 4, then path 3, path 2, and finally path 1. (c) The thermal energy added to the gas is also greatest for path 4, then path 3, path 2, and finally path 1.

RE 19-7: (a) For any cyclic process, $Q - W = 0$ or $Q = W$. (b) Because the net work that the gas does on its environment is negative here, that means that the net thermal energy Q transfer to the system is also negative and has the same value as the work. Thus thermal energy equal in magnitude to the work done by the system must be transferred *from the gas to the environment* each cycle.

RE 19-8: (a) Plates 2 and 3 will be tied for the largest increase in their vertical heights, followed by plate 1 and then plate 4. (b) Plate 3 will have the greatest increase in area, followed by plate 2, with plates 1 and 4 tied for last place.

RE 19-9: The hole gets larger as the plate's temperature increases, as would a circle drawn on the plate.

RE 19-10: The pressure at the base of the rod decreases, since the weight of the rod remains constant while the area of the base supporting that weight increases a bit.

RE 19-11: The greater the thermal conductivity, the smaller is the temperature difference between the two faces of samples of the same thickness. Since the temperature differences here, going from left to right, are 10 C°, 5 C°, 15 C°, and 5 C°, slabs b and d tie for greatest thermal conductivity, followed by slab a, with slab c having the smallest thermal conductivity.

Problems

1. (a) 320 °F; (b) −12.3 °F **3.** (a) Dimensions are inverse time. **5.** (a) 523 J/kg·K; (b) 26.2 J/mol·K; (c) 0.600 mole **7.** 42.7 kJ **9.** 1.9 times as great **11.** (a) 33.9 Btu; (b) 172 F° **13.** 160 s **15.** 2.8 days **17.** 742 kJ **19.** 73 kW **21.** 33 g **23.** (a) 0°C; (b) 2.5°C **25.** A: 120 J, B: 75 J, C: 30 J **27.** −30 J **29.** (a) 6.0 cal; (b) −43 cal; (c) 40 cal; (d) 18 cal, 18 cal **31.** 348 K **33.** (a) −40°; (b) 575°; (c) Celsius and Kelvin cannot give the same reading **35.** 960 μm **37.** 2.731 cm **39.** 29 cm³ **41.** 0.26 cm³ **43.** 360°C **47.** 0.68 s/h, fast **49.** 7.5 cm **51.** (a) 0.13 m; (b) 2.3 km **53.** 1660 J/s **55.** (a) 16 J/s; (b) 0.048 g/s **57.** 0.50 min **59.** (a) 17 kW/m²; (b) 18 W/m² **61.** 0.40 cm/h

Chapter 20

RE 20-1: Processes (a), (b), (d), and (e) start and end on the same isotherm because each has $PV = 12$ units.

RE 20-2: (a) The average translational kinetic energy doubles when the temperature in kelvins of the gas doubles. (b) The average translational kinetic energy would be zero if the temperature of the gas were 0 K. However, all real gases condense into liquids before reaching 0 K.

RE 20-3: (a) The average kinetic energy of each of the three types of molecules is the same. (b) Since that is true, the rms speed of each is inversely related to its molecular mass, so type 3 has the greatest rms speed, followed by type 2, with type 1 the smallest.

RE 20-4: Since the internal energy of an ideal gas depends only on its temperature, path 5 has the greatest change in E^{int}, followed by the other four paths, all of which tie for second place.

Problems

1. 0.933 kg **3.** 6560 **5.** (a) 5.47×10^{-8} mol; (b) 3.29×10^{16} **7.** (a) 0.0388 mol; (b) 220°C **9.** (a) 106; (b) 0.892 m³ **11.** $A(T_2 - T_1) - B(T_2^2$

$- T_1^2)$ **13.** 5600   J **15.** 100   cm³ **17.** 2.0×10^5 Pa **19.** 180   m/s **21.** 9.53×10^6 m/s **23.** 1.9 kPa **25.** 3.3×10^{-20} J **27.** (a) 6.75×10^{-20} J; (b) 10.7 **31.** (a) 6×10^9 km **33.** 15 cm **35.** (a) 3.27×10^{10}; (b) 172 m **37.** (a) 6.5 km/s; (b) 7.1 km/s **39.** (a) 1.0×10^4 K; (b) 1.6×10^5 K; (c) 440 K, 7000 K; (d) hydrogen, no; oxygen, yes **41.** (a) 7.0 km/s; (b) 2.0×10^{-8} cm; (c) 3.5×10^{10} collisions/s **43.** (a) $\frac{2}{3}v_0$; (b) $N/3$; (c) $122v_0$; (d) $1.31v_0$ **45.** RT $\ln(V_f/V_i)$ **47.** $(n_1C_1 + n_2C_2 + n_3C_3)/(n_1 + n_2 + n_3)$ **49.** (a) 6.6×10^{-26} kg; (b) 40 g/mol **51.** 8000 J **53.** (a) 6980 J; (b) 4990 J; (c) 1990 J; (d) 2990 J **55.** (a) 14 atm; (b) 620 K **59.** 1.40 **61.** (a) In joules, in the order Q, ΔE^{int}, W: $1 \rightarrow 2$: 3740, 3740, 0; $2 \rightarrow 3$: 0, −1810, 1810; $3 \rightarrow 1$: −3220, −1930, −1290; cycle: 520, 0, 520; (b) $V_2 = 0.0246$ m³, $p_2 = 2.00$ atm, $V_3 = 0.0373$ m³, $p_3 = 1.00$ atm

Chapter 21

RE 21-1: As the putty falls to the floor, gravitational potential energy is converted into translational kinetic energy. When the putty hits the floor it looks at first as if all the mechanical energy is somehow destroyed. But a closer look at the putty after the fall reveals that the putty and the floor beneath it have warmed up a bit. How so? As the putty collides with the floor, the floor does work on the putty. Since the putty doesn't bounce back, the work done on the putty system serves to raise its internal energy. Now the putty's temperature rises and it begins transferring microscopic thermal energy to its surroundings (air and floor) until thermal equilibrium is achieved. Even more careful observations show, in fact, that all of the mechanical energy present in the system before the fall is still there after the putty hits the floor, just in other forms, primarily as thermal energy.

RE 21-2: Driving a nail into a board, letting your hot coffee cool, and saying hello to your friend are all examples of irreversible processes in the sense that if you saw a movie of them running backward you would know something was wrong.

RE 21-3: Process (c) and (b) involve the same amount of heat energy transfer to the water, while (a) adds twice as much heat to the water. Process (a) also happens at the lowest average temperature, so it involves the greatest entropy change of the water, followed by (b) and then (c).

RE 21-4: Equation 21-11 relates the efficiency of a Carnot engine to the two thermodynamic temperatures between which it operates. Applying this to these three cases yields Carnot efficiencies of (a) 0.20, (b) 0.25, and (c) 0.33, so ranking the efficiencies, greatest first, yields (c), then (b), then (a).

RE 21-5: (a) Raising the lower temperature T_L by (δT) increases the numerator of Eq. 21-14 by (δT) and simultaneously decreases the denominator by the same amount. This yields the *greatest increase* in the coefficient of performance of the refrigerator. (b) Lowering the lower temperature T_L by (δT) decreases the numerator of Eq. 21-14 by (δT) and simultaneously decreases the denominator by the same amount. This yields the *greatest decrease* in the coefficient of performance of the refrigerator. (c) Increasing the higher temperature T_H by (δT) makes the denominator bigger by (δT) with no change in the numerator, decreasing the coefficient of performance of the refrigerator, but not as much as in (b). (d) Decreasing the higher temperature T_H by (δT) makes the denominator smaller by (δT) with no change in the numerator, increasing the coefficient of performance of the refrigerator, but not as much as in (a). So, from greatest to least, the changes in the coefficient of performance of the refrigerator are (a), (d), (c), and finally (b).

RE 21-6: If we had, say, 6 molecules, then the number of microstates corresponding to 3 molecules in each half of the box would be $6!/(3!)^2 = 20$. Generalizing Eq. 21-18 to three bins in the box with 2 molecules in each bin would have $6!/(2!)^3 = 90$. In this case a greater number of microstates is associated with dividing the box up into a larger number of equally populated equal subvolumes. This remains true as the number of molecules is increased, so (b) has more microstates than (a).

Problems

1. 14.4 J/K **3.** (a) 9220 J; (b) 23.0 J/K; (c) 0 **5.** (a) 5.79×10^4 J; (b) 173 J/K **7.** (a) 14.6 J/K; (b) 30.2 J/K **9.** (a) 57.0°C; (b) −22.1 J/K; (c) +24.9 J/K; (d) +2.8 J/K **13.** (a) 320 K; (b) 0; (c) +1.72 J/K **15.** +0.75 J/K **17.** (a) −943 J/K; (b) +943 J/K; (c) yes **19.** (a) $3p_0V_0$; (b) $\Delta E^{\text{int}} = 6RT_0$, $\Delta S = \frac{3}{2}R \ln 2$; (c) both are zero **21.** (a) 31%; (b) 16 kJ **23.** (a) 23.6%; (b) 1.49×10^4 J **25.** 266 K and 341 K **27.** (a) 1470 J; (b) 554 J; (c) 918 J; (d) 62.4% **29.** (a) 2270 J; (b) 14800 J; (c) 15.4% (d) 75.0%, greater **31.** (a) 78%; (b) 81 kg/s **33.** (a) $T_2 = 3T_1$, $T_3 = 3T_1/4^{\gamma-1}$, $T_4 = T_1/4^{\gamma-1}$, $p_2 = 3p_1$, $p_3 = 3p_1/4^\gamma$, $p_4 = p_1/4^\gamma$; (b) $1 - 4^{1-\gamma}$ **35.** 21 J **37.** 440 W **39.** 0.25 hp **41.** $[1 - (T_2/T_1)]/[1 - (T_4/T_3)]$ **45.** (a) $W = N!/(n_1!n_2!n_3!)$; (b) $[(N/2)! (N/2)!]/[(N/3)! (N/3)! (N/3)!]$; (c) 4.2×10^{16}

Photo Credits

Index

Page references followed by italic *table* indicate material in tables.
Page references followed by italic *n* indicate material in footnotes.

Mathematical Formulas*

Quadratic Formula

If $ax^2 + bx + c = 0$, then $x = \dfrac{-b \pm \sqrt{b^2 - 4ac}}{2a}$

Binomial Theorem

$(1 + x)^n = 1 + \dfrac{nx}{1!} + \dfrac{n(n-1)x^2}{2!} + \cdots \qquad (x^2 < 1)$

Products of Vectors

Let θ be the smaller of the two angles between $\vec{a}$ and $\vec{b}$. Then

$\vec{a} \cdot \vec{b} = \vec{b} \cdot \vec{a} = a_x b_x + a_y b_y + a_z b_z = |\vec{a}||\vec{b}|\cos\theta$

$\vec{a} \times \vec{b} = -\vec{b} \times \vec{a} = \begin{vmatrix} \hat{i} & \hat{j} & \hat{k} \\ a_x & a_y & a_z \\ b_x & b_y & b_z \end{vmatrix}$

$= \hat{i}\begin{vmatrix} a_y & a_z \\ b_y & b_z \end{vmatrix} - \hat{j}\begin{vmatrix} a_x & a_z \\ b_x & b_z \end{vmatrix} + \hat{k}\begin{vmatrix} a_x & a_y \\ b_x & b_y \end{vmatrix}$

$= (a_y b_z - b_y a_z)\hat{i} + (a_z b_x - b_z a_x)\hat{j} + (a_x b_y - b_x a_y)\hat{k}$

$|\vec{a} \times \vec{b}| = |\vec{a}||\vec{b}|\sin\theta$

Trigonometric Identities

$\sin\alpha \pm \sin\beta = 2\sin\tfrac{1}{2}(\alpha \pm \beta)\cos\tfrac{1}{2}(\alpha \mp \beta)$

$\cos\alpha + \cos\beta = 2\cos\tfrac{1}{2}(\alpha + \beta)\cos\tfrac{1}{2}(\alpha - \beta)$

Derivatives and Integrals

$\dfrac{d}{dx}\sin x = \cos x$ $\qquad \displaystyle\int \sin x\, dx = -\cos x$

$\dfrac{d}{dx}\cos x = -\sin x$ $\qquad \displaystyle\int \cos x\, dx = \sin x$

$\dfrac{d}{dx}e^x = e^x$ $\qquad \displaystyle\int e^x\, dx = e^x$

$\displaystyle\int \dfrac{dx}{\sqrt{x^2 + a^2}} = \ln(x + \sqrt{x^2 + a^2})$

$\displaystyle\int \dfrac{x\, dx}{(x^2 + a^2)^{3/2}} = -\dfrac{1}{(x^2 + a^2)^{1/2}}$

$\displaystyle\int \dfrac{dx}{(x^2 + a^2)^{3/2}} = \dfrac{x}{a^2(x^2 + a^2)^{1/2}}$

Cramer's Rule

Two simultaneous equations in unknowns x and y,

$a_1 x + b_1 y = c_1 \qquad$ and $\qquad a_2 x + b_2 y = c_2,$

have the solutions

$x = \dfrac{\begin{vmatrix} c_1 & b_1 \\ c_2 & b_2 \end{vmatrix}}{\begin{vmatrix} a_1 & b_1 \\ a_2 & b_2 \end{vmatrix}} = \dfrac{c_1 b_2 - c_2 b_1}{a_1 b_2 - a_2 b_1}$

and

$y = \dfrac{\begin{vmatrix} a_1 & c_1 \\ a_2 & c_2 \end{vmatrix}}{\begin{vmatrix} a_1 & b_1 \\ a_2 & b_2 \end{vmatrix}} = \dfrac{a_1 c_2 - a_2 c_1}{a_1 b_2 - a_2 b_1}.$

* See Appendix E for a more complete list.

The Greek Alphabet

Alpha	A	α	Iota	I	ι	Rho	P	ρ
Beta	B	β	Kappa	K	κ	Sigma	Σ	σ
Gamma	Γ	γ	Lambda	Λ	λ	Tau	T	τ
Delta	Δ	δ	Mu	M	μ	Upsilon	Y	υ
Epsilon	E	ϵ	Nu	N	ν	Phi	Φ	ϕ, φ
Zeta	Z	ζ	Xi	Ξ	ξ	Chi	X	χ
Eta	H	η	Omicron	O	o	Psi	Ψ	ψ
Theta	Θ	θ	Pi	Π	π	Omega	Ω	ω